The Dental Pulp

Michel Goldberg
Editor

The Dental Pulp

Biology, Pathology, and Regenerative Therapies

Springer

Editor
Michel Goldberg, DDS, PhD
Department of Oral Biology
Institut National de la Santé et de la
Recherche Médicale
Université Paris Descartes
Paris
France

ISBN 978-3-642-55159-8 ISBN 978-3-642-55160-4 (eBook)
DOI 10.1007/978-3-642-55160-4
Springer Heidelberg New York Dordrecht London

Library of Congress Control Number: 2014944383

Printed on acid-free paper

Springer is part of Springer Science+Business Media (www.springer.com)

Preface

The dental pulp has the characteristic of being exclusively the nonmineralized part of a mineralized tooth. This tissue is surrounded by a robust shell-like complex structure that includes dentin and enamel in the crown and cementum in the root. Like most connective tissues, the dental pulp is vascularized and innervated, which constitute important functional differences between the crown and the root. It is also a reservoir of structural fibroblasts (named pulpoblasts long ago by Louis Baume). The pulp complexity is increased by the presence of progenitors (or stem cells) implicated in pulp repair and regeneration.

Inflammatory immune cells are concerned with the destruction of pathogens, cell debris (apoptotic bodies), and/or adverse molecules. Altogether, the heterogeneous cell colonies restore the reparative functions of the dental pulp and consequently the biological approaches of pulp therapies. As an alternative for the surgical or chemical ablation of this tissue and for a limited size of the lesion, pulp capping with biomolecules has been successful. In the case of a more advanced dental destruction, it comes out nowadays that vital pulp regeneration is our next goal. The formation/regeneration of an artificial pulp may be followed by the construction of a homogeneous mineralization, sealing the root canal.

For future prospects in pulp therapies, it is important to take into account three major points: (1) Usually, the dental pulp is not mineralized, and some specificities of its composition may explain why it remains a soft tissue. This point should be further explored and clarified. (2) In a very few cases, the pulp may eventually mineralize, namely, when genetic pathologies such as *dentinogenesis imperfecta* or *dentin dysplasia* disturb the tissue. In this respect, gene expression seems crucial for the understanding of such processes. (3) It should also be noted that in the aging pulp, some loci initiate gradually localized or diffuse mineralized areas. They also contribute to the formation of laminated pulp stones. Eventually, mineral diffusion occurs and totally fills the dental pulp.

Therefore, these different aspects lead to a series of questions: (1) Why, in "normal" physiological conditions, does the dental pulp remain a nonmineralized tissue? (2) Why, in some pathological cases, does mineralization of the dental pulp occur? Can we identify the factors that influence the mineralization even if the pulp was initially a soft tissue? Different studies shed some light on questions that are still open. As a matter of fact, the answers obtained from pulp studies may also expand to other tissues and/or to ectopic mineralization. We may expect some answers to these unanswered questions. Atheroma plaques, atherosclerosis, skin, eyes, kidney stones or large areas of mineralizations, and other pathological calcifications are still under examination. Identification of some extracellular matrix molecules may contribute to identify which molecules are in the lead. They may be mineralization inhibitors, as it is the case for α_2HS glycoprotein. Indeed, enlarged dental pulps are seen when DMP1 is mutated. In contrast, such studies can contribute to our understanding of the role of other molecules or ECM peptides. They may shed light on normal regulation or pathological processes leading to the initiation and development of pulp mineralization.

Finally, we should also pinpoint that dental caries lead to chronic or acute pulp reactions. They may be limited and controlled or expand within the totality of the pulp tissue. Alternatively, pulp necrosis may occur, or pulp healing can be observed. For a long period of time, endodontic treatment was the privileged therapy. Biotherapies or biological approaches with mineral components (e.g., calcium hydroxide), ECM, or bioactive peptide-derived molecules have substantially reinforced the cascade of events occurring in the wounded pulp. The healing process involves stem cell recruitment, their proliferation and differentiation, and ECM synthesis and secretion, altogether contributing to pulp recovery. As a consequence of forming an artificial pulp, pulp regeneration is occurring. In terms of biology, stem cells contribute to the construction of a nonmineralized pulp. The pulp will further mineralize, and the final step of pulp biotherapy mimicks what can be expected from an endodontic substitute. Tooth engineering is a promising tool for the near future, sealing the lumen of the pulp root.

The selection of the contributors to this book was based on their unanimously well-recognized knowledge in this field and on their important publications in the field of pulp biology or physiopathology. They have explored new concepts, new areas of research, and consequences of new data, which may have some impact on the future of pulp therapies. The list of authors that contributed to this book is based mostly on the warm friendship developed during hot discussions, drinking many glasses of beer or wine and exchanging arguments on burning points in many international conferences. We were delighted to be present in such meetings and to show our latest findings or raise some discussions on unanswered points. We were most happy to meet again or to become connected with new friends.

After all these years, some of the coauthors really became my closest friends, and we became coconspirators. This is the case especially for Art Veis, who is for me an older brother and a wonderful example of what should be done through a life span. In this Noah's ark, I feel sometimes that I am in a floating boat, ensuring our temporary survival in a scientific domain. It is certainly the reason why in this ship full of friends, we have launched—if we keep a certain sense of humor—what may be called a solid friendship.

During all these years, altogether we have built a kind of family, and it is a pleasure to be associated with a group of researchers forming, so to say, a "pulp knot"—a new organizing center, which is reflected in this book.

Paris, France Michel Goldberg, DDS, PhD

Contents

Contributors

Imad About, PhD Aix-Marseille Université, Centre National de la Recherche Scientifique, Marseille, France

Lars Bjørndal, DDS, PhD Department of Odontology, Section of Cariology and Endodontics/Pediatric and Clinical Genetics, University of Copenhagen, Copenhagen, Denmark

Agnès Bloch-Zupan, HDR, PhD, DDS Department of Sciences Biologiques, Faculty of Dentistry, University of Strasbourg, Strasbourg, France

Reference Center for Orodental Manifestations of Rare Diseases, Pôle de Médecine et Chirurgie Bucco-Dentaires, Hôpitaux Universitaires de Strasbourg, Strasbourg, France

Institute of Genetics and Molecular and Cellular Biology, INSERM, Illkirch, France

Eastman Dental Institute, UCL, London, UK

Paul R. Cooper, PhD Department of Oral Biology, School of Dentistry, University of Birmingham, Birmingham, B4 6NN, UK

Pamela Den Besten, DDS, MS Division of Pediatric Dentistry, Department of Orofacial Sciences, School of Dentistry, University of California, San Francisco, San Francisco, CA, USA

Sasha Dimitrova-Nakov, DDS, PhD Department of Oral Biology, Institut National de la Santé et de la Recherche Médicale, Université Paris Descartes, Paris, France

Kaj Fried, DDS, PhD Department of Neuroscience, Karolinska Institutet, Stockholm, Sweden

Kerstin M. Galler, DDS, PhD Department of Restorative Dentistry and Periodontology, University Hospital Regensburg, Regensburg, Germany

Jennifer Lynn Gibbs, MAS, DDS, PhD Department of Endodontics, New York University College of Dentistry, New York, NY, USA

Michel Goldberg, DDS, PhD Department of Oral Biology, Institut National de la Santé et de la Recherche Médicale, Université Paris Descartes, Paris, France

Koichiro Iohara, DDS, PhD Department of Dental Regenerative Medicine, Center of Advanced Medicine for Dental and Oral Diseases, National Center for Geriatrics and Gerontology, Research Institute, Obu, Japan

Mina Mina, DMD, MSD, PhD Division of Pediatric Dentistry, School of Dental Medicine, University of Connecticut Health Center, Farmington, CT, USA

Yukiko Nakano, DDS, PhD Department of Orofacial Sciences, School of Dentistry, University of California, San Francisco, San Francisco, CA, USA

Misako Nakashima, PhD, DDS Department of Dental Regenerative Medicine, Center of Advanced Medicine for Dental and Oral Diseases, National Center for Geriatrics and Gerontology, Research Institute, Obu, Japan

Jacques Eduardo Nör, DDS, MS, PhD Department of Cariology, Restorative Sciences, and Endodontics, University of Michigan, Ann Arbor, MI, USA

Tracy L. de Peralta, DMD, PhD, MClinEd Department of Cariology, Restorative Sciences, and Endodontics, University of Michigan, Ann Arbor, MI, USA

Domenico Ricucci, MD, DDS Private Practice, Cetraro, Italy

Gottfried Schmalz, DDS, DMD, PhD Department of Operative Dentistry and Periodontology, University Hospital Regensburg, University Clinics, Regensburg, Germany

School of Dental Medicine - ZMK Bern, University of Bern, Switzerland

Stéphane Simon, DDS, PhD Department of Oral Biology and Endodontics, Dental School of the University of Paris Diderot, Hoptial de la Pitié Salpétrière, Paris, France

Anthony J. Smith, PhD Department of Oral Biology, School of Dentistry, University of Birmingham, Birmingham, UK

Arthur Veis, PhD Department of Cell and Molecular Biology, Feinberg School of Medicine, Northwestern University, Chicago, IL, USA

Part I

Pulp Biology

1 Pulp Development

Sasha Dimitrova-Nakov and Michel Goldberg

1.1 Introduction

The induction and the human dentition development takes place during embryonic, fetal, neonatal, and postnatal childhood stages of development.

Human tooth development begins with the induction of the primary dentition during the fifth week of gestation (embryogenesis). Biomineralization starts during the fourteenth week of gestation, and the permanent dentition is completed at the end of adolescence.

The tooth is composed of different tissues. The enamel, dentin, and cementum are mineralized dental tissues, whereas dental pulp is the only non-mineralized dental tissue. The dental pulp is a specialized loose connective tissue localized in the central part of the tooth.

Anatomically and functionally, the dentin (synthesized by odontoblasts) and dental pulp are considered a single entity. Both tissues are often associated as the "dentin-pulp complex." But, biologically, this anatomical entity has no consistency.

Understanding odontogenesis is a prerequisite to be able to understand the processes involved in dentin repair. Many studies underline that genes and signaling pathways involved in the early stages of odontogenesis also play a role in the dental pulp repair process in adults [1, 2].

1.2 Tooth Development: The Initial Steps

The odontogenesis is associated with the initial stages of craniofacial development and is regulated by epithelial-mesenchymal interactions. The epithelium may be ectodermal or endodermal. The mesenchyme in the first branchial arch is termed ectomesenchyme because neural crest cells have migrated in it [3–5].

In mammals, the ectoderm is at the origin of the oral epithelium which gives rise to ameloblasts, responsible for dental enamel formation. Odontoblasts, cells secreting dentin, derive from the ectomesenchyme.

The neural crest cells (NCC) of the rostral hindbrain (rhombomeres 1 and 2) and caudal midbrain migrate and colonize the first branchial arch, forming the presumptive territories of the teeth, mandible, and maxilla. Combinatory expression of homeobox genes (Hox) assigns an identity to the branchial arches after NCC migration. Prior to tooth bud formation, these cells

S. Dimitrova-Nakov, DDS, PhD
Department of Oral Biology, Institut National de la Santé et de la Recherche Médicale, UMR-S 1124, Université Paris Descartes, 45 Rue des Saints Pères, Paris 75006, France
e-mail: dimitrova.sasha@gmail.com

M. Goldberg, DDS, PhD (✉)
Department of Oral Biology, Institut National de la Santé et de la Recherche Médicale, Université Paris Descartes, 45 Rue des Saints Pères, Paris 75006, France
e-mail: mgoldod@gmail.com, michel.goldberg@parisdescartes.fr

M. Goldberg (ed.), *The Dental Pulp*,
DOI 10.1007/978-3-642-55160-4_1, © Springer-Verlag Berlin Heidelberg 2014

already express the LIM-homeobox-containing genes, Lhx 6 and Lhx 7, which are hallmarks of the odontogenic lineage [6].

In the mouse embryo, Hoxa2 appears to be the only homeobox gene expressed in rhombomere 2, while HOXa2 is absent in the NCC of rhombomere 1. Furthermore, the "knockout" (KO) of HOXa2 induces transformation of the skeletal elements of the second arc in those of the first arc [7]. It was noted as the absence of expression of homeotic genes in the first branchial arch [8]. This absence of expression suggests that cell fate is not "determined" at this stage, which, in term, would promote morphogenesis/differentiation of the elements of the jaw during the later stages of development.

NCCs of the first branchial arch are at the origin of the odontogenic ectomesenchyme that will interact with the oral epithelium to form presumptive territories of the incisors, canines, and molars (in humans) in each quadrant of the two jaws. The early expression of FGF8 and BMP4 in the oral epithelium allows the induction of the homeobox gene expression (Barx1, Dlx1/2, Msx1, Msx2, Alx3) in the cells of the underlying ectomesenchyme and establishes a Hox gene expression pattern specifying separate territories [9, 10].

This combinatory Hox gene expression creates a "dental homeocode" that will control the morphogenesis/differentiation. This "homeocode" assigns an identity to these "pools" of progeny cells which will form the tooth germs specific to different types of teeth and thus plays a crucial role in the spatiotemporal regulation of odontogenesis.

Tooth morphogenesis is similar to other organ's morphogenesis formed by the cells deriving from the neural crest (tooth, hair, feathers, salivary glands, mammary glands) [11]. During the initiation of these organs, the ectoderm thickens and forms the epithelial placode that buds in the underlying mesenchyme. The interaction between the ectoderm and underlying mesenchyme provokes the condensation of mesenchyme around the epithelial bud. During morphogenesis, the mesenchyme directs the folding and the ramification of the epithelium, a crucial step for the morphogenesis of the organ.

The teeth have been used as a model extensively to illustrate the importance of ecto-mesenchymal interactions and particularly the role of these interactions during the morphogenesis of different types of teeth.

The molecular signals mediating these interactions belong to several conserved signalization families. Many growth factors such as FGFs (fibroblast growth factors), Wnt(s), BMPs (bone morphogenetic proteins), the Hh(s) (Hedgehog), Notch, and EDA (Ectodysplasin-A) are involved in the dental development [8, 12–30], but their exact roles are not yet clear.

Specific spatial and temporal expression of a number of homeotic genes, such as Pitx2, Pax9, Msx1/2, Lhx6, Lhx7, Dlx1/2, and Barx1, marks the induction of odontogenesis and can be used as markers of tooth development [23, 31–44]. Recently, it was suggested that Sox2 regulates the progenitor state of dental epithelial cells and that the expression patterns of Sox2 support the hypothesis that dormant capacity for continuous tooth renewal exists in mammals [45].

MicroRNAs (miRNAs) are emerging as important regulators of the various aspects of embryonic development, including the odontogenesis. The small noncoding RNA function is a transcriptional and posttranscriptional regulation of gene expression. It was admitted that miRNAs have different roles in the epithelium and mesenchyme during odontogenesis. Furthermore, "microarray" and hybridization in situ analysis have identified several miRNAs having a differential expression between the incisors and molars [46–48].

Finally, although the spatiotemporal gene expression pattern was determined in the mouse embryo, the precise role of each of these actors in the development program of the tooth is far from being fully elucidated.

Next, we describe briefly the different stages during odontogenesis, without details on molecular level. There are many extensive reviews related to this topic [30, 49, 50].

1.3 Stages of Tooth Development

It is well established that the basic steps of tooth morphogenesis are similar in all vertebrates. After 5 weeks of development, continuous bands of thickened epithelium, horseshoe shaped, are formed around the mouth in the presumptive upper and lower jaws. These epithelium bands, named primary epithelial bands, will give rise to the dental lamina. The establishment of the dental lamina, the area that forms the teeth, precedes the initiation of individual teeth.

The key event for the initiation of tooth development is the formation of localized thickenings or *dental placodes* (sixth week) within the primary epithelial bands, at the site of the future dental arches in the embryonic mandible and maxilla. The *basement membrane* (BM) separates, even at this early stage, the epithelium from the underlying ectomesenchyme. The BM controls the epithelial-mesenchymal interactions and exchanges. The interactions between the surface epithelium and underlying ectomesenchyme are crucial both for the formation of dental placodes and during various stages of odontogenesis (Fig. 1.1).

The first evidence of the future teeth appears when the epithelial cells near the basement membrane begin to multiply (four to five cell layers) and invaginate into the underlying ectomesenchyme, giving rise to the *dental lamina*. The ectomesenchyme starts to change composition in response and becomes more condensed. Thus, this initial epithelial invagination clearly marks the apparition of the tooth crown area and will develop through several distinct stages (bud, cap, bell stage). Tooth development is a continuous process, so clear distinction between the transition stages is not possible.

Each dental lamina is at the origin of a tooth *bud* (Fig. 1.2). Tooth buds of the deciduous canines and incisors are apparent in the 8-week-old human embryo, and buds of the deciduous molars are formed during the ninth week. The *bud stage* is characterized by the progression of ectodermal invagination in the underlying ectomesenchyme, in which cells are packed closely around the epithelial bud. This will be followed by the changes in the shape of the dental bud and formation of the *dental cap* (Figs. 1.2 and 1.3). The *cap stage* is characterized by a concavity of the epithelium that partially envelops the underlying mesenchyme. During the cap stage, the epithelial outgrowth is referred widely as the *enamel organ* and is related to the differentiation of the outer dental epithelium, inner dental epithelium, and the *appearance of the enamel knot*. Also, there is a condensation of the ectomesenchyme in

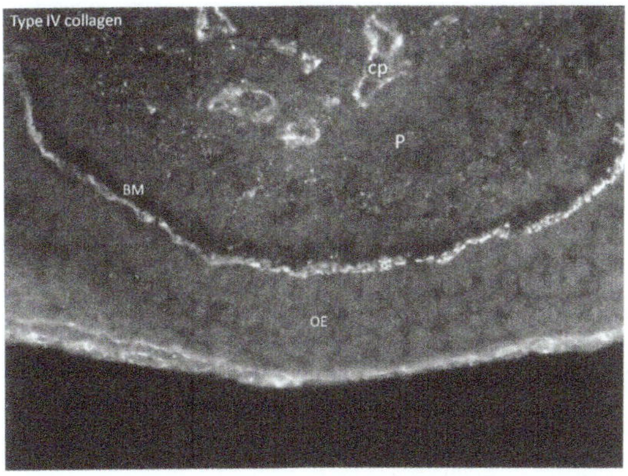

Fig. 1.1 At the initial stage of tooth germ formation, a basement membrane (*BM*) separates the outer epithelium (epithelial placode) (*OE*) from the subjacent mesenchyme. The future pulp (*P*) (condensation zone) displays type IV collagen immunostaining limiting capillaries (*cp*)

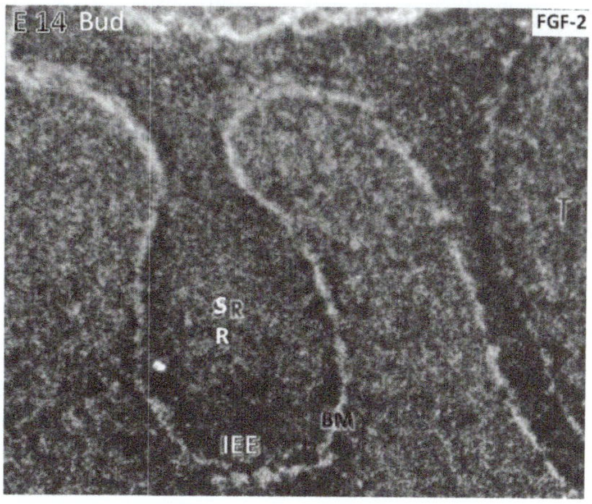

Fig. 1.2 At the embryonic day, E14 binding of the radioactive FGF-2 occurs along the basement membrane (*BM*) limiting the inner epithelial epithelium (*IEE*). The central part of the tooth bud contains the stratum reticulum (*SR*)

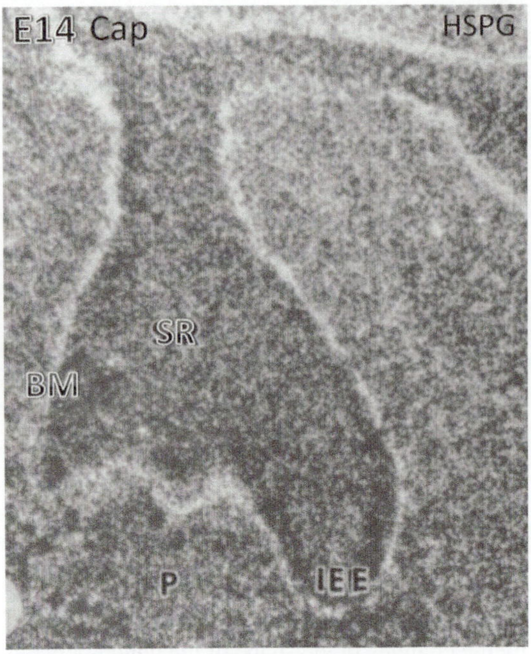

Fig. 1.3 Heparan sulfate proteoglycan (*HSPG*) binding is mostly situated along the basement membrane (*BM*) located at the early cap stage along the inner enamel epithelium (*IEE*) above the stratum reticulum (*SR*). In the embryonic pulp (*P*), numerous cells are immunostained

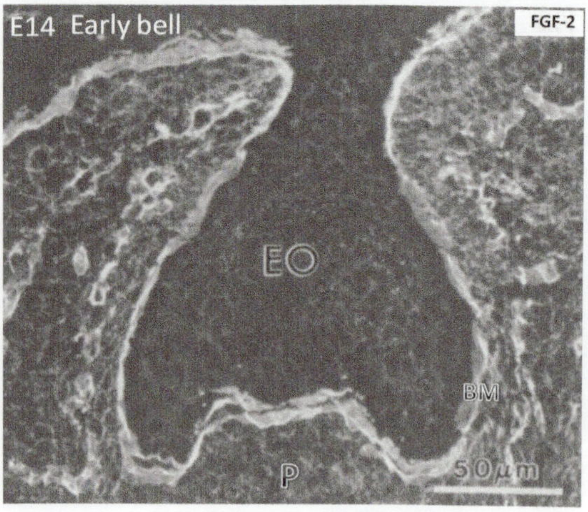

Fig. 1.4 At a more advanced cap stage (day 14), dense FGF-2 binding occurs along the basement membrane (*BM*). The binding is weaker in the enamel organ (*EO*), limited to the cell surface. The binding is stronger in the pulp (*p*), capillaries, and the early stages of the trabecular bone

the concavity of the enamel organ forming the *dental papilla*, at the origin of the odontoblasts and the dental pulp [51, 52] (Figs. 1.4 and 1.5).

Starting from the late cap stage and through the transition from cap to bell stage of tooth development, many developmental changes are observed. All the elements of the *enamel organ* are well distinguished (*histodifferentiation*):

- The outer enamel epithelium located at the periphery of the cap in contact with the peridental mesenchyme.
- The inner enamel epithelium formed by cells which are precursors of ameloblasts and which are separated from the future dental pulp by a basement membrane.
- The stellate reticulum and the stratum intermedium, two intermediate layers of the enamel organ involved in the transcellular and intercellular transfer of precursors of enamel proteins, and in the provision of energy for these transfers (synthesis and degradation of glycogen). For some authors, the stratum intermedium differentiates during the bell stage.
- The primary enamel knot, a transient structure situated in the center of the enamel organ that will control the morphogenesis of dental cusps and determine the final shape of the tooth [53]. Subsequently, secondary enamel knots will be formed, contributing to the formation of molar cusps. The enamel knots within epithelium are described as organizing centers composed by "clusters" of cells that secrete many morphogen signals like Shh, Wnts, FGFs, and BMPs, whose roles are not yet fully defined.

Besides the role of the enamel knots to regulate the size and shape of the teeth, the signals from the mesenchyme are also necessary for the formation and maintenance of epithelial compartments.

The cap stage is followed by the *bell stage*, during which the dental crown acquires its final shape (*morphodifferentiation*) and the formation of the cusp pattern is observed (Figs. 1.6 and 1.7).

The outer and inner enamel epithelia are continuous, and they meet at the rim of the enamel organ known as the zone of reflection or cervical loop. Extended in Hertwig's epithelial root

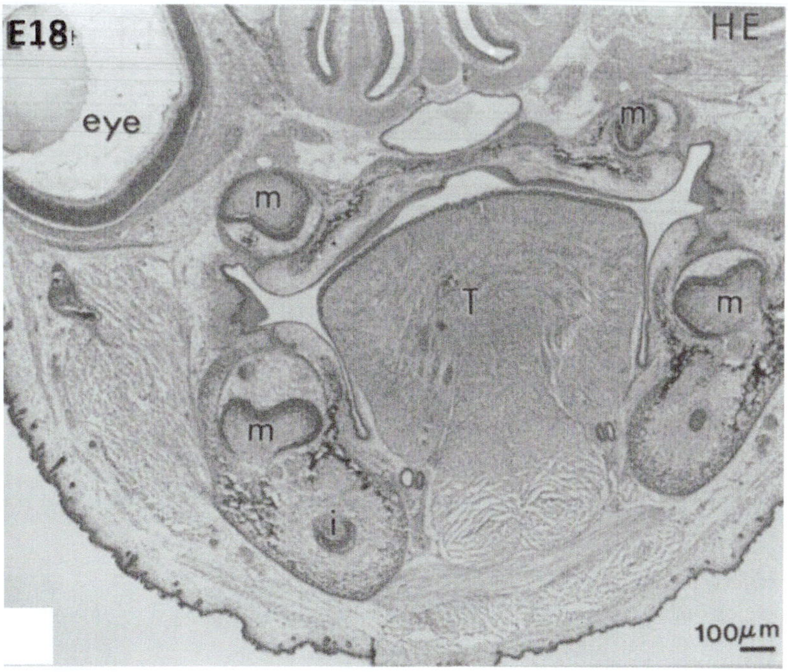

Fig. 1.5 At embryonic day 18, hematoxylin-eosin staining reveals early bell stages of molars (*m*) in the mandible (*lower part* of the figure) and maxillary (*upper part*, near the nasal cavities, the mineralizing palatal layer, and the eye). The tongue (*T*) occupies the central part of the mouse head. Beneath the molars, the incisors are seen in the transverse sections of the mandible

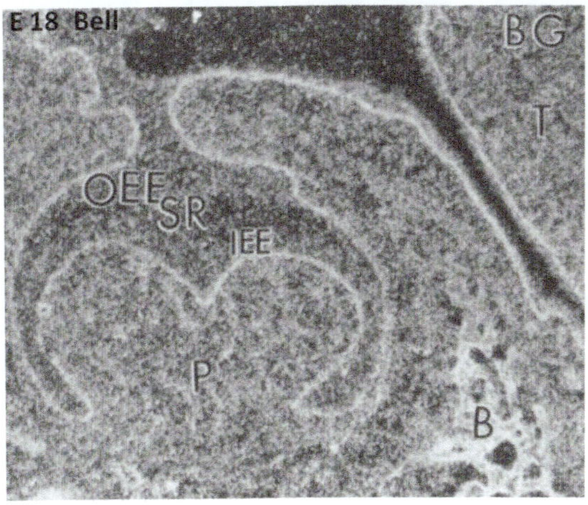

Fig. 1.6 E18. At the bell stage, the binding of FGF-2 (*BG*) concentrate along the basement membrane in the outer enamel epithelium (*OEE*), stellate reticulum (*SR*), and inner enamel epithelium (*IEE*). The pulp (*P*) is heavily labeled and the forming bone (*B*) as well. *T* tongue

sheath, this cervical loop will control the formation of the root, including the root odontoblast differentiation (Figs. 1.8 and 1.9). The cervical loop progresses in apical direction, and many cell divisions sustain this preeruptive crown growth, thus delimiting increasingly the dental papilla area. In the pluricuspid teeth, the secondary enamel knots appear at the top of each cusp.

In the late bell stage, tooth morphogenesis is followed by a phase of cell differentiation of the inner enamel epithelium and of the ectomesenchymal cells at the epithelial-mesenchymal interface with the basement membrane (*histodifferentiation*). These cells will differentiate in pre-ameloblasts and pre-odontoblasts in order to become polarized and secreting ameloblasts and odontoblasts to form the enamel and dentin, respectively. The first layers of the enamel and dentine are visible at the end of the coronary morphogenesis. Thus, during embryogenesis, morphogenesis and differentiation are coupled. Cells acquire the competence to differentiate according to their position. Differentiation of pre-ameloblasts and odontoblasts is pre-coupled spatiotemporally to these morphogenetic movements that provide the pattern formation of the crown and the beginning of the root formation.

The condensed ectomesenchyme situated at the periphery of the enamel organ and dental papilla is referred as the *dental follicle or dental sac* and will give rise to the supporting dental

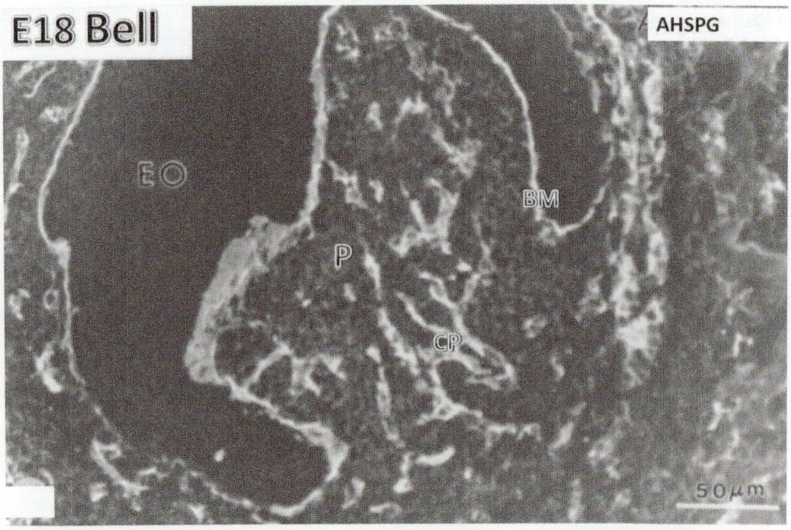

Fig. 1.7 E18. Bell stage. Anti-heparan sulfate proteoglycan (*AHSPG*) immunolabeling. No labeling is detectable in the enamel organ (*EO*). The basement membrane (*BM*) is densely immunostained. In the pulp (*P*), the capillaries (*CP*) are well stained. At the stage of crown formation, endothelial cells of capillaries proliferate, elongate, and form a dense vascular network

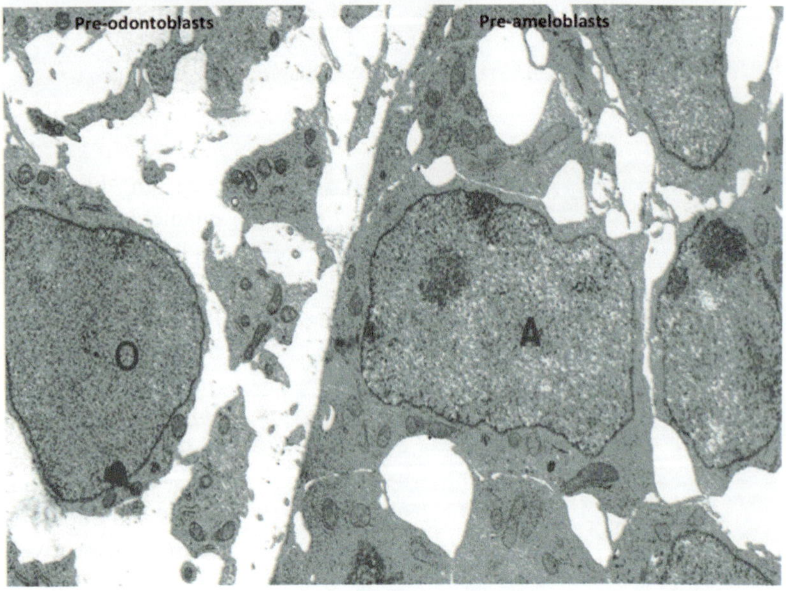

Fig. 1.8 At a later stage, pre-odontoblasts are facing pre-ameloblasts. Postmitotic polarizing ameloblasts form a cell network, with large intercellular spaces. Presecretory pre-polarized odontoblasts (*O*) are located in an electron-lucent dental pulp

tissues as the tooth cementum, periodontal ligament, and alveolar bone. Thus, the dental follicle is involved in the formation of the root and tooth eruption (Figs. 1.5 and 1.10).

The enamel organ, the dental papilla, and the dental follicle form the *dental organ or tooth germ*.

In humans, there are two dentitions, deciduous and permanent (primary, temporary). The development of the deciduous dentition begins around the sixth gestational week. Then quickly, there is coexistence of deciduous and permanent dental germs, and odontogenesis ends around the age of 18–25 years by formation of the dental root and the eruption of the third permanent molars.

The development of the permanent teeth begins during the odontogenesis of deciduous teeth. The deciduous teeth are formed from the primary dental lamina. In humans, the permanent teeth develop in two ways: (1) sequentially at the lingual region of the enamel organ of each temporary tooth (successional teeth) or (2) permanent molars grow from an extension of the

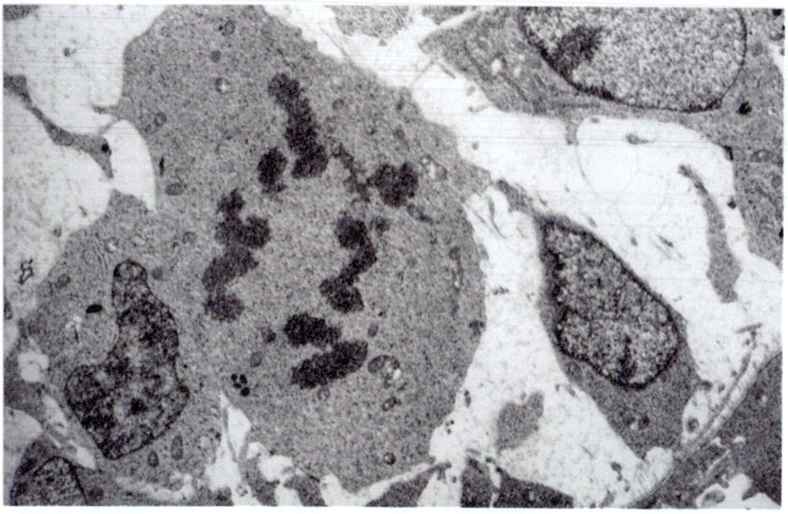

Fig. 1.9 The ultimate asymmetric division allows pre-odontoblasts to become postmitotic odontoblasts implicated in dentin formation

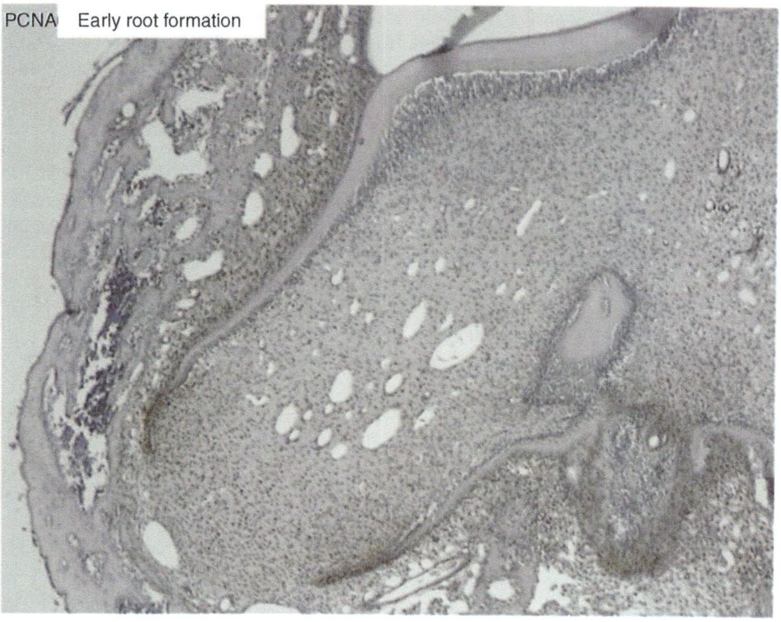

Fig. 1.10 Proliferating cell nuclear antigen (*PCNA*) labeling of the apical cells of the forming root. The epithelial Hertwig's root sheath divides and elongates, contributing to root elongation and to the tooth eruption

initial dental lamina. These are non-successional teeth.

The spatiotemporal control of dental development is orchestrated by epithelial-mesenchymal interactions. At the molecular level, these interactions provide reciprocal and sequential exchanges of signals through the basement membrane. These signals/morphogens are associated with the "determination" (commitment) of the cells in different territories during the morphogenesis.

References

1. Kitamura C, Kimura K, Nakayama T, Terashita M. Temporal and spatial expression of c-jun and jun-B proto-oncogenes in pulp cells involved with reparative dentinogenesis after cavity preparation of rat molars. J Dent Res. 1999;78(2):673–80.
2. Mitsiadis TA, Fried K, Goridis C. Reactivation of Delta-Notch signaling after injury: complementary expression patterns of ligand and receptor in dental pulp. Exp Cell Res. 1999;246(2):312–8.

3. Tucker A, Sharpe P. The cutting-edge of mammalian development; how the embryo makes teeth. Nat Rev Genet. 2004;5(7):499–508.
4. Soukup V, Epperlein HH, Horacek I, Cerny R. Dual epithelial origin of vertebrate oral teeth. Nature. 2008;455(7214):795–8.
5. Fraser GJ, Hulsey CD, Bloomquist RF, Uyesugi K, Manley NR, Streelman JT. An ancient gene network is co-opted for teeth on old and new jaws. PLoS Biol. 2009;7(2):e31.
6. Mandler M, Neubuser A. FGF signaling is necessary for the specification of the odontogenic mesenchyme. Dev Biol. 2001;240(2):548–59.
7. Rijli FM, Mark M, Lakkaraju S, Dierich A, Dolle P, Chambon P. A homeotic transformation is generated in the rostral branchial region of the head by disruption of Hoxa-2, which acts as a selector gene. Cell. 1993;75(7):1333–49.
8. Cobourne MT, Sharpe PT. Tooth and jaw: molecular mechanisms of patterning in the first branchial arch. Arch Oral Biol. 2003;48(1):1–14.
9. McCollum MA, Sharpe PT. Developmental genetics and early hominid craniodental evolution. Bioessays. 2001;23(6):481–93.
10. McCollum M, Sharpe PT. Evolution and development of teeth. J Anat. 2001;199(Pt 1–2):153–9.
11. Pispa J, Thesleff I. Mechanisms of ectodermal organogenesis. Dev Biol. 2003;262(2):195–205.
12. Jernvall J, Aberg T, Kettunen P, Keranen S, Thesleff I. The life history of an embryonic signaling center: BMP-4 induces p21 and is associated with apoptosis in the mouse tooth enamel knot. Development. 1998;125(2):161–9.
13. Kettunen P, Karavanova I, Thesleff I. Responsiveness of developing dental tissues to fibroblast growth factors: expression of splicing alternatives of FGFR1, -2, -3, and of FGFR4; and stimulation of cell proliferation by FGF-2, -4, -8, and -9. Dev Genet. 1998;22(4):374–85.
14. Trumpp A, Depew MJ, Rubenstein JL, Bishop JM, Martin GR. Cre-mediated gene inactivation demonstrates that FGF8 is required for cell survival and patterning of the first branchial arch. Genes Dev. 1999;13(23):3136–48.
15. Dassule HR, Lewis P, Bei M, Maas R, McMahon AP. Sonic hedgehog regulates growth and morphogenesis of the tooth. Development. 2000;127(22): 4775–85.
16. Hjalt TA, Semina EV, Amendt BA, Murray JC. The Pitx2 protein in mouse development. Dev Dyn. 2000;218(1):195–200.
17. Jackman WR, Draper BW, Stock DW. Fgf signaling is required for zebrafish tooth development. Dev Biol. 2004;274(1):139–57.
18. Mitsiadis TA, Regaudiat L, Gridley T. Role of the Notch signalling pathway in tooth morphogenesis. Arch Oral Biol. 2005;50(2):137–40.
19. Jarvinen E, Salazar-Ciudad I, Birchmeier W, Taketo MM, Jernvall J, Thesleff I. Continuous tooth generation in mouse is induced by activated epithelial Wnt/beta-catenin signaling. Proc Natl Acad Sci U S A. 2006;103(49):18627–32.
20. Chen S, Gluhak-Heinrich J, Martinez M, Li T, Wu Y, Chuang HH, Chen L, Dong J, Gay I, MacDougall M. Bone morphogenetic protein 2 mediates dentin sialophosphoprotein expression and odontoblast differentiation via NF-Y signaling. J Biol Chem. 2008;283(28):19359–70.
21. Klein OD, Lyons DB, Balooch G, Marshall GW, Basson MA, Peterka M, Boran T, Peterkova R, Martin GR. An FGF signaling loop sustains the generation of differentiated progeny from stem cells in mouse incisors. Development. 2008;135(2):377–85.
22. Liu F, Chu EY, Watt B, Zhang Y, Gallant NM, Andl T, Yang SH, Lu MM, Piccolo S, Schmidt-Ullrich R, Taketo MM, Morrisey EE, Atit R, Dlugosz AA, Millar SE. Wnt/beta-catenin signaling directs multiple stages of tooth morphogenesis. Dev Biol. 2008;313(1):210–24.
23. Chen J, Lan Y, Baek JA, Gao Y, Jiang R. Wnt/beta-catenin signaling plays an essential role in activation of odontogenic mesenchyme during early tooth development. Dev Biol. 2009;334(1):174–85.
24. Jackman WR, Yoo JJ, Stock DW. Hedgehog signaling is required at multiple stages of zebrafish tooth development. BMC Dev Biol. 2010;10:119.
25. Liu F, Millar SE. Wnt/beta-catenin signaling in oral tissue development and disease. J Dent Res. 2010;89(4):318–30.
26. Lohi M, Tucker AS, Sharpe PT. Expression of Axin2 indicates a role for canonical Wnt signaling in development of the crown and root during pre- and postnatal tooth development. Dev Dyn. 2010;239(1):160–7.
27. Mitsiadis TA, Graf D, Luder H, Gridley T, Bluteau G. BMPs and FGFs target Notch signalling via jagged 2 to regulate tooth morphogenesis and cytodifferentiation. Development. 2010;137(18):3025–35.
28. Li J, Huang X, Xu X, Mayo J, Bringas Jr P, Jiang R, Wang S, Chai Y. SMAD4-mediated WNT signaling controls the fate of cranial neural crest cells during tooth morphogenesis. Development. 2011;138(10): 1977–89.
29. Haara O, Harjunmaa E, Lindfors PH, Huh SH, Fliniaux I, Aberg T, Jernvall J, Ornitz DM, Mikkola ML, Thesleff I. Ectodysplasin regulates activator-inhibitor balance in murine tooth development through Fgf20 signaling. Development. 2012;139(17): 3189–99.
30. Thesleff I. Current understanding of the process of tooth formation: transfer from the laboratory to the clinic. Aust Dent J. 2013 doi: 10.1111/adj.12102.
31. Chen Y, Bei M, Woo I, Satokata I, Maas R. Msx1 controls inductive signaling in mammalian tooth morphogenesis. Development. 1996;122(10):3035–44.
32. Mucchielli ML, Mitsiadis TA, Raffo S, Brunet JF, Proust JP, Goridis C. Mouse Otlx2/RIEG expression in the odontogenic epithelium precedes tooth initiation and requires mesenchyme-derived signals for its maintenance. Dev Biol. 1997;189(2):275–84.
33. Neubuser A, Peters H, Balling R, Martin GR. Antagonistic interactions between FGF and BMP

signaling pathways: a mechanism for positioning the sites of tooth formation. Cell. 1997;90(2):247–55.
34. Mitsiadis TA, Mucchielli ML, Raffo S, Proust JP, Koopman P, Goridis C. Expression of the transcription factors Otlx2, Barx1 and Sox9 during mouse odontogenesis. Eur J Oral Sci. 1998;106 Suppl 1:112–6.
35. Tucker AS, Matthews KL, Sharpe PT. Transformation of tooth type induced by inhibition of BMP signaling. Science. 1998;282(5391):1136–8.
36. Tucker AS, Sharpe PT. Molecular genetics of tooth morphogenesis and patterning: the right shape in the right place. J Dent Res. 1999;78(4):826–34.
37. Zhao Y, Guo YJ, Tomac AC, Taylor NR, Grinberg A, Lee EJ, Huang S, Westphal H. Isolated cleft palate in mice with a targeted mutation of the LIM homeobox gene lhx8. Proc Natl Acad Sci U S A. 1999;96(26): 15002–6.
38. Ferguson CA, Tucker AS, Sharpe PT. Temporospatial cell interactions regulating mandibular and maxillary arch patterning. Development. 2000;127(2):403–12.
39. Aberg T, Wang XP, Kim JH, Yamashiro T, Bei M, Rice R, Ryoo HM, Thesleff I. Runx2 mediates FGF signaling from epithelium to mesenchyme during tooth morphogenesis. Dev Biol. 2004;270(1):76–93.
40. Tucker AS, Headon DJ, Courtney JM, Overbeek P, Sharpe PT. The activation level of the TNF family receptor, Edar, determines cusp number and tooth number during tooth development. Dev Biol. 2004;268(1):185–94.
41. Chen S, Rani S, Wu Y, Unterbrink A, Gu TT, Gluhak-Heinrich J, Chuang HH, Macdougall M. Differential regulation of dentin sialophosphoprotein expression by Runx2 during odontoblast cytodifferentiation. J Biol Chem. 2005;280(33):29717–27.
42. Denaxa M, Sharpe PT, Pachnis V. The LIM homeodomain transcription factors Lhx6 and Lhx7 are key regulators of mammalian dentition. Dev Biol. 2009;333(2): 324–36.
43. Venugopalan SR, Li X, Amen MA, Florez S, Gutierrez D, Cao H, Wang J, Amendt BA. Hierarchical interactions of homeodomain and forkhead transcription factors in regulating odontogenic gene expression. J Biol Chem. 2011;286(24):21372–83.
44. Sharpe PT. Homeobox genes and orofacial development. Connect Tissue Res. 1995;32(1–4):17–25.
45. Juuri E, Jussila M, Seidel K, Holmes S, Wu P, Richman J, Heikinheimo K, Chuong CM, Arnold K, Hochedlinger K, Klein O, Michon F, Thesleff I. Sox2 marks epithelial competence to generate teeth in mammals and reptiles. Development. 2013;140(7): 1424–32.
46. Michon F, Tummers M, Kyyronen M, Frilander MJ, Thesleff I. Tooth morphogenesis and ameloblast differentiation are regulated by micro-RNAs. Dev Biol. 2010;340(2):355–68.
47. Jheon AH, Li CY, Wen T, Michon F, Klein OD. Expression of microRNAs in the stem cell niche of the adult mouse incisor. PLoS One. 2011;6(9):e24536.
48. Oommen S, Otsuka-Tanaka Y, Imam N, Kawasaki M, Kawasaki K, Jalani-Ghazani F, Anderegg A, Awatramani R, Hindges R, Sharpe PT, Ohazama A. Distinct roles of MicroRNAs in epithelium and mesenchyme during tooth development. Dev Dyn. 2012;241(9):1465–72.
49. Thesleff I. Epithelial-mesenchymal signalling regulating tooth morphogenesis. J Cell Sci. 2003;116(Pt 9):1647–8.
50. Thesleff I. Developmental biology and building a tooth. Quintessence Int. 2003;34(8):613–20.
51. Goldberg M. La dent normale et pathologique. Bruxelles: De Boeck University; 2001.
52. Nanci A. Ten cate's oral histology. Development, structure, and function. Philadelphia: Mosby Elsevier; 2008.
53. Jernvall J, Thesleff I. Reiterative signaling and patterning during mammalian tooth morphogenesis. Mech Dev. 2000;92(1):19–29.

Pulp Anatomy and Characterization of Pulp Cells

2

Michel Goldberg

2.1 Introduction

A dental tooth is formed by a series of mineralized tissues, including three outer mineralized layers located at the periphery of the teeth: enamel, dentin, and cementum. They are surrounding a non-mineralized dental soft tissue located in the inner part of the tooth, the dental pulp (Figs. 2.1a–d, 2.2, 2.3, and 2.4).

2.2 General Organization

In adult teeth, odontoblasts and Hoehl's cells form a superficial layer at the periphery of the pulp, contributing to the configuration an outer border lining the dental pulp. Odontoblasts are implicated in the production of the extracellular predentin/dentin matrix and, subsequently, they are involved in the dentin mineralization process. Odontoblasts and Hoehl's cells take origin from the neural crests. The progenitors migrate toward the first branchial arch and contribute to the formation of tooth germs. They differ substantially from the pulp cells, both from a developmental point of view, displaying specific composition and functionality. Additional differences were suggested between the cell-free zone presumably located at the pulp surface, beneath the subodontoblastic cell layer, now recognized as a fixation artifact, and the cell-rich zone, containing progenitor cells that display plasticity and pluripotent capabilities.

This points out the complexity of the odontoblast, Hoehl's cells, and pulp layers. Although reference is often made in the literature to the existence of a so-called dentino-pulpal complex, much evidence denies this notion. This working hypothesis refers apparently to physiopathologic pain perception, pulpitis, and dental treatments but is not actually based on any biological proof. This statement is clearly questionable both with respect to anatomical and biological specificities, gene and transcription factors expression, and with respect to the embryological origin of the tissues, the neural crest-derived tissues differing from mesenchymal branchial arch [1].

The dental pulp is formed by cells which are implicated in the secretion and reorganization of a collagen-rich extracellular matrix (ECM). Pulp cells play a crucial role in the synthesis and in the ECM molecules remodeling:

(i) In the dental pulp, a series of characteristic cells have been identified. The pulp stromal fibroblasts, also named pulpoblasts by Baume [2], constitute the most abundant pulp cell population. In addition, other cell lineages have been recognized in the pulp: specifically progenitors (also named stem

M. Goldberg, DDS, PhD
Department of Oral Biology,
Institut National de la Santé et de la Recherche Médicale, Université Paris Descartes,
45 Rue des Saints Pères, Paris 75006, France
e-mail: mgoldod@gmail.com,
michel.goldberg@parisdescartes.fr

M. Goldberg (ed.), *The Dental Pulp*,
DOI 10.1007/978-3-642-55160-4_2, © Springer-Verlag Berlin Heidelberg 2014

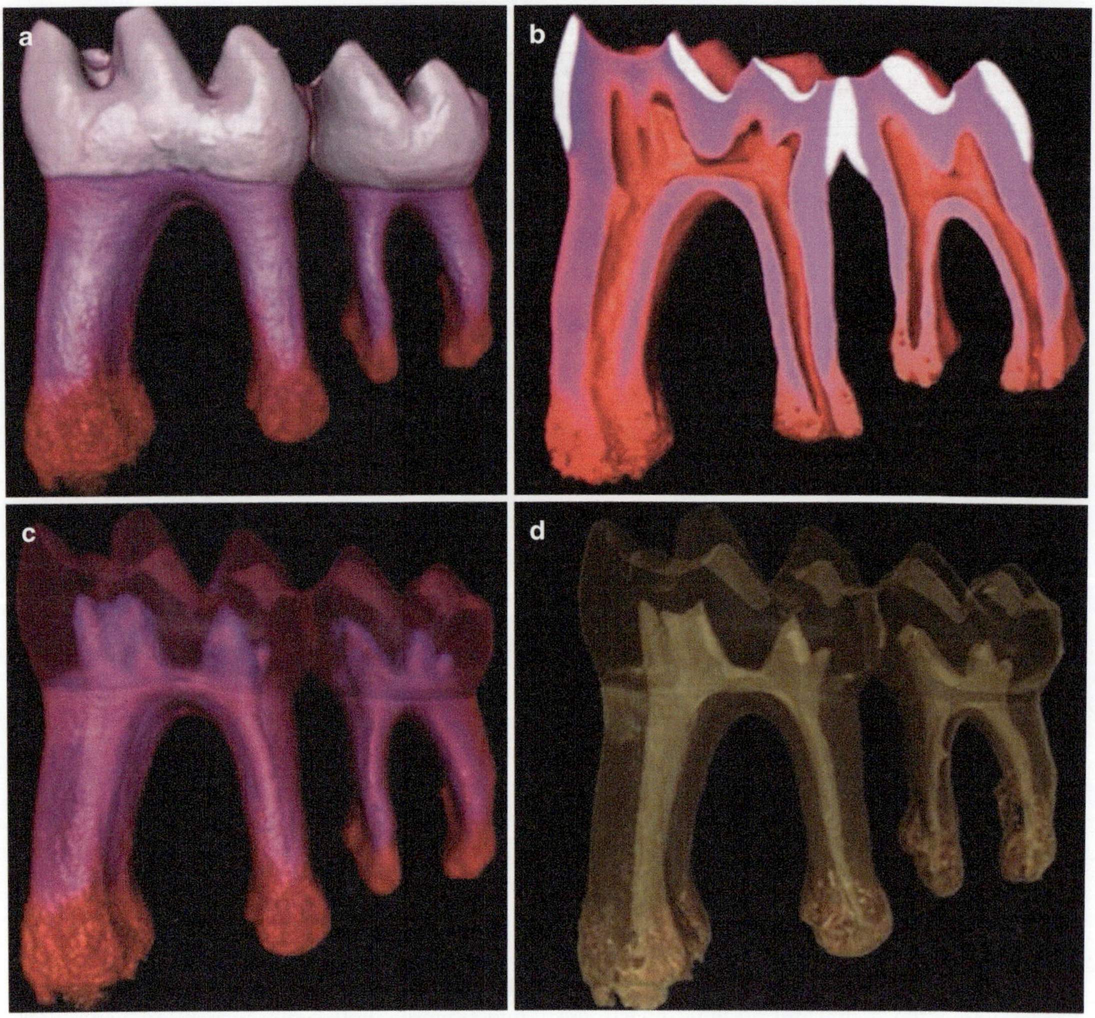

Fig. 2.1 (**a–d**) MicroCT reconstruction of a rodent mandibular molar showing in (**a**) the outer surface of the first mandibular molar; in (**b**) a virtual section showing, namely, the enamel, dentin, pulp, and cementum; and in (**c**, **d**) the 3D reconstruction

cells, or side population) and neuronal, vascular, and immune system cells. *Resident structural cells* are permanently subjected to renewal and apoptosis. This cell population includes stem cells, which are acting as progenitors. According to some researchers, in the dental pulp, a ~9 % positivity was found for STRO-1, a stem cell marker, whereas according to Kenmotsu et al. [3] between 0.11 and 0.40 % of the pulp cells are stem cells (Figs. 2.5 and 2.6).

(ii) In addition, a few *nonresident cells* are found. They migrate from the blood or take origin in the bone marrow and/or other non-dental tissues. They migrate and penetrate within the dental pulp through the apical foramen.

Different adjacent domains contribute to the pulp organization, each domain bearing its own specificity. The dental pulp includes a mosaic of territories, varying in the crown and the root, and in the central part versus the peripheral pulp as well.

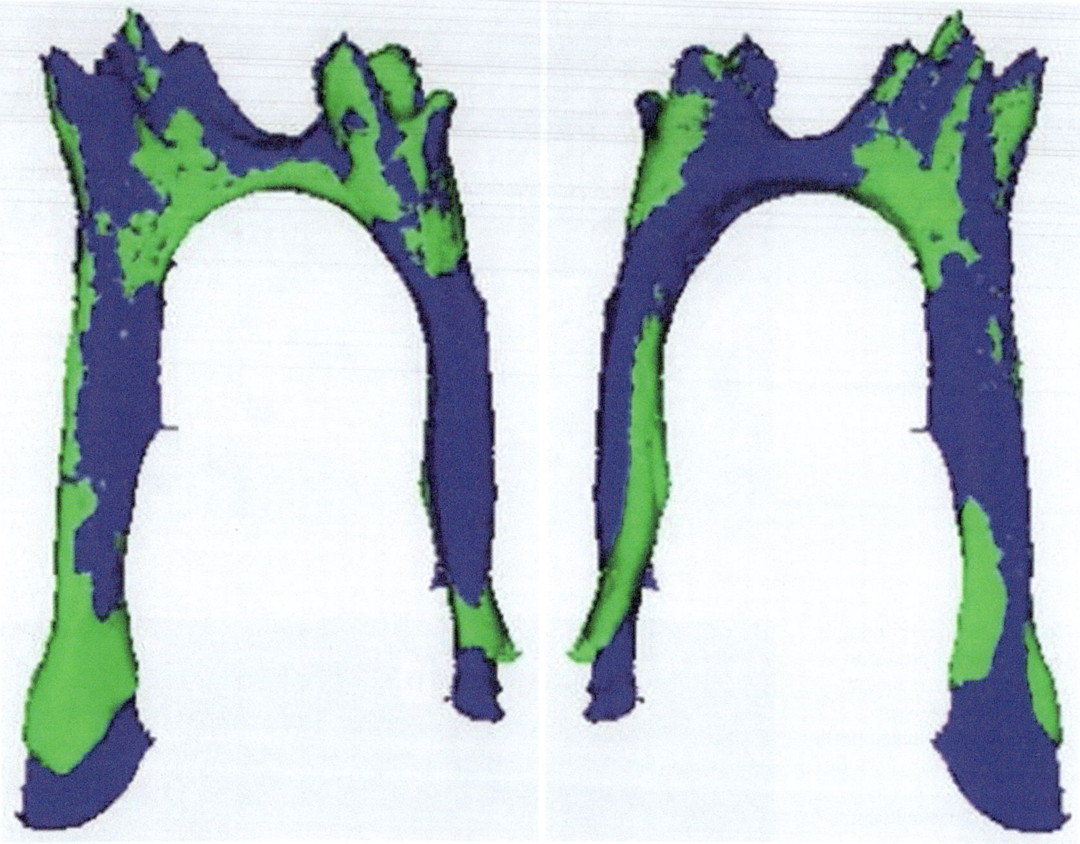

Fig. 2.2 3D reconstitution of the pulp of a wild-type (in *green*) molar compared to the pulp of a $5HT_{2B}R$ KO mouse in *purple*. The pulp of the KO mice is longer and wider. The comparison between the age-matched WT and KO mice shows that dentinogenesis is altered by the receptor deletion

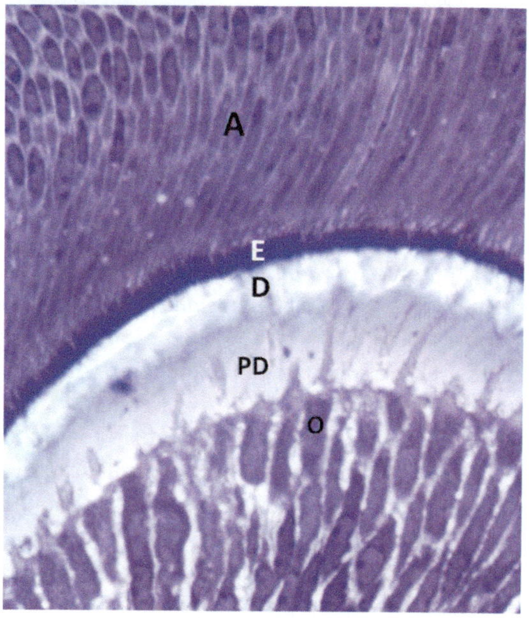

Fig. 2.3 Postmitotic secretory odontoblasts (*O*) are implicated in the secretion of predentin (*PD*) and in dentin (*D*) mineralization. At this early stage of odontogenesis, a thin layer of forming enamel is secreted by the secretory ameloblasts (*A*)

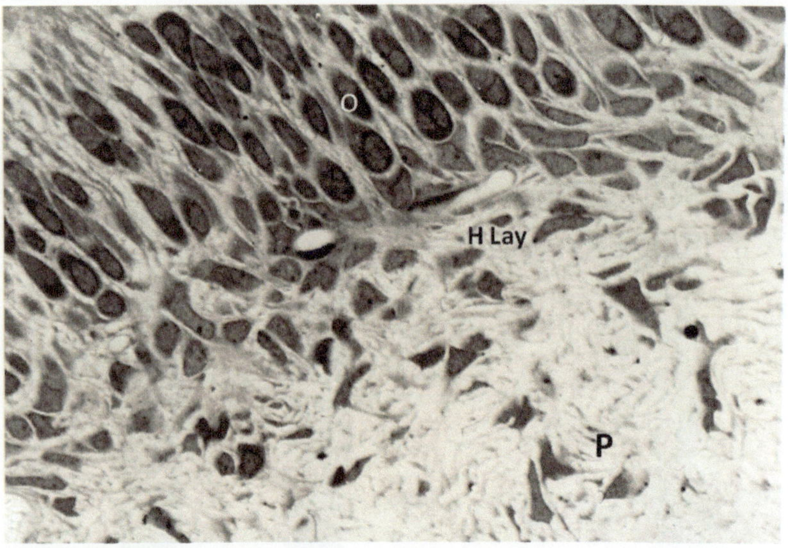

Fig. 2.4 From the outer layer (*upper part*) to the inner zone (*lower zone*), odontoblast cell bodies (*O*) and Hoehl's cell layer (*H Lay*) are located at the pulp periphery (*P*)

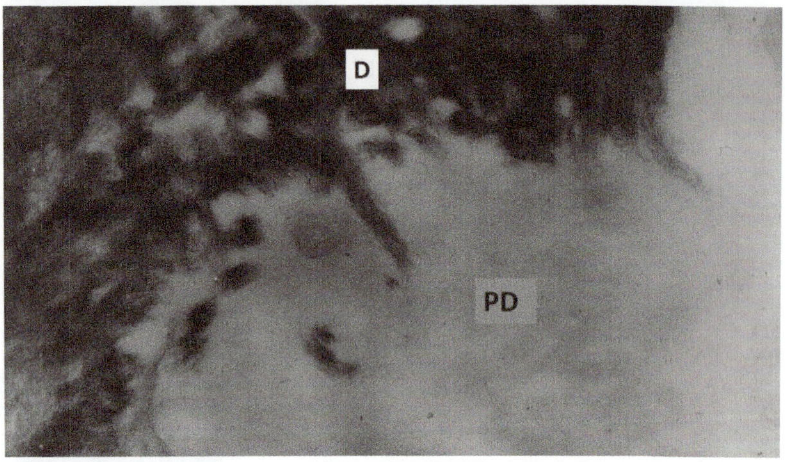

Fig. 2.5 Unstained ultrathin section. Dentin-predentin junction. Hydroxyapatite needlelike crystals are located along collagen fibrils at the mineralizing front in dentin (*D*), whereas no mineral is detectable in the predentin (*PD*)

2.3 Odontoblasts

The postmitotic *odontoblasts* are implicated in dentinogenesis. A cell body and a cell process form each odontoblast. Cell bodies are grouped in four to five rows parallel to the tooth surface at the periphery of the pulp.

Each *cell body* comprises a basal third containing an abundant rough endoplasmic reticulum (RER) and mitochondria. A nucleus is also located in this basal third. Many cells display cilia. Cytoskeletal proteins direct the shape and functionality of the cell bodies. In the central part, dictyosomes are located in the supranuclear area. Multivesicular vesicles and lysosomal electron-dense vesicles, with variable content, are also detected. The RER located along the lateral borders contributes to the MEC synthesis. The Golgi apparatus and multivesicular bodies play role in the terminal steps of ECM synthesis and, after re-internalization of molecules cleaved by metalloproteases (MMPs), in the control of degradation. In the distal cell body, the RER is interrupted at halfway. Small mitochondria are grouped near the place where the processes take origin.

The *odontoblast main processes* display bundles of cytoskeletal proteins such as microtubules,

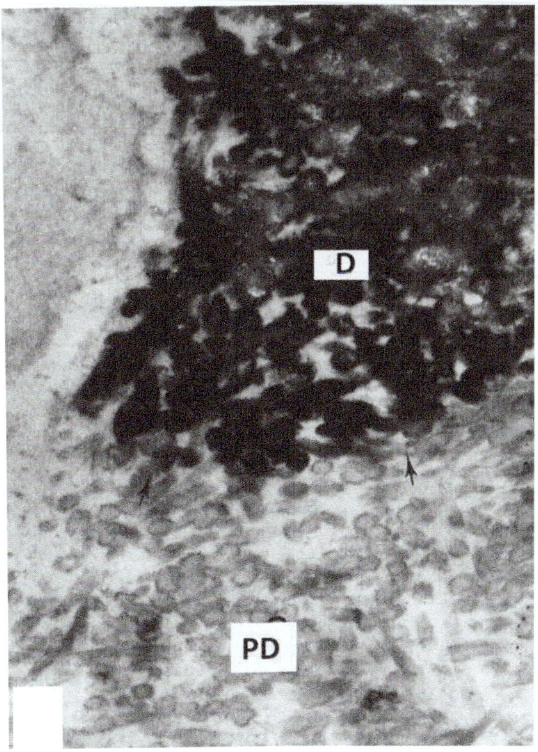

Fig. 2.6 Junction between dentin (*D*) and predentin (*PD*) stained with the phosphotungstic acid/chromic acid mixture. Dentin loaded by the dentin sialophosphoprotein is heavily electron dense, whereas in the predentin (PD), thin rings around individual collagen fibrils revealed electron-dense phosphorylated proteins

intermediary filaments identified as vimentin and nestin, and actin microfilaments. These later contribute to a sub-plasmalemmal undercoat. Secretory vesicles and acid phosphatase-rich endocytotic vesicles (coated vesicles) are implicated in active secretion and/or reabsorption. Odontoblast processes cross the predentin and penetrate inside dentin tubules either in the inner third or along the whole dentin length, up to the dentinoenamel junction. The diameter of odontoblast lateral branchings is thinner. They establish connections between tubules, penetrating in minute tubules and crossing the whole thickness of the hypermineralized peritubular dentin. The lateral branches do not contain nestin, but only vimentin and actin.

Odontoblasts have a limited lifespan and when they mature and become aged, they start to be loaded by lysosomes and autophagic vacuoles. Then, they become apoptotic cells [4]. The number of odontoblast inside the outer cell layers is gradually reduced. The cells become smaller, and eventually they are reduced to a single layer. Differentiating Hoehl's cells, which behave as odontoblast second generation, presumably renews them [5].

During the teeth formation, odontoblasts are implicated in the synthesis and secretion of the dentin extracellular matrix (Figs. 2.7 and 2.8a–c).

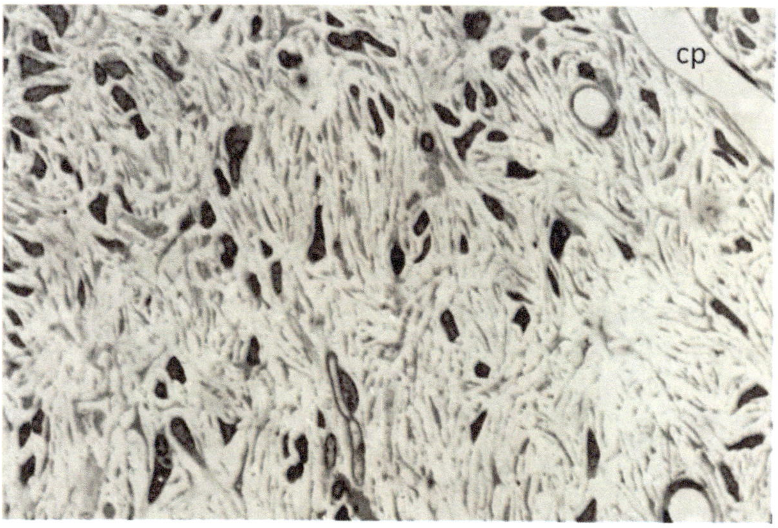

Fig. 2.7 Pulp cells include a heterogeneous population of fibroblasts (pulpoblasts) and endothelial cells of capillaries (*cp*)

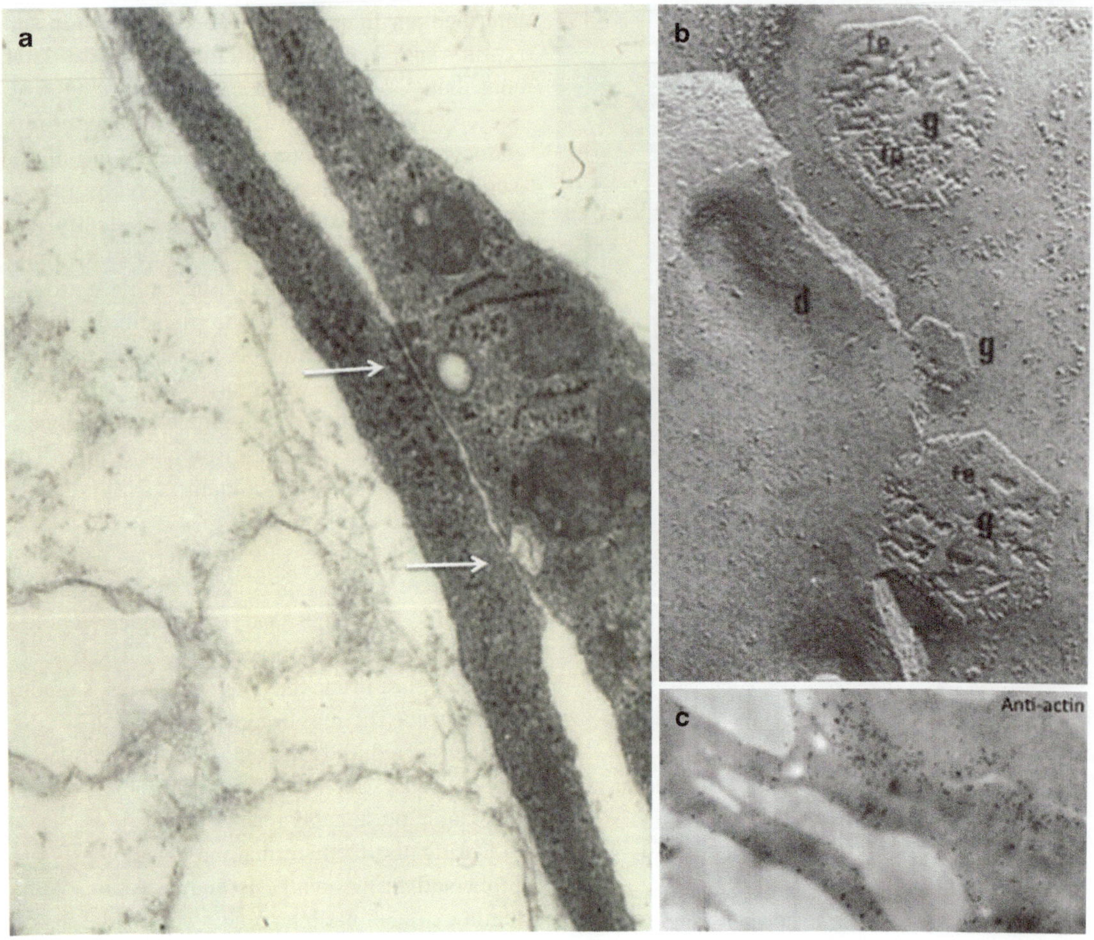

Fig. 2.8 (**a**) Intercellular junctions of the gap junction and desmosome-like type between pulpoblasts (*white arrows*). Thin collagen fibrils are detectable in the extracellular spaces. (**b**) Freeze fracture replica obtained after rapid freezing-freeze substitution. *G* gap junction (*e* or *f* faces), *d* desmosome-like structure. (**c**) Plasmalemmal undercoat after immunogold labeling with an anti-actin antibody

The cells issued from the neural crest migrate toward the first branchial arch and settle near the dental lamina of the mandibular, maxillary, and nasofrontal buds. In front of the epithelial dental lamina, the pre-odontoblasts divide and migrate from the central to the outer part of the pulp. The number of mitosis is fixed depending on the species. During the last division of pre-odontoblast, an asymmetric division occurs. The taller cells establish a limited contact with the basement lamina (BM). These pre-odontoblasts become presecretory pre-polarized odontoblasts. The smaller cells are located some distance away from the basement membrane. They are clustered in Hoehl's cell layer. At an early stage, odontoblasts contribute to the formation of the coronal dentin. They are implicated in the synthesis of an extracellular matrix, implicated in dentin mineralization. The first outer dentin layer formed is atubular. This layer is also named mantle dentin in the crown. Afterward they are involved in the physiologic primary and secondary dentin formation.

After crown completion, during the next stage, the root starts to be formed. Pre-odontoblasts migrate from the central dental pulp toward the periphery, just beneath Hertwig's epithelial root sheath. Differentiation of pre-odontoblasts precedes the terminal differentiation of these cells into radicular odontoblasts. There is still a debate to clarify if the inner layer of Hertwig's epithelial layer is susceptible to transdifferentiate. Due to phenotypic changes, epithelial cells become cementoblasts and afterward may become cementocytes. As an alternative possibility, pre-cementoblasts issued from the dental follicle may slide between the unbound cells of Hertwig's epithelial root sheath. Intercellular junctions are disrupted, and intercellular spaces enlarge. Mesenchymal cells of the dental follicle infiltrate the spaces of the root sheath and come in contact with the outer dentin surface of the root, where they acquire the final cementoblast phenotype. The dental follicle contributes to the formation of the bonny socket and of the dental ligament. Subsequently the root formation starts, preceding tooth eruption and pulp lengthening. In the dental pulp, stem cells or odontoblast progenitors differentiate and contribute to root dentinogenesis. After the outer dentin layer(s) formation (Tomes' granular layer and/or amorphous Hopewell-Smith layers), the root circumpulpal dentin starts its initial construction either as tubular dentin or appearing as a fibrodentin structure.

After tooth formation, odontoblasts are located at the periphery of the pulp keeping a pseudostratified palisade structure. Primary dentinogenesis occurs during the early secretory period of tooth formation. The synthesis and secretion of ECM are gradually reduced and autophagic activities increase. Thereafter comes a period of decreasing activity for the odontoblasts. The *primary* dentinogenesis starts just after the formation of the mantle dentin and stops when the teeth become functional, with occlusal pressures. As postmitotic cells, odontoblasts are implicated in the formation and maintaining of dentin. Quiescent odontoblasts are implicated in the formation of *secondary* dentin during all life span. During aging, odontoblast develops an autophagic-lysosomal system organized in large vacuoles, which are acid phosphatase positive. The lysosomal markers LC3 and LAMP2 are indicative of a dynamic autophagic activity, implicated in the turnover and degradation machinery. Accumulation of lipofuscin was seen within lysosomes [4]. *Reactionary* (*or tertiary*) dentin is produced in response to a carious lesion, to abrasion, or a noxious reaction to dental materials.

2.3.1 Subodontoblastic Layer/ Hoehl's Cell Layer

Presumably Hoehl's cells take origin from the last pre-odontoblast cell division (Figs. 2.5 and 2.6). Odontoblasts form originally a structured layer including about four rows of cells. During tooth maturation and aging, the number of odontoblasts is gradually reduced, due to apoptotic events, and finally they appear as a thin cell monolayer. There is high probability that odontoblasts have a limited life span. In this context, subodontoblastic cells are implicated in cell replacement. Their terminal differentiation is activated, and they become what have been named second-generation odontoblasts. This hypothesis is reinforced by the fact that Hoehl's cells express high alkaline phosphatase activity. The majority of subodontoblastic cells express Thy-1, a cell surface marker of stem cells and progenitors. The capacity of Thy-1 to be expressed by the subodontoblastic cells was evaluated following stimulation with BMP-2. Thy-1 positive cells showed accelerated induction of ALP activity. They formed alizarin red-positive mineralized nodules and induced the formation of bone-like matrix. Hosoya et al. [6] concluded that the subodontoblastic cells have the ability to differentiate into hard tissue-forming cells, and consequently they may serve as a source of odontoblastic cells. To conclude, odontoblasts and the subodontoblastic layer are both involved in dentinogenesis. These cells contribute to reactionary dentin formation, whereas pulp cells are implicated in reparative dentin formation.

For reviews on the biology of odontoblasts, see Refs. [5, 7].

2.4 Stromal Fibroblasts (or Pulpoblasts)

2.4.1 Resident Cells

2.4.1.1 Phenotypic Characterization

Most of the pulp cells are *resident cells*. They play a structural role, shaping the construction of the pulp and acting as feeder cells. These cells

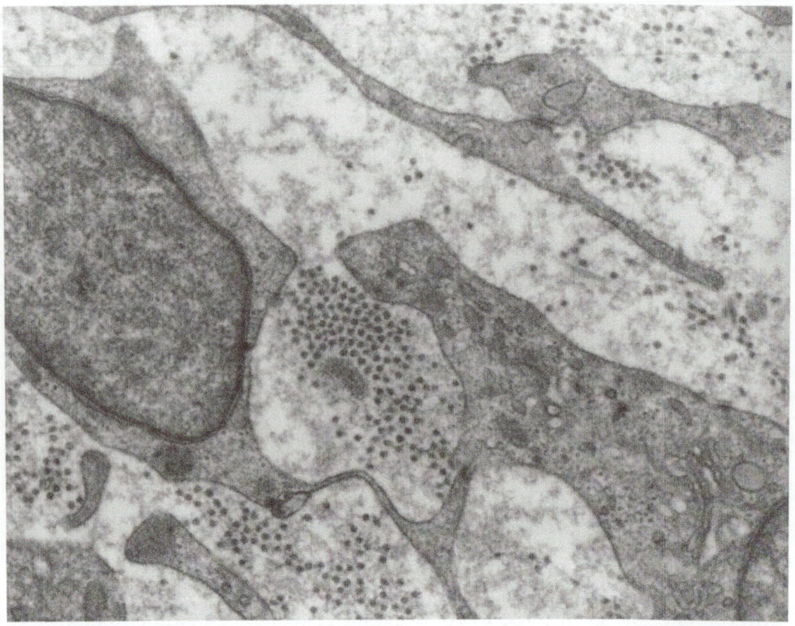

Fig. 2.9 Pulp fibroblasts. In the extracellular spaces, different types of collagen fibrils are found, corresponding to type I (aggregates of round structures) and type III collagens (ramified thin and thick fibrils less electron dense)

were identified on the basis of their morphology. Pulp fibroblasts (or pulpoblasts) are elongated cells, with narrow diameter and long protracted processes. In the dental pulp, fibroblasts (or pulpoblasts) are fusiform cells, bound by intercellular desmosome-like, gap junctions and a few tight junctions. A small number of stem cells, or pulp progenitors, are included in this group (Figs. 2.9, 2.10, and 2.11).

A few other cells take origin and migrate from blood circulation, bone marrow, and other nondental tissues. This second group constitutes the *nonresident cells*, also designed as migrating mesenchymal cells. They penetrate the pulp by the apical part and invade the tissue.

There is much evidence that resident and nonresident cells have a limited life span. In order to keep a continuous volume and maintain the diverse functions of the tissue, a constant renewal of pulp cells is mandatory. An experimental approach was conducted on young rats, using a essential fatty acid-deficient diet (EFAD) from day 0 to day 21 (Group I), whereas another group received after birth a normal diet, followed at day 21 for 4 weeks by a deficient diet (Group II). These two groups were compared with a group of rats receiving a normal diet (Group III) and with a per-fed group (a group of rats receiving a reduced food intake) (Group IV) [8]. The number of cells/mm^2 was scored in the central part and in the subodontoblastic lateral areas. In groups I and II, the cell density was related to the experimental period of time selected. In the EFAD rats, cell accumulation was seen initially in the central pulp. This was followed by the sliding of pulp cells located in the central coronal pulp, migrating toward the lateral subodontoblastic area. Near this limit, pulpoblasts were subjected to apoptosis [9]. Dendritic cells and macrophages seem to be implicated in the destruction of pulp fibroblasts. Apoptotic bodies were engulfed by macrophages and destroyed inside lysosomes in an outer subodontoblastic region. This pointed to equilibrium between the generation of new pulpoblasts emerging in the central pulp and the cell destruction occurring mostly in the outer pulp border (Figs. 2.12 and 2.13).

Human dental pulp cells display pluripotency. BMP-2 treatment produced alkaline phosphatase in the cells, which produce and secrete osteocalcin in the culture medium. They also overexpress Sox2, Col2, and ColX. They express two adipocyte-specific transcripts PPARg2 and LPL. Koyama et al. [10] concluded

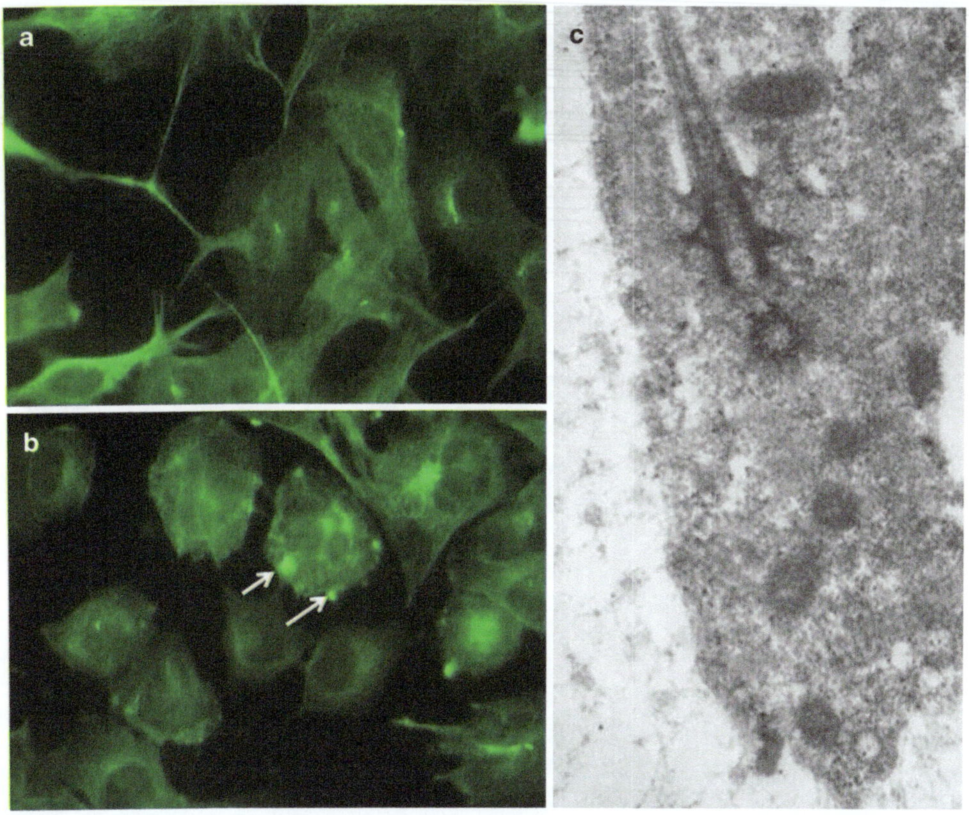

Fig. 2.10 (**a**–**c**) Immunostaining of cilium using an alpha acetyl tubulin antibody (*white arrows*) (**a**, **b**). In (**c**), a cilium and basal body in an odontoblast cell body. This structure formed by bundles of microtubules contributes to act as a receptor in the odontoblast cell bodies

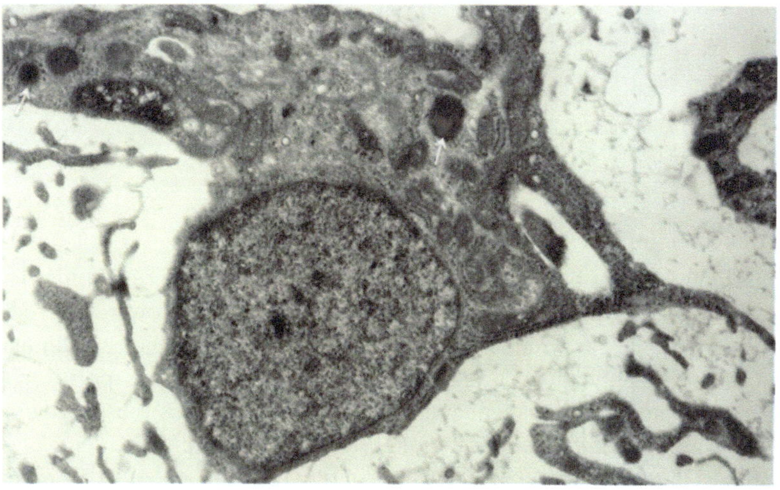

Fig. 2.11 Pulp fibroblasts are also implicated in lysosomal (*white arrow*) and catabolic activities

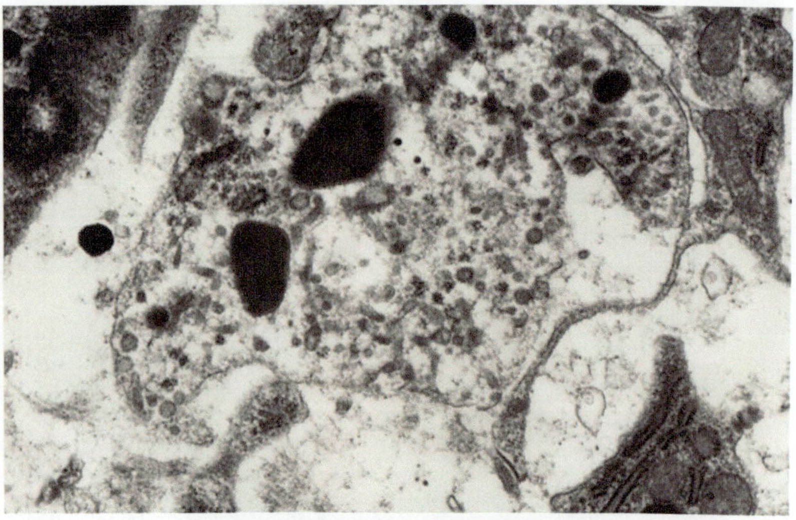

Fig. 2.12 Histiocyte/macrophage pulp cell displaying tubulovesicular structures and electron-dense lysosomes

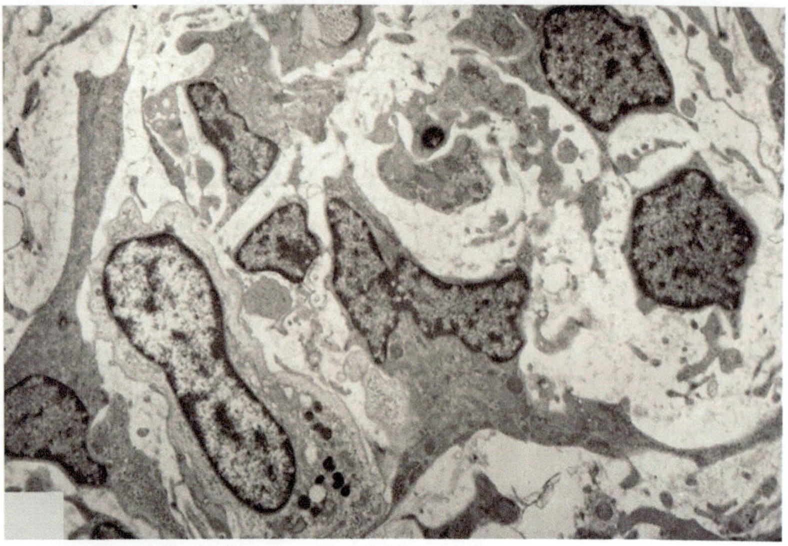

Fig. 2.13 Heterogeneous cell population in the dental pulp. In addition to fibroblasts, macrophages are seen, loaded with electron-dense lysosomes. Many apoptotic bodies are present in the intercellular spaces

that pluripotent pulp cells differentiate into osteoblasts, chondrocytes, and adipocytes. The dental pulp may provide enough cells for potential clinical applications.

Multipotent and unipotent dental pulp progenitors are currently detected among pulp stem cells. Three clonal pulp precursor lines were established from embryonic ED18 first molars of mouse transgenic for a recombinant plasmid adeno-SV40. The clones were cultured on 2D and 3D scaffolds. They were induced to differentiate toward the odontogenic/osteogenic, chondrogenic, or adipogenic lineage. The A4 clone has the capacity to be recruited toward at least three mesodermal lineages, contributing to dentin-like or bone formation. The A4 lines appeared to be multipotent cells. In contrast C5 and H8 displayed a more restricted potential and were monopotent. In the dental pulp, progenitor cells demonstrate the coexistence of multipotent

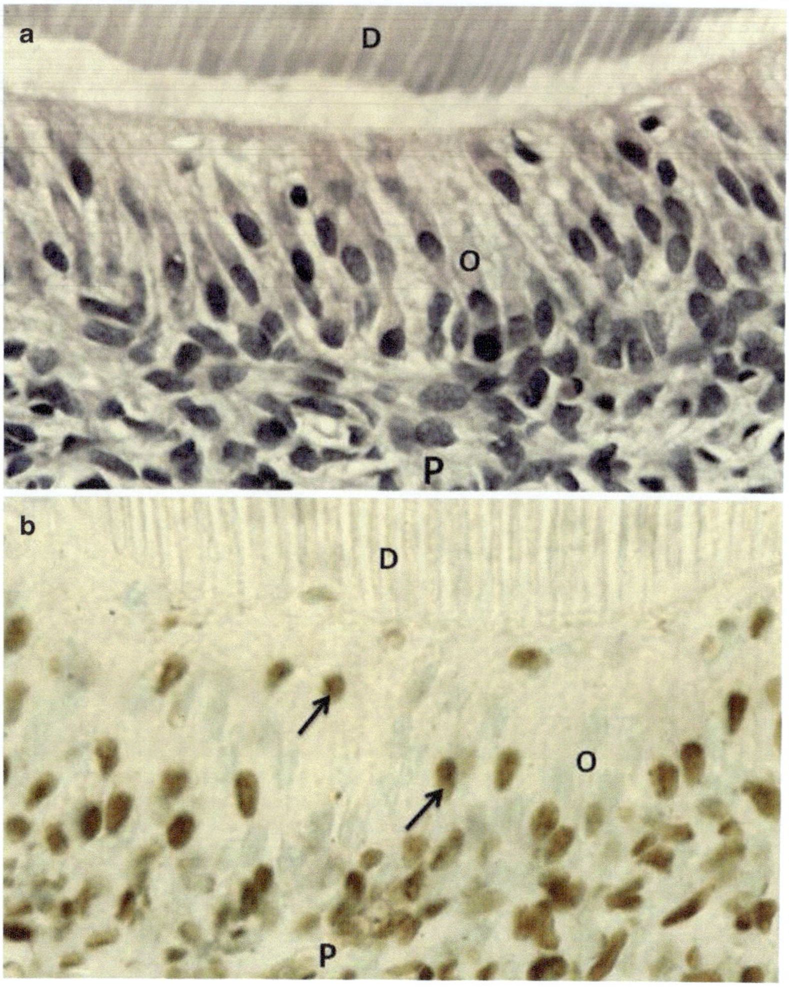

Fig. 2.14 (**a**) Hematoxylin/eosin staining of a section including the inner part of dentin (*D*), odontoblasts (*O*) cell bodies, and pulp (*P*). (**b**) Staining the section with the TUNEL method allows visualization of a few apoptotic cells (*arrow*) within the odontoblast layer, and many cells in the so-called Hoehl's cell layer above the pulp

cells and cells with restricted lineages [11]. The A4 cells express DMP-1 after 4 days of culture, reaching a maximum at 7 days. Then, the expression of the protein was decreased. Fourteen days are needed for A4 cells to reach a maximal expression of DSPP.

2.4.2 Intracellular Proteins

In addition to the series of constitutive proteins, found in most cellular compartments, some intracellular molecules seem to be more specifically expressed by the *stromal pulp fibroblasts*. They are implicated both in cell differentiation, in the synthesis of ECM molecules, and after endocytosis, in their degradation by MMPs (collagenases, MMP-3, MMP-2, and MMP-9). Pulp cells are also involved in ECM lipids endocytosis [12] (Figs. 2.14a, b, 2.15, and 2.16a, b).

Senescent human dental pulp cells display increased autophagic activity. Senescence-β-galactosidase, microtubule-associated protein light chain 3, and Beclin 1 staining were higher in old cells compared with young pulp. More autophagic vacuoles were seen in senescent cells [13]. Apoptosis of dental pulp cells was investigated in the rat incisor by the TUNEL method. It was reported that the number of apoptotic cells was increased in the incisal direction. OX6, ED1, and ED2 antibodies were used to detect macrophage and dendritic cells. Apoptotic bodies were

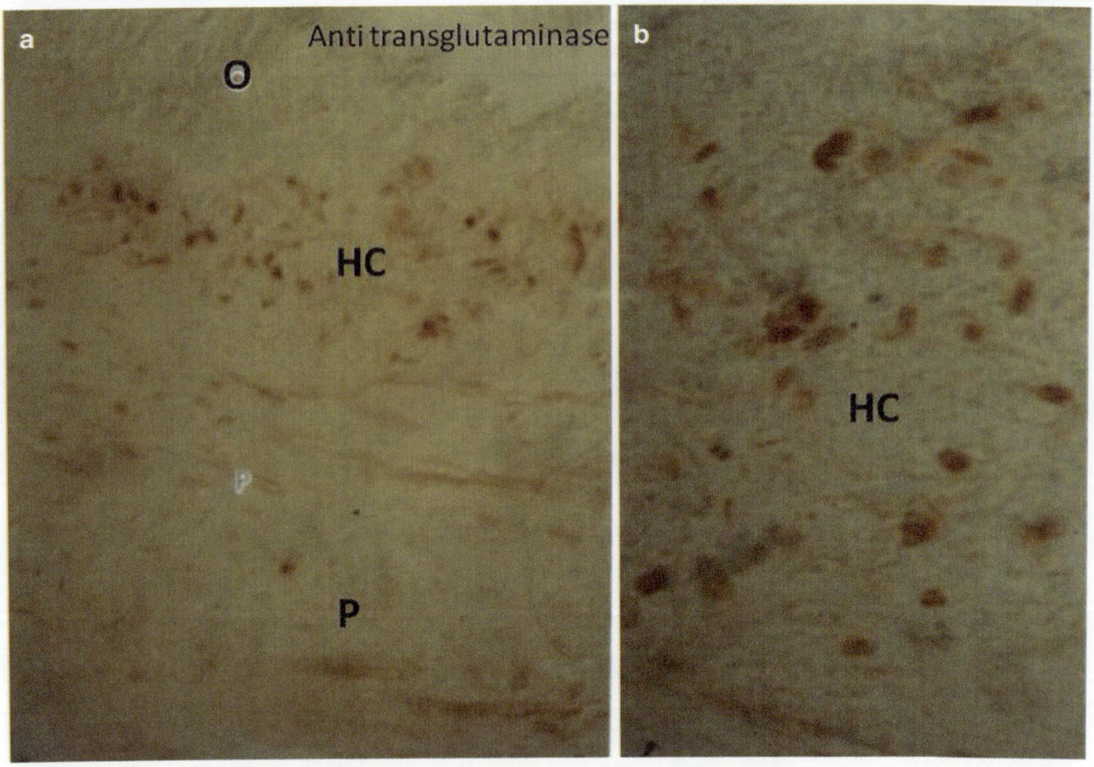

Fig. 2.15 (**a**, **b**) Immunostaining with an anti-transglutaminase labels mostly the cells located beneath the subodontoblastic Hoehl's cell layer (*HC*). *O* odontoblasts, *P* pulp

eliminated mainly by MHC class II-expressing cells [14] (Figs. 2.17 and 2.18).

Positive immunostaining was seen for *vimentin* and *α1 type 1 collagen*. In contrast, the staining for CD45, CD34, and cytokeratin was negative as evidenced by flow cytometry. The cytoskeleton was well developed. *Connexin 43* is a 43 kDa gap junction protein forming hexametric structures creating intercellular channels and allowing cell-to-cell diffusion of small molecules. Pulp cells transmit information as a syncytium via gap junctions. Lucifer yellow penetrates between odontoblasts and stains the superficial pulp cells [15]. Specific immunolabeling revealed *α-smooth muscle actin* and intermediary filaments, namely, vimentin. Observed with the TEM, the cells looked like myofibroblasts or pericytes with stress fibers, fibronexus, indented nuclei, and gap junctions linking the pulp cells [16]. TGF-β enhances the expression of α-smooth muscle actin in cultured human pulpal fibroblasts [17]. *Nestin* is also an intermediary filament. It was reported to be specific of the odontoblast lineage. This seems to be true for some species but erroneous for some other groups of animals where vascular endothelial cells, nerves, and other types of cells reacted also positively to nestin antibodies.

Smad 1/5 is involved in BMP-2-induced odontoblastic differentiation in human dental pulp cells. BMP-2 induces the differentiation of human dental pulp cells (DPCs) into odontoblast-like cells. BMP-2 induces the phosphorylation and nuclear translocation of Smad 1/5. *Noggin*, a BMP signaling inhibitor, inhibits alkaline phosphatase activity and odontoblast differentiation. DPCs treated with BMP-2 display a reduced formation of mineralized nodules [18]. *Histone deacetylase inhibitors* (HDACIs) alter the homeostatic balance between two groups of

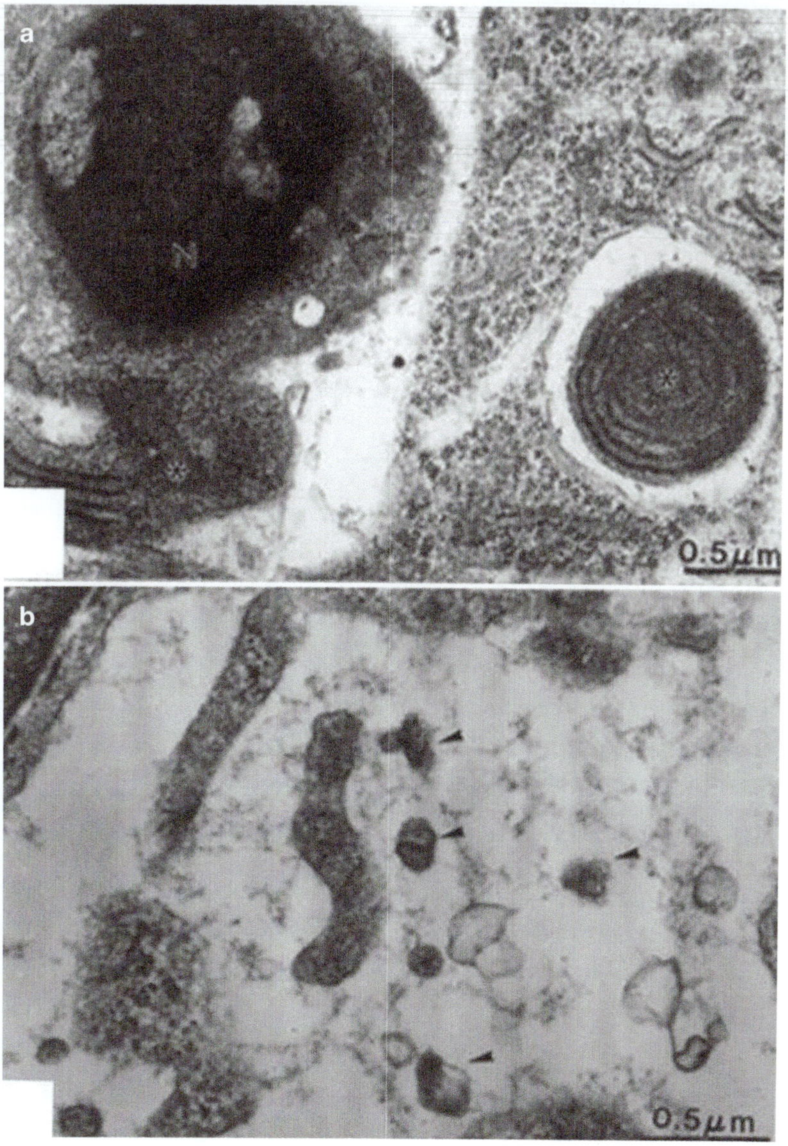

Fig. 2.16 (**a**) Apoptotic cell with marginal chromatin concentration in the nucleus (*N*). A cell-derived fragment including rough endoplasmic reticulum is degraded in an electron-lucent vesicle. In (**b**) apoptotic bodies are seen in the intercellular space (*arrowheads*)

cellular enzymes, *histone deacetylases* (HDACs) and *histone acetyltransferases* (HATs), increasing transcription and influencing cell behavior. Epigenetically modulating pulp cell behavior has direct implications on pulp regeneration [19].

A small number of mesenchymal cells are derived from human dental pulp. Pulp stem cells are implicated in stemness. Small molecules (*Pluripotin* (*SC1*), *BIO*, *and rapamycin*) are implicated in the maintenance of stem cells. They proliferate, display expression of pluripotent and mesenchymal stem cell markers, and exhibit differentiation capacities and intracellular signaling activation. The expressions of STRO-1, NANOG, OCT4, and SOX2 are increased, while the cell differentiation into odontogenic/osteogenic, adipogenic, and neurogenic lineages was diminished in vitro [20].

The RER occupied the lateral borders, with flat cisternae, intermediate elements, and spherical distensions. In the central part of the cell, near the nucleus, Golgi saccules and secretory granules

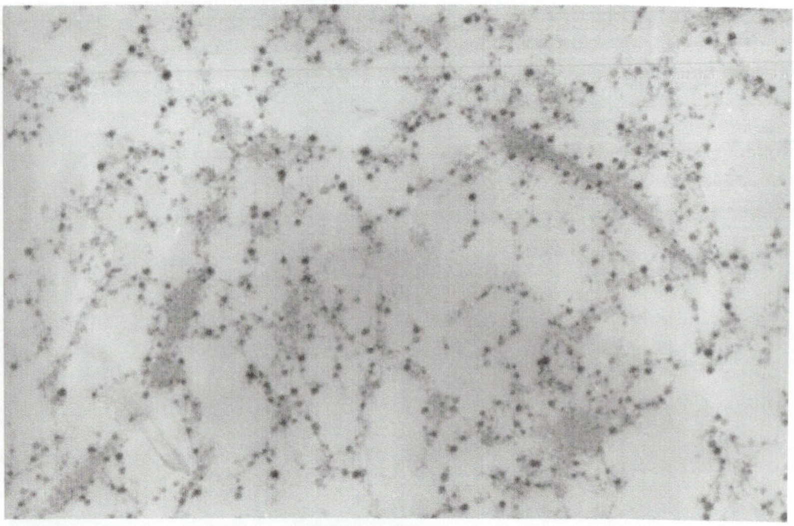

Fig. 2.17 Fibrillar collagen and the associated electron-dense proteoglycan granules are seen in the intercellular spaces

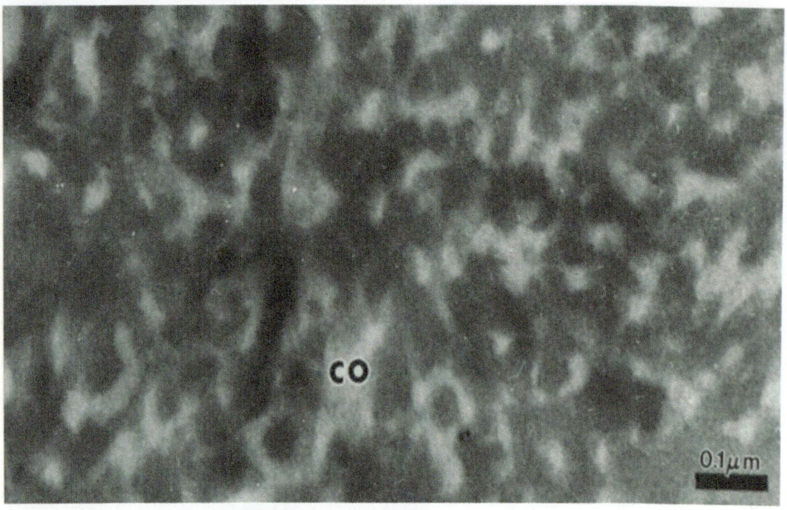

Fig. 2.18 After rapid freezing and freeze substitution, the collagen fibrils (*co*)appear as electron-lucent areas, whereas the inter-collagenous spaces are filled with a homogeneous amorphous expanded proteoglycan gel

play a crucial role in the assembly of ^{3}H-labeled precursors and in the maturation and secretion of the collagen-rich extracellular matrix (ECM) [21]. Pulp cells and odontoblasts incorporate [^{3}H] proline as a collagen precursor. Pulp cells are implicated in the synthesis and secretion of a network of distended thin collagen fibrils. [^{33}] P and [^{3}H] serine are synthesized and incorporated in dentin as *phosphoprotein precursors* but also in the pulp within *phosphorylated ECM proteins* [22, 23]. *Tenascin C* promotes the differentiation of rat dental pulp cells in vitro by activation of Notch 1. In culture, pulp cells form mineralizing nodules, observed until the fifth passage. Mineralization was confirmed by the presence of calcium and phosphate. *DMP-1* (10 ng/mL) added to lipopolysaccharide stimulates the production of IL-6 and IL-8 from pulp fibroblasts. Inhibition of the p38 mitogen-activated protein kinase pathway blocked the proinflammatory effect of DMP-1 on pulp fibroblasts [24]. DMP-1 has a let-7 binding site in its 3′-untranslated region, directly regulated by the members of let-7. DMP-1 is regulated posttranscriptionally by let-7 during odontoblast differentiation.

2.4.3 Nonresident Pulp Cells

Nonresident pulp cells, also referred to adult mesenchymal stromal cells (MSCs), are undifferentiated multipotent cells, which reside primarily in the bone marrow. They have the potential to differentiate into three skeletal phenotypes such as osteoblasts, chondroblasts, and adipocytes. MSCs are lacking expression of the hematopoietic markers CD45, CD14, CD34, HLA-DR, and CD31, an endothelial marker. They are expressing CD105, CD73, and CD90 [25, 26].

2.5 Extracellular Matrix and Cell Differentiation

The ECM is composed by thin *collagen* fibrils (Fig. 2.17). After aqueous aldehyde fixation incorporating cationic dye, *glycosaminoglycans* (GAGs) appear as small aggregates or dots punctuating the fibrils. After physical fixation (rapid freezing followed by freeze substitution), GAGs form an expanded amorphous gel [27] (Figs. 2.17 and 2.18). This organization favors the transport of the cells from the place where they emerge as progenitors and differentiate and migrate toward the place where they underwent apoptosis. Cells are fragmented and disappear, leaving within the pulp, cell remnants and/or apoptotic bodies. Type I (56 %), III (41 %), and V (2 %) *collagens* are found in the sound pulp. *BSP* and *OPN* are the main phosphorylated matrix proteins. *Fibronectin*, *versican*, and *glycosaminoglycans (GAGs)* (CS-4 and CS-6; DS and KS) are linked to core proteins. This association contributes to *proteoglycan* formation. In contrast, the expanded *hyaluronic acid* is not bound to any protein and stays as GAGs. *Fibrillins*, the major constituents of microfibrils, facilitate the release of active transforming growth factor-β (TGF-β). They are involved in barrier formation in exposed human dental pulps. Some factors seem to be associated with fibrillin degradation and cytodifferentiation. Fibrillin-1 immunoreactivity was seen until day 7 after a surgical exposure and becomes undetectable at 14 days, just beneath the exposure site. MMP-3 was transiently detected at day 14. At 42 days, fibrillin-1 immunoreactivity and fibrillin-1 expression were downregulated. Degradation and downregulation of fibrillin take place during the formation of a mineralized tissue barrier [28].

The effects of recombinant human DSP (rh-DSP) on proliferation, migration, and odontoblast differentiation in human dental pulp cells demonstrate that rh-DSP markedly increase ALP activity, calcium nodule formation, and the levels of odontoblastic marker mRNA. Rh-DSP increased BMP-2 expression and Smad1/5/8 phosphorylation. The BMP antagonist noggin blocked phosphorylation. Rh-DSP phosphorylates extracellular signal-regulated kinase (ERK), c-Jun N-terminal kinase (JNK), Akt, and IkB-α and induced the nuclear translocation of the NF-kBp65 subunit. Altogether rh-DSP may have therapeutic utility in dentin regeneration or dental pulp tissue engineering [29].

Biglycan, decorin, versican, and link protein mRNAs have been identified as pulp proteoglycans. As small molecular weight PGs, biglycan and decorin have a regulatory effect on collagen fibrillogenesis. By contrast, expression of link protein and versican is related to larger proteoglycans in the dental pulp, forming large PGs aggregates.

MMP-2, MMP-9, TIMP-2, and myeloperoxidase protein (MPO) were found in the clinically sound pulp and inflamed pulp tissue specimens. MMP-1 and MMP-3 were localized in the infiltrating neutrophils, macrophages, and ECM of the pulpitis group. MMP-8 is expressed by neutrophils, human chondrocytes, cultured fibroblasts, and endothelial cells. TGF-β-1 downregulated the MMP-8 mRNA. The metalloproteinase-8 participates in the organization of the dentin ECM prior to mineralization [30]. BMPs, type IA and II receptors for TGF-β, activin, MMPs, and tissue inhibitors of metalloproteases (TIMPs) are detectable in the ECM. Some non-membranous phospholipids are currently also present in the pulp.

It is of note that dental pulp cells transplanted subcutaneously form mineralized tissues. It has been shown that types I and type II collagen surround the implanted cells. Osteonectin, osteocalcin, and dentin matrix protein-I were identified in the transplanted cell aggregates at day 7. Dentin sialophosphoprotein was detected after 28 days [31]. This capacity demonstrates the influence of local environment on pulp cells.

In addition, genes coding for DMP-1, SLIT 2, period 2 (PER 2), period 3 (PER 3), osteo-adherin, glypican-3, midkine, activin receptor-interacting protein-1 (AIP 1), growth hormone receptor (GRH), adrenomedullin (ADM), interleukin-11, BSP, matrix Gla protein (MGP), endothelial cell growth factor-1 (ECGF 1), inhibin beta A, and oromucoid-1 were identified in diseased pulp [32].

2.6 Immune Defense Mechanisms of the Dental Pulp [33]

Antigen-presenting cells (APC) are, namely, dendritic cells, macrophages, B lymphocytes, and endothelial cells. They express class II MHC molecules. Using an antihuman leukocyte antigen (HLA)-DR monoclonal antibody, immunopositive cells appear with a spindle-like or a dendritic profile. They are densely distributed throughout the dental pulp. They exhibit various sizes of vesicles, enclosing electron-lucent or electron-dense contents, multivesicular bodies, and characteristic tubulovesicular structures as well. HLA-DR-immunopositive cells were predominantly located at the pulp periphery, in the subodontoblastic layer. Contacts between cell processes and odontoblasts were observed. They contribute to the initial defense reaction after a tooth injury [34]. The cells identified as immunocompetent in the normal dental pulp comprise all the peripheral T cells, helper/inducer T cells, cytotoxic/suppressor T cells, macrophages, and class II antigen-expressing cells. However, B cells were not detected in any of the pulp samples examined [35]. Only one type of immunolabeled cells displays a pronounced dendritic appearance. They are located mostly at the periphery of the pulp. The other type has morphological characteristics similar to those of macrophage. They are observed mainly in the central part of the pulp. A ratio of 1:4 was established between the two cell types [36].

Two major populations of *lymphocytes* were identified, the B lymphocytes and T lymphocytes. The latter are themselves subdivided into T helper cells ($CD4^+$) and cytotoxic T cells ($CD8^+$). These cells are natural killers (NK), producing chemokines and inflammatory cytokines such as interferon-γ and tumor necrosis factor-α. NK and NKT (natural killer T) are usually located in the peripheral blood, spleen, and bone marrow. However, they are also present in the dental pulp. Immunohistochemical investigation shows that positively labeled cells are situated in the central portion of the coronal pulp and in the subodontoblastic zone. After activation, T cells secrete several cytokines that regulate the intensity and/or duration of the immune response by stimulating or inhibiting the action of various target cells. CD4+ T lymphocytes are further classified into Th1 and Th2 cells. Th1 cells produce cytokines such as IL-2 and interferon-gamma. In contrast Th2 cells are primarily involved in the activation of IL-4, IL-5, and IL-6. They stimulate the proliferation and differentiation of B lymphocytes. T lymphocytes are recognized in the dental pulp as essential normal residents. They are present along blood vessels in the inner portion of the pulp. Stem cells are significantly targets of the NK cell cytotoxicity. B lymphocytes are not found in the sound pulp, despite the fact that a few researchers have shown the occasional presence of B lymphocytes in the human pulp. In normal dental pulp, NK and NKT represent an essential line of defense, together with macrophages and dendritic cells [37].

Pulpal dendritic cells (DCs) provide the necessary signals, which activate T lymphocytes. They are characterized by their dendritic morphology, the constitutive expression of class II molecules, their high motility, and a limited phagocytic activity. DCs have elongated cytoplasmic processes and contain a few lysosomal structures. Apparently, no distinct phagosomes are present.

Macrophages, initially identified as members of the mononuclear phagocyte system (MPS), then form a group of cells including a series of bone-marrow-derived cells comprising monoblasts, promonocytes taking origin in the bone marrow, monocytes in the peripheral blood, and diverse populations of recognized functionally as macrophages. They display prominent phagocytic capacities. They produce (i) microbicidal enzymes and reactive oxygen species (ROS); (ii) several cytokines such as IL-I, IL-6, and tumor necrosis factor (TNF); and (iii) growth factors promoting pulp repair. Macrophages are implicated in phagocytosis and are important for the elimination of antigen immune complexes [36]. The rat molar contains many OX6-positive cells, and a large number of ED2 (anti-tissue macrophages)-positive and/or OX35 (anti-macrophage and CD4+lymphocytes)-positive cells. Macrophages act as phagocytes or antigen-presenting cells, implicated in the initiation of the immune response [38]. Most of the dental pulp macrophages proliferate and differentiate in the dental pulp without the supply of precursor cells from the blood stream. Myeloid colony-stimulating factor (M-CSF) contributes to stimulate macrophage development in the dental pulp, whereas serum factors do not affect the development of macrophages [39]. Labeling with Mac-1, Mac-2, and MOMA-2, three antibodies raised against macrophages revealed that Mac-2 appeared in the dental pulp at the postnatal day 0 (PN0), but not near or around odontoblasts, and disappeared at PN3. A few MOMA-2 positive cells were detected during the period examined, extending from PN0 to PN5. The labeling detected with the F4/80 antibody raised against dendritic cells and macrophages include immature dendritic cells during the early stage preceding eruption and exposure to antigenic stimuli.

Mast cells are mobile bone-marrow-derived, granule-containing immune cells, which are found in the dental pulp. Each mast cell contains between 80 and 300 granules. When they are activated, degranulation occurs. The key mediators that are released are chymase and tryptase (two serine proteases), histamine, TNF, and cathepsin G; all these mediators are packaged separately within the granules. Mast cell-derived histamine mediates changes in vascular permeability (endothelial contraction and intercellular gap formation) and adhesion of platelets. Mast cell TNF contributes to the control of inflammation, acting on T lymphocytes, Il-1, and Il-4 [40].

Polymorphonuclear leukocytes (*PMNs*) are implicated in the synthesis and secretion (azurophilic granules) of PMN elastase, cathepsin G, and alpha-2 macroglobulin (alpha-2M).

2.7 Microvasculature

Based on the morphology of vascular structures, muscular arterioles, terminal arterioles, precapillaries arterioles, capillaries, postcapillaries venules, and collecting or muscular venules constitute six different segments (Figs. 2.19a, b, 2.20, and 2.21). They are in close association with smooth muscle cells and pericytes, which appear as spiderlike cells. Vasoconstriction and vasodilatation regulate the blood flow [41]. Capillaries are fenestrated in the subodontoblastic plexus. In normal pulp, α-SMA and STRO-1 are expressed in vascular smooth muscle cells, pericytes, and endothelial cells.

The vasculature differs in the crown and root parts of the pulp, although arteries and veins are both located centrally. The vessels located in the two parts of the pulp appear quite distinct as a result of their formation at different stages of tooth development. In the root, blood vessels penetrate the apical area of the pulp, infiltrate the tissue, and form a network of tiny branches. As seen in corrosion resin casts of dog pulp, in the crown, the subodontoblastic and odontoblastic layers are fed by capillaries forming successive glomerular well-individualized structures that feed limited areas about 100–150 μm wide. This implies that a pulp lesion provokes controlled diffusion in the crown. Bypass or arteriovenous anastomoses favor in the crown the healing of a restricted inflammatory pulp area. After a lesion or in case of infection, the radicular pulp underwent necrosis, irreversibly. In the root, resin casts indicate a continuous network of subodontoblastic capillaries. The inflammatory process diffuses and invades the whole radicular pulp, following a fishnet pathway [42].

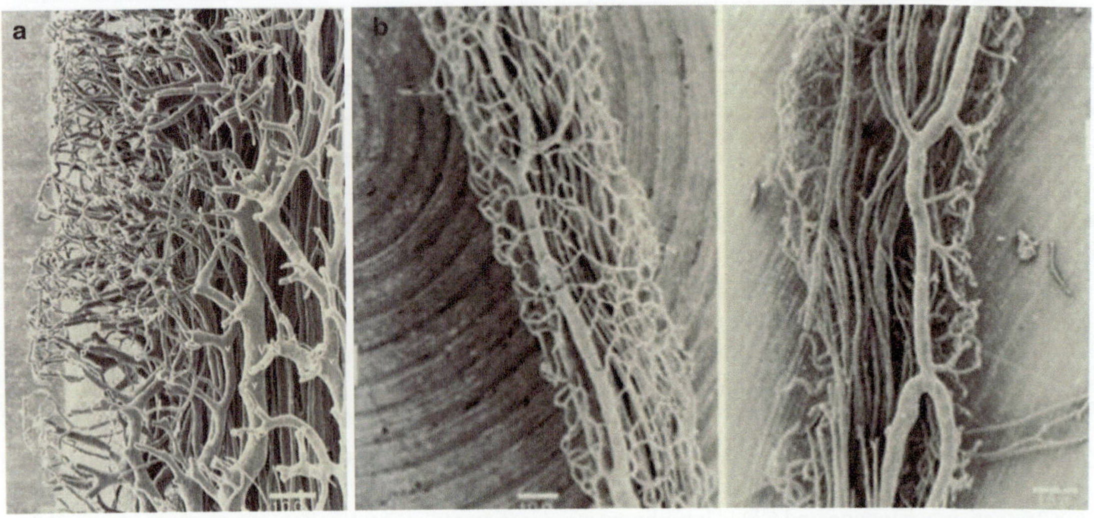

Fig. 2.19 (**a**, **b**) After vascular perfusion with resin, the casts obtained in the coronal part of the pulp (**a**) indicate the presence of successive domains about 150 μm in diameter. In (**b**), in the root, a fisherman network establishes the continuity of vascularization at the pulp periphery (Used with permission from Takahashi [42])

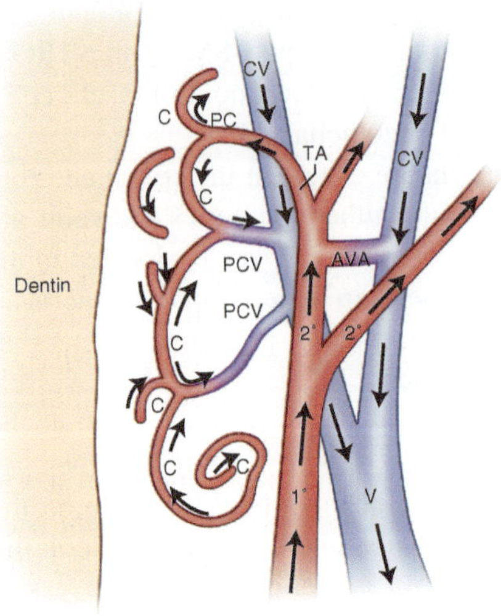

Vessel type	Diameter (μm)	Vm (mm/s)	Q ($10^{-4}mm^3/s$)	$\frac{Qa}{Q}$
1 Feeding arteriole	35–45	1.46 ± 0.11	16.68 ± 1.79	1.0
2 Feeding arteriole	24–24	1.08 ± 0.09	6.31 ± 0.71	2.6
Terminal arteriole (TA)	16–23	0.58 ± 0.06	1.71 ± 0.21	9.8
Pre-capillary (PC)	12–15	0.48 ± 0.13	0.82 ± 0.27	20.3
Capillary (C)	>8	0.27 ± 0.03	0.16 ± 0.01	104.2
Post-cap. venule (PCV)	12–23	0.20 ± 0.02	0.57 ± 0.06	29.4
Collecting venule (CV)	24–50	0.37 ± 0.03	3.56 ± 0.54	4.7
Venule (V)	>50	0.57 ± 0.05	16.83 ± 1.75	1.0

$V_m=\frac{V_{CL}}{1.6}$ (Mean velocity) Mean ± SEM

$Q=V_m \cdot \pi D^2$ (Volumetric flow rate)

Fig. 2.20 Vascular continuity between the collecting venules, capillaries, and feeding arterioles

In healthy impacted third molars processed for TEM examination, capillaries composed of endothelial and peri-endothelial cells were in a 4:1 ratio. Endothelial cells contained typical Weibel-Palade bodies. Three types of peri-endothelial cells were observed: pericytes, transitional cells, and fibroblasts. Pericytes were embedded within the capillary basement membrane. Transitional cells were partly surrounded by basement membrane but were separated from the endothelium by collagen fibrils. Pericytes and transitional cells, but not peri-endothelial fibroblasts, contained low numbers of endothelial Weibel-Palade bodies. It comes out from this

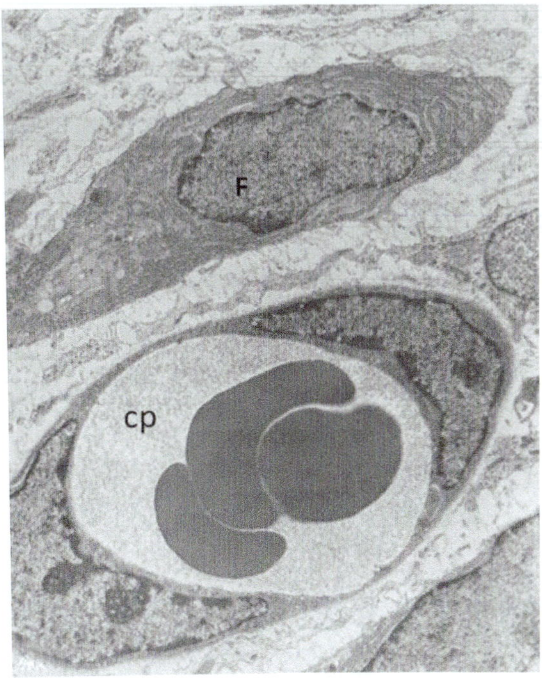

Fig. 2.21 Fusiform fibroblasts (*F*) share pulp territory with endothelial cells of capillaries (*cp*)

observation that during normal tissue turnover, some pericytes may originate from endothelium and migrate from the vessel wall and undergo transition to a fibroblastic phenotype [43]. Nitric oxide synthases (NOS) are important enzymes present in endothelial cells and in macrophages. They are responsible for vasodilatation, blood pressure regulation, platelet aggregation, cardiac contractility, and the mediation of immunity during infection or inflammation. By immunohistochemistry neurokinin 1 (NK1) receptors were located within odontoblasts and endothelial cells of capillaries and postcapillaries venules. The subodontoblastic layer was also labeled. This suggests a vasoactive function, regulation of vascular permeability, and a modulation of pain transmission [44].

The lymphatic system plays a crucial role in the immune barrier function and for tissue fluid balance. The transport of filtered fluid, proteins, and immune cells stimulates lymphangiogenesis. Lymphatic capillaries were identified under local anesthesia after injection in the pulp horn of 0.2–0.3 cc of sterile colloidal carbon. Lymphatic capillaries are characterized by occasional large intercellular clefts, absence or incompleteness of basement membrane, absence of pericytes, absence of luminal red blood cells, and presence of a filamentous material between the endothelium and the surrounding collagen fibrils [45]. Martin et al. denied this [46], using anti-Prox 1 an antibody specific for lymphatic vessels. They were unable to detect any vascular structure lined by lymphatic endothelium. In contrast, Berggreen et al. [47] found positive immunostaining using antibodies that rose against the lymphatic endothelial hyaluronan receptor-1 (LYVE-1) and vascular endothelial growth factor receptor-3 (VEGFR-3), the initial lymphatics in the coronal molar pulp, whereas in the incisor, immunostaining was found only in the apical part. Immunopositive cells were found both in the molar and incisor pulp. These observations suggest that macrophages contribute directly to the formation of lymphatic vessels after pulp exposure (see Chap. 5 on vascularization).

2.8 Innervation

See Chap. 6 and references on pulp innervations.

Nerves penetrate through the apex. They accompanied venules and arterioles in the central root pulp. They fan out in the crown forming the subodontoblastic Raschkow's plexus. Nerve terminal endings establish close contact with the plasma membrane of the odontoblasts and subodontoblastic layer. Junctions of the gap type provide continuity between nerve ending containing synaptic vesicles (small electron-lucent and electron-dense vesicles and/or large vesicles) and the plasma membrane of the odontoblasts. Some axons penetrate the predentin between odontoblast cell bodies. They cross the predentin and, in humans, enter in the inner third of dentin tubules, associated in the lumen with odontoblast processes. Immunohistochemical staining established also that some axons located at 150 μm from the mineralization front are horizontally oriented and connect a few tubules.

References

1. Goldberg M, Lasfargues J-J. The pulpo-dentinal complex revisited. J Dent. 1995;23:15–20.
2. Baume LJ. Biology of pulp and dentin. In: Monographs in oral science. Basel: S. Karger AG; 1980.
3. Kenmotsu M, Matsuzaka K, Kokubu E, Azuma T, Inoue T. Analysis of side population cells derived from dental pulp tissue. Into Endod J. 2010;43:1132–42.
4. Couve E, Schmachtenberg O. Autophagic activity and aging in human odontoblasts. J Dent Res. 2011;90: 523–8.
5. Goldberg M, Smith A. Cells and extracellular matrices of dentin and pulp. A biological basis for repair and tissue engineering. Crit Rev Oral Biol Med. 2004;15:13–27.
6. Hosoya A, Hiraga T, Ninomiya T, Yukita A, Yoshiba K, Yoshiba N, Takahashi M, Ito S, Nakamura H. Thy-1-positive cells in the subodontoblastic layer possess high potential to differentiate into hard tissue-forming cells. Histochem Cell Biol. 2012;137:733–42.
7. Goldberg M, Kulkarni AB, Young M, Boskey A. Dentin: structure, composition and mineralization. – the role of dentin ECM in dentin formation and mineralization. Front Biosci (Elite Ed). 2011;3: 711–35.
8. Vermelin L, Ayanoglou C, Septier D, Carreau JP, Bissila-Mapahou P, Goldberg M. Effects of essential fatty acid deficiency on rat molar pulp cells. Eur J Oral Sci. 1995;103:219–24.
9. Vermelin L, Lécolle S, Septier D, Lasfargues J-J, Goldberg M. Apoptosis in human and rat dental pulp. Eur J Oral Sci. 1996;104:547–53.
10. Koyama N, Okubo Y, Nakao K, Bessbo K. Evaluation of pluripotency in human dental pulp cells. J Oral Maxillofac Surg. 2009;67:501–6.
11. Lacerda-Pinheiro S, Dimitrova-Nakov S, Harichane Y, Souyri M, Petit-Cocault L, Legrès L, Marchadier A, Baudry A, Ribes S, Goldberg M, Kellermann O, Poliard A. Concomitant multipotent and unipotent dental pulp progenitors and their respective contribution to mineralised tissue formation. Eur Cell Mater. 2012;23:371–86.
12. Stanislawski L, Carreau J-P, Pouchelet M, Chen ZHJ, Goldberg M. In vitro culture of human dental pulp cells: some aspects of cells emerging early from the explant. Clin Oral Investig. 1997;1:131–40.
13. Li L, Zhu YQ, Jiang L, Peng W. Increased autophagic activity in senescent human dental pulp cells. Int Endod J. 2012;45:1074–9.
14. Nishikawa S, Sasaki F. Apoptosis of dental pulp cells and their elimination by macrophages and MHC Class II-expressing dendritic cells. J Histochem Cytochem. 1999;47:303–11.
15. Ikeda H, Suda H. Rapid penetration of lucifer yellow into vital teeth and dye coupling between odontoblasts and neighbouring pulp cells in the cat. Arch Oral Biol. 2006;51:123–8.
16. Alliot-Licht B, Hurtrel D, Gregoire M. Characterization of smooth muscle actin positive cells in mineralized human dental pulp cultures. Arch Oral Biol. 2001; 46:221–8.
17. Martinez EF, Araújo VC, Sousa SOM, Arana-Chavez VE. TGF-1 enhances the expression of alpha–smooth muscle actin in cultured human pulpal fibroblasts: immunochemical and ultrastructural analyses. J Endod. 2007;33:1313–8.
18. Qin W, Yang F, Deng R, Li D, Song Z, Tian Y, Wang R, Ling J, Lin Z. Smad 1/5 is involved in bone morphogenetic protein-2-induced odontoblastic differentiation in human dental pulp cells. J Endod. 2012;38:66–71.
19. Duncan HF, Smith AJ, Fleming GJ, Cooper PR. Histone deacetylase inhibitors induced differentiation and accelerated mineralization of pulp-derived cells. J Endod. 2012;38:339–45.
20. Alhabib M, Yu Z, Huang GT. Small molecules affect human dental pulp stem cell properties via multiple signaling pathways. Stem Cells Dev. 2013;22(17): 2402–13.
21. Marchi F, Leblond CP. Radioautographic characterization of successive compartments along the rough endoplasmic reticulum-Golgi pathway of collagen precursors in foot pad fibroblasts of [3 H]proline-injected rats. J Cell Biol. 1984;98:1705–9.
22. Weinstock M, Leblond CP. Radioautographic visualization of the deposition of a phosphoprotein at the mineralization front in the dentin of the rat incisor. J Cell Biol. 1973;56:838–45.
23. Weinstock M, Leblond CP. Synthesis, migration, and release of precursor collagen by odontoblasts as visualized by radioautography after [^{3}H] proline administration. J Cell Biol. 1974;60:92–127.
24. Abd-Elmeguid A, Yu DC, Kline LW, Moqbel R, Vliagoftis H. Dentin matrix protein-1 activates dental pulp fibroblasts. J Endod. 2012;38:75–80.
25. Dominici M, Le Blanc K, Mueller I, Slaper-Cortenbach I, Marini FC, Krause DS, Deans RJ, Keating A, Prockop DJ, Horwitz EM. Minimal criteria for defining multipotent mesenchymal stromal cells. The International Society for Cellular Therapy position statement. Cytotherapy. 2006;8:315–7.
26. Valtieri M, Sorrentino A. The mesenchymal stromal cell contribution to homeostasis. J Cell Physiol. 2008;217:296–3000.
27. Goldberg M, Escaig F. The appearance of the proteoglycans phase in TEM of predentine is fixation dependant. J Microsc. 1984;134:161–7.
28. Yoshiba N, Yoshiba K, Ohkura N, Hosoya A, Shigetani Y, Yamanaka Y, Izumi N, Nakamura H, Okiji T. Expressional alterations of fibrillin-1 during wound healing of human dental pulp. J Endod. 2012;38:177–84.
29. Lee SY, Kim SY, Park SH, Kim JJ, Jang JH, Kim EC. Effects of recombinant dentin sialoprotein in dental pulp cells. J Dent Res. 2012;91:407–12.
30. Palosaari H, Wahlgren J, Larmas M, Rönkä H, Sorsa T, Salo T, Tjäderhane L. The expression of

MMP-8 in human odontoblasts and dental pulp cells is down-regulated by TGF-beta1. J Dent Res. 2000;79:77–84.
31. Ikeda-Isogai M, Ohtsuka T, Baba K, Nonaka N, Nakamura M. Calcified tissue formation of subcutaneously transplanted mouse dental pulp. Acta Histochem. 2012;114:55–61.
32. McLachlan JL, Smith AJ, Bujalska IJ, Cooper PR. Gene expression profiling of pulpal tissue reveals the molecular complexity of dental caries. Biochim Biophys Acta. 2005;1741:271–81.
33. Jontell M, Okiji T, Dahlgren U, Bergenholtz G. Immune defense mechanisms of the dental pulp. Crit Rev Oral Biol Med. 1998;9:179–200.
34. Ohshima H, Maeda T, Takano Y. The distribution and ultrastructure of class II MHC-positive cells in human dental pulp. Cell Tissue Res. 1999;295:151–8.
35. Jontell M, Gunraj MN, Bergenholtz G. Immunocompetent cells in the normal dental pulp. J Dent Res. 1987;66:1149–53.
36. Jontell M, Bergenholtz G, Scheynius A, Ambrose W. Dendritic cells and macrophages expressing Class II antigens in the normal rat incisor pulp. J Dent Res. 1988;67:1263–6.
37. Kawashima N, Wongyaofa I, Suzuki N, Kawanishi HN, Suda H. NK and NKT cells in the rat dental pulp tissues. Oral Surg Oral Med Oral Pathol Oral Radiol Endod. 2006;102:558–63.
38. Okiji T, Kawashima N, Kosaka T, Matsumoto A, Kobayashi C, Suda H. An immunohistochemical study of the distribution of immunocompetent cells, especially macrophages and Ia antigen- expressing cells of heterogeneous populations, in normal rat molar pulp. J Dent Res. 1992;71:1196–202.
39. Iwasaki Y, Otsuka H, Yanagisawa N, Hisamitsu H, Manabe A, Nonaka N, Nakamura M. In situ proliferation and differentiation of macrophages in dental pulp. Cell Tissue Res. 2011;346:99–109.
40. Walsh LJ. Mast cells and oral inflammation. Crit Rev Oral Biol Med. 2003;14:188–98.
41. Iijima T, Zhang J-Q. Three-dimensional wall structure and the innervation of dental pulp blood vessels. Microsc Res Tech. 2002;56:32–41.
42. Takahashi K. Vascular architecture of dog pulp using corrosion resin cast examined under a scanning electron microscope. J Dent Res. 1985;64: 579–84.
43. Carlile MJ, Sturrock MG, Chisholm DM, Ogden GR, Schor AM. The presence of pericytes and transitional cells in the vasculature of the human dental pulp: an ultrastructural study. Histochem J. 2000;32:239–45.
44. Kido MA, Ibuki T, Danjo A, Kondo T, Zhang JQ, Yamaza T, Yamashita Y, Higuchi Y, Tanaka T. Immunocytochemical localization of the neurokinin 1 receptor in rat dental pulp. Arch Histol Cytol. 2005;68:259–65.
45. Frank RM, Wiedemann P, Fellinger E. Ultrastructure of lymphatic capillaries in the human dental pulp. Cell Tissue Res. 1977;178:229–38.
46. Martin A, Gasse H, Staszyk C. Absence of lymphatic vessels in the dog dental pulp: an immunohistochemical study. J Anat. 2010;217:609–15.
47. Berggreen E, Haug SR, Mkonyi LE, Bletsa A. Characterization of the dental lymphatic system and identification of cells immunopositive to specific lymphatic markers. Eur J Oral Sci. 2009; 117:34–42.

Pulp Extracellular Matrix

3

Arthur Veis and Michel Goldberg

3.1 Introduction

Many features of the pulp anatomy and pulp cells have already been described in Chaps. 1 and 2 but a few general remarks are in order here to clearly define the scope of this discussion of the pulp extracellular matrix. At the earliest stage of each tooth's development, the epithelial neural crest derived cells at each tooth-defined position along the dental lamina fold forming a complex three-dimensional bud which invaginates into the underlying mesenchyme. The epithelial dental lamina remains as a continuous layer, with an intact basement membrane but, for each tooth the lamina bud grows into a different and particular cap-like shape. Cells in the interior of the folded bud become the enamel organ, The epithelial cells lining the concave part of the interface with the opposing mesenchymal cells divide and spread forming a layer of "inner enamel epithelium (IEE)" cells. This layer expands to surround and enclose the mesenchymal cells destined to become the dental papilla and defines the three dimensional shape of the tooth crown. The cells of the IEE ultimately become preameloblasts, which upon maturation, become secretory ameloblasts and form the tooth enamel, which mineralizes. The "trapped" mesenchyme is destined to become the "dental papilla" and in the fully formed tooth, the "dental pulp." The cuboidal ectomesenchymal neural crest cells opposing the basement membrane at the IEE interface form a layer and develop into pre-odontoblasts which mature into secretory odontoblasts (OD) and ultimately form the mineralized dentin. The extension of the IEE into the mesenchymal tissue ceases upon reaching the position of the tooth cervix, at which the cervical loop forms and merges on the outer aspect of the enamel organ and the layer of epithelial cells is now called the outer enamel epithelium (the OEE). In three dimensions the formation of the cervical line marks the limit of the tooth crown. However, tooth growth does not stop. The layer of ectomesenchymal neural-crest derived odontoblastic cells continues to divide and extend to form the root dentin. The OD comes to face the mesodermal-derived mesenchyme containing a variety of cells of different character, but including bone-like cementoblasts. These cementoblasts produce mineralized cementum in opposition to the mineralized radicular dentin. Tooth growth ceases when the root apex is reached.

Why begin a discussion of the tooth pulp with a description of overall tooth development?

A. Veis, PhD
Department of Cell and Molecular Biology, Feinberg School of Medicine, Northwestern University, 303 East Chicago Avenue, Chicago, IL 60611, USA
e-mail: aveis@northwestern.edu

M. Goldberg, DDS, PhD (✉)
Department of Oral Biology, Institut National de la Santé et de la Recherche Médicale, Université Paris Descartes, 45 Rue des Saints Pères, Paris 75006, France
e-mail: mgoldod@gmail.com, michel.goldberg@parisdescartes.fr

M. Goldberg (ed.), *The Dental Pulp*,
DOI 10.1007/978-3-642-55160-4_3, © Springer-Verlag Berlin Heidelberg 2014

It is well known [1, 2] that tooth development is directed and regulated by a series of reciprocal epithelial–mesenchymal signaling molecule transfers that pass locally between the epithelium and mesenchyme. As described previously, the tooth coronal (ameloblast–odontoblast) and radicular (mesoderm–ectomesenchyme) interfaces are distinctly different. In an early, but classic, paper [3], Sharpe pointed out that almost all the developmentally regulated genes are expressed in the tooth germ, and posited that it is the localized variation in timing of expression of these genes and their relative concentrations that create overlapping "expression fields" which combine to determine the overall local pathways for development of each tooth's shape and structure. The coronal ameloblasts mature and lose their secretory functions, become apopotic, degrade at the tooth surface, and have no further impact on the deposited enamel. On the other hand, the polarized secretory odontoblasts secrete their collagenous dentin matrix at the DEJ and push the main body of the OD into the space of the dental pulp, but the OD still remain in contact with the dentin through their odontoblastic processes as they create the dentinal tubular structure. Since the secretory OD is joined by tight junctions, the basal pole of the cell presents a coronal ectomesenchymal cell face to the cells of the trapped pulpal mesoderm. The role of the pulp is to keep the dentin alive throughout the life of the tooth. Thus, all the factors controlling the OD, their growth rate, feeding, metabolism and removal of metabolic wastes, and transmission of the signals and factors from other parts of the body via the circulation pass through the pulp to the OD and dentin. Conversely, external inputs are equally possible. Any breach in the enamel exposing the dentin and tubular network may reciprocally affect the pulp and other organs of the body. Immediately under the tightly organized odontoblast layer is a loose, disorganized, less dense layer of cells also of ectomesenchymal origin designated as Hoehl's cells which appear to have the capacity to transform to odontoblasts and may be a pool of reserve cells that could be used to repair or replace damaged or aged odontoblasts. This loose layer of Hoehl's cells is probably best thought of as part of the odontoblast–dentin complex and represents the true boundary which interfaces the dental pulp. Although it is of great interest and complexity here we exclude the odontoblast-dentin complex from further direct discussion, and focus discussion on the loose connective tissue of the dental pulp [4].

See Table 3.1 for a listing of extracellular matrix components of the dental pulp.

3.2 Isolation of the Dental Pulp

The dental pulp is comprised of fibroblast-like cells and mesenchymal stem cells (MSC) as well as more specialized cells and the regulatory factors they secrete. These reside in a more or less typical loose highly hydrated viscous extracellular connective tissue matrix rich in hyaluronan, glycosaminoglycans, and proteoglycans all held together in a network of thin collagen fibrils, reticular fibrils, and fibronectin. The pulp becomes highly vascularized [5, 6] and the vascular bed serves both as the pathway for entry of circulating factors produced elsewhere, and for the removal of metabolic degradation products. The coronal pulp cavity is also richly innervated [7]. The pulp fibrillar network gives the pulp a fixed shape and it can be removed intact if the confining mineralized dentin is carefully removed. Van Amerongen et al. [8] collected the pulp from freshly extracted human teeth, which were washed with 4 °C water, then immediately rinsed and frozen to −20° with no further treatment. A diamond disc saw, cooled with 4 °C water, was used to cut each tooth along the lingual, buccal, and both occlusal sides without entering the pulp. The teeth were then split with a chisel along each cut to expose the pulp. The intact pulp could then be plucked out with tweezers retaining its hydrated shape. Alternatively, an intact premolar could be cut into three portions perpendicular to the incisal edge-apical root direction, yielding a clear coronal portion cut at the cervical line, "limited by the cemento-enamel junction," and two radicular sections of

Table 3.1 Extracellular matrix components of the dental pulp

Dental pulp collagens	Type I collagen	56 %
	Type III collagen	41–42.6 %, associated with fibronectin
	Type V collagen	2 %
	Type VI collagen	Thin fibrils -100 nm periodicity
	Fibrillin-containing microfibrils	
Phosphorylated non-collagenous ECM proteins	Small integrin-binding ligand N-linked glycoproteins (SIBLINGs)	Dentin sialophosphoprotein (DSPP)
		Cleaved into dentin sialoprotein (DSP) and dentin phosphoprotein (DPP)
		Dentin matrix protein 1 (DMP1)
		Bone sialoprotein (BSP)
		Osteopontin
		Matrix extracellular phosphoglycoprotein (MEPE)
Non-phosphorylated ECM protein	Osteocalcin	
	Osteonectin (secreted protein, acidic and rich in cysteine –SPARC)	
	Tenascin	
	Fibronectin	
Glycosaminoglycans and proteoglycans	Small leucine-rich proteoglycans (SLRPs)	*Class I*: CS/DS PGs – decorin and biglycan (collagen fibrillogenesis)
		Class II: KSPGs – fibromodulin and lumican (collagen fibrillogenesis)
		Keratocan and PRELP osteoadherin
		Class III: CS/DS/KS. Epiphycan and mimecan/osteoglycin
	Large GAGs and proteoglycans	Versican
		GAG chains
Cellular and pericellular pulp molecules	Growth factors	Hepatocyte growth factor
		Fibroblast growth factor 2
		TGF beta 1
		Tumor necrosis factor α
		Lymphocyte enhancer-binding factor 1 (LEF1)
		Bone morphogenetic protein 2 (BMP2)
	Receptors of growth factors: Smad1, Smad7, BMPs	
	Enzymes	Alkaline phosphatase (nonspecific alkaline phosphatase, TNAP)
		Acid phosphatases
		Metalloproteinases
		Collagenases: MMP-1, MMP-8, MMP-13
		Collagenase inhibitors
		Disintegrin and metalloproteinase 28 (ADAM28)
		Cysteine cathepsin
		Dipeptidyl peptidase II
		PMN elastase
		Cathepsin G
	Matrilin-2, matrilin-4, CD44, interleulin-1 β	

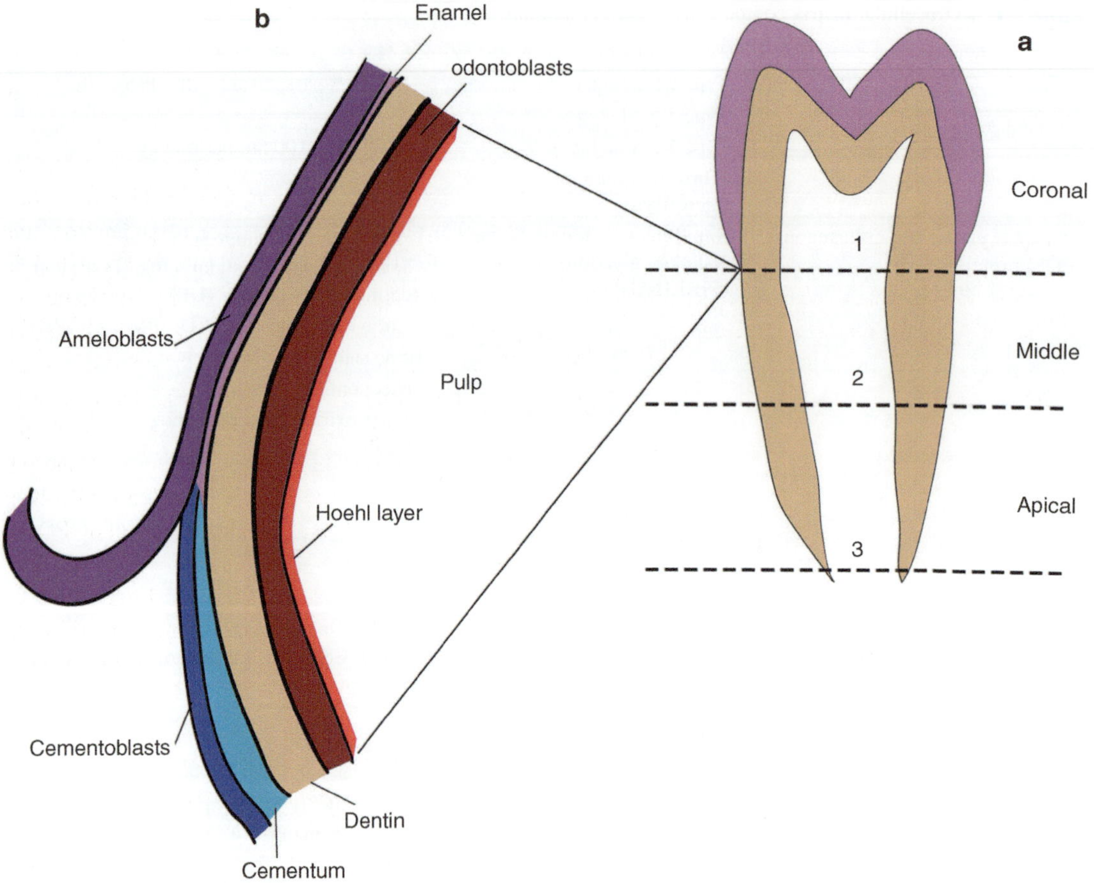

Fig. 3.1 (**a**) The plan for separating pulp zones in a human premolar tooth according to van Amerongen et al. [8]. Cuts are made along each of the *dashed lines*. *Line 1* is at the cervical line, *line 3* is at the tooth apex, *line 2* is midway between 1 and 3. The pulp has sufficient mechanical strength so that the pulp segments can be plucked from the surrounding dentin with tweezers. Although there are no physical barriers, the three pulp segments have, on average, different structures and significantly different compositions. (**b**) Detail of the arrangement of the structures at the cervical line early in development. The ameloblasts in the IEE stop secreting enamel meet and form a tight junction with the cementoblasts and developing cementum layer, which covers the root dentin. The dentin–enamel junction (*DEJ*) is replaced by cementum–dentin junction (*CEJ*). However, the thin, approximately 4-cell thick layer of loosely arranged Hoehl's cells remains interfaced between the dentin, odontoblasts, and the dental pulp cells. Hoehl's cell region, depicted in *red*, is densely vascularized. The reciprocal enamel–dentin and cementum–dentin signaling interactions are probably also communicated locally and reciprocally through Hoehl's cells to the pulp cells accounting for the differences seen in the compositions of the different pulp regions

equal length representing the middle radicular pulp and the apical section (Fig. 3.1a, b).

The collection of the pulp via mechanical removal and the definition of the pulp as not including the dentin–odontoblast complex odontoblasts leave open the question of the sharpness of the mechanical separation. In the coronal region, the odontoblastic processes go deeply into the dentinal tubules, in some cases 2/3 of the total tubule length. Byers and Sugaya [9] used a fluorescent carbocyanine dye, Di-I (1,1′-dioctadecyl-3, 3,3′,3′ tetramethylindo-carbocyanine perchlorate), to stain the organelles and cytosol within the dentinal tubules. While there was no specificity to the staining, the packing of the individual tubules was readily seen, as was the high variability of the length of penetration of the odontoblast processes within the tubules. Control experiments were conducted to verify that the staining was in the processes and not in the surrounding dentin. The density, length, degree of convolution, and extent of branching of the tubules was different in the coronal, central, and apical radicular regions. In our experience mechanical removal of the pulp is not entirely clean, some odontoblasts or odontoblast precursors (the Hoehl's cells) may adhere to the extracted pulp, or some of the contents of the torn and contracted odontoblast processes may bind to the pulp surface in spite of the macroscopically clean surface appearance. Thus, locally the composition of the pulp at the pulp-odontoblast interface may appear to be artifactually high in type I collagen and associated components [10].

3.3 Composition of the Dental Pulp: Type III Collagen, Type V Collagen

The composition of the intact premolar pulp was then compared to the compositions of the three sub-compartments of similar teeth. It was also possible to compare the compositions of equivalent premolar and more mature third molar pulps. There was a statistically significant larger content of collagen on a percentage dry weight basis in the third molar pulp than in the premolars. But the collagen content, based on hydroxyproline determinations, was not the same within the premolar pulp fractions. The radicular middle and apical sections had significantly higher contents of total collagen than the coronal pulp. DNA analysis of the coronal and apical pulp sections showed significantly higher DNA content (cellularity) than did the intervening middle radicular section. Very little of the collagen was soluble in neutral salts or 0.15 M acetic acid, but pepsin digestion did yield soluble collagen. SDS-gel electrophoresis of the pepsin soluble collagen showed a major peak at the 3-chain γ-component position, and that peak was converted to a mixture of α and β chains upon the addition of DTT. Differential denaturation and renaturation experiments [11] showed that the γ-component was essentially type III collagen, and this was confirmed by analysis of the compositions of the cyanogen bromide peptides [12]. By comparing the area under the well-resolved $\alpha1(III)CB_{3,6,8}$ peak to the total integrated areas of all peaks, it was determined that type III collagen comprised ~42.6 % of the total pulp collagen, a surprisingly high concentration compared to the <3 % found as type III in mature skin and bone collagens. Lechner and Kalnitsky [13] and Tsuzaki et al. [14] subsequently showed in detail that the same was true in bovine dental pulp obtained from more mature molars and in teeth with multiple roots. The more detailed isolation of the collagens from bovine pulps showed overall compositions of 56 % Type I, 41 %Type III, and 2 % Type V. The small amount of type V was heterogeneous: a mixture of two isoforms of Type V, $[\alpha1(V)_2\ \alpha2(V)_1]$, and $[\alpha1(V)\ \alpha2(V)\ \alpha3(V)]$. The type III molecule was present as high molecular weight aggregates, which could be disassociated by treatment with 2-mercaptoethanol to yield α1(III) chains, indicating that the type III molecules were disulfide-linked homotrimers. Unfortunately, the more recent papers quantitating the relative contents of the pulp type I and III collagens all used whole homogenized pulps rather than the coronal and apical portions used by Amerongen et al. [8]. Thus, there is little fresh information concerning the differences in the type III contents in the apical and coronal pulp domains. Nevertheless, the analytical methodologies used by Amerongen et al. were essentially equivalent to those used later and there is no reason to dispute the regional composition differences reported in the direct and simultaneous comparison of pulps from different regions. In fact, perhaps, that should have been expected based on the fact discussed earlier that the pulp cells, including pulp-MSCs, are probably subjected to different ratios of factors stimulating MSC gene expression in different parts of the cervical and apical regions, and the recent findings that ectomesenchymal stem cells derived from the pulp (DP-MSC) are most effective and active [15–17]. In particular, DP-MSCs demonstrated in culture the highest osteogenic potential compared with that of bone marrow-derived, adipose tissue-derived and umbilical cord-derived MSC's [17], and increased in activity with higher passage numbers, whereas the BM-MSC, AD-MSC and UBC-MSC cultures decreased in activity with increasing cell passage. Thus, DP-MSCs increase in activity with tissue aging, whereas the others decrease in activity with aging. Stanko et al. [17] noted that Snail, a zinc finger transcriptional repressor that downregulates ectodermal genes within the mesoderm, including E-cadherin, is expressed at lower levels in the DP-MSCs. The lower expression of Snail allows the E-cadherin level to remain high and favors the epithelial-mesenchymal transition so important for tooth development and osteogenic induction. Bone morphogenic protein BMP 4 is also higher in pulp than in bone and adipogenic tissues.

When type III was discovered to be present in small amounts in mature tissue and somewhat larger percentages were present in fetal tissues it was thought that type III, with its thin fibrils and different cross-linking chemistry when compared to type I collagen, was confined to a transient role with its expression decreasing with age. That is obviously not the case within the dental pulp, where type III collagen remains a major component in mature pulp ECM, suggesting that it may be important in maintaining the pulp structural integrity by stabilizing the highly hydrated network of thin reticular type III collagen fibrils in the ECM. Type VI collagen microfibrils with a characteristic periodicity of about 100 nm have also been detected, appearing as long thin and flexible filaments. They aggregate by lateral association. Microfibrils of 10–14 nm in diameter contained fibrillin. Fibrillin-containing microfibrils were abundant in the newly formed pulp [18]. The collagen network associates thin fibrils and fibers with a larger diameter. It has been estimated that about 14 % of the volume of the pulp is made up of blood vessels [19] but the rate of blood flow is different in different regions. The blood flow in the coronal half of a pulp is about twice that in the apical half.

3.4 Other Structural Proteins of the Dental Pulp

Dental pulp cells in culture can, depending on the culture conditions and the factors added to the culture, produce almost all of the ECM proteins found in other tissues, but only a few are present in normal pulp tissues. Of these fibronectin is a principal noncollagenous protein of the dental pulp, associated principally, but not exclusively, with type III collagen in argyrophillic fibrils. Fibronectin (FN) is abundant in the odontoblast layer around blood vessels and in the core of the pulp. A fucosylated glycoprotein, fibronectin is a disulfide linked dimer of two similar ~220,000 mass subunits. Other species of glycoproteins were also detected. Fibronectin is also associated with proteoglycans (PGs). The molecule is also detectable as the free dimer in the water extract of the pulp (0.030 %), in the collagenase extract (0.094 %), and in the hyaluronidase extract (0.109 %). The reappearance of collagen type III is parallel to the advanced vascularization of the dental papilla. The expression of fibronectin is dependent of the calcium-ions dose, but not on magnesium ions [20]. Twice as much fibronectin was extracted from the apical pulp as from the coronal and middle parts, in accord with the higher concentration of type III collagen also found in the apical region, suggesting a 1 to 1 collagen III to fibronectin complex formation.

3.5 Non-collagenous Proteins

3.5.1 The Phosphorylated Proteins, SIBLINGS

The small integrin-binding ligand N-linked glycoprotein family (SIBLING) is a family of phosphoproteins, present in dentin, in bone, and in many pathologic mineralized tissues (kidney, salivary glands, tumors etc.). Immunostaining of the SIBLING molecules shows much lower reactions in the pulp compared with the dentin staining, but they still remaining at a detectable level.

Among the phosphorylated NCP, dentin sialophosphoprotein (DSPP) is the most abundant and is a specific factor of odontoblast differentiation and dentin mineralization. Just after the molecule is released in the tissue, DSPP is proteolytically cleaved into dentin sialoprotein (DSP) and dentin phosphoprotein (DPP). BMP-1 cleaves DSPP to form DSP. Matrix metalloproteinases, MMP-2 and MMP-20, are implicated in cleavage of the COOH-terminal portion of DSPP yielding DPP. BMP-1/Tolloid-like metalloproteinases are also implicated in this cleavage. There is no evidence that DPP, the highly phosphorylated component created in the same molar concentration as DSP, has any bearing on the potential stimulation of pulp matrix mineralization.

The 53 kDa sialic acid-rich glycoprotein DSP, the NH2-terminal portion cleavage product of DSPP, is synthesized by dental pulp cells. It may be an effective stimulator for dental pulp stem/progenitor cell differentiation, dental tissue repair, and pulp regeneration [21], but it does not appear to be involved in pulp matrix mineralization as implicated in the dentin matrix.

Dentin matrix protein 1 (DMP1) is another SIBLING found in the dental pulp. It may be mutated, but in contrast with mild dentin defects, there is no apparent change in the pulp structure. However, deletion of the DMP-1 gene shows a postnatal failure of maturation of predentin into dentin, accompanied by an increased width of the predentin but reduced width of the dentin also leading to enlarged pulp chambers. This tooth phenotype is very similar to the human dentinogenesis imperfecta. Altogether this demonstrates that DMP-1 is essential for the late dentinogenesis development especially during the postnatal period [22].

Bone sialoprotein (BSP) is the sialylated bone analog of DSP. Some is expressed by pulp cells and has been detected by immunohistochemistry and in situ hybridization (ISH) in the pulp. Pulp capping with BSP leads to the formation of a robust homogeneous mineralized reparative dentinal bridge [23].

Osteopontin is synthesized and secreted by pulp cells. This SIBLING binds to cells via the following: (1) an adhesion sequence that recognize the alpha V beta 3 integrin, (2) regulation of the formation and remodeling of mineralized tissues, (3) recruitment and stimulation of macrophages and lymphocytes as part of a nonspecific response to microbial infections, (4) interaction with Ca^{2+}-mediated or Ca^{2+}-dependent processes, and (5) inhibition of the growth of oxalate crystals by disruption of the growing crystal lattice [24].

Matrix extracellular phosphoglycoprotein (*MEPE*) is expressed within the dental pulp. The C-terminal fragment of the molecule keeps the odontoblasts in an immature condition. Dentonin, a 23 amino acid peptide derived from MEPE, stimulates pulp cell proliferation and differentiation, with a potential role in pulp repair [25, 26].

3.5.2 Non-phosphorylated ECM Proteins

Osteocalcin mRNA (OCN) was identified by histochemistry in the pulp tissue. Osteocalcin-coated type I collagen was seen in predentin, but not elsewhere. In addition, no Jun D was found in the dental pulp cells. OCN is considered as a reparative molecule inside the dental pulp. OCN RNA was detected in young and old samples of the dental pulp, with a marked decrease in the aged group [27].

A light to moderate staining of *osteonectin* (ONEC) has been observed in the dental pulp. Also named secreted protein, acidic, and rich in cysteine (SPARC), ONEC was present mostly at sites of rapid collagen remodeling [28]. *Tenascin* (TN) and ONEC identified on pulpal fibroblasts may play a role in the differentiation into hard-tissue-forming cells and in the formation of mineralized tissues. TN-C promoted the differentiation of dental pulp cells by activation of Notch. Mostly present in the forming tooth germ, the molecule disappears gradually, becomes residual, and is not present in the adult.

3.6 Glycoaminoglycans and Proteoglycans

Small leucine-rich proteoglycans (*SLRPs*) act as tissue organizers and influence cell growth and maturation. They are biological filters and modulate growth-factor activities. They regulate collagen fibrillogenesis [29]. The small leucine-rich proteoglycan (SLRP) family comprises three classes. Class I includes decorin and biglycan, which display a high homology (~57 % identity), the only SLRPs that contain a pro-peptide, highly conserved across species. The terminal domain is usually substituted with one (decorin) or two (biglycan) chondroitin/dermatan sulfate side chains that provide polyanionic properties. The C-terminal domain comprises about 50 amino acid residues and two disulfide-linked cysteine residues separated by ~32 amino acids. Both decorin and biglycan densely immunostained the predentin but displayed a weaker reaction as a collagen-associated network within the pulp. Collagen fibrillation is regulated by these PGs.

Class II comprises five members, subdivided in three subfamilies. Fibromodulin and lumican form the first subfamily and exhibit ~48 % protein sequence identity, keratocan and PRELP the second, and osteoadherin constitute a third distinct subfamily. Keratan sulfate chains, and polylactosamine, an unsulfated KS, substitute them.

Class III includes epiphycan and mimecan/osteoglycin. They contain either CS or DS and can be secreted as a glycoprotein [30].

Decorin and fibromodulin interact with collagen and regulate collagen fibrillogenesis. PGs are implicated in ECM assembly. They are TGF-β blockers. They control cell proliferation by interacting with the receptor tyrosine kinase.

Biglycan (BGN), decorin (DCN), fibromodulin (Fmod), and lumican are involved in dentinogenesis. Dental pulps contain sulfated glycosaminoglycans (GAGs) such as chondroitin 4-sulfate, dermatan sulfate, and chondroitin 6-sulfate. Versican-like large proteoglycan, decorin- and biglycan-like small proteoglycans, and a small amount of sulfated protein have been identified in the culture medium, released by pulp cells.

Fibromodulin restricted the diameter of the collagen fibrils and compensatory mechanisms occurred in the Fmod-deficient mice leading to an increased immunolabeling for DSP, DMP1, and BSP in the pulp of incisors. BSP and DMP1 were also enhanced in the molar, whereas DSP and OPN were decreased [31]. Fibromodulin (Fmod) is a keratan sulfate member of the small leucine-rich proteoglycan (SLRP) family.

The leucine-rich keratan sulfate proteoglycan lumican was detected by intense immunolabeling in predentin around odontoblasts and in the pulp with a fibrillar distribution [32]. This is also the case for osteoadherin (OSAD). OSAD is restricted to mineralized tissues and in close association with collagen fibers. Immunohistochemical staining revealed the presence of OSAD mostly in predentin, but not within the dental pulp [33, 34].

3.7 Large Proteoglycans

Versican has a molecular mass of approximately 800 kDa. Versican and mimecan (also named osteoglycin) are 25-kDa corneal keratan sulfate proteoglycans and accumulate in the subodontoblastic layer in the coronal pulp, with co-expression of MEPE and mimecan in predentin. The versican large proteoglycan exhibits heterogeneity, ranging from 250 to 400 kDa, and the GAG chains have a molecular mass of approximately 42 kDa. Digestion with chondroitinase ABC indicates the presence of 10–13 GAG chains per core protein. The GAG chains contain approximately 63 % 4-sulfated disaccharides. Strong immunoreaction appeared in the subodontoblastic layer in the crown. A strong immunoreaction is shown for hyaluronan (HA), whereas a weak reaction is observed for versican in the subodontoblastic layer of the radicular pulp. Hence, regional differences are detected between the coronal and radicular pulp [35, 36].

3.8 Incidence of Cellular and Pericellular Molecules on ECM (Enzymes and Growth Factors (GF), Differentiation Factors)

Dental pulp cells were also used to investigate the migration of pulp stem cells in response to chemotactants and extracellular matrix proteins (EMPs). Chemotactants were sphingosine-1 phosphate, fibroblast growth factor, epidermal growth factor, and transforming growth factor beta-1. EMPs were collagen-1, collagen-IV, laminin, and fibronectin. Some EMPs, namely, laminin, and some chemotactants, are promoters of pulp cell migration, and they mediate the process of pulp regeneration after a tooth injury [37].

Hepatocyte growth factor (HGF) mediates wound repair. The GF is sequestered within the dentin matrix and as chemokine, excerpt some action on the differentiation and mineral deposition of pulp-derived cells. Its receptor, c-Met, was also detected in close relation with human dental pulp cells. HGF is therefore an important molecule implicated in pulp repair and healing [38]. Other growth factors have been identified in the dental pulp. Fibroblast growth factor 2 (FGF2) enhances the proliferation and differentiation of pulp stem cells. FGF2 and TGF beta1 initiate an odontoblast-like differentiation of dental pulp stem cells (DPSCs). Tumor necrosis factor-alpha (TNF-alpha) stimulates the differentiation of dental pulp cells toward an odontoblastic phenotype via p38.

Receptors of growth factors are present in the dental pulp. The expression of mRNAs for Smad1, Smad7, BMPs, ALP, and osteocalcin was enhanced by laser irradiation, whereas Smad6 mRNA was inhibited. The lymphocyte enhancer-binding factor 1 (LEF1) plays a role in inducing tissue interaction during tooth development, contributing to the formation of mineralization nodules by stimulating the expression of DSPP, osteocalcin, and alkaline phosphatase. Finally, pulp vasculogenesis is under the control of the bone morphogenetic protein 2 (Bmp2).

3.9 Collagenases and Other Proteolytic Enzymes

3H-proline-labeled collagens showed that the turnover of pulp collagens occurred at the same rate as the NCP turnover. This suggests that the whole pulp matrix undergoes rapid remodeling, with more than one metabolic pool of collagen. Collagenases, collagenase inhibitors, and other proteolytic enzymes are involved in the process [39]. The action of collagenases converges in patients with chronically inflamed dental pulp tissue. MMP-1, MMP-8, and MMP-13 metalloproteinases have the capacity to degrade most ECM proteins, as well as basement membrane components [40].

A potent collagenase inhibitor was associated with a latent collagenase. It is a sialoglycoprotein containing approximately 20 % carbohydrate and together with complex-type oligosaccharides. The addition of the inhibitor to the activated collagenase resulted in a dose-dependent inhibition of the enzyme activity [41]. A disintegrin and metalloproteinase 28 (ADAM28) has been identified in dental pulp stem cells. This molecule is implicated in proliferation, differentiation, and apoptosis of human dental pulp stem cells [42].

Cysteine cathepsin was identified in pulp tissue, with a minor response to TGF-beta. Dipeptidyl peptidase II, also known as dipeptidyl aminopeptidase II, is present in high concentrations in bovine dental pulp. It hydrolyzes prolyl bonds at acidic pH [43]. Human PMN elastase and cathepsin-G are also active in the dental pulp, the two enzymes being probably implicated in the tissue destruction that occurs during inflammation. Matrilin-2 showed significant downregulation during odontogenic differentiation of dental pulp cells. The molecule was observed throughout the pulp. Matrilin-4 was identified in odontoblasts, but not in dental pulp cells, except under a deep carious lesion. The messenger RNAs were detected by RT-PCR during the odontogenic differentiation of dental pulp cells, and matrilin-4 was increased in a time-dependent manner [44]. CD44, a transmembrane glycoprotein, is expressed in odontoblasts and is involved in the mineralization of dental pulp cells [45].

Interleukin-1 beta (IL-1beta) is implicated in inflammatory processes, and this cytokine is also found in the healthy tissue.

3.10 Summary and Conclusions

Comparisons of the differences between the two closely related tissues, dentin and pulp, allow some insights on the mechanisms regulating mineralization. Altogether, the extracellular matrix of the dental pulp forms an irregular loose network of thin and thick collagen fibrils, associated with non-collagenous proteins either acting as part of the stabilizing fibril network or participating as mineralization stimulators or inhibitors. Abundant highly hydrated GAGs and/or PGs provide an amorphous gel allowing easy cell movement and differentiation moving in the pulp from the apex of the root to the crown part. This is contained within the more solid dense collagenous matrix of dentin, which is mineralized. Small leucine-rich PGs are associated with collagen fibrils in dentin, serving as nucleating agents, whereas GAGs, small and large LRPGs, control the collagen fibril structure and the three-dimensional cell environment needed by pulp fibroblasts to move and renew within the dental pulp. All the biological effects reported here implicate growth factors, transcription factors, and a series of enzymes that are released by pulp cells, implicated in molecule cleavage and/or degradation. Although the dental pulp is a non-mineralized tissue, in some circumstances such as a genetic alteration of ECM molecules involving SIBLINGs deletions or mutations (e.g., DSPP, DMP-1), the pulp may display aberrant mineralization (dentinogenesis imperfecta or dentin dysplasia). In other circumstances, such as in the presence of dental carious lesions and/or aging effects, pulp calcospherites or large mineralized areas may alter the pulp matrix properties. The pulp, although isolated for the large part, is never out of touch with the rest of the animal because of the rich vascular network entering through the apical foramen and providing for easy blood flow and communication with the blood for supply of nutrients and clearance of metabolic waste products. The tooth pulp keeps the tooth as a living part of the host.

This chapter has two clear objectives. Obviously it catalogs the various types of proteins and other components of the pulp and notes what their roles and distribution within the pulp might be. Viewing this listing, it is evident that the contents of the dental pulp are not unique, it is very much like a listing of the components of any of the other loose connective tissues of the body, except that certain proteins such as type III collagen and fibronectin and large proteoglycans such as versican are more prominent. The one factor that seems to make the pulp unique is that there are no physical mechanical forces operating to stress the tissue in any way other than changes in hydraulic pressure and blood flow and unavoidable small gravitational forces. In each tooth these cells may diffuse and interact and ultimately organize with other pulp cells to form regions, such as in the apical pulp, to form aggregates with special function. In a way, the pulp cells may be a model for cell behavior in bone under weightless conditions where the bone demineralizes. On the other hand, unmineralized is the normal pulp condition.

The second objective of this discussion is to emphasize the independence of each tooth from its neighbors. The initial group of mesenchymal cells trapped and enclosed in the early phase of coronal development is not homogeneous and may in fact differ in composition as each tooth forms independently and becomes an independent entity. The external forces on each tooth are also position dependent. The remaining chapters in this book may be better appreciated from the perspective of the complexities of pulp development and behavior.

Acknowledgment All of the work carried out in the Veis laboratory on mineralization and tooth behavior has been supported by the National Institutes of Health and National Institute for Dental and Craniofacial Research Grant DE01374. The French Foundation has supported the work in the Goldberg laboratories for dental research and STEM Pole for funding this research. A personal note of appreciation is in order from AV for the privilege of working with Professor Michel Goldberg, a gifted clinician scientist. He has taught me much about dentistry and the real world of clinical science as compared to the strict physical chemistry approach I brought to my studies of dentin 60 years ago. I have not been the best student, but he has been a great teacher.

References

1. Thesleff I, Vaahtokari A, Kettunen P, Aberg T. Epithelial-mesenchymal signaling during tooth development. Connect Tissue Res. 1995;32:9–15.
2. Ruch JV. Tooth crown morphogenesis and cytodifferentiations: candid questions and critical comments. Connect Tissue Res. 1995;32:1–8.

3. Sharpe PT. Homeobox genes and orofacial development. Connect Tissue Res. 1995;32:17–25.
4. Kuboki Y, Takagi T, Sasaki S, Saito S, Mechanic GL. Comparative collagen biochemistry of bovine periodontium, gingiva, and dental pulp. J Dent Res. 1981;60:159–63.
5. Bishop MA. An investigation of pulp capillaries and tight junctions between odontoblasts in cats. Anat Embryol (Berl). 1987;177:131–8.
6. Bishop MA. Extracellular fluid movement in the pulp; the pulp/dentin permeability barrier. Proc Finn Dent Soc. 1992;88 Suppl 1:331–5.
7. Egan CA, Bishop MA, Hector MP. An immunohistochemical study of the pulpal nerve supply in primary human teeth: evidence for the innervation of deciduous dentine. J Anat. 1996;188:623–31.
8. Van Amerongen JP, Lemmens AG, Tonino GJM. The concentration, extractability and characterization of the collagen in human dental pulp. Arch Oral Biol. 1983;28:339–45.
9. Byers MP, Sugaya A. Odontoblast processes in dentin revealed by fluorescent Di-1. J Histochem Cytochem. 1995;43:159–69.
10. Lukinmaa PL, Waltimo J. Immunohistochemical localization of types I, V, and VI collagen in human permanent teeth and periodontal ligament. J Dent Res. 1992;71:391–7.
11. ChandraRajan J. Separation of type III collagen from type I collagen and pepsin by differential denaturation and renaturation. Biochem Biophys Res Commun. 1978;83:180–6.
12. Miller EJ, Epstein EH, Piez KA. Identification of three genetically distinct collagens by cyanogen bromide cleavage of insoluble human skin and cartilage collagen. Biochem Biophys Res Commun. 1971;42: 1024–9.
13. Lechner JH, Kalnitsky G. The presence of large amounts of type III collagen in bovine dental pulp and its significance with regard to the mechanism of dentinogenesis. Arch Oral Biol. 1981;26:265–73.
14. Tsuzaki M, Yamauchi M, Mechanic GL. Bovine dental pulp collagens: characterization of types III and V collagen. Arch Oral Biol. 1990;35:195–200.
15. Xing Y, Haiyan Q, Cunye Q, Tuan RS, Shi S, George T, Huang J. iPS cells reprogrammed from human mesenchymal-like stem/progenitor cells of dental tissue origin. Stem Cells Dev. 2010;19:469–80.
16. Makino Y, Yamaza H, Akiyama K, Ma L, Hoshino Y, Nonaka K, Terada Y, Kukita T, Shi S, Yamaza T. Immune therapeutic potential of stem cells from human supernumerary teeth. J Dent Res. 2013;92: 609–15.
17. Stanko P, Kaiserova K, Altanerova V, Altaner C. Comparison of human mesenchymal stem cells derived from dental pulp, bone marrow, adipose tissue, and umbilical cord tissue by gene expression. Biomed Pap Med Fac Univ Palacky Olomouc Czech Repub. 2013;157(XX):1–5.
18. Shuttleworth CA, Berry L, Kielty CM. Microfibrillar components in dental pulp: presence of both type VI collagen- and fibrillin-containing microfibrils. Arch Oral Biol. 1992;37:1079–84.
19. Suda H, Ikeda H. The circulation of the pulp. In: Hargreaves KM, Goodis HE, editors. Seltzer and bender's dental pulp. 3rd ed. Carol Stream: Quintessence Publishing; 2002.
20. van Amerongen JP, Lemmens IG, Tonino GJ. Immunofluorescent localization and extractability of fibronectin in human dental pulp. Arch Oral Biol. 1984;29:93–9.
21. Butler WT, Ritchie HH, Bronckers AL. Extracellular matrix proteins of dentine. Ciba Found Symp. 1997;205:107–15.
22. Ye L, MacDougall M, Zhang S, Xie Y, Zhang J, Li Z, Lu Y, Mishina Y, Feng JQ. Deletion of dentin matrix protein-1 leads to a partial failure of maturation of predentin into dentin, hypomineralization, and expanded cavities of pulp and root canal during postnatal tooth development. J Biol Chem. 2004;279: 19141–8.
23. Decup F, Six N, Palmier B, Buch D, Lasfargues J-J, Salih E, Goldberg M. Bone sialoprotein-induced reparative dentinogenesis in the pulp of rat's molar. Clin Oral Investig. 2000;4:110–9.
24. Denhardt DT, Guo X. Osteopontin: a protein with diverse functions. FASEB J. 1993;7:1475–82.
25. Liu H, Li W, Gao C, Kumagai Y, Blacher RW, DenBesten PK. Dentonin, a fragment of MEPE, enhanced dental pulp stem cell proliferation. J Dent Res. 2004;83:496–9.
26. Six N, Septier D, Chaussain-Miller C, Blacher R, DenBesten P, Goldberg M. Dentonin, a MEPE fragment, initiates pulp healing response to injury. J Dent Res. 2007;86:780–5.
27. Muramatsu T, Hamano H, Ogami K, Ohta K, Inoue T, Shimono M. Reduction of osteocalcin expression in aged human dental pulp. Int Endod J. 2005;38: 817–21.
28. Salonen J, Domenicucci C, Goldberg HA, Sodek J. Immunohistochemical localization of SPARC (osteonectin) and denatured collagen and their relationship to remodelling in rat dental tissues. Arch Oral Biol. 1990;35:337–46.
29. Iozzo RV. Matrix proteoglycans: from molecular design to cellular function. Ann Rev Biochem. 1998; 67:609–52.
30. Iozzo RV. The biology of the small leucine-rich proteoglycans- functional network of interactive proteins. J Biol Cell. 1999;274:18843–6.
31. Goldberg M, Septier D, Oldberg A, Young MF, Ameye LG. Fibromodulin-deficient mice display impaired collagen fibrillogenesis in predentin as well as altered dentin mineralization and enamel formation. J Histochem Cytochem. 2006;54:525–37.
32. Hall RC, Embery G, Lloyd D. Immunochemical localization of the small leucine-rich proteoglycan lumican in human predentine and dentine. Arch Oral Biol. 1997;42:783–6.
33. Buchaille R, Couble ML, Magloire H, Bleicher F. Expression of the small leucine-rich proteoglycan

osteoadherin/osteomodulin in human dental pulp and developing rat teeth. Bone. 2000;27:265–70.
34. Nikdin H, Olsson M-L, Hultenby K, Sugars RV. Osteoadherin accumulates in the predentin towards the mineralization front in the developing tooth. PLOS One. 2012;7:e31525.
35. Yoneda S, Shibata S, Yamashita Y, Yanagishita M. Biosynthesis of versican by rat dental pulp cells in culture. Arch Oral Biol. 2002;47:435–42.
36. Shibata S, Yoneda S, Yanagishita M, Yamashita Y. Isolation of proteoglycan (versican) aggregate from rat dental pulp. Arch Oral Biol. 2000;45:563–8.
37. Howard C, Murray PE, Namerow KN. Dental pulp stem cell migration. J Endod. 2010;36:1963–6.
38. Tomson PL, Lumley PJ, Alexander MY, Smith AJ, Cooper PR. Hepatocyte growth factor is sequestered in dentine matrix and promotes regeneration-associated events in dental pulp cells. Cytokine. 2013; 61:622–9.
39. Orlowski WA. The turnover of collagen in the dental pulp of rat incisors. J Dent Res. 1977;56:437–40.
40. Lee YH, Kim GE, Cho HJ, Yu MK, Bhattarai G, Lee NH, Yi HK. Aging of in vitro pulp illustrates change of inflammation and dentinogenesis. J Endod. 2013; 39:340–5.
41. Kishi J, Hayakawa T. Purification and characterization of bovine dental pulp collagenase inhibitor. J Biochem. 1984;96:395–404.
42. Zhao Z, Liu H, Wang D. ADAM28 manipulates proliferation, differentiation, and apoptosis of human dental pulp stem cells. J Endod. 2011;37:332–9.
43. McDonald JK, Schwabe C. Dipeptidyl peptidase II of bovine dental pulp. Initial demonstration and characterization as a fibroblastic, lysosomal peptidase of the serine class active on collagen-related peptides. Biochim Biophys Acta. 1980;616:68–81.
44. Chen C, Wei X, Ling J, Xie N. Expression of matrilin-2 and -4 in human dental pulps during dentin-pulp complex wound healing. J Endod. 2011;37:642–9.
45. Chen KL, Huang YY, Lung J, Yeh YY, Yuan K. CD44 is involved in mineralization of dental pulp cells. J Endod. 2013;39:351–6.

Strategies for Tracking the Origin and Fate of Odontoblasts and Pulp Cell Progenitors

4

Mina Mina

4.1 Odontoblasts and Dentinogenesis

Dentinogenesis is regulated by a single layer of highly differentiated postmitotic odontoblasts originating from the cranial neural crest-derived cells of the dental papilla [1, 2]. The terminal differentiation of odontoblasts from the dental papilla occurs independently in each cusp and is dependent on inductive signals derived from the inner dental epithelium and its associated basement membrane [1, 2]. This differentiation involves a series of change in the morphology, transcriptional profile, and expression of proteins secreted by cells in the odontoblast lineage.

During this process, the dental papilla cells in close proximity to the epithelial–mesenchymal interface first form cuboidal pre-odontoblasts. The differentiation of pre-odontoblasts occurs at the bell stage of tooth development and is regulated by signals from inner enamel epithelium and secondary enamel knots [3, 4]. Further differentiation proceeds in a graded fashion from cusp tips toward the intercuspal areas and cervical loops and includes the withdrawal of pre-odontoblasts from cell cycle and the formation of polarized odontoblasts in close contact with the epithelial–mesenchymal interface. The formation polarizing odontoblasts that display cytological polarization with nucleus occupying the proximal part of the cell body is followed by the formation of secretory/functional odontoblasts and finally the formation of mature and terminally differentiated odontoblasts [5].

Functional/secretory odontoblasts are engaged in the secretion of unmineralized predentin matrix, composed primarily of type I collagen (Col1a1) secreted at their apical end [6]. As odontoblasts continue their differentiation, they secrete many non-collagenous proteins (NCP) including SIBLINGs proteins (osteopontin, bone sialoprotein, dentin matrix protein 1 (DMP1), dentin sialophosphoprotein, osteonectin, and MEPE) and the proteoglycans (biglycan, decorin, fibromodulin, lumican, and osteoadherin) shown to be essential for initiating the mineralization of the type 1 collagen network in the predentin matrix [7].

Many of the NCPs of dentin are also found in the bone. However, dentin is characterized by the presence of two dentin-specific NCPs, dentin sialoprotein (DSP) and dentin phosphoprotein (DPP) [8]. These two dentin-specific NCPs are encoded by a single gene and are specific cleavage products of a larger protein called dentin sialophosphoprotein (DSPP) [8]. Although low levels of *Dspp* have been detected in the bones by RT-PCR, high level of *Dspp* expression is specific to odontoblasts [8].

At the same time, odontoblasts recede toward pulp and leave behind cell processes that extend

M. Mina, DMD, MSD, PhD
Division of Pediatric Dentistry,
School of Dental Medicine,
University of Connecticut Health Center,
263 Farmington Ave, Farmington,
CT 06030-4081, USA
e-mail: mina@nso.uchc.edu

M. Goldberg (ed.), *The Dental Pulp*,
DOI 10.1007/978-3-642-55160-4_4, © Springer-Verlag Berlin Heidelberg 2014

into the mineralized dentin and give the dentin matrix its characteristic tubular morphology that makes dentin morphologically different from the atubular bone matrix [6].

Therefore, the differentiation of odontoblasts from dental papilla is a long process involving several intermediate steps and shares many similarities to that of the bone. However, although much has been learned about the cellular and molecular mechanisms that regulate the progression of osteoprogenitor cells into fully differentiated osteoblasts, not much is known about the regulatory mechanisms involved in the differentiation program of odontoblasts. This has been due in part to the lack of suitable markers for identifying the intermediate stages of odontoblast differentiation.

For example, in vivo studies have proven to be difficult for understanding the molecular mechanisms of odontoblast differentiation because of the close proximity of cells in varying stages of differentiation. In addition, in mice, the animal model used extensively in tooth research, the steps between the formation of pre-odontoblasts and mature odontoblasts occur fast and are completed within 6–10 h [5, 9]. The differentiation of odontoblasts from cultured dental pulps and cell outgrowth from cultured explants has made dental pulp cells a valuable model for examining the mechanisms regulating the sequential steps involved in odontoblast differentiation. The differentiation of odontoblasts in all these cultures has been characterized by a number of standard biochemical and molecular methods including the expression of extracellular matrix components such as Col1a1 and various NCPs such as *OC*, *Dmp1*, and *Dspp* that are the hallmark of differentiated odontoblasts.

Most often, published results have been based on observations in whole primary dental pulp cultures, which are composed of a heterogeneous population of cells composed of fibroblasts, macrophages, endothelial cells, and lymphocytes and contain nodules at various stages of differentiation/mineralization. Thus, investigation of the mechanisms regulating differentiation of progenitor cells into odontoblasts has been hampered by the lack of availability of stage-specific molecular markers for cell lineage studies and the inability to identify and isolate relatively pure populations of cells from dental pulp during progression into the odontoblast lineage.

4.2 Methods of Studying Lineage Progression

Key to the isolation of cells at intermediate stages during lineage progression is the ability to identify and isolate the cells at specific stages of differentiation. Techniques most often employed include isolation of cells by fluorescence-activated cell sorting (FACS), based on the expression of cell surface antigens, and laser capture microscopy based on location of recognizable anatomical markers.

The field of hematology, by isolating and identifying cell surface receptors and intracellular proteins that are differentially expressed between hematopoietic stem cells and their differentiated progeny, has made significant advances in understanding the progression of stem cell and progenitors into different progenies [10]. This not only has led to an increased understanding of hematopoietic stem cell biology but has also become a gold standard for other fields.

In recent years, GFP under the control of specific gene regulatory elements has become one of the most widely used, noninvasive protein markers for studying fate mapping and lineage determination and progression in vivo and in vitro [11, 12]. Many new GFP-based techniques have been developed including the generation of transgenic animals carrying GFP coding sequences under the control of tissue-specific or stage-specific promoters. The availability of such transgenic animals and the utilization of FACS-mediated cell isolation and enrichment have provided powerful experimental tools for developmental and lineage studies [13, 14] osteogenesis [15–22].

GFP-based reporters linked to promoters that activate at defined stages of development within the odontoblast lineage have been used [15, 23–25]. The key to the use of this strategy is to show that the cells identified by the expression of a transgene (GFP signal) in a transgenic line represent a cell with reproducible properties of a defined stage of differentiation. In our studies we routinely characterize the temporal and spatial expression of a transgene in vivo and in vitro and then examine the proliferation and differentiation of the cells expressing the transgene (GFP+).

Considering the abundance of Col1a1 in dentin (approximately 86–90 %), we started our studies using pOBCol3.6GFP (referred to as 3.6-GFP) and pOBCol2.3GFP (referred to as 2.3-GFP) transgenic animals as models to distinguish and identify populations of cells at early stages of odontoblast differentiation [23]. In these animals 3.6- and 2.3-kb fragments of rat type I collagen promoter drive the expression of GFP, respectively. Our in vivo studies in the developing teeth (molars and incisors) of these transgenic animals [22, 26–28] showed low but detectable levels of both transgenes in differentiating odontoblasts at the tip of the mesiolingual cusp of the first mandibular molar at the late bell stage (E18) of tooth development. Both 3.6-GFP and 2.3-GFP transgenes were expressed at high intensity in secretory and differentiated odontoblasts expressing high levels of *Dspp* [26, 27] (Fig. 4.1a–d). The patterns of expression of Col1a1-GFP in differentiating odontoblasts are consistent with immunostaining for CAT protein in developing teeth of transgenic line that carries a CAT reporter gene fused to 3.6 and 2.3 kb of rat Col1a1 regulatory sequences. These observations indicated that these promoter fragments contain sufficient regulatory elements to direct the expression of GFP to odontoblasts. Thus, Col1a1-GFP transgenes provide an excellent noninvasive marker for examining the progression of odontoblast differentiation from progenitor cells. This possibility was supported by transplantation studies in which pieces of dental pulp isolated from 2.3-GFP mice were transplanted under the kidney capsule. In the explanted dental pulps dentin-like and bone-like mineralized tissues were formed [27]. Dentin-like matrices were composed of tubular matrix (characterized by extended expression of 2.3-GFP into the tubular matrices) lined with cells expressing high levels of 2.3-GFP, and *Dspp* (Fig. 4.2a, b) and bone-like matrices were composed of atubular matrix with cells embedded within the matrix expressing high levels of 2.3-GFP and lacking the expression of *Dspp*.

By multiplexing the promoter-GFP colors in which the donor and recipient carry a different color, GFP driven by the same promoter is particularly useful in interpreting a transplantation experiment in that it allows distinction in the contributions of donor vs. host cells.

We also analyzed the temporal and spatial expression of these transgenes during in vitro mineralization and odontoblast differentiation in primary cultures derived from the coronal portions of dental pulp of molars from these transgenic animals. Our studies showed the expression of these transgenes in scattered cells at day 7 before initiation of mineralization and expression of *Dspp* indicating that, during odontoblasts differentiation, 2.3-GFP and 3.6-GFP are activated at early stages of differentiation [23].

With the ability of FACS to separate cells based on GFP expression, we obtained relatively homogenous subpopulations of Col1a1-GFP + cells and analyzed their proliferation and dentinogenic potentials. Our results showed that 2.3-GFP + and 3.6-GFP + populations contain highly proliferative cells enriched in progenitors committed to mineralization and dentinogenesis shown by sheets of mineralized tissue expressing *Dspp* in these cultures. Most interesting was the differences in levels of *Dspp* in different populations. These studies showed these transgenes are activated before the onset of matrix deposition and in cells at intermediate stages of odontoblast differentiation. The 3.6-GFP transgene was activated in cells in early stages of polarization, whereas the 2.3-GFP transgene was activated at a later stage of polarization just before or at the time of transition into secretory odontoblast. These differences were not appreciated in the developing teeth in vivo because of the close proximity of cells in the early and late stage of polarization. Further experiments are in progress to examine the expression of these two transgene in dual GFP reporter mice in which different color GFPs (topaz and cyan) are driven by 3.6 and 2.3 kb of the Col1a1 promoter, respectively. See Fig. 4.3a, b.

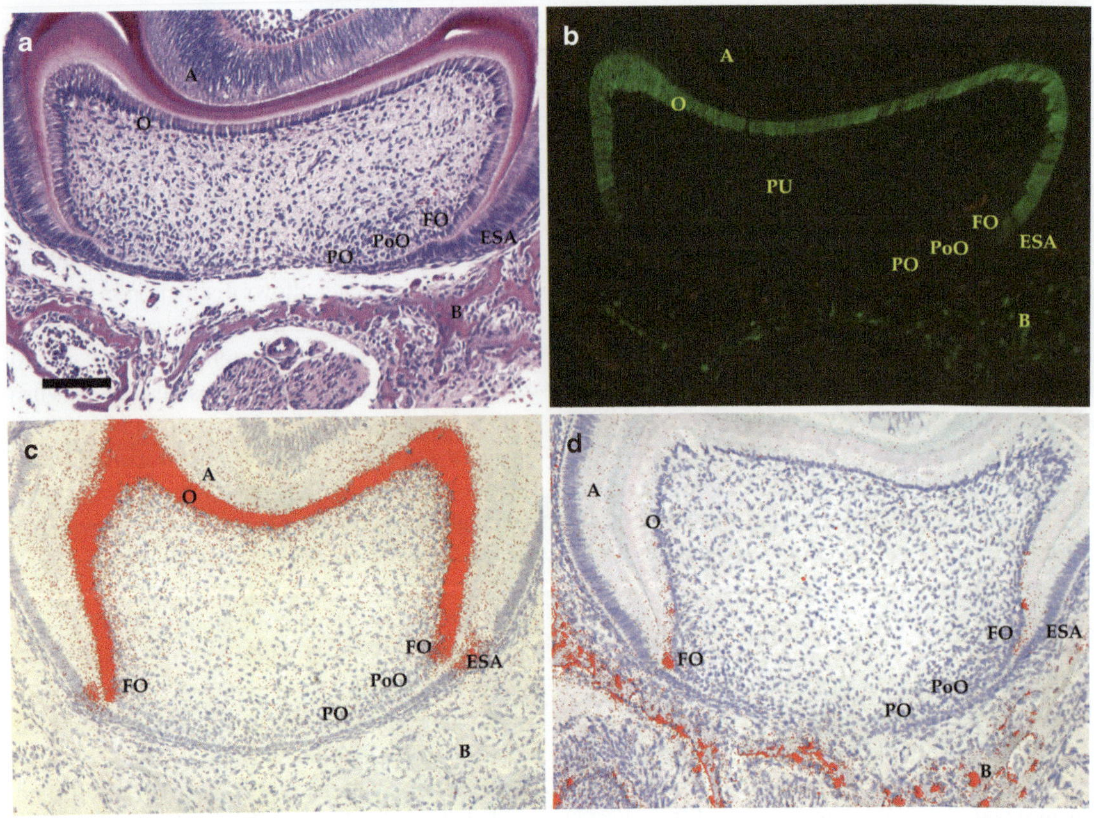

Fig. 4.1 (**a–d**) The expression 2.3-GFP, *Dspp*, and *DMP-1* in the developing maxillary first molars. Bright field (**a**), epifluorescence (**b**) pseudo-colored bright field (**c**, **d**) images of the frontal sections through first maxillary molars from postnatal day 4. (**a**) During the secretory stage of crown development, there is a gradient of odontoblast differentiation in the developing cusp of the first maxillary molars. The terminally differentiated odontoblasts (*O*) associated with predentin and dentin opposed by ameloblasts (*A*) secreting enamel are present at the tip of the cusps. Functional odontoblasts (*FO*) secreting only predentin, pre-odontoblasts (*PO*), and polarizing odontoblasts (*PoO*) are present at the cervical loops. (**b**) 2.3-GFP is not expressed in pre-odontoblasts (*PO*) or polarizing odontoblasts (*PoO*), but is expressed at high levels in the functional odontoblasts (*FO*), in terminally differentiated odontoblasts, and in osteocytes of the developing alveolar bone. (**c**) *Dspp* is expressed at high levels by functional odontoblasts (*FO*) secreting predentin and terminally differentiated odontoblasts (*O*) associated with predentin and dentin. *Dspp* is also expressed by a group of early secretory ameloblasts (ESA) located at the cervical loops opposing the functional odontoblasts (*FO*) expressing *Dspp*. Pre-odontoblasts (*PO*), polarizing odontoblasts (*PoO*), or the osteoblasts of the developing bone (*B*) do not express *Dspp*. (**d**) Unlike *Dspp*, *DMP-1* is expressed at low levels only by functional odontoblasts (*FO*) located at the cervical loop and not by terminally differentiated odontoblast (*O*) at the tips of the cusps. *DMP-1* is not expressed by pre-odontoblasts (*PO*) and polarizing odontoblasts (*PoO*). Also note the high levels of *DMP-1* expression by the osteoblasts of the developing bone (*B*). Scale bar=200 um

We also extended these studies to identify markers for later stages of odontoblast differentiation. Given the important roles of DMP1 in dentinogenesis, we used DMP1-GFP transgenic mice in which GFPtpz is under the control of 8 kb upstream regulatory sequences of DMP-1 [25]. The in vivo studies showed that DMP1-GFP was first expressed at E19 in the secretory/functional odontoblasts prior to the expression of *Dspp*. The expression of DMP1-GFP intensified in odontoblasts synthesizing primary dentin and was decreased in highly differentiated odontoblasts synthesizing secondary dentin.

Our in vitro studies using pulp cultures from DMP1-GFP transgenic animal showed that during mineralization of primary pulp cultures,

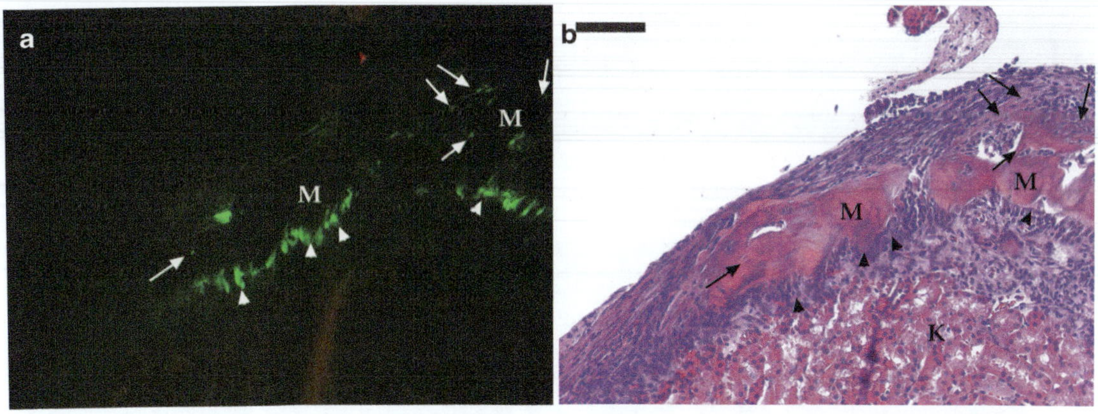

Fig. 4.2 (**a**, **b**) Analysis of pulp from pOBCol2.3GFP transgenic mice after grafting under the kidney capsule. Epifluorescent (**a**) and bright field (**b**) images of a piece of the coronal portions of pulp isolated from mandibular first molars of pOBCol2.3GFP transgenic mice transplanted for 10 days under the kidney capsule of CD-1 hosts. (**a**) The formation of odontoblast-like cells (indicated by *arrowheads*) expressing high levels of 2.3-GFP that are lining the surface of the mineralized matrix (*M*). Also note the formation of osteocyte-like cells (indicated by *arrows*) embedded within the mineralized matrix. Scale bar = 200 um

DMP1-GFP is expressed in cells in odontogenic and osteogenic lineage. In the osteogenic areas of cultures (identified by the lack of DSP expression), DMP1-GFP was expressed at high intensity. On the other hand, in the dentinogenic areas (identified by the expression of DSP), DMP1-GFP was expressed at two different intensities. The expression of DMP1-GFP at high intensity was associated with secretory/functional and newly differentiated odontoblasts, whereas the expression at very low intensity was associated with highly differentiated odontoblasts [25].

Our in vitro studies also showed that the behavior of DMP1-GFP + population was different from 2.3-GFP + populations. Unlike the 2.3-GFP + population, DMP1-GFP + cells exhibited poor proliferation and survival, suggesting that this population contains postmitotic cells in advanced stages of differentiation (Fig. 4.4). The lack of detectable levels of *Dspp* in DMP1-GFP + population isolated at day 7 (early time point) from pulp cultures suggested that DMP1-GFP is activated in secretory odontoblasts prior to the expression of *Dspp* [25].

It is important to note that the 2.3-GFP, 3.6-GFP, and DMP1-GFP transgenes, similar to expression of the endogenous proteins, are expressed by both odontoblasts and osteoblasts and therefore make it difficult to distinguish between the two cell types. To overcome this difficulty, we have generated Dspp-Cerulean transgenic mice that show restricted expression of transgene in odontoblasts.

These studies together showed the heterogeneity of pulp cultures and the necessity for development of markers for identification and isolation of more homogenous population for careful lineage analysis. In these cultures the 3.6-GFP transgene was activated in cells in early stages of polarization, whereas the 2.3-GFP transgene was activated at a later stage of polarization just before or at the time of formation of secretory/functional odontoblast. This is followed by the activation of DMP1-GFP transgene in secretory/functional odontoblasts (Fig. 4.5). These studies also indicate that these transgene can be used for identification and isolation of cells at different stages of odontoblast differentiation from the heterogeneous population of dental pulp cells. The ability to isolate relatively homogenous populations using GFP strategy during odontoblast differentiation by FACS sorting has the potential to overcome shortcoming of microarray analysis of whole cultures and should

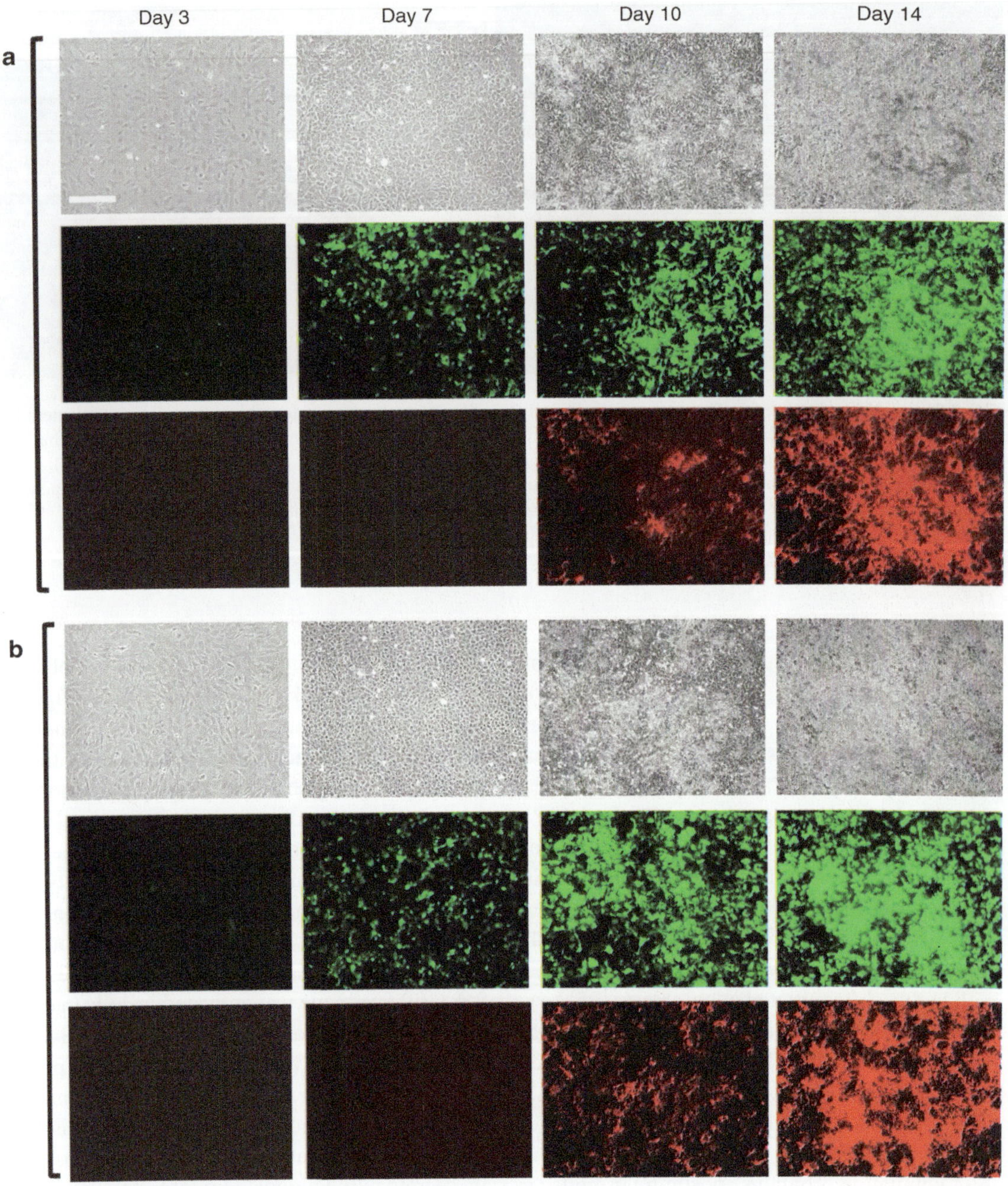

Fig. 4.3 (**a**, **b**) Expression of 3.6-GFP and 2.3-GFP transgenes in primary dental pulp cultures during in vitro mineralization and dentinogenesis. Primary dental pulp cultures obtained from pOBCol3.6GFP (**a**) and pOBCol2.3GFP (**b**) transgenic animals grown for 14 days in culture conditions supporting their mineralization and dentinogenesis. Images of the same areas in cultures at different time points analyzed under phase contrast (*upper rows* in **a** and **b**), epifluorescent light using filters for GFPtpz and GFPemd for detection of GFP (*middle rows* in **a** and **b**), and epifluorescent light using TRITC red filter for detection of xylenol orange (XO) staining in mineralized matrix (*lower rows* in **a** and **b**). Note the presence of GFP + cells early in the culture (day 3), in cell clusters at day 7, and in differentiating and differentiated nodules between days 10 and 14. Scale bars = 200 μm

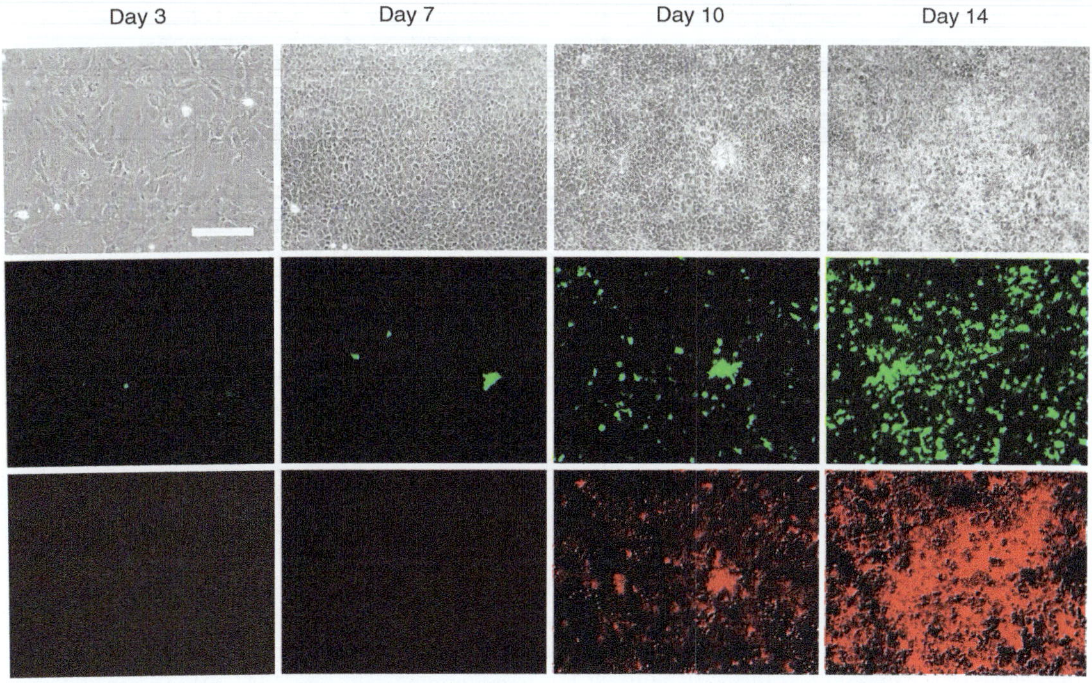

Fig. 4.4 Expression of DMP1-GFP transgenes in primary dental pulp cultures during in vitro mineralization and dentinogenesis. Primary dental pulp cultures obtained from DMP1-GFP transgenic animal grown for 14 days in culture conditions supporting their mineralization and dentinogenesis. Images of the same areas in cultures at different time points analyzed under phase contrast (*upper row*), epifluorescent light using filters for GFPtpz and GFPemd for detection of GFP (*middle row*), and epifluorescent light using TRITC red filter for detection of xylenol orange (XO) staining in mineralized matrix (*lower rows*). Scale bars = 200 μm

provide new expression profiles of stage-specific genes during normal and abnormal dentinogenesis in mice with various genetic mutations.

4.3 Dental Pulp Progenitor and Stem Cells

Dentin secreted by odontoblasts during tooth development and before the completion of root formation is defined as primary dentin and is deposited at a rate of 4–20 μm/day. Following primary dentinogenesis, secondary dentin is secreted throughout life at a much slower rate (0.4 μm/day) and results in a decrease in the size of the pulp chamber. Primary and secondary dentins secreted by odontoblasts are characterized by closely packed dentinal tubules that span the entire thickness of the dentin. However, the function of odontoblasts is not limited to primary and secondary dentinogenesis, but includes maintenance of dentin thorough the life of the tooth.

The dentin–pulp complex has a regenerative potential leading to the formation of tertiary dentin. In response to mild environmental stimuli (attrition or early caries), preexisting live odontoblasts upregulate their secretory activity and secrete a tubular reactionary dentin matrix [29, 30]. This layer shows many anatomical, biochemical, and functional similarities to the primary and secondary dentins and contributes to the protection of the pulp tissue. Reactionary dentin formation is promoted by low amounts of pro-inflammatory cytokines and/or growth factors and extracellular matrix components sequestered in dentin that are released following dentin demineralization by cariogenic bacteria.

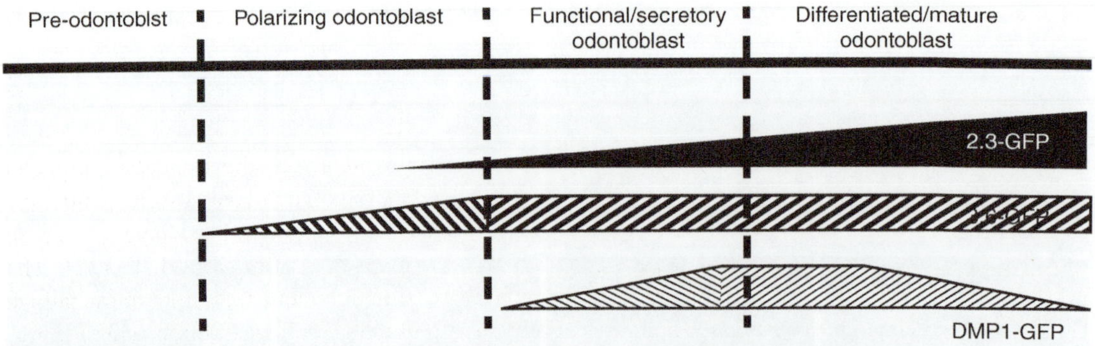

Fig. 4.5 Schematic representation of proposed stages of activation of 3.6-GFP and 2.3-GFP transgenes during odontoblast differentiation. *DSPP* was used as a marker of early and later stages of mineralization

On the other hand, trauma of greater intensity that causes the death of the preexisting odontoblasts leads to reparative dentinogenesis involving the recruitment and proliferation of progenitor and/or stem cells to the site of injury and their differentiation into a second generation of odontoblasts or "odontoblast-like cells" [29, 30]. These odontoblast-like cells synthesize an atubular dentin-like mineralized matrix immediately below the site of damage to preserve pulp vitality [29, 30]. The reparative dentin also referred to as osteodentin contains cells trapped within the matrix that forms in the absence of inner dental epithelium and basement membrane. Reparative dentinogenesis is thought to be dependent on multiple signaling molecules sequestrated in the dentin matrix [29, 30].

Potential populations of cells within dental pulp capable of giving rise to the new generation of odontoblast-like cells during reparative dentinogenesis are many and include the cell-rich layer of Höhl adjacent to the odontoblasts, fibroblasts, and progenitors/mesenchymal stem cells (MSCs) [29, 30]. Available evidence suggests that the dental pulp contains several niches of potential progenitors and stem cells involved in reparative dentinogenesis.

The discovery of stem cells from human dental pulp of impacted third molars in the year 2000 (DSTP) and later in the pulp of exfoliated primary teeth (SHED) suggested that these populations are among the potential populations involved in reparative dentinogenesis [31, 32]. Transplantation of in vitro expanded DPSCs and SHED mixed with hydroxyapatite/tricalcium phosphate particles into immunocompromised mice resulted in the formation of pulp–dentin-like tissue complexes including vascularized pulp-like tissue, surrounded by a layer of odontoblast-like cells without an active hematopoietic marrow [31, 32]. Transplantation of SHED seeded in tooth slice/scaffolds into the subcutaneous space of immunodeficient mice also resulted in the generation of tissue-like human dental pulps [33]. The odontoblast-like cells in these experiments expressed dentin sialoprotein (DSP), a marker for odontoblastic differentiation [31–33].

A recent study provided strong evidence for the differentiation of SHED into functional odontoblasts by demonstrating that these cells were able to generate new tubular dentin in vivo, as determined by tetracycline staining [34]. These studies suggested that human adult dental pulp contained a small population of self-renewing, highly proliferative multipotent mesenchymal stem cells (MSCs) that reside within a larger population of more committed progenitors [35]. These cells exhibit similar characteristic with bone marrow stromal stem cells (BMSCs), including fibroblast-like morphology, adherent colony-forming units.

MSCs, first described in postnatal bone marrow [36], constitute a rare population of cells of non-hematopoietic origin that were identified by their adherence to plastic tissue culture dish and their fibroblastic morphology. When transplanted into an animal, MSC from bone marrow form bone, cartilage, hematopoietic marrow, fat cells, and the stroma that supports blood formation [37]. On the basis of their origin and their multipotency, these cells were originally referred to as "bone marrow stromal stem cells" (BMSCs) and more recently MSC [37].

Since their discovery in the bone marrow, MSCs have been identified in almost every organ in the body using criteria established for BMSCs including dental pulps. Furthermore, different populations of MSCs have been isolated from dental pulp of various teeth in humans and animal models.

All of these populations are heterogeneous in that cell colonies derived from these cells exhibit morphological and functional (cell proliferation and differentiation) similarities and differences. These cells are all capable of regenerating dentin and pulp-like tissues after transplantation [35]. The fraction of multipotent stem cells in the dental pulp is small as defined by their in vitro ability to undergo odontogenic, angiogenic, adipogenic, chondrogenic, neurogenic, or myogenic differentiation.

Comparative studies demonstrated that MSC isolated from different organs, including dental pulps, exhibit many differences including differences in the expression of stem cell markers, proliferation, and osteogenic differentiation [35]. Furthermore, there are significant differences in their multipotency and expression of markers of pluripotency and in their differentiation potential that is related to the developmental stage at which a stem cell is isolated.

Although these studies have provided evidence that the cells are able to form the dentin pulp complex, the origin/identity of the progenitor cells and signaling pathways involved in reparative dentinogenesis in vivo remain elusive. This is, in part, due to the lack of suitable markers for the identification of potential progenitors and appropriate animal models to trace the fate of potential progenitors during reparative dentinogenesis.

Later studies showed that MSCs in almost all organs, including in dental pulp and bone marrow, reside in the perivascular area [38–42]. Several studies have shown that perivascular cells can be identified/isolated based on the expression of several markers including CD146, chondroitin sulfate proteoglycan 4 (NG2), platelet-derived growth factor receptor-beta (PDGF-Rβ), and alpha-smooth muscle actin (αSMA) [43].

4.4 Methods Used to Identify and Characterize Stem Cells

Over the last few decades, several techniques have been utilized to identify stem cells including the use of genetically modified mouse models. Proliferation kinetics and lineage-tracing assays based on transgenic animal models have provided crucial insights into the stem and progenitor cells in many organs including dental pulp [44–46].

Because stem cells divide infrequently, studying them and their niches in vivo in the different organs including dental pulp requires injury to the organ. The most common method of injury for studying dental pulp stem cells has been the creation of experimental pulp exposure. Pulp exposures are created in vivo on incisors and molars in animal models including rats and mice. In these models cavities are prepared on specific surfaces of a tooth using a dental burr. Exposure of pulp is obtained by pushing a fine instrument into the deepest part of the cavity. Exposed pulps are capped with appropriate materials such as mineral trioxide aggregate (MTA) and restored with cement. Animals are usually sacrificed by intracardiac perfusion for best histological analysis at various time points after injury, and then the fate and behavior of dental pulp stem/progenitor cells are studies using various methods described next. Thus, the experimental animal models for cavity preparation could be helpful for better understanding the molecular mechanisms regulating the odontoblast differentiation after tooth injury.

Using an injury model on the occlusal surface of the 3.6-GFP transgenic mice, we showed the absence of odontoblasts expressing 3.6-GFP at the injury site immediately after pulp exposure. Evidence of reparative dentinogenesis was apparent at 4 weeks in 3.6-GFP mice in CD1 background and at 8 weeks in 3.6-GFP mice with

C57/Bl6 background. The reparative dentin in both groups contained newly formed atubular-mineralized tissue resembling a dentin bridge and/or osteodentin that was lined by cells expressing 3.6-GFP as well as 3.6-GFP expressing cells embedded within the atubular matrix [47].

4.4.1 Label-Retaining Cells

This slow-cycling characteristic of stem cells has allowed the use of DNA-labeling techniques with detectable nucleotide analogs to track stem cells in situ. Bromodeoxyuridine (BrdU) is the common nucleotide analog that has been used in pulse-chase experiments to track label-retaining cells (LRCs) after a prolonged "washout" period that dilutes the label within the more rapidly cycling transient-amplifying (TA) cells.

Recent studies showed that dense LRCs mainly reside in the center of the dental pulp are associated with the blood vessel and co-express mesenchymal stem cell markers such as STRO-1 and CD146, a marker of pericytes. Using experimental pup exposure these authors showed the active of migration of these dense LRCs toward the site of the injury and that reparative dentinogenesis contained a few dense LRCs that expressed nestin and *Dspp* suggesting of their odontoblast-like phenotype [48–52].

More recently, a pulse-chase strategy on the basis of a transgenic mouse model that allows the expression of a GFP fused to the histone 2B (H2B–GFP) in a doxycycline-dependent manner has been developed. In these mice, in the absence of doxycycline, H2B–GFP transgene is expressed at high and homogenous level in all proliferative cells. When doxycycline is given to the mice, H2B–GFP expression is completely repressed. However, cells labeled before administration of doxycycline remains H2B–GFP+. In this model, in rapidly dividing cells the GFP signal is diluted, while slowly dividing and/or postmitotic cells remain labeled (green). This approach enables the isolation of live cells on the basis of their slow or rapidly cycling properties as well as fate mapping of slow-cycling cells and their progeny and has been successfully used in several tissues.

4.4.2 In Vivo Lineage Tracing

In recent years, a labeling strategy has been developed using transgenic animal models that allows the identification of the progeny of individually marked cells. In mice, the most popular labeling technique is genetic inducible fate mapping based on a drug-inducible Cre recombinase together with a reporter system (The *Cre-ER-loxP* system). In this system, a drug-inducible Cre recombinase (i.e., Cre–estrogen receptor, Cre-ER) is placed under the control of a gene-specific promoter (a putative stem cell marker). The transient activation by drug administration (e.g., tamoxifen) leads to the excision of a stop cassette and the permanent expression of a reporter construct in targeted cells and their progeny. The use of different cell-specific promoters allows different cell subpopulations to be labeled.

The *Cre-ER-loxP* system in mice has provided a powerful technique for lineage tracing, and it has significantly increased our understanding of the identity and behaviors of stem cells in numerous tissues. The real power of inducible genetic-labeling assays lies in their capacity to resolve individual cell fate behavior and has been used to trace cells in many organisms.

To date, a few studies have utilized in vivo lineage tracing to study adult stem cells in dental pulp. Cre-mediated genetic lineage tracing of pericytes, using tamoxifen-inducible NG2creER crossed with reporter lines, showed that some pericytes differentiate into odontoblasts. However, the pericyte-derived mesenchymal cell contribution to odontoblast differentiation was low and could not account for all odontoblasts suggesting the presence of multiple populations of stem cells and stem cell niches in dental pulp [39]. Further studies should provide additional insight into this mater.

Furthermore, in addition to the roles of resident progenitor populations within dental pulp in reparative dentinogenesis, the possible roles and involvement of the nonresident populations including progenitor/stem cells from other sites and the hematopoietic cell population in reparative dentinogenesis have not been examined. The parabiosis model in which two mice are surgically joined to allow the establishment of common blood circulation has provided an excellent system for studying various biological processes including the roles and involvement of nonresident progenitors/stem and hematopoietic cells migrating to the site of injury from circulating blood during tissue remodeling and repair. By using parabiotic pairs between mice expressing (EGFP) (referred to as GFP+) and nonfluorescent wild-type mice, we studied the contributions of the nonresident progenitor/stem cells and hematopoietic cells to reparative dentinogenesis. After 2 weeks of parabiosis, pulp exposures were created in the maxillary first right molar of the GFP mice to stimulate reparative dentinogenesis. Epifluorescence analysis of molars with pulp exposures in the recipient parabiont GFP mice 4 and 8 weeks after pulp exposure showed an influx of GFP + cells through the apical foramen. At 8 weeks after pulp exposure, there were numerous GFP + cells in close contact with the newly synthesized matrix. The location and the size of the GFP + cells and the lack of TRAP staining (a specific histochemical marker for odontoclast/osteoclasts) in the majority of the GFP + cells suggested that the GFP + cells associated with reparative dentin represent cells such as macrophages within the mononuclear phagocyte lineage originating from the hematopoietic system of the GFP + donor animals [53].

Acknowledgments I would like to thank Drs. Barut, Balic, Frozoni, and Sagomonyants for their hard work in generating the data reported here. I also would like to thank all the individuals who provided reagents, valuable input, and technical assistance in various aspects of these studies including Drs. Rowe and Kalajzic and Mrs. Rodgers for critical review of the manuscript. This work was supported by a grant from National Institutes of Health (NIDCR) to MM (DE016689).

References

1. Lisi S, Peterkova R, Peterka M, Vonesch JL, Ruch JV, Lesot H. Tooth morphogenesis and pattern of odontoblast differentiation. Connect Tissue Res. 2003;44 Suppl 1:167–70.
2. Tziafas D, Kodonas K. Differentiation potential of dental papilla, dental pulp, and apical papilla progenitor cells. J Endod. 2010;36(5):781–9.
3. Thesleff I, Keranen S, Jernvall J. Enamel knots as signaling centers linking tooth morphogenesis and odontoblast differentiation. Adv Dent Res. 2001;15:14–8.
4. Arana-Chavez VE, Massa LF. Odontoblasts: the cells forming and maintaining dentine. Int J Biochem Cell Biol. 2004;36(8):1367–73.
5. Ruch JV, Lesot H, Begue-Kirn C. Odontoblast differentiation. Int J Dev Biol. 1995;39(1):51–68.
6. Nanci A. Ten Cate's oral histology: development, structure and function. 7th ed. Maryland Heights, Missouri: Mosby; 2008.
7. Bleicher F. Odontoblast physiology. Exp Cell Res. 2013. (Epub ahead of print).
8. Qin C, Baba O, Butler WT. Post-translational modifications of sibling proteins and their roles in osteogenesis and dentinogenesis. Crit Rev Oral Biol Med. 2004;15(3):126–36.
9. Lesot H, Lisi S, Peterkova R, Peterka M, Mitolo V, Ruch JV. Epigenetic signals during odontoblast differentiation. Adv Dent Res. 2001;15:8–13.
10. Bryder D, Rossi DJ, Weissman IL. Hematopoietic stem cells: the paradigmatic tissue-specific stem cell. Am J Pathol. 2006;169(2):338–46.
11. Abe T, Fujimori T. Reporter mouse lines for fluorescence imaging. Dev Growth Differ. 2013;55(4):390–405.
12. Hoffman RM. Fluorescent proteins as visible in vivo sensors. Prog Mol Biol Transl Sci. 2013;113:389–402.
13. Hadjantonakis AK, Macmaster S, Nagy A. Embryonic stem cells and mice expressing different GFP variants for multiple non-invasive reporter usage within a single animal. BMC Biotechnol. 2002;2:11.
14. Hadjantonakis AK, Nagy A. FACS for the isolation of individual cells from transgenic mice harboring a fluorescent protein reporter. Genesis. 2000;27(3):95–8.
15. Balic A, Mina M. Analysis of developmental potentials of dental pulp in vitro using GFP transgenes. Orthod Craniofac Res. 2005;8(4):252–8.
16. Bilic-Curcic I, Kronenberg M, Jiang X, Bellizzi J, Mina M, Marijanovic I, et al. Visualizing levels of osteoblast differentiation by a two-color promoter-GFP strategy: Type I collagen-GFPcyan and osteocalcin-GFPtpz. Genesis. 2005;43(2):87–98.
17. Jiang X, Kalajzic Z, Maye P, Braut A, Bellizzi J, Mina M, et al. Histological analysis of GFP expression in murine bone. J Histochem Cytochem. 2005;53(5):593–602.
18. Rowe DW. Viewing problems in bone biology from the perspective of lineage identification. J Musculoskelet Neuronal Interact. 2005;5(4):350–2.
19. Kalajzic I, Braut A, Guo D, Jiang X, Kronenberg MS, Mina M, et al. Dentin matrix protein 1 expression during osteoblastic differentiation, generation of an osteocyte GFP-transgene. Bone. 2004;35(1):74–82.
20. Kalajzic I, Kalajzic Z, Kaliterna M, Gronowicz G, Clark SH, Lichtler AC, et al. Use of type I collagen

green fluorescent protein transgenes to identify subpopulations of cells at different stages of the osteoblast lineage. J Bone Miner Res. 2002;17(1):15–25.
21. Kalajzic Z, Li H, Wang LP, Jiang X, Lamothe K, Adams DJ, et al. Use of an alpha-smooth muscle actin GFP reporter to identify an osteoprogenitor population. Bone. 2008;43(3):501–10.
22. Kalajzic Z, Liu P, Kalajzic I, Du Z, Braut A, Mina M, et al. Directing the expression of a green fluorescent protein transgene in differentiated osteoblasts: comparison between rat type I collagen and rat osteocalcin promoters. Bone. 2002;31(6):654–60.
23. Balic A, Aguila HL, Mina M. Identification of cells at early and late stages of polarization during odontoblast differentiation using pOBCol3.6GFP and pOBCol2.3GFP transgenic mice. Bone. 2010;47(5):948–58.
24. Balic A, Mina M. Characterization of progenitor cells in pulps of murine incisors. J Dent Res. 2010;89(11):1287–92.
25. Balic A, Mina M. Identification of secretory odontoblasts using DMP1-GFP transgenic mice. Bone. 2011;48(4):927–37.
26. Mina M, Braut A. New insight into progenitor/stem cells in dental pulp using Col1a1-GFP transgenes. Cells Tissues Organs. 2004;176(1–3):120–33.
27. Braut A, Kollar EJ, Mina M. Analysis of the odontogenic and osteogenic potentials of dental pulp in vivo using a Col1a1-2.3-GFP transgene. Int J Dev Biol. 2003;47(4):281–92.
28. Braut A, Kalajzic I, Kalajzic Z, Rowe DW, Kollar EJ, Mina M. Col1a1-GFP transgene expression in developing incisors. Connect Tissue Res. 2002;43(2–3):216–9.
29. Sloan AJ, Smith AJ. Stem cells and the dental pulp: potential roles in dentine regeneration and repair. Oral Dis. 2007;13(2):151–7.
30. Sloan AJ, Waddington RJ. Dental pulp stem cells: what, where, how? Int J Paediatr Dent. 2009;19(1):61–70.
31. Gronthos S, Mankani M, Brahim J, Robey PG, Shi S. Postnatal human dental pulp stem cells (DPSCs) in vitro and in vivo. Proc Natl Acad Sci U S A. 2000;97(25):13625–30.
32. Miura M, Gronthos S, Zhao M, Lu B, Fisher LW, Robey PG, et al. SHED: stem cells from human exfoliated deciduous teeth. Proc Natl Acad Sci U S A. 2003;100(10):5807–12.
33. Cordeiro MM, Dong Z, Kaneko T, Zhang Z, Miyazawa M, Shi S, et al. Dental pulp tissue engineering with stem cells from exfoliated deciduous teeth. J Endod. 2008;34(8):962–9.
34. Sakai VT, Zhang Z, Dong Z, Neiva KG, Machado MA, Shi S, et al. SHED differentiate into functional odontoblasts and endothelium. J Dent Res. 2010;89(8):791–6.
35. Huang GT, Gronthos S, Shi S. Mesenchymal stem cells derived from dental tissues vs. those from other sources: their biology and role in regenerative medicine. J Dent Res. 2009;88(9):792–806.
36. Friedenstein AJ, Chailakhjan RK, Lalykina KS. The development of fibroblast colonies in monolayer cultures of guinea-pig bone marrow and spleen cells. Cell Tissue Kinet. 1970;3(4):393–403.
37. Bianco P, Robey PG, Simmons PJ. Mesenchymal stem cells: revisiting history, concepts, and assays. Cell Stem Cell. 2008;2(4):313–9.
38. Dimarino AM, Caplan AI, Bonfield TL. Mesenchymal stem cells in tissue repair. Front Immunol. 2013;4:201.
39. Feng J, Mantesso A, De Bari C, Nishiyama A, Sharpe PT. Dual origin of mesenchymal stem cells contributing to organ growth and repair. Proc Natl Acad Sci U S A. 2011;108(16):6503–8.
40. Feng J, Mantesso A, Sharpe PT. Perivascular cells as mesenchymal stem cells. Expert Opin Biol Ther. 2010;10(10):1441–51.
41. Kerkis I, Caplan AI. Stem cells in dental pulp of deciduous teeth. Tissue Eng Part B Rev. 2012;18(2):129–38.
42. Crisan M, Corselli M, Chen WC, Peault B. Perivascular cells for regenerative medicine. J Cell Mol Med. 2012;16(12):2851–60.
43. Crisan M, Corselli M, Chen CW, Peault B. Multilineage stem cells in the adult: a perivascular legacy? Organogenesis. 2011;7(2):101–4.
44. Blanpain C, Simons BD. Unravelling stem cell dynamics by lineage tracing. Nat Rev Mol Cell Biol. 2013;14(8):489–502.
45. Blanpain C. Tracing the cellular origin of cancer. Nat Cell Biol. 2013;15(2):126–34.
46. Liu X, Driskell RR, Engelhardt JF. Stem cells in the lung. Methods Enzymol. 2006;419:285–321.
47. Frozoni M, Balic A, Sagomonyants K, Zaia AA, Line SR, Mina M. A feasibility study for the analysis of reparative dentinogenesis in pOBCol3.6GFPtpz transgenic mice. Int Endod J. 2012;45(10):907–14.
48. Ishikawa Y, Ida-Yonemochi H, Suzuki H, Nakakura-Ohshima K, Jung HS, Honda MJ, et al. Mapping of BrdU label-retaining dental pulp cells in growing teeth and their regenerative capacity after injuries. Histochem Cell Biol. 2010;134(3):227–41.
49. Mutoh N, Nakatomi M, Ida-Yonemochi H, Nakagawa E, Tani-Ishii N, Ohshima H. Responses of BrdU label-retaining dental pulp cells to allogenic tooth transplantation into mouse maxilla. Histochem Cell Biol. 2011;136(6):649–61.
50. Onshima H, Nakagawa E, Ida-Yonemochi H. O39-establishment of in vitro culture system for evaluation of the dentin-pulp complex regeneration with special reference to differentiation capacity of the BrdU-label-retaining dental pulp cells. Bull Group Int Rech Sci Stomatol Odontol. 2010;49(3):92.
51. Saito K, Ishikawa Y, Nakakura-Ohshima K, Ida-Yonemochi H, Nakatomi M, Kenmotsu S, et al. Differentiation capacity of BrdU label-retaining dental pulp cells during pulpal healing following allogenic transplantation in mice. Biomed Res. 2011;32(4):247–57.

52. Saito K, Nakatomi M, Ohshima H. Dynamics of bromodeoxyuridine label-retaining dental pulp cells during pulpal healing after cavity preparation in mice. J Endod. 2013;39(10):1250–5.

53. Frozoni M, Zaia AA, Line SR, Mina M. Analysis of the contribution of nonresident progenitor cells and hematopoietic cells to reparative dentinogenesis using parabiosis model in mice. J Endod. 2012;38(9):1214–9.

Pulp Vascularization and Its Regulation by the Microenvironment

5

Imad About

5.1 Introduction

Like any other tissue, pulp vasculature brings oxygen supply and nutrients to the dental pulp and provides means to remove waste and toxic materials. However, unlike any other tissue, the dental pulp is highly vascularized and located within a rigid environment as it is enclosed within calcified dentin walls. Indeed, when the pulp is subjected to an inflammation which occurs under carious injuries, the vasodilatation leading to an increased blood flow may be detrimental to the dental pulp due to the inextensible environment (Fig. 5.1a–c). However, the vascularization provides a protection to the pulp by different mechanisms [1]. This is why dental pulp vascularization can be considered as a unique example.

Recent advancements revealed some interesting aspects demonstrating a local regulation of pulp angiogenesis that may be valuable in future tissue engineering investigations for regenerating a vascularized pulp.

For a better understanding of the pulp vasculature, it is important to know how it is formed during tooth development and in mature teeth. It is also important to understand how it can be re-established after pulp injury and during tissue engineering.

I. About, PhD
Aix-Marseille Université, Centre National de la Recherche Scientifique, ISM UMR 7287, 13288, Marseille cedex 09, France
e-mail: imad.about@univ.amu.fr

5.2 Establishment of the First Capillary Network (Vasculogenesis)

The main cell-forming unit of the vascular system is the endothelial cell which forms the internal lining of blood vessels. These cells are derived from mesoderm stem cells which give rise to haemangioblast precursor cells which in turn give rise to the haematopoietic stem cells and angioblasts: the progenitors of endothelial cells [2]. Thus, blood cells and endothelial cells share a common origin and remain tightly linked to each other during adult life (Fig. 5.2).

During the tooth development and in the absence of blood vessels, oxygen and nutrients may reach the embryonic tissues by simple diffusion [3]. The first evidence of mesodermally derived cell migration was observed within the dental papilla at the early bell stage [4]. This migration was demonstrated on tooth slice cultures of E14.5 mouse mandibles comprising the first molar tooth germ at the cap stage. CD31-positive endothelial cells were present in the dental follicle surrounding the forming dental papilla. After 4 days in culture, they entered the dental papilla, and the first capillary network of blood vessels rises de novo at the late bell stage [5]. At this stage, the metabolic activity and oxygen requirements of both epithelial and mesenchymal cells are high. Endothelial cells connect to each other and organize into hollow and interconnecting structures forming the blood vessels (Fig. 5.2) of the tooth by a process called vasculogenesis [6].

M. Goldberg (ed.), *The Dental Pulp*,
DOI 10.1007/978-3-642-55160-4_5,

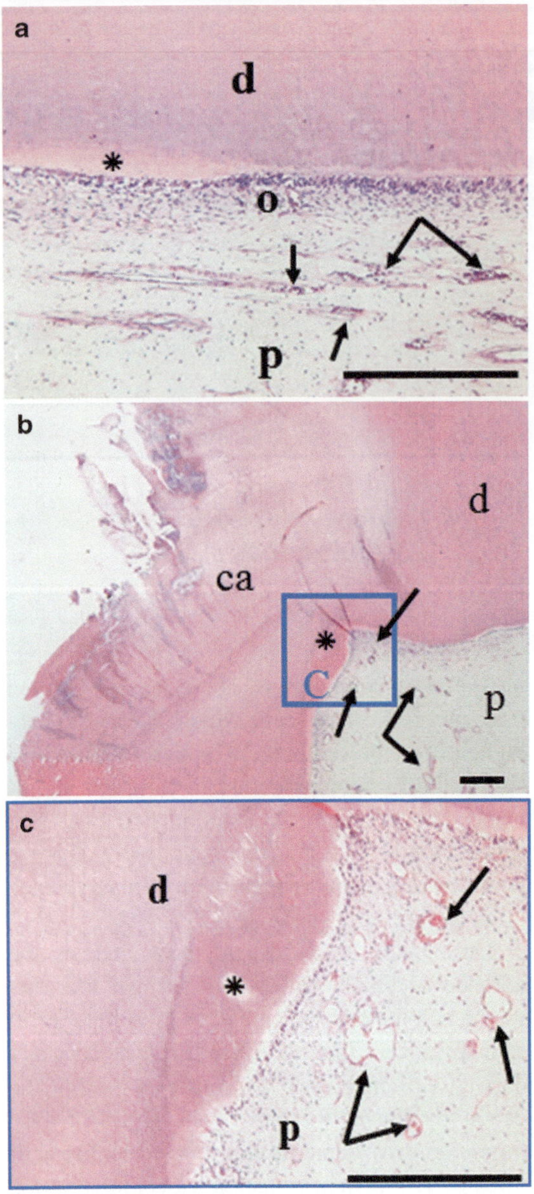

Fig. 5.1 Dental pulp vascularization. The dental pulp is highly vascularized. This vascularization is re-established after pulp healing and dentin regeneration. (**a**) Inflammation induces dilatation of blood vessels. These can be easily distinguished under a carious injury at low (**b**) and high magnifications (**c**). *Arrows* blood vessels, *P* pulp, *D* dentin, *O* odontoblasts, *star* tertiary dentin scale bar 100 μm

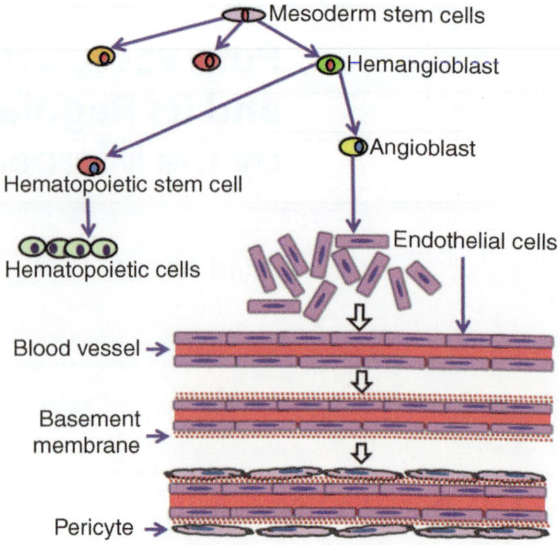

Fig. 5.2 Origin of endothelial cells and vasculogenesis. Endothelial cells derive from mesoderm stem cells and share a common origin with haematopoietic cells. During the embryonic development, angioblasts differentiate into endothelial cells which establish well-organized networks of blood vessels through the vasculogenesis process

This is followed by the recruitment of smooth muscle-like pericytes to cover capillaries [7, 8]. During angiogenesis which occurs at later developmental stages, this primary network is remodelled, adjusted and specialized into arterial and vein capillary beds [9]. Expression analysis of vascular endothelial growth factor (VEGF) and its receptors (VEGFR) demonstrated that VEGF signalling pathway plays a central role in the migration of angioblasts during vasculogenesis [10]. The importance of VEGF signalling was further supported by the fact that heterozygote deletion of VEGF-A, VEGFR1 and VEGFR2 in knockout mice led either to a strong alteration or to a nearly complete lack of blood vessels in mouse embryos [2, 11, 12].

Although vasculogenesis is believed to be the main process involved in blood vessel formation until the late bell stage of tooth development, investigations using CD34 as marker of endothelial cells revealed that vasculogenesis can still be functional at later stages in mature teeth. The presence of CD34+ endothelial cells in permanent teeth suggests that vasculogenesis persists into adult life, where it contributes to continuous adjustment of vessel and network structures in response to functional needs and dental tissue homeostasis [13].

5.3 Angiogenesis

At later developmental stages and in the mature pulp, endothelial cells form blood vessels by another process called angiogenesis. This process implies the formation of new blood vessels by "sprouting" from pre-existing blood vessels. This occurs via the extension or remodelling from existing capillaries (Fig. 5.3a–c). This process is encountered during chronic inflammation, menstrual cycle and tumour growth. It is a critical part of the wound healing process in all tissues, and the local pulpal angiogenesis is a prerequisite for successful repair in the tooth [14]. It is also a key process in tissue engineering procedures. If blood supply is not established rapidly into the transplanted cells/tissues, a necrosis occurs due to a lack of oxygen and nutrient supply [15].

Angiogenesis is a complex process. It implies extracellular matrix secretion and remodelling, secretion of proteolytic enzymes to degrade the vessel wall extracellular matrix, endothelial cell migration and proliferation, capillary differentiation and anastomosis [16]. This process is regulated by many inductive and inhibitory signals [17, 18]. Among all the pro-angiogenic factors, VEGF is considered the most essential for the differentiation of the vascular system. It induces endothelial cell proliferation, migration and survival [12]. Basic fibroblast growth factor (FGF-2) stimulates angiogenesis in vivo and plays a significant role in vascularization of damaged or traumatized tissue [19]. Platelet-derived growth factor (PDGF) is a smooth muscle cell mitogen [20] and plays an essential role in the formation of new vessels and in maintaining their stability [21]. Angiopoietins (Ang1, Ang2) and their receptor (Tie-2) play a role in destabilizing the blood vessel and establishment of contacts between endothelial cells. Transforming growth factor β1 (TGF-β1) is involved in endothelial cell proliferation [22]. On the other hand, anti-angiogenic signals regulate the vasculature and inhibit overgrowth of blood vessels. Among these, endostatin inhibits endothelial proliferation and angiogenesis [23]. Angiostatin is considered as a potent inhibitor of angiogenesis by selectively inhibiting endothelial cell proliferation [24]. Thus, angiogenesis appears as a balance between pro- and anti-angiogenic signals, and a "switch" occurs under pathological conditions.

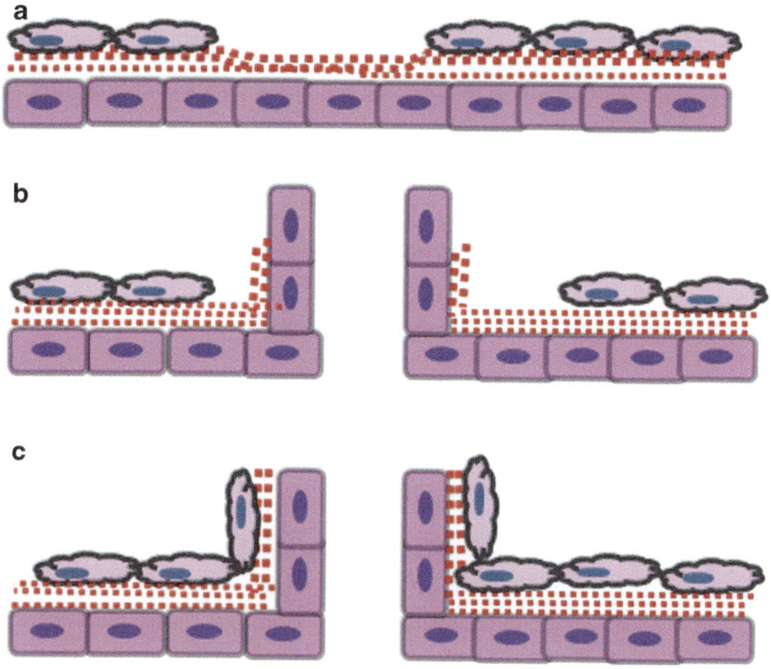

Fig. 5.3 Angiogenesis by sprouting from pre-existing vessels. This is a multistep process which implies the following: (**a**) pre-existing vessel destabilization (angiopoietin 2, VEGF), extracellular matrix degradation (MMP, chymases, heparanases) and release of matrix-sequestered molecules (VEGF, FGF-2, IGF1); (**b**) proliferation (VEGF, FGF-2, EGF, HGF, PLGF, TGF-β, TNF-α) and anastomosis (VEGF, angiopoietin 1, VE cadherin, ephrin B2/B4); and (**c**) pericyte recruitment (angiopoietin 1, PDGF, TGF-β, VE cadherin)

The endothelial cells in the blood vessels are closely related to the vascular basement membrane extracellular matrix. This membrane inhibits endothelial cell proliferation, but upon degradation by proteolytic enzymes, it allows endothelial cell proliferation and angiogenesis to proceed. The basement membrane also plays critical roles in most of the processes of blood vessel formation and stabilization. This role is best demonstrated by extracellular matrix molecule null mutations. Indeed, in α-1(I) collagen knockouts, mutants die at E12–E14 due to blood vessel rupture. This suggests that fibrillar collagen type I may play an important role in the blood vessel stability [25]. In fibronectin knockout mice, a range of phenotypes including deformed heart and embryonic vessels was observed. This may be due to mesenchymal defects mainly in cell proliferation, adhesion and migration [26, 27]. Perlecan is a proteoglycan synthesized by both vascular endothelial and smooth muscle cells and deposited in the extracellular matrix. It has been shown that half of perlecan knockout mice die at E10 or E12 due to pericardial haemorrhage and to death from respiratory failure at birth. This may be due to rupture of basement membranes [28]. These data clearly demonstrate that the composition of the basement membrane of blood vessels and its structure is a key element in blood vessel stabilization and in maintaining its physiological function.

5.3.1 Intussusceptive or Non-sprouting Angiogenesis

Intussusceptive angiogenesis does not require cell proliferation but rather endothelial cell reorganization to form new blood vessels from preexisting ones. During this reorganization, an individual blood vessel splits into two separate compartments leading to the formation of two individual blood vessels [29]. There is no evidence of the existence of this process within the mature dental pulp where angiogenesis occurs mainly by sprouting angiogenesis.

5.4 The Dental Pulp Is a Highly Vascularized Tissue

The development and establishment of pulp vascularization by these processes in embryonic and mature teeth lead to the formation of a densely vascularized pulp tissue. In mature teeth, the maxillary artery brings blood supply to the pulp through the dental artery which feeds each pulp via arterioles. These arterioles enter the apical foramina and ascend the central region of the coronal pulps and branches off to form a rich capillary network at the periphery of the pulp [30, 31]. During primary dentinogenesis, this vasculature develops in the odontogenic zone of the dentine-pulp complex, showing increased vessel fenestration for odontoblast nutrition. With the completion of dentine formation, a decrease in fenestration and a withdrawal of vessels from the odontoblast layer can be seen so that in the mature tooth of limited growth, the capillary network is confined to the subodontoblastic region [32]. A parallel network drains blood into the central pulp venules and leaves out the pulp via the apical foramina [30].

Thus, the dental pulp becomes a highly vascularized tissue both due to vasculogenesis and angiogenesis. It has been shown that the dental pulp has an average capillary density higher than in most other tissues [33]. The pulp blood flow is also high (around 50 ml/min/100 g of pulp tissue) [34] as compared to that of other tissues [35].

5.5 The Pulp Vascularization Has Unique Properties

1. The dental pulp is surrounded by inextensible calcified dentin walls. Because of this low-compliance encasement, intrapulpal tissue pressure is of prime importance in pulpal physiology. During acute inflammation, capillary dilatation may lead to a significant swelling of the dental pulp which, in turn, increases the local blood pressure that stimulates pulp nerves leading to pain sensation. The measurements of the local pressure indicate that

its increase due to inflammation is a localized event which is not transmitted to the rest of the pulp. This was demonstrated by measuring interstitial fluid pressure in cat dental pulp 7 days after inducing pulpitis in vivo. Under pulpitis, the pressure was significantly high (16 mmHg) as compared with controls (5.5 mmHg). However, the pressure measurements at a site 1–2 mm distant to the induced inflammation site averaged 7.0 mm indicating that the increase in pulp blood pressure is a very localized event [36].

2. The combination of both the pulp resilient ground substance properties and the localized pulpal pressure does not transmit the deleterious effects of increased pressure to the pulp and limits its transmission throughout the pulp [37]. Thus, the thin-walled veins and venules collapse only in the area of the affected pulp tissue, leading to a local vascular stasis and cell death.
3. The dental pulp has arteriovenous anastomoses providing a direct communication between arterioles and venules. Although the function of these anastomoses is not well elucidated [31], these shunt vessels may open up to counterbalance any significant rise in pulp blood pressure. When the intrapulpal pressure rises during inflammation, these shunt vessels that are abundant at the apical half of the pulp may communicate to reduce the intrapulpal pressure and maintain normal blood flow [38]. This suggests a local vascular regulation in the pulp [39]. Indeed, recent works allow a better understanding of this regulation which seems to occur at different levels:

5.6 Neural Control of Blood Flow

The sensory nerves predominate in the dental pulp which is innervated by large numbers of unmyelinated and small myelinated axons of autonomic and sensory origin [40]. However, there is no evidence for pulpal blood flow control by sympathetic or parasympathetic nerve activity to meet specific requirements of the pulp tissue, and neural control seems to be operative on a local scale. It has been shown that large terminal arterioles in the rat incisors receive a dense nerve supply, suggesting an important role of neuronal regulation on the vessels [41]. Perivascular sympathetic nerve fibres liberate noradrenaline and neuropeptide Y leading to a reduction of pulp blood flow [42], whereas intradental sensory nerves liberate neuropeptides which increase pulp blood flow [43]. The perivascular nerve endings are either adrenergic containing noradrenaline or somatosensory nerve fibres containing substance P or calcitonin gene-related peptides [41]. These nerve fibres participate in the pulp blood flow regulation by affecting vascular smooth muscle tone and vessel diameter. Thus pulp blood flow is under the influence of local nerve impulses. This is why pulp blood flow is considered to be predominantly under neural local control [44].

5.7 Neo-angiogenesis Is a Requirement for Regeneration and Healing

While an increased vasodilatation brings more blood supply to the inflamed tissue, sprouting of capillaries leads to an increase of their density which increases blood perfusion of hypoxic tissue to restore local oxygen and nutrition supply. Thus, neo-angiogenesis appears as a basic requirement under hypoxia.

5.8 Hypoxia Triggers Pulp Angiogenesis

After the initial vasodilatation of acute inflammatory reaction which increases blood supply to the pulp, an adaptive formation of capillaries by neo-angiogenesis is initiated as a reaction to hypoxia in ischemic tissues during the regeneration process [45, 46]. The regulation of angiogenesis by hypoxia ensures an adequate oxygen and nutrient supply required to meet the needs of the hypoxic

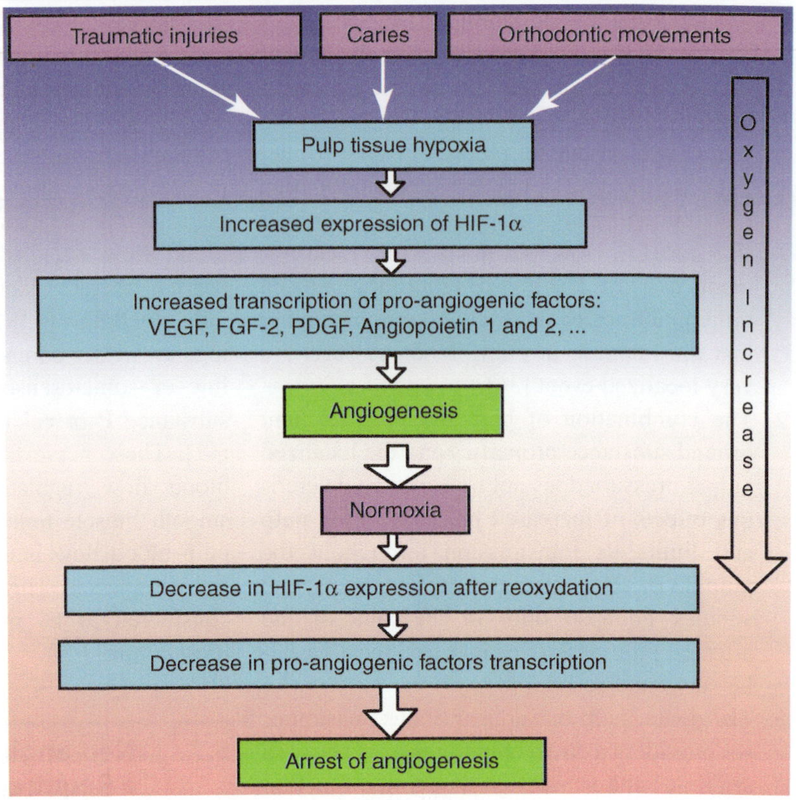

Fig. 5.4 Pulp hypoxia induces neo-angiogenesis. Hypoxia up-regulates hypoxia-inducible transcription factor-1α (Hif-1α) which induces pro-angiogenic factor transcription and neo-angiogenesis. This increases oxygen supply to the hypoxic tissue. After reoxidation, Hif-1α expression decreases and neo-angiogenesis is arrested. *Arrow* indicates the increase in oxygen supply

tissue. Hypoxia can be frequently encountered in the dental pulp under different clinical situations such as carious or traumatic pulp injuries, orthodontic movements or replanted avulsed teeth. Maintenance of oxygen homeostasis is essential to all living tissues. It is regulated through the activity of hypoxia-inducible transcription factor-1 (Hif-1) [47]. This transcription factor is a hetero-dimeric protein of a constitutive subunit (Hif-1β) and an oxygen-sensitive subunit (Hif-1α). Under hypoxic conditions, Hif-1α expression increases rapidly in the cells and binds to specific enhancer elements in the promoter region of pro-angiogenic genes [48, 49]. Hif-1α plays a key role in angiogenesis by activating the transcription of genes encoding pro-angiogenic growth factors including VEGF, FGF-2, angiopoietin 1 and 2 (Ang1, Ang2), placental growth factor (PGF), PDGF and angiogenic receptors [50–53]. These factors induce different steps of the neo-angiogenesis process (Fig. 5.4). The strong mitogen VEGF-A, for example, has induced not only proliferation of endothelial cells but also permeabilization in the initiation of the neo-angiogenesis process.

Given the fact that TGF-β, FGF-2 and PDGF growth factors have been reported to up-regulate VEGF expression [54], the capillary network formation may be triggered by a combined action of both hypoxia and locally secreted factors [49].

In the dental pulp, induction of Hif-1α was demonstrated after plating pulp cells under anaerobic conditions for 6 h [55]. Immunocytochemistry revealed that Hif-1α was localized primarily in the cell nuclei under hypoxic conditions [53]. This induction was also reported, in the dental pulp, in an orthodontic tooth movement model.

Hif-1α induction was also demonstrated both in dental pulp stem cells and pulp fibroblasts cultured in a hypoxia chamber at 37 °C containing 1 % O_2. In both cell types, Hif-1α expression peaked at 4 h of hypoxia and then decreased gradually through 24 h under normoxia. This indicates that this protein subunit is degraded after re-oxygenation. This degradation was also associated with a decrease in pro-angiogenic factor transcription, and this led to an arrest of neo-angiogenesis (Fig. 5.4). Interestingly, the fact that Hif-1α expression increases in dental pulp stem cells implies that these cells may be activated under hypoxia not only to regenerate dentin but also to regenerate the pulp vasculature.

5.9 Pulp Fibroblasts Induce Endothelial Network Formation

In an attempt to understand the effect of both hypoxia and locally secreted pro-angiogenic factors within the pulp, a coculture system was used. Both endothelial and pulp fibroblasts were transduced with fluorescent proteins and plated as a mix in an 80–20 % ratio on the surface of a Matrigel extracellular matrix to simulate the pulp extracellular matrix. The coculture of endothelial cells with pulp fibroblasts induced time-independent morphological changes in the endothelial cells. At 3–5 h, both cell types had a spherical or a fibroblastic morphology. The endothelial cells spread and started to organize after 24 h, while fibroblasts had the typical fibroblastic appearance. At 48 h, most endothelial cells formed tubular and closed structures. After 6 days, most endothelial cells became elongated, forming thin cords of interconnecting cells, and exhibited elongations and branching to form a 2D network of capillary-like structures corresponding to neo-angiogenesis in vivo [56, 57]. Blood vessels are clearly visible after 14 days (Fig. 5.5a–d). This indicates that, in addition to hypoxia, pulp fibroblasts are essential in maintaining pulp vasculature homeostasis.

In order to understand the consequences of hypoxia on the secretion of angiogenic factors, pulp fibroblasts were cultured in vitro, and injuries were performed mechanically to disrupt the fibroblast monolayer, simulating the hypoxia under traumatic pulp injury. The culture medium obtained after 5 h of contact with intact or injured pulp fibroblasts was used as a culture medium to feed a separate culture of endothelial cells on Matrigel. These media induced marked changes in endothelial cell morphology, with structural rearrangements leading to the organization of endothelial cells into capillary-like networks in 24 h. The dimensions of the newly formed tubes were larger in the medium obtained from injured cells when compared to the dimensions obtained with the medium from intact cells. This organization is due to the secretion of higher level of angiogenic factors from injured pulp fibroblasts as demonstrated by quantitative ELISA measurements of FGF-2, VEGF and PDGF [50]. This increased secretion of pro-angiogenic factors and higher stimulation of endothelial organization by injured pulp fibroblasts highlights the direct involvement of pulp fibroblasts in angiogenesis under hypoxia.

5.10 Pulp Angiogenesis Is Tightly Controlled by the Microenvironment

Several lines of evidence suggest that pulp vascularization is under a strong local control [57, 58] via pro- and anti-angiogenic factors and neuropeptides that can be liberated from the microenvironment [59]. The sources of these signals include the dentin and the different cell types within the dental pulp including pulp fibroblasts, endothelial cells and pulp nerves (Fig. 5.6).

5.10.1 Dentin and Pulp Fibroblasts

The dentin acts as a reservoir of signalling molecules. Pro-angiogenic factors such as VEGF and FGF-2 were identified in the dentin and may play a role in angiogenesis when released under carious injuries [60].

Pulp fibroblasts express several pro-angiogenic growth factors such as angiogenin, angiopoietin 2, epidermal growth factor (EGF), heparin-binding epidermal growth factor, hepatocyte growth factor, leptin and placental growth factor [59]. They also secrete FGF-2, VEGF and PDGF under normal culture conditions, and this secretion in the culture medium increases when injuries were performed to the cell layer. When the culture medium obtained from injured pulp fibroblasts was pre-incubated with neutralizing antibodies to both VEGF and FGF-2 and used for the culture of endothelial cells on Matrigel extracellular matrix, a significant decrease of capillary formation

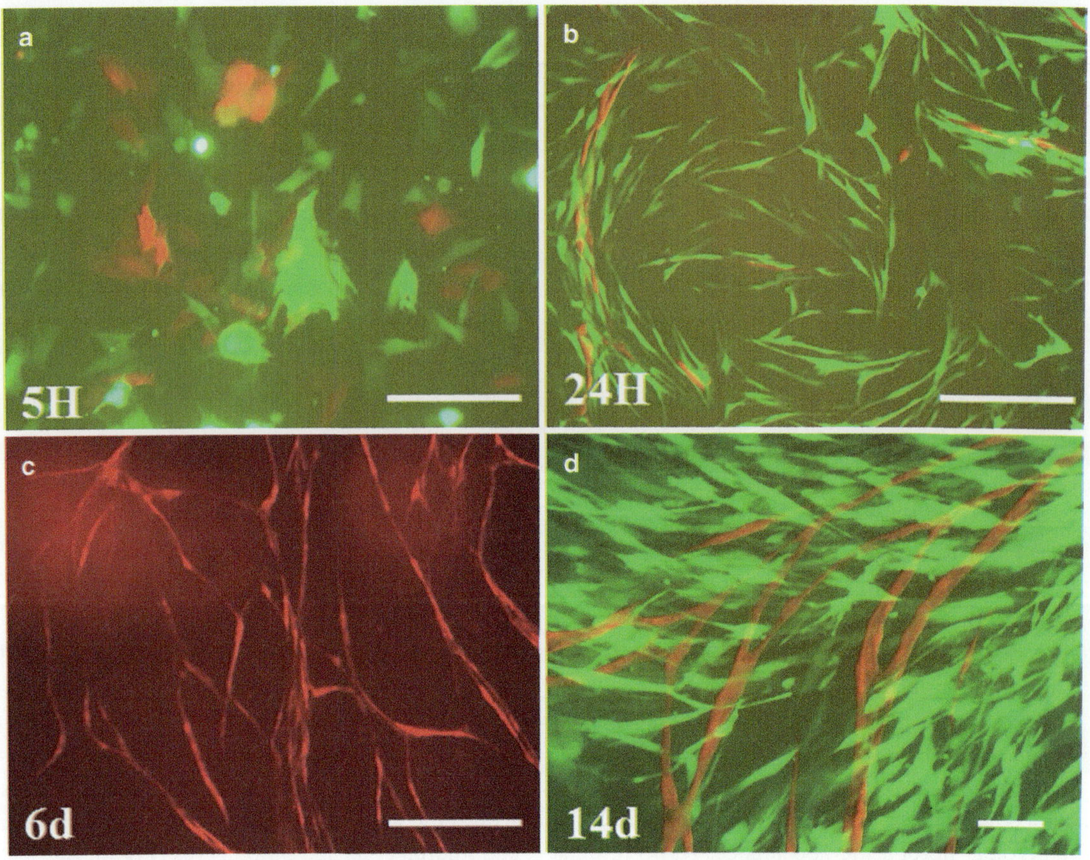

Fig. 5.5 Pulp fibroblasts induce neo-angiogenesis. Pulp fibroblasts and endothelial cells were transduced with fluorescent proteins and cocultured on Matrigel. Pulp fibroblasts induced formation of vascular network by endothelial cells. The cells are separated from each other after 5 h of culture (**a**). After 24 h, endothelial cells elongate and form anastomosis (**b**). After 6 days, a 2D network is formed (**c**). Blood vessels are seen after 14 days (**d**). Scale bar = 200 μm

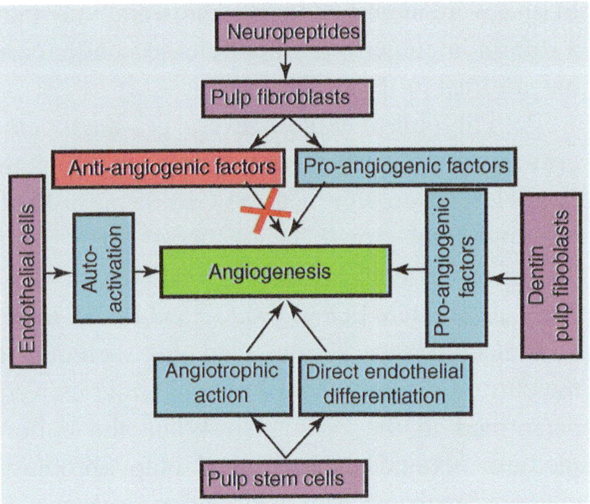

Fig. 5.6 Regulation of pulp angiogenesis by the microenvironment

capacity was obtained [56]. This result demonstrated that both factors were involved in pulp neo-angiogenesis. These data highlight the fact that the locally secreted VEGF and FGF-2 under hypoxia play a major role in neovascularization [61]. Several works strongly suggest the involvement of these pro-angiogenic factors under pathological pulp conditions. VEGF expression has been reported in inflamed pulp tissues [62], and its secretion has recently been reported in MDPC-23 cells in response to adhesive resins [63] and resinous monomers [50] or after lipoteichoic acid application suggesting the implication of this growth factor in angiogenesis under pathological conditions [64]. The dental pulp fibroblasts express and secrete the pro-angiogenic factor

TGF-ß1. The secretion of this factor is modulated by the restorative materials. Its secretion increases when pulp cells are placed in contact with bioactive tricalcium-based restorative materials such as the mineral trioxide aggregate or Biodentine. On the opposite, this secretion decreases when the cells were placed in contact with a resin-based adhesive [65].

5.10.2 Endothelial Cells

Endothelial cells also produce pro-angiogenic growth factors for the regulation of their own activity. It has been well established that injured endothelial cells release signalling molecules that initiate the inflammatory reaction and the healing process [66]. Cultured endothelial cells, which are known to express VEGF receptors, secrete VEGF and FGF-2. When physical injuries were performed on these cells, endothelial cells increased VEGF and FGF-2 secretion. The secretion level of VEGF significantly peaked 5 h after the injury and returned to baseline values 24 h later. Similarly, a significant increase in FGF-2 secretion was observed after 24 h and returned to baseline values after 48 h [58].

The expression of these factors in healthy tissues is relevant to physiological angiogenesis, while its expression in pulp cells under hypoxia is suggestive of a possible role in angiogenesis under pathological conditions. This was demonstrated in tooth slice cultures prepared from human molars and incubated with rhVEGF or rhFGF-2 for 7 days. This experiment showed that both angiogenic factors enhanced pulp microvessel density in vitro. When the slices were pretreated with rhVEGF for 1 h prior to subcutaneous implantation into immunodeficient mice, an increase in microvessel density was observed in vivo [61].

5.10.3 Neuropeptides Regulate the Local Expression of Angiogenic Growth Factors

Pulp vascularization is also under the control of locally secreted neuropeptides which have been implicated in mediating angiogenesis [67]. The pulpal nerve release of neuropeptides such as substance P (SP), calcitonin gene-related peptide (CGRP), neuropeptide Y (NPY) and vasoactive intestinal polypeptide (VIP) is up-regulated in carious teeth [68–70]. Furthermore, these neuropeptides modulated angiogenic growth factor secretion in cultures of human fibroblasts. Indeed, the neuropeptides CGRP and VIP induced substantial increases in placenta growth factor (PIGF) secretion after 24 h, while VIP and SP substantially decreased EGF levels after 48 h of treatment. NPY had an inhibitory effect on PDGF secretion from 24 h. These data suggest that neuropeptides are involved in the local regulation of angiogenic growth factors in human dental pulp fibroblasts [59].

5.11 Pulp Express Anti-angiogenic Factors

It has been demonstrated that pulp cells secrete anti-angiogenic factors such as angiostatin and endostatin [18]. Endostatin is a 20 kDa C-terminal fragment of collagen XVIII. This protein specifically inhibits endothelial proliferation and potently inhibits angiogenesis and tumour growth [23]. Angiostatin, an internal fragment of plasminogen, is a potent inhibitor of angiogenesis, which selectively inhibits endothelial cell proliferation. When given systemically, angiostatin potently inhibits tumour growth and can maintain metastatic and primary tumours in a dormant state defined by a balance of proliferation and apoptosis of the tumour cells [71]. These anti-angiogenic factors may be involved in the regulation of pro-angiogenic factor secretion in function of hypoxia.

5.12 Locally Produced Pro-angiogenic Factors Induce Pulp Stem Cell Direct Differentiation into Endothelial Cells

Dental pulp stem cells seem to contribute to blood vessel formation by differentiating themselves into endothelial cells and, thus, become actively incorporated into the newly formed blood vessels. Dental pulp stem cell differentiation into endothelial cells has been reported in many studies where these cells were incubated with differentiation culture media supplemented with angiogenic growth factors [72, 73].

Recent data demonstrated that the secreted FGF-2 and VEGF from injured endothelial cells and pulp fibroblasts directly affect this pulp stem cell differentiation into endothelial cells. These factors are involved in the differentiation of the progenitors expressing CD34+ and VEGFR2+/FLK+, but not CD31 or CD146 into endothelial cells [74]. Incubation of these cells with FGF-2 and VEGF led to an up-regulation of endothelial cell markers such as von Willebrand factor (vWF), CD31, CD34 and CD105 and the formation of blood vessel-like structures on Matrigel extracellular matrix in 10 days. When these cells were injected into an ischemic site of a mouse hindlimb, they led to the re-establishment of blood flow and vascularization after 14 days [74]. Similarly, when stem cells from exfoliated deciduous teeth were seeded in biodegradable scaffolds and transplanted into immunodeficient mice, they differentiated into endothelial-like cells [75]. Interestingly, when the same cells were seeded in tooth slices/scaffolds and implanted subcutaneously into immunodeficient mice, they differentiated into functional odontoblasts and endothelial cells only after the addition of VEGF [76]. These data provide further support to the idea of a local regulation of pulp vasculature by the locally secreted VEGF. Although this is not demonstrated yet, these observations indicate that pulp stem cells may be involved in neo-angiogenesis by the process of vasculogenesis which predominates in the embryonic tooth development.

5.13 Vascularization Is a Basic Requirement for Tissue Regeneration

Angiogenesis is a key process in tissue regeneration as blood supply is essential to provide the regenerating tissue with oxygen and nutrient transport to ensure both cell viability and function. The involvement of angiogenesis in regeneration has been demonstrated during bone regeneration where a soluble, neutralizing VEGF receptor decreased angiogenesis, bone formation and callus mineralization in mice femoral fractures. Inhibition of VEGF also dramatically suppressed healing of a tibial cortical bone defect, while exogenous VEGF enhanced blood vessel formation, ossification and new bone maturation in mouse femur fractures [77].

In the dental pulp, this has been demonstrated in experiments where stem cells from exfoliated deciduous teeth were seeded in tooth slices/scaffolds and implanted subcutaneously into immunodeficient mice. These cells differentiated into functional odontoblasts and endothelial cells. This response may be due to synergistic effects between endothelial and pulp cells as demonstrated by investigating cocultures of human pulp with endothelial cells at different ratios. This revealed greater alkaline phosphatase activity and alizarin red staining quantification of calcification as compared with cultures of pulp cells alone. Also, the expression levels of alkaline phosphatase, bone sialoprotein and dentin sialophosphoprotein genes confirmed the greater odontogenic potential of cocultured cells as compared to those of pulp cell cultures alone [76]. These data show clearly that angiogenesis is a requirement for tissue regeneration.

5.14 Pulp Stem Cells Exert a Trophic Angiogenic Action

In tissue engineering, if blood supply cannot reach the implanted tissue to provide it with oxygen and nutrients, this will lead to a failure and necrosis of the implanted tissue [15].

The transplanted pulp stem cells contribute to blood vessel formation either by a direct differentiation into endothelial cells as described previously or via a trophic angiogenic action by secreting angiogenic factors that induce angiogenesis from pre-existing host endothelial cells [78].

Dental pulp stem cells express numerous pro-angiogenic factors as demonstrated by mRNA and protein expression and secretion. The production of high amounts of angiogenic molecules such as VEGF and monocyte chemotactic protein-1 (MCP-1) enables stem cells to significantly induce endothelial cell migration as demonstrated in a transwell migration assay in vitro. This migration was inhibited by adding the phosphatidylinositol 3 kinase (PI3K) inhibitor LY294002 and the extracellular-signal-regulated kinase (MEK) inhibitor U0126. This suggests that pulp stem cells have the capacity to stimulate endothelial cell migration in vitro by activating the PI3K/AKT and MEK/ERK pathways. Additionally, in the chicken chorioallantoic membrane assay in ovo, the transplantation of pulp stem cells significantly induced blood vessel formation [18]. This might have a therapeutic application in restoring vascularization in ischemic sites in vivo. Indeed, intramyocardial injection of GFP-transduced pulp stem cells in a myocardial infarction rat model improved cardiac function and led to a reduction of the infarct size and to a higher neovascularization in cell-treated animals as compared with controls [79]. Transplantation of a CD31-negative side population of porcine dental pulp stem cells into a mouse hindlimb ischemia model produced a higher blood flow and capillary density compared with animals treated with bone marrow CD31- mesenchymal stem cells [80, 81].

When pulp stem cells expressing CD34+ and VEGFR2+/FLK+, but not CD31 or CD146, were labelled with fluorescent DiI dye and autotransplanted into the amputated pulp of dog teeth in a scaffold of type I and type III collagen and the cavity was sealed with zinc phosphate cement and a composite resin, pulp tissue with capillaries extending on the top of the tissue beneath the site of the filling cement was observed after 14 days. The DiI-labelled cells were seen in the amputated area closely related to newly formed capillaries and expressed several pro-angiogenic factors such as VEGF-A, granulocyte colony-stimulating factor (G-CSF), granulocyte-macrophage colony-stimulating factor (GM-CSF) and matrix metalloproteinase-3 (MMP3), implying trophic action on endothelial cells [82].

5.15 Conclusions

Taken together, the previously mentioned data show that the pulp has a unique example of tissue vascularization due to its confined localization within rigid dentin walls, high vascularization and high blood flow rate. It has a unique regulation system via both nervous and angiogenic local controls. Dentin and different pulp cell types produce pro- and anti-angiogenic factors which maintain pulp vasculature under normoxic conditions and affect the local pulp angiogenic capacity under hypoxia. In case of severe injuries, this chapter highlights the fact that pulp stem cells influence the vascular system directly either by differentiating themselves into endothelial cells and establishing neo-angiogenesis or by exerting trophic actions on endothelial cells via soluble factors to induce angiogenesis from preexisting vessels. This interaction between stem cells and the vascular system is of prime importance for establishing future pulp tissue regeneration and engineering strategies (Fig. 5.6).

References

1. Yu C, Abbott P. An overview of the dental pulp: its functions and responses to injury. Aust Dent J. 2007;52:S4–6.
2. Shalaby F, Rossant J, Yamaguchi TP, Gertsenstein M, Wu X-F, Breitman ML, et al. Failure of blood-island formation and vasculogenesis in Flk-1-deficient mice. Nature. 1995;376(6535):62–6.
3. Noden DM. Embryonic origins and assembly of blood vessels. Am Rev Respir Dis. 1989;140(4):1097–103.
4. Nait Lechguer A, Kuchler-Bopp S, Hu B, Haikel Y, Lesot H. Vascularization of engineered teeth. J Dent Res. 2008;87(12):1138–43.
5. Rothová M, Feng J, Sharpe PT, Peterková R, Tucker AS. Contribution of mesoderm to the developing dental papilla. Int J Dev Biol. 2011;55(1):59–64.
6. Fong G-H, Rossant J, Gertsenstein M, Breitman ML. Role of the Flt-1 receptor tyrosine kinase in regulating the assembly of vascular endothelium. Nature. 1995;376(6535):66–70.
7. Davis S, Aldrich TH, Jones PF, Acheson A, Compton DL, Jain V, et al. Isolation of angiopoietin-1, a ligand for the TIE2 receptor, by secretion-trap expression cloning. Cell. 1996;87(7):1161–9.
8. Suri C, Jones PF, Patan S, Bartunkova S, Maisonpierre PC, Davis S, et al. Requisite role of angiopoietin-1, a ligand for the TIE2 receptor, during embryonic angiogenesis. Cell. 1996;87(7):1171–80.

9. Jain RK. Molecular regulation of vessel maturation. Nat Med. 2003;9(6):685–93.
10. Schmidt A, Brixius K, Bloch W. Endothelial precursor cell migration during vasculogenesis. Circ Res. 2007;101(2):125–36.
11. Carmeliet P, Ferreira V, Breier G, Pollefeyt S, Kieckens L, Gertsenstein M, et al. Abnormal blood vessel development and lethality in embryos lacking a single VEGF allele. Nature. 1996;380(6573):435–9.
12. Ferrara N. Vascular endothelial growth factor. Eur J Cancer Oxf Engl. 1996;32A(14):2413–22.
13. Trubiani O, Tripodi D, Fratte TD, Caputi S, Primio RD. Human dental pulp vasculogenesis evaluated by CD34 antigen expression and morphological arrangement. J Dent Res. 2003;82(9):742–7.
14. Baume LJ. The biology of pulp and dentine. A historic, terminologic-taxonomic, histologic-biochemical, embryonic and clinical survey. Monogr Oral Sci. 1980;8:1–220.
15. Laschke MW, Harder Y, Amon M, Martin I, Farhadi J, Ring A, et al. Angiogenesis in tissue engineering: breathing life into constructed tissue substitutes. Tissue Eng. 2006;12(8):2093–104.
16. Folkman J, Shing Y. Angiogenesis. J Biol Chem. 1992;267(16):10931–4.
17. Carmeliet P, Jain RK. Molecular mechanisms and clinical applications of angiogenesis. Nature. 2011;473(7347):298–307.
18. Bronckaers A, Hilkens P, Fanton Y, Struys T, Gervois P, Politis C, et al. Angiogenic properties of human dental pulp stem cells. PLoS One. 2013;8(8):e71104.
19. Gerwins P, Sköldenberg E, Claesson-Welsh L. Function of fibroblast growth factors and vascular endothelial growth factors and their receptors in angiogenesis. Crit Rev Oncol Hematol. 2000;34(3):185–94.
20. Betsholtz C, Karlsson L, Lindahl P. Developmental roles of platelet-derived growth factors. Bioessays. 2001;23(6):494–507.
21. Carmeliet P. Angiogenesis in health and disease. Nat Med. 2003;9(6):653–60.
22. Ross R. Atherosclerosis — an inflammatory disease. N Engl J Med. 1999;340(2):115–26.
23. O'Reilly MS, Boehm T, Shing Y, Fukai N, Vasios G, Lane WS, et al. Endostatin: an endogenous inhibitor of angiogenesis and tumor growth. Cell. 1997;88(2):277–85.
24. Ribatti D. Endogenous inhibitors of angiogenesis: a historical review. Leuk Res. 2009;33(5):638–44.
25. Löhler J, Timpl R, Jaenisch R. Embryonic lethal mutation in mouse collagen I gene causes rupture of blood vessels and is associated with erythropoietic and mesenchymal cell death. Cell. 1984;38(2):597–607.
26. George EL, Georges-Labouesse EN, Patel-King RS, Rayburn H, Hynes RO. Defects in mesoderm, neural tube and vascular development in mouse embryos lacking fibronectin. Development. 1993;119(4):1079–91.
27. Rhodes JM, Simons M. The extracellular matrix and blood vessel formation: not just a scaffold. J Cell Mol Med. 2007;11(2):176–205.
28. Costell M, Gustafsson E, Aszódi A, Mörgelin M, Bloch W, Hunziker E, et al. Perlecan maintains the integrity of cartilage and some basement membranes. J Cell Biol. 1999;147(5):1109–22.
29. Burri PH, Djonov V. Intussusceptive angiogenesis—the alternative to capillary sprouting. Mol Aspects Med. 2002;23(6, Suppl):1–27.
30. Kishi Y, Shimozato N, Takahashi K. Vascular architecture of cat pulp using corrosive resin cast under scanning electron microscope. J Endod. 1989;15(10):478–83.
31. Takahashi K, Kishi Y, Kim S. A scanning electron microscope study of the blood vessels of dog pulp using corrosion resin casts. J Endod. 1982;8(3):131–5.
32. Yoshida S, Ohshima H. Distribution and organization of peripheral capillaries in dental pulp and their relationship to odontoblasts. Anat Rec. 1996;245(2):313–26.
33. Vongsavan N, Matthews B. The vascularity of dental pulp in cats. J Dent Res. 1992;71(12):1913–5.
34. Meyer MW. Pulpal blood flow: use of radio-labelled microspheres. Int Endod J. 1993;26(1):6–7.
35. Kim S. Microcirculation of the dental pulp in health and disease. J Endod. 1985;11(11):465–71.
36. Tønder KJH, Kvinnsland I. Micropuncture measurements of interstitial fluid pressure in normal and inflamed dental pulp in cats. J Endod. 1983;9(3):105–9.
37. Van Hassel HJ. Physiology of the human dental pulp. Oral Surg Oral Med Oral Pathol. 1971;32(1):126–34.
38. Kim S, Lipowsky HH, Usami S, Chien S. Arteriovenous distribution of hemodynamic parameters in the rat dental pulp. Microvasc Res. 1984;27(1):28–38.
39. Yu CY, Boyd NM, Cringle SJ, Su EN, Alder VA, Yu DY. An in vivo and in vitro comparison of the effects of vasoactive mediators on pulpal blood vessels in rat incisors. Arch Oral Biol. 2002;47(10):723–32.
40. Bradley RJ. Dental sensory receptors. In: Smythies JR, Bradley RJ (eds.). Int Rev Neurobiol. Academic Press, Budapest; 1984;25:39–94.
41. Zhang J-Q, Nagata K, Iijima T. Scanning electron microscopy and immunohistochemical observations of the vascular nerve plexuses in the dental pulp of rat incisor. Anat Rec. 1998;251(2):214–20.
42. Kim SK, Ang L, Hsu YY, Dörscher-Kim J, Kim S. Antagonistic effect of d-myo-inositol-1,2,6-trisphosphate (PP56) on neuropeptide Y-induced vasoconstriction in the feline dental pulp. Arch Oral Biol. 1996;41(8–9):791–8.
43. Olgart LM, Edwall B, Gazelius B. Neurogenic mediators in control of pulpal blood flow. J Endod. 1989;15(9):409–12.
44. Olgart L. Neural control of pulpal blood flow. Crit Rev Oral Biol Med. 1996;7(2):159–71.
45. Carmeliet P. Cardiovascular biology: creating unique blood vessels. Nature. 2001;412(6850):868–9.
46. Fukumura D, Xu L, Chen Y, Gohongi T, Seed B, Jain RK. Hypoxia and acidosis independently up-regulate vascular endothelial growth factor transcription in brain tumors in vivo. Cancer Res. 2001;61(16):6020–4.

47. Jiang B-H, Rue E, Wang GL, Roe R, Semenza GL. Dimerization, DNA binding, and transactivation properties of hypoxia-inducible factor 1. J Biol Chem. 1996;271(30):17771–8.
48. Semenza GL. Signal transduction to hypoxia-inducible factor 1. Biochem Pharmacol. 2002;64(5–6):993–8.
49. Heil M, Eitenmüller I, Schmitz-Rixen T, Schaper W. Arteriogenesis versus angiogenesis: similarities and differences. J Cell Mol Med. 2006;10(1):45–55.
50. Tran-Hung L, Laurent P, Camps J, About I. Quantification of angiogenic growth factors released by human dental cells after injury. Arch Oral Biol. 2008;53(1):9–13.
51. Kelly BD, Hackett SF, Hirota K, Oshima Y, Cai Z, Berg-Dixon S, et al. Cell type–specific regulation of angiogenic growth factor gene expression and induction of angiogenesis in nonischemic tissue by a constitutively active form of hypoxia-inducible factor 1. Circ Res. 2003;93(11):1074–81.
52. Vincent KA, Feron O, Kelly RA. Harnessing the response to tissue hypoxia: HIF-1α and therapeutic angiogenesis. Trends Cardiovasc Med. 2002;12(8): 362–7.
53. Aranha AMF, Zhang Z, Neiva KG, Costa CAS, Hebling J, Nör JE. Hypoxia enhances the angiogenic potential of human dental pulp cells. J Endod. 2010;36(10):1633–7.
54. Pertovaara L, Kaipainen A, Mustonen T, Orpana A, Ferrara N, Saksela O, et al. Vascular endothelial growth factor is induced in response to transforming growth factor-beta in fibroblastic and epithelial cells. J Biol Chem. 1994;269(9):6271–4.
55. Römer P, Wolf M, Fanghänel J, Reicheneder C, Proff P. Cellular response to orthodontically-induced short-term hypoxia in dental pulp cells. Cell Tissue Res. 2014;355(1):173–80.
56. Tran-Hung L, Mathieu S, About I. Role of human pulp fibroblasts in angiogenesis. J Dent Res. 2006;85(9):819–23.
57. About I. Dentin–pulp regeneration: the primordial role of the microenvironment and its modification by traumatic injuries and bioactive materials. Endod Top. 2013;28(1):61–89.
58. About I. Dentin regeneration in vitro the pivotal role of supportive cells. Adv Dent Res. 2011;23(3):320–4.
59. El Karim IA, Linden GJ, Irwin CR, Lundy FT. Neuropeptides regulate expression of angiogenic growth factors in human dental pulp fibroblasts. J Endod. 2009;35(6):829–33.
60. Roberts-Clark DJ, Smith AJ. Angiogenic growth factors in human dentine matrix. Arch Oral Biol. 2000;45(11):1013–6.
61. Mullane EM, Dong Z, Sedgley CM, Hu JC-C, Botero TM, Holland GR, et al. Effects of VEGF and FGF2 on the revascularization of severed human dental pulps. J Dent Res. 2008;87(12):1144–8.
62. Artese L, Rubini C, Ferrero G, Fioroni M, Santinelli A, Piattelli A. Vascular Endothelial Growth Factor (VEGF) expression in healthy and inflamed human dental pulps. J Endod. 2002;28(1):20–3.
63. Mantellini MG, Botero T, Yaman P, Dennison JB, Hanks CT, Nör JE. Adhesive resin and the hydrophilic monomer HEMA induce VEGF expression on dental pulp cells and macrophages. Dent Mater. 2006;22(5):434–40.
64. Telles PDS, Hanks CT, Machado MA, Nör JE. Lipoteichoic acid up-regulates VEGF expression in macrophages and pulp cells. J Dent Res. 2003;82(6):466–70.
65. Laurent P, Camps J, About I. BiodentineTM induces TGF-β1 release from human pulp cells and early dental pulp mineralization. Int Endod J. 2012;45(5):439–48.
66. Martin P. Wound healing–aiming for perfect skin regeneration. Science. 1997;276(5309):75–81.
67. Hökfelt T, Bartfai T, Bloom F. Neuropeptides: opportunities for drug discovery. Lancet Neurol. 2003;2(8):463–72.
68. Caviedes-Bucheli J, Muñoz HR, Azuero-Holguín MM, Ulate E. Neuropeptides in dental pulp: the silent protagonists. J Endod. 2008;34(7):773–88.
69. Awawdeh L, Lundy FT, Shaw C, Lamey P-J, Linden GJ, Kennedy JG. Quantitative analysis of substance P, neurokinin A and calcitonin gene-related peptide in pulp tissue from painful and healthy human teeth. Int Endod J. 2002;35(1):30–6.
70. El Karim IA, Lamey P-J, Linden GJ, Awawdeh LA, Lundy FT. Caries-induced changes in the expression of pulpal neuropeptide Y. Eur J Oral Sci. 2006;114(2):133–7.
71. O'Reilly MS, Holmgren L, Chen C, Folkman J. Angiostatin induces and sustains dormancy of human primary tumors in mice. Nat Med. 1996;2(6):689–92.
72. D' Aquino R, Graziano A, Sampaolesi M, Laino G, Pirozzi G, De Rosa A, et al. Human postnatal dental pulp cells co-differentiate into osteoblasts and endotheliocytes: a pivotal synergy leading to adult bone tissue formation. Cell Death Differ. 2007;14(6).
73. Marchionni C, Bonsi L, Alviano F, Lanzoni G, Di Tullio A, Costa R, et al. Angiogenic potential of human dental pulp stromal (stem) cells. Int J Immunopathol Pharmacol. 2009;22(3):699–706.
74. Iohara K, Zheng L, Wake H, Ito M, Nabekura J, Wakita H, et al. A novel stem cell source for vasculogenesis in ischemia: subfraction of side population cells from dental pulp. Stem Cells. 2008;26(9): 2408–18.
75. Cordeiro MM, Dong Z, Kaneko T, Zhang Z, Miyazawa M, Shi S, et al. Dental pulp tissue engineering with stem cells from exfoliated deciduous teeth. J Endod. 2008;34(8):962–9.
76. Sakai VT, Zhang Z, Dong Z, Neiva KG, Machado MA, Shi S, et al. SHED differentiate into functional odontoblasts and endothelium. J Dent Res. 2010;89(8):791–6.
77. Street J, Bao M, deGuzman L, Bunting S, Peale FV, Ferrara N, et al. Vascular endothelial growth factor stimulates bone repair by promoting angiogenesis and bone turnover. Proc Natl Acad Sci. 2002;99(15): 9656–61.

78. Sieveking DP, Ng MK. Cell therapies for therapeutic angiogenesis: back to the bench. Vasc Med. 2009;14(2):153–66.
79. Gandia C, Armiñan A, García-Verdugo JM, Lledó E, Ruiz A, Miñana MD, et al. Human dental pulp stem cells improve left ventricular function, induce angiogenesis, and reduce infarct size in rats with acute myocardial infarction. Stem Cells. 2008;26(3): 638–45.
80. Murakami M, Horibe H, Iohara K, Hayashi Y, Osako Y, Takei Y, et al. The use of granulocyte-colony stimulating factor induced mobilization for isolation of dental pulp stem cells with high regenerative potential. Biomaterials. 2013;34(36):9036–47.
81. Ishizaka R, Hayashi Y, Iohara K, Sugiyama M, Murakami M, Yamamoto T, et al. Stimulation of angiogenesis, neurogenesis and regeneration by side population cells from dental pulp. Biomaterials. 2013;34(8):1888–97.
82. Iohara K, Zheng L, Ito M, Ishizaka R, Nakamura H, Into T, et al. Regeneration of dental pulp after pulpotomy by transplantation of $CD31^-/CD146^-$ side population cells from a canine tooth. Regen Med. 2009;4(3):377–85.

Dental Pulp Innervation

6

Kaj Fried and Jennifer Lynn Gibbs

6.1 Introduction

The mammalian dentition is of ultimate importance for survival in the animal kingdom. It is thus not surprising that teeth are equipped with an abundant, sophisticated, protective neurosensory system that mediates the sensation of pain (see [1]). Impressive progress in the understanding of this system has made it evident that it differs in many ways from nociceptive (i.e., pain detecting) networks at other body sites. Despite this, fundamental issues regarding the formation, structure, reaction to injuries, and especially the transduction mechanisms of the sensory system within the dental pulp remain elusive. From a functional standpoint, it appears enigmatic why most or all stimuli that excite pulpal nerve fibers, whether noxious cold or noxious heat to a fully intact tooth, or extremely light mechanical forces or subtle thermal, osmotic, or chemical changes to exposed dentin, result only in the sensation of pain, with no mechanism for discrimination (see [2]). In this context, it is of interest to consider the tooth from an evolutionary perspective. Hence, it may not be that teeth are simply mineralized feeding and fending structures incidently provided with highly sensitive nerves. Rather, they may have evolved from primitive electroreceptor organs that ultimately accumulated a calcified shield. Accordingly, it has been proposed that cartilage, bone, dentin, and enamel-like tissues evolved in association with new vertebrate sense organs and only secondarily provided mechanical support [3, 4]. This may have been possible through an evolution of cranial neural crest populations with mixed neurogenic, osteogenic, and odontogenic potentials [5]. Intriguingly, teeth could then be regarded as vestigial sensors that have gradually adapted to synthesize mineralized matrix and eventually changed fate to become neurosensory organs for mastication [6].

To maintain an efficient afferent transduction system in highly mineralized teeth, there is a need for a low-threshold sensory apparatus that will be able to detect stimuli through a hard shell of calcified tissue. Activation of highly sensitive intradental mechanoreceptors would alert to potentially endangering hardness and texture of food or other intraoral objects [7–9]. This, in turn, would provide input for coordination and reflex activity of the masticatory muscle complex [10, 11]. Nerve fibers with higher thresholds would also be required to record and report on inflammatory threats. The pulp of the tooth seems to possess both these nerve fiber types. At odds with the current general concepts of pain transduction, the low-threshold mechanosensory

K. Fried, DDS, PhD (✉)
Department of Neuroscience, Karolinska Institutet, Retzius väg 8, Stockholm SE-171 77, Sweden
e-mail: kaj.fried@ki.se

J.L. Gibbs, MAS, DDS, PhD
Department of Endodontics, New York University College of Dentistry, 345 E 24th Street, Clinic 7W, New York, NY 10010, USA
e-mail: jlg15@nyu.edu

M. Goldberg (ed.), *The Dental Pulp*,
DOI 10.1007/978-3-642-55160-4_6,

fibers apparently connect to central pain-mediating, rather than tactile-mediating, pathways. In fact, intrapulpal nerves are probably the main source of tissue-damaging stimulus signaling from the dentition, while periodontal afferents serve to provide information on tooth load when subjects contact and gently manipulate food [12].

6.2 Development of Tooth Pulp Innervation

The ingrowth of trigeminal ganglion (TG) nerve fibers to the neural crest-derived condensed mesenchyme that will form the dental pulp occurs at a comparatively late developmental stage. This is in contrast to the surrounding mesenchymal tissue, which has a well-developed neural supply much earlier (for review, see [13, 14]). Thus, already at embryonic day 13.5 (E13.5), the mandibular molar tooth germ of the mouse has buccal and lingual nerve branches that surround the dental mesenchyme in basketlike formations. However, they remain in that position for a considerable developmental period [13, 15]. Only after the crown shape is set and mineralization of both enamel and dentin has commenced, around postnatal day 3–4 in the mouse and rat, do pioneer nerve fibers enter the apical region of the tooth germ [16, 17] (Fig. 6.1). The functional explanation for this delay, which cannot be accounted for by any obvious physical boundary such as an epithelial barrier, is not clear. In the dental papilla, neurotrophic factor genes are expressed long before pulpal innervation is established. However, the dental papilla/pulp cells also express neurite growth inhibitory factors at early stages [15, 18, 19], whose effects most likely dominate over the neurotrophic ones. Accordingly, early fetal dental mesenchyme repels neurites from TG explants of corresponding stages in vitro [19]. Among several putative neurorepelling factors that could be active during odontogenesis, the semaphorin (Sema) group of molecules has received the most attention. A number of Sema gene family members are present in tooth-related mesenchyme from embryonic and postnatal mice. The expression of some of them, namely, 3A, 3C, 3F, 4F, 5B, 6A, 6B, and 6C, is high early in development and then decreases in a temporal pattern that correlates with neurite inhibitory/repulsive effects of dental mesenchyme [19]. Of particular interest is Sema3A, which shows a spatiotemporal expression pattern in restricted dental mesenchyme areas in areas where axons appear to be unable to enter. Furthermore, in Sema3A mutant embryos, nerve fibers grow into the dental mesenchyme prematurely and ectopically, suggesting that Sema3A has a major role in preventing axonal ingrowth to early tooth anlagen [15]. Interestingly, the tooth-instructive oral and dental epithelia, as well as epithelial Wnt4, induce Sema3a expression in the dental mesenchyme at early developmental stages. At the bud stage, epithelial Wnt4 and Tgfβ1, which both are pivotal in odontogenesis, regulate Sema3a expression in the dental mesenchyme. This suggests that a coordinating axis exists between epithelial-mesenchymal interactions that lead to tooth formation and the control of the subsequent innervation of the dental organ [15]. Sema3A continues to exert important functions during postnatal innervation of the dental pulp. In addition to a continued axon-repelling effect which demarcates and directs ingrowing nerve fibers to appropriate sites, it also affects the structural development of the axonal pathways. This is evident by the fact that in the molars of mice deficient for Sema3A, nerves become defasciculated and thinner and form a premature, abnormal, enlarged nerve plexus at the pulp-dentin border [20]. Another member of the Sema family, Sema3F, may serve additional functions as a tooth target-derived axonal chemorepellent to control the establishment of the local nerve supply [21]. A functional role for Semas in tooth-nerve interactions is underpinned by the fact that the relevant Sema receptors, Npn1, plexinA3, and -A4, are expressed in trigeminal ganglion neurons during development [19, 21].

As seen from this discussion, a shift in expression from neurorepulsive to neuroattractive dental papilla/pulpal factors apparently takes place during odontogenesis. In tissue culture,

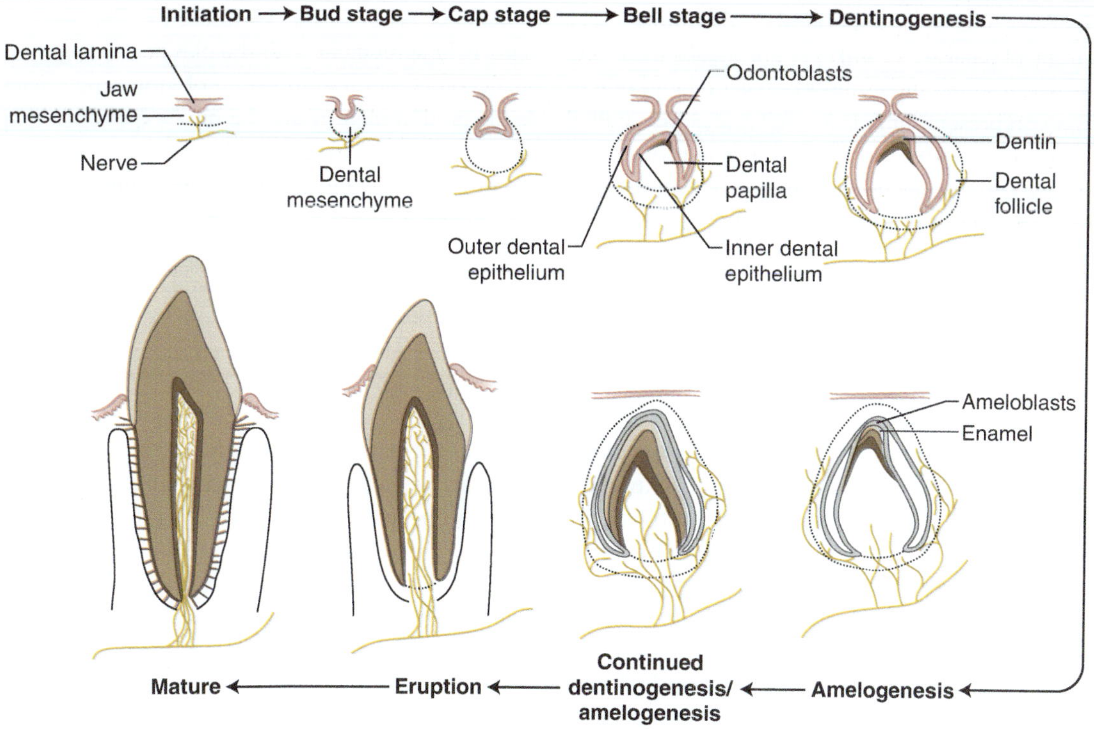

Fig. 6.1 This schematic drawing shows the relationship between tooth development and pulpal innervation. At early stages, nerve fibers are located below the dental lamina. Axons then form a plexus underneath the tooth organ and innervate the dental follicle but do not enter the dental papilla. Later, when the formation of mineralized tissue is already initiated, nerve fibers invade the tooth pulp, apparently as a result of a shift from secretion of pulpal neurorepelling to pulpal neurotrophic factors (Used with permission of Elsevier from Fried et al. [16])

late embryonic or early postnatal dental mesenchyme strongly attracts TG neurites [19]. The main molecular candidate for this effect is nerve growth factor (NGF). NGF in the developing tooth pulp has been demonstrated with a variety of methods [22–24]. In support of this, mutant mice which lack the high-affinity NGF receptor trkA do not develop a pulpal nerve supply [25]. In addition, glial cell line-derived neurotrophic factor (GDNF) and its receptor GFR-α1 mRNAs are expressed in patterns that suggest that GDNF contributes to the establishment of pulpal innervation [24, 26, 27]. However, in vitro, neutralizing antibodies against NGF, brain-derived neurotrophic factor (BDNF), and GDNF applied to cocultures of pulpal and TG explants do not fully block neurite outgrowth. This could be due to growth-stimulating activities of other GDNF-related factors such as neurturin (NRTN), artemin (ARTN), and/or persephin (PSPN), which are expressed in pulpal mesenchymal cells [28]. It may also be explained by effects from other hitherto largely unexamined pulpal neurotrophic factors, e.g., neuregulins [29].

Once having entered the dental pulp, it is likely that local extracellular matrix (ECM) proteins help guide and promote the growth of axons toward their final targets. Among them, laminins, a group of heterotrimeric αβγ proteins, display a clear-cut specificity in this zone. Pulpal nerves seem to use defined laminin substrates for growth and likely also nerve terminal integrity. Tooth pulp nerves express the laminin chains α2, α4, β1, and γ1, as reported for other peripheral nerves. Larger, but not smaller, nerve fascicles also express α5 [30]. In addition, and unexpectedly, laminin α1 chain immunoreactivity is present in tooth pulp nerve bundles. Nerve trunks display marked immunoreactivity for laminin integrin receptors INTα3, INTα6, INTβ1, and

INTβ4 chains. Importantly, laminins 211 (α2β1γ1) and 411 (α4β1γ1) are synthesized and secreted from pulpal fibroblasts and could potentially represent important substrates for pulpal nerve fibers. However, when TG neurons were cultured on isolated laminin-211 or laminin-411 surfaces, only 411 promoted neurite outgrowth. Conversely, 211 exerted minimal, if any, neuritogenic activity and seems rather to be involved in mineralization [31]. Thus, in the tooth pulp stroma, laminin-411 may promote the migration of nerves during development and/or regeneration after injury. Another ECM glycoprotein, reelin, which is important for axon development in the central nervous system, is strongly expressed in fully differentiated human odontoblasts. In vitro cocultures with rat TG neurons have indicated that neurites contact odontoblasts at sites of reelin expression. Consequently, since reelin receptors ApoER-2, VLDLR, CNR, and Disabled-1 are expressed in the trigeminal ganglion, it has been suggested that reelin might be an ECM molecule that is involved in the terminal innervation of the dentin-pulp complex [32]. Other nervous system-related signaling molecules such as glutamic acid, phosphatidylcholine, phosphatidylserine, and phosphatidylinositol are present in the mineralized matrix of the peritubular dentin that encapsulates odontoblast processes [33], where they may interact with axons.

In diphyodont species, the primary dentition is eventually replaced by the permanent dentition. The developmental anatomy of the intradental axons is similar in primary and permanent teeth, although the formation of a sensory innervation is more rapid in deciduous than in permanent teeth [34].

6.3 The Structure of Pulpal Axons

When mature, the innervation of primary teeth is structurally identical to that of permanent teeth, although axon numbers are smaller due to size differences [1]. Within the root pulp of permanent teeth in experimental animals and humans, ~70–90 % of axons are unmyelinated, and most of the remainder seem to be Aδ-fibers [1, 23, 34–36]. This is in agreement with the classical concept of nociceptors and has appeared obvious since pain is the predominant if not the only experience that can be evoked when pulpal nerves are excited. However, the parent axons of most pulp afferents are myelinated and have larger diameters, usually in the Aβ-fibers range [36]. They often have rapid extradental conduction velocities as found in large-diameter fibers, for example, in the cat reaching up to almost 60 m^{-s}, while Aδ axons usually conduct in the order of 25 m^{-s} [37]. Their trigeminal cell bodies are of medium or large sizes and have a number of cytochemical characteristics that are specific for the category of primary sensory neurons usually associated with low-threshold mechanoreceptors (LTMs) (see [38]). These observations suggest that a very large number of pulpal axons are end branches of larger or much larger parent axons that branch, taper, and lose their myelin sheaths. Thus, in the rat, EM analysis has shown that whereas 95.6 % of the parent nerve fibers innervating the dental pulp are myelinated, a minority of all axons in the apical part of the radicular pulp have myelin coverings [36]. Further, within the tooth, the unmyelinated axons show immunoreactivity to specific neurofilament antibodies that are conventional markers for myelinated, medium-sized, and large primary sensory neurons [39, 40] (Figs. 6.2a–c and 6.3a, b). Nonetheless, there is no reason to doubt that some unmyelinated pulpal axons are "true" C-fibers and belong to a restricted proportion of pulp-innervating trigeminal ganglion neurons that are small sized and express heat-sensitive TRPV1 and cold-sensitive TRPA1 receptors [41]. These nerve fibers likely terminate in the coronal pulp and convey thermo-induced pain sensations. Similarly, some thinly myelinated pulpal fibers are most probably genuine Aδs with properties and cell soma sizes that are typical for this category of primary sensory neurons.

A subset of intradental sensory nerves is involved in the local control of blood flow. By virtue of their neuropeptide content, these afferent fibers cause vasodilation and inhibit sympathetic vasoconstriction in response to painful stimulation of the tooth [42].

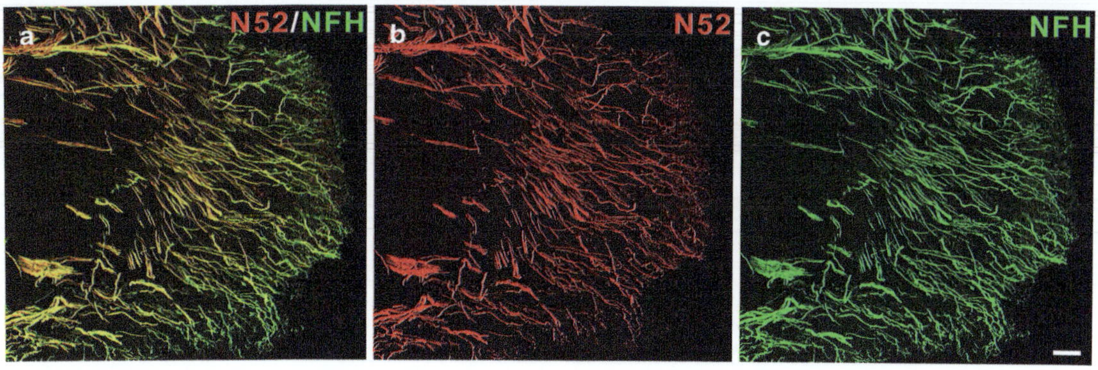

Fig. 6.2 (**a–c**) Neurofilament 200 kDa expression is prominent in the human dental pulp. Confocal micrographs showing nerve fibers identified by two different neurofilament 200 kDa antibodies [**b** N52-mouse monoclonal; **c** neurofilament heavy (*NFH*)-chicken monoclonal] in the pulp horn of a normal human dental pulp. The overlapping of the N52 and NFH immunoreactivity appears *yellow* in the merged image (**a**). Scale bar, 50 μm (Used with permission from Henry et al. [39])

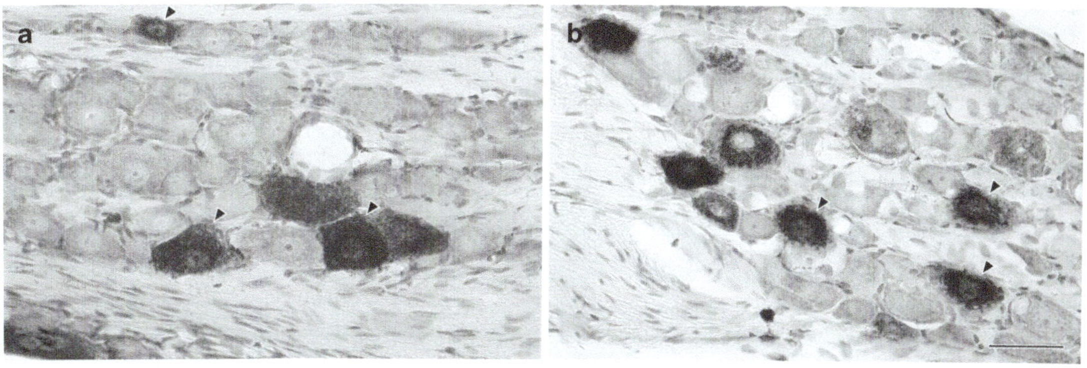

Fig. 6.3 (**a**, **b**) Light micrographs showing HRP-labeled somata in the TG that innervate the upper molar (**a**) and lower incisor (**b**) pulp. Both large- and medium-sized neurons were frequently labeled. The *arrowheads* indicate labeled cells with clear nucleoli, selected for measurements of cross-sectional area. Scale bar = 50 μm (Used with permission of Elsevier from Paik et al. [36])

As axons traverse the radicular canal to reach the coronal regions of the pulp, they give off a few collaterals, taper, and those that still are myelinated have progressively thinner and shorter internodes [1]. Up to 90 % of the myelinated axons lose their myelin within the short intradental course from the radicular to the coronal pulp [36, 40]. In the pulpal horn, there is an extensive axonal arborization. The sensitivity of the tooth is also most intense here and then gradually declines in parallel with a decrease in nerve fiber density at the pulp-dentin border toward the crown-root transition [43]. Many axons terminate below or in the odontoblast layer region. Near the terminals, they lose their Schwann cell ensheathment altogether, assuming intimate relationships with odontoblasts as well as with specific sub- and periodontoblastic cells with features similar to central nervous system glia. These cells are associated with the local microcirculation in what seems to be analogous to a blood-barrier system [6]. Some axon terminals proceed beyond this site and continue along odontoblast processes into dentinal tubules to innervate the inner segment (0.1 mm) of the dentin. A single intrapulpal axon might branch and innervate more than 100 dentinal tubules ([44]; for further references, see [1, 34]). The fact that mature odontoblast processes and associated nerve fibers are embedded in mineralized dentin limits their accessibility for structural as well as functional studies. Consequently, many aspects of the complex nerve-odontoblast architecture and possible interactions remain obscure. A number of electron microscopical studies on odontoblast-axon relationships have yielded inconclusive results (for references, see [45]). This is probably to some extent caused by inadequate preservation techniques, which fail to maintain the native morphology. Thus, samples from the pulp-dentin junction usually have to be decalcified, which removes the peritubular dentin and distorts estimations of tubule and peri-odontoblastic size and content [46]. Even more important though is a lack of reliable markers in existing reports to determine the identity of cellular elements in ultrathin sections. In what seems to be a singular exception, an anterogradely transported neuronal tracer was used to examine odontoblast-predentin-dentin innervation. Here, it was concluded that clear-cut ultrastructural signs of synaptic formations were absent from this region [47].

6.4 Neuropeptides in Pulpal Afferents

The neurons innervating the dental pulp express numerous biologically active neuropeptides that are released from both the peripheral terminal of the neurons (within the pulp) and the central terminal located within the trigeminal nuclear complex in the medulla. Some of the neuropeptides identified in pulpal afferents include substance P, calcitonin gene-related peptide (CGRP), vasoactive intestinal peptide (VIP), neuropeptide Y (NPY), and somatostatin. In the periphery, these neuropeptides have multiple varied effects including regulating blood flow, recruitment and modulation of activity of immune cells, and finally proliferation of and secretion of bioactive molecules from pulpal fibroblasts [48, 49]. Sensory neurons themselves express receptors for neuropeptides; thus, peripherally and centrally released neuropeptides bind to membrane-bound neuronal receptors, either increasing or decreasing neuronal activity, and thus modulating inflammatory pain states.

Small diameter C-fiber neurons expressing the neuropeptides CGRP and substance P represent an anatomically and functionally distinct class of sensory neurons than those without peptides, which typically express a different set of markers including the IB4-lectin binding site, the purinergic P2X3 receptor, and the Mrgprd receptor [50, 51]. The peptidergic and non-peptidergic C-fibers are responsive to different growth factors with the non-peptidergic fibers responding to GDNF and the peptidergic to NGF via the trkA receptor. Interestingly, the dental pulp appears to mostly lack the non-peptidergic C-fiber population, but is well populated by the peptidergic fiber types, both with and without myelin. Other "deep" tissues, including the knee joint and intestines, also have very low levels or even no innervation by non-peptidergic neurons, in contrast to superficial tissues such as the skin in which these fibers are plentiful [52, 53]. The biological consequence of this unique property of neurons innervating the dental pulp is not fully understood, but it could be relevant to the quality and persistence of pain states produced in the setting of injury to pulpal tissues [54, 55].

The neurotransmitter CGRP is expressed in many neurons that innervate the dental pulp, more so than other functionally important neurotransmitters like substance P. Further, the CGRP-expressing pulpal afferents are likely anatomically and functionally unique relative CGRP-expressing afferents innervating other tissues [56–59]. The expression of CGRP in pulpal afferents is dynamic, with increased expression observed after pulpal injury [60–62]. Anatomical studies demonstrate that CGRP-expressing axons will sprout adjacent to an area of a dentinal damage and this sprouting precedes the observation of reparative dentin deposition [63] (Fig. 6.4a–d). After artificial mechanical exposure of the dental pulp to the oral environment, sprouting of CGRP-expressing axons is observed in the remaining vital pulp tissues, adjacent to abscesses where no vital tissue is found [64]. Although in these experiments CGRP was primarily used as an anatomical marker of pulpal axons, there is good evidence that CGRP mediates numerous effects on resident cells of the pulp, supporting the

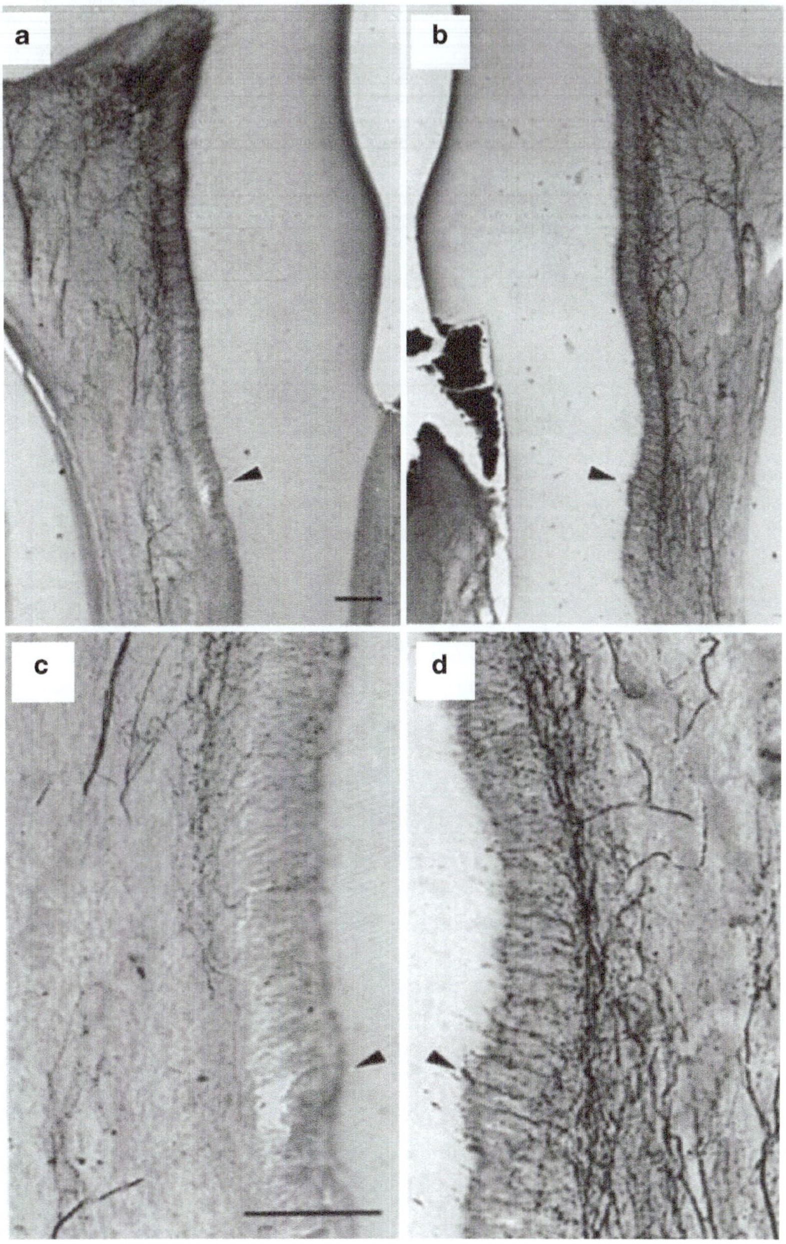

Fig. 6.4 (**a–d**) Sprouting of CGRP fibers in response to dentinal injury (Used with permission of Elsevier from Taylor et al. [63])

hypothesis that CGRP release from sensory neurons is an important component of healing and repair processes. The function of CGRP has been more thoroughly studied in the context of bone physiology, where it plays an important role in bone healing and remodeling, in part by inducing osteoblast proliferation and differentiation of stem cells into osteoblasts [65–67]. Similarly, in the dental pulp, CGRP can promote the proliferation of fibroblasts, causing BMP-2 production, and thus could potentially stimulate dentin formation [68–71]. Further in vivo experiments are needed to determine if this is a mechanism that can be utilized to promote dentin bridge formation and pulpal healing after injury.

In addition to influencing healing and repair via fibroblasts, CGRP release from sensory neurons mediates several aspects of inflammatory

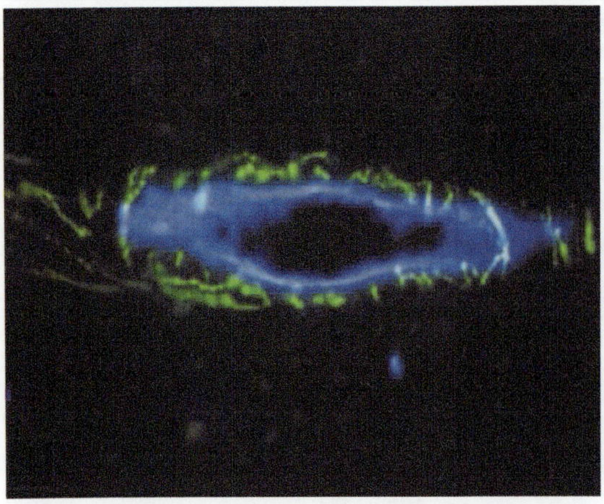

Fig. 6.5 Substance P-expressing fibers (*green*) forming a plexus around a blood vessel (*blue*) (Used with permission of John Wiley and Sons from Rodd and Boissonade [75])

processes. CGRP is a potent vasodilator and also causes plasma extravasation [72]. In fact, activation of sensory neurons in the pulp produces an overall vasodilatory effect and increases vascular permeability [73]. In contrast, activation of sympathetic neurons produces vasoconstriction, mediated by both monoamine sympathetic neurotransmitters as well as the peptide NPY [74]. CGRP, substance P, and sympathetic NPY-expressing nerve fibers are found in abundance in close approximation to arterioles [75] (Fig. 6.5). Like CGRP, substance P also causes vasodilation, and the magnitude of their individual vasodilatory effects is augmented when they are co-administered [76].

CGRP and substance P also produce several effects on the immune system. Although there are contradictory findings, the effects of CGRP are found to generally inhibit the immune responses, while substance P is an immune system stimulant [77, 78]. However, in vivo experiments in rats show that denervating the pulp results in reduced immune cell recruitment in response to experimental cavity preparation, suggesting an overall immunostimulatory effect of sensory neuron activation. Both CGRP and substance P cause cytokine release from pulpal fibroblasts [79]. Relevant to inflammatory mechanisms in the dental pulp, CGRP was recently shown to inhibit the release of bacterially stimulated TNF-α release from macrophages, and reduce lymphadenopathy in vivo, after acute bacterial exposure [80].

The immunomodulatory mechanisms of neuropeptides released from dental pulp afferents are complex, and many questions regarding these processes remain. The more we learn about inflammation, the more difficult it is to interpret findings relating to very specific immunomodulatory effects on overall disease processes. From a high level perspective, it's important to recognize that pulpal sensory neurons are a critical player in the defense mechanisms of the pulp, as pulpal necrosis proceeds more rapidly in denervated teeth that receive a pulp exposure, than in teeth with intact innervation [81]. As this protective effect is likely related to neurosecretions, manipulation of neuropeptide signaling represents an important potential point of therapeutic intervention in the inflamed pulp. Currently, the options for pulpal therapeutic interventions are expanding to include the promotion of biological repair and regenerative processes; thus, a fundamental understanding of the role of neuropeptides in these processes is needed.

The receptors for neuropeptides are found on peripheral sensory neurons, in the trigeminal nucleus, where processing of sensory signaling occurs, as well as other more rostral neuronal structures involved in pain/sensory perception. Endogenous release of neuropeptides can thus modulate sensory neuron activity and pain. Increased levels of neuropeptides, including CGRP, substance P, and NKA, are found in pulps from carious teeth versus non-carious teeth [56]. However, only substance P expression levels were found to be elevated in symptomatic versus non-symptomatic pulps and as well as elevated in pulpal tissues of patients with irreversible pulpitis [82, 83, 84] (Fig. 6.6a, b). Multiple preclinical studies have supported a role for substance P, via the NK1 receptor, to be an important mechanism for maintaining inflammatory and neuropathic pain states. However, an NK1 antagonist was not successful in demonstrating pain relief in clinical studies [85]. On the other hand, CGRP antagonists have demonstrated clinical efficacy in treating migraine pain [86]. Preclinical studies using animal models of pain have also suggested that

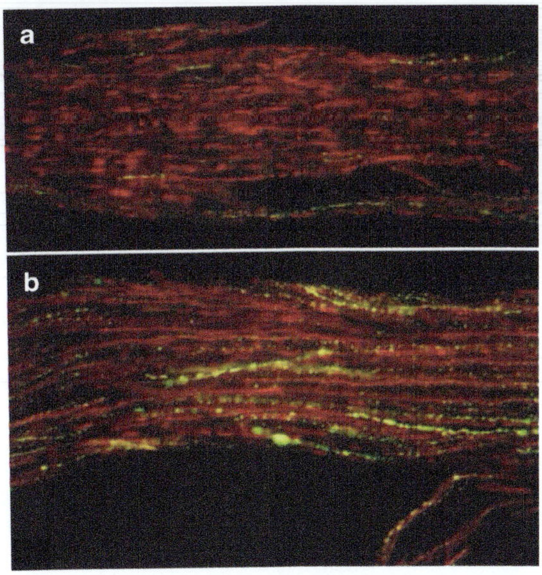

Fig. 6.6 (**a**, **b**) Substance P upregulated in carious human teeth (Used with permission of John Wiley and Sons from Rodd and Boissonade [84])

CGRP receptors have value as a therapeutic target for neuropathic pain. Interestingly, there may be some specificity toward the trigeminal system for the anti-hyperalgesic effects of CGRP antagonist after nerve injury [87]. NPY was shown to produce anti-hyperalgesic effects via the Y1 receptor in animal models in the pulpal tissues as well as in the spinal system [88]. NPY is highly expressed in the spinal cord and trigeminal nucleus and appears to be an important component of endogenous pain relief [89]. In sum, the receptors for neuropeptides expressed in afferents innervating dental pulp are attractive targets for manipulating pain of pulpal origin.

6.5 TRP Channels

Our current understanding of how peripheral neurons detect and transmit thermal, mechanical, and chemical stimuli is greatly influenced by the characterization of a family of cation-permeable channels, termed the transient receptor potential channels or TRPs [90]. The molecular basis for the specificity of populations of peripheral neurons to detect distinct stimuli (e.g., noxious cold or low pH) can be attributed, in part, to their expression of TRP receptors. Interestingly, the expression of TRPs and other sensory receptors differs by the target tissue being innervated; thus, tissues with unique sensory capacity, such as dental pulp, likely demonstrate unique expression of sensory receptors including TRPs [91].

The most studied TRP channel to date is the TRPV1 receptor. It was the first cloned and is notable for being activated by heat in the noxious range, low pH, and capsaicin, the pungent chemical found in chili peppers that causes a warm or burning sensation when ingested [92, 93]. Interestingly, this channel appears to be underrepresented in neurons innervating the dental pulp relative to its expression in other tissues innervated by trigeminal nerves, including the skin and periodontal tissues [41, 94, 95]. As TRPV1 is required for normal heat detection, the underrepresentation of TRPV1 in dental pulp afferents may be one reason why heat is an unreliable stimulus to evaluate pulpal vitality in a clinical setting [96]. Although TRPV1 may not play an important role in sensation in normal pulp, it is very likely involved in pulpal pain in the setting of inflammation, as the TRPV1 receptor is an important site for the integration of signaling pathways from several inflammatory mediators. Also, TRPV1 appears to be upregulated in inflamed human dental pulp [97, 98]. Finally, as capsaicin, a specific agonist for TRPV1, can stimulate the release of neurotransmitters such as CGRP from rodent and human dental pulp, the TRPV1 receptor is clearly functional in the dental pulp [99, 100].

It is of interest that the TRPV2 receptor, which, like TRPV1, was originally described as heat responsive, is highly expressed in the neurons innervating the dental pulp [41, 94, 95] (Fig. 6.7a–d). The neuronal population expressing TRPV2 does not overlap with those neurons expressing TRPV1. The neurons expressing TRPV2 are larger in diameter and myelinated and thus more likely to be low-threshold mechanosensitive neurons than classical nociceptors [43]. Although the TRPV2 receptor was originally described as heat responsive, this characteristic has only been demonstrated in vitro, and

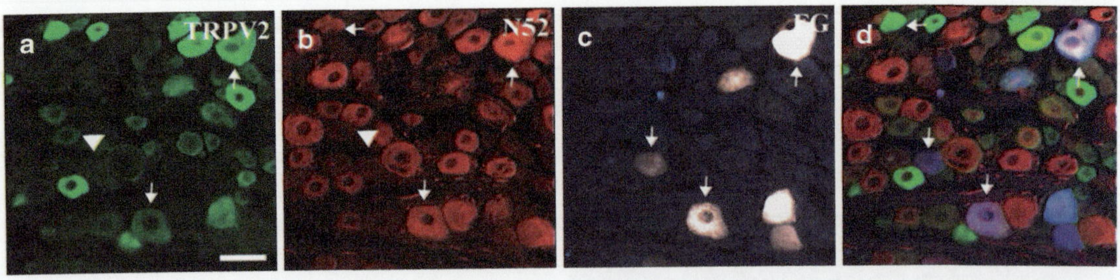

Fig. 6.7 TRPV2 (**a**) and neurofilament (**b**) expression in trigeminal ganglion neurons innervating the dental pulp after retrograde labeling with Fluoro-Gold (*FG*) (**c**). *Tailed arrows* point at *FG* cell bodies that show immunoreactivity for the indicated antigen. *Arrowheads* highlight cells that contain *FG* but are immunonegative for the indicated antigen. (**d**) merged images, *arrows* indicate FG labeled pulpal neurons. Scale bar = 200 μm. (Used with permission from Gibbs et al. [41])

further studies suggest the TRPV2 receptor is likely not involved in thermodetection. However, TRPV2 is clearly expressed on both high-threshold and low-threshold mechanosensitive fibers [101, 102]. Whether TRPV2 is a marker for this class of sensory neurons or is functionally involved in transducing mechanosensation has yet to be clearly demonstrated.

Perhaps of more relevance to the dental pulp are the cold-responsive channels. Cold allodynia is a common complaint in persons experiencing odontalgia of several etiologies, including pulpitis and dentin hypersensitivity [103, 104]. In fact, an abnormal lingering response to cold is considered the most important diagnostic test for irreversible pulpitis, the clinical diagnosis used when a root canal or extraction is deemed necessary to relieve pain [105]. Two TRP receptors have thus far been identified as molecular sensors for cold, TRPM8, and TRPA1. The TRPM8 receptor is responsive to cool temperatures in the non-noxious range, as well as chemicals that produce a cooling sensation such as menthol and icilin [106, 107]. It has been identified in neurons that innervate dental pulp, both in humans and rodents, although its expression was not correlated with cold sensitivity in humans [108–110]. The TRPA1 receptor is activated by cold temperatures in the noxious spectrum and is also a detector of environmental irritants and pungent compounds such as mustard oil [111, 112]. Like TRPV1, the TRPA1 receptor activity can be modulated by the signaling of several inflammatory mediators including bradykinin [113]. TRPA1 is highly expressed in neurons innervating the dental pulp and may be upregulated in teeth with painful pulpitis [109, 114, 115]. Although both TRPM8 and TRPA1 are interesting novel targets for treating the pain of pulpitis, further work is needed to understand their role in pain transduction within the dental pulp.

6.6 Sodium Channels

Voltage-gated sodium and potassium channels are needed for the generation of action potentials to convey peripheral sensory input into the central nervous system. These channels are termed "voltage gated" as the channels undergo a conformational change in response to application of a voltage, leading to sodium influx and membrane depolarization. There are several subtypes of sodium channels, some of which are expressed in specific subclasses of sensory fibers, including pain fibers, which make them potentially favorable targets for prospective therapeutics [116, 117]. Sodium channels are characterized as being either tetrodotoxin resistant (TTX-R) or tetrodotoxin sensitive (TTX-S), with the TTX-R current mediated by the $Na_v1.8$ and $Na_v1.9$ channels [118, 119]. Studies utilizing mouse genetics to knock out the receptor completely, or to make the neurons expressing the receptor susceptible toxins and subsequent ablation, have shown that the $Na_v1.8$ channel is required for the transmission of painful cold stimuli, mechanical pain, and mechanical and thermal hypersensitivity after inflammation [118, 120, 121]. The channel is also expressed at higher levels under inflammatory conditions, and increased expression of $Na_v1.8$ has been demonstrated in human dental pulp in persons experiencing painful pulpitis [122–125]. Importantly, the channel has also been shown to reduce the efficacy of lidocaine to block nerve transduction. Thus, the upregulation of $Na_v1.8$ within nerves innervating the dental pulp during pulpitic states could contribute to the clinical challenge of achieving adequate local anesthesia during dental procedures.

Another interesting molecular target in the sodium channel family is the TTX-S channel $Na_v1.7$. The importance of this channel to pain was convincingly demonstrated by the identification of genetic mutations of this channel in humans that led to either a gain in function or loss of function of the receptor that was clearly linked to very unique pain symptomatology [126]. Persons with a loss of function mutation were found to demonstrate congenital insensitivity to pain, i.e., they are unable to detect any type of painful stimulus [127]. These patients highlight the importance of pain perception to survival, as they tend to have shortened life spans due to gross injuries sustained because of their inability to detect tissue damage. Moreover, persons found to have a gain in function mutation were found to suffer from chronic ongoing spontaneous pain with an intense burning characteristic. The channel $Na_v1.7$ is found to be upregulated in many animal models of inflammatory pain and also in humans with painful pulpitis [128, 129]. Based on these findings, both the $Na_v1.8$ and $Na_v1.7$ channels are appealing targets for further investigation of the pain mechanisms originating from the dental pulp.

6.7 Autonomic Innervation

The autonomic nerves of the dental pulp belong to the sympathetic division of the autonomic nervous system. Parasympathetic fibers do not seem to innervate the tooth pulp [130]. The sympathetic axons of the dental pulp have their cell bodies in the superior cervical ganglion (SCG). They mainly project to the radicular pulp and form plexa along the blood vessels, while the odontoblast and subodontoblast layers seem to lack a sympathetic innervation [131, 132]. The main sympathetic functional output in the pulp is related to blood vessel constriction. Thus, stimulation of these nerves, or injections of sympathetic transmitters, causes a robust fall in pulpal blood flow [74, 133].

The distribution and density of pulpal sympathetics in mammalian teeth has been estimated with different methods and with varying results. It is conceivable that the extent of sympathetic innervation of the pulp varies between species. Thus, when monoamines have been targeted as markers of sympathetic transmitters using formaldehyde-induced fluorescence, positive nerve fibers were observed in pulps of humans, rabbits, and cats but not rats [132, 134]. Accordingly, the proportion of unmyelinated axons in rat molar pulps was not altered by sympathectomy [35], and retrograde tracer studies demonstrated that very few neurons in the ipsilateral superior cervical ganglion of the rat had projections to the rat molar pulp [135]. Immunohistochemistry has shown that antibodies against neuropeptide Y (NPY), a well-known marker of the sympathetic nervous system, label nerve fibers that line the blood vessels of normal human [75, 131], cat, and rat pulps [74, 131, 132]. Another sympathetic nerve marker, tyrosine hydroxylase (TH), is expressed in both rat [17] and human [39] pulps. Nonetheless, these data should be interpreted with some care, since TH is expressed also in a population of sensory nerves [136]. This is true for NPY as well, which is upregulated in sensory pulpal nerves as a response to challenges such as injury [137] or neuropathy [138]. To conclude, sympathetic stimulation of the dental pulp provides effective vasoconstrictor machinery in mammalian tooth pulps, although the numbers of intrapulpal sympathetic axons involved seem to vary between types of teeth as well as between species. Furthermore, it cannot be excluded that in some cases this mechanism is partly executed through sympathetic fibers on extrapulpal blood vessels, which would escape detection in structural studies of the pulp.

The sympathetic nervous system has an influence on the immune system, through local release of various molecules (see [139]). In sympathectomized rat pulpal tissue, granulocyte recruitment was impaired during experimental orthodontic tooth movement [140]. In line with this, electrical sympathetic nerve stimulation recruited such cells to the pulp. Moreover, immunoglobulin-producing cells were recruited to normal uninflamed dental pulps bilaterally after unilateral sympathectomy. Consequently, pulpal sympathetic nerves appear to play an important role in monitoring and influencing immunocompetent cells in states of infectious/inflammatory challenges to the dental pulp. However, it seems to be unclear as to whether sympathetic activity increases or reduces the severity of different types of inflammation. Thus, resection of the SCG in rats reduced abscess formation after molar pulp damage, but only at short time points. After longer periods, there was no difference in extent or severity of inflammation when compared to controls [141]. Similarly, conflicting results exist with regard to the degree of reparative dentin formation in sympathectomized inflamed teeth [141, 142].

6.8 Generator Mechanisms of Sensory Pulp Nerves

Weak mechanical stimuli such as air puffs and water spray, which are innocuous when applied to, e.g., the skin, evoke intense pain when directed at exposed dentin [143]. It appears unlikely that this is due to direct stimulation of dentinal nerve endings, since these terminate far away in the initial pulp-adjacent segment of the dentin. The hydrodynamic theory holds that force applied at the outermost end of dentinal tubules is transmitted to the sensory transduction apparatus deep inside by mechanical displacement, i.e., flow, of the fluid that the tubules contain [144, 145]. A prerequisite is then that the nerves that are stimulated by these very weak forces are LTMs (provided that they are not sensitized by, e.g., inflammation), since no obvious amplification mechanism is present. This fits well with the data that many if not most dentinal afferents are not classical nociceptors, but rather LTMs. An overwhelming majority is probably A-fibers, but low-threshold C-fibers could theoretically also contribute. The mechanical detection of dentinal fluid movement would require mechanosensory membrane receptors/ion channels in the dental LTM afferents. A number of such molecules have by now been identified in pulpal primary nerve cells. Among these are epithelial sodium channels (ENaCs), ASIC3, TREK1, and TREK2 [114, 146]. Furthermore, members of the TRP family of ion channels, which have been implicated in mechanosensation, are expressed in pulp-innervating trigeminal ganglion neurons, including TRPV2 and TRPA1 (see [147–149]. However, the individual contribution and possible coordinated action of these and perhaps additional membrane sensors remain to be elucidated.

6.9 The Odontoblast as a Putative Pulpal Transducer Cell

The hydrodynamic theory, propagated more than 40 years ago, still provides an attractive model to explain the mechanism behind the sharp and immediate pain that is elicited by various stimuli on dentin. However, it leaves several issues with regard to, e.g., hot and cold sensitivity, in the pulp unresolved. In some cases there seems to be no relationship between pain sensation and movement of dentinal fluid after cold stimulation [150], although some authors claim that distal movement of the fluid in response to cold stimulation is more rapid than proximal movement by hot stimuli, which could affect sensory thresholds [151]. This raises the possibility that additional mechanisms might be activated to convey sensations when teeth are challenged by thermal and perhaps also other stimuli. Very recently, several lines of evidence have pointed to the likelihood that the odontoblast has a role in sensory transduction from teeth, although this is not yet conclusively shown. Thus, calcium imaging studies have demonstrated that human odontoblasts express functional TRPM8, TRPA1, and TRPV1 channels [152]. This indicates that odontoblasts could mediate thermal stress, in concert with sensory nerves in teeth. Furthermore, odontoblasts also express mRNA or protein for mechanosensitive ion channels such as the TREK-1 and K_{Ca} potassium channels, which could suggest a mechanosensory function as well [153–155]. An additional electrogenic sensor of stretch activation function of odontoblasts might be accomplished by the recently characterized primary cilia of these cells [156]. Finally, and importantly, odontoblasts express functional voltage-gated sodium channels, which would enable them to become electrically excitable. They also express mRNA for major subunits of ionotropic glutamate receptors (NMDARs), which potentially might be used to generate action potentials [157]. Other sensory cell-related genes present in odontoblasts, again with putative roles in stimulus transduction, include those that code for parvalbumin, the membrane adaptor protein harmonin, the neuronal calcium sensor-1 [6], and synaptic vesicle protein 2b [158].

As seen from this discussion, there is mounting evidence that odontoblasts can respond to sensory stimuli and become electrically excited. However, there is still no reliable proof for the presence of a system, synaptic or other, that translates odontoblast activity into afferent nerve fiber signaling. An interaction that involves ATP is conceivable since purinergic pulp nerve fibers [2, 159] seem to become sensitized by ATP from pulpal cells following inflammation or injury [160, 161]. This may well involve odontoblasts, but is apparently not a sensory cell/nerve-specific mechanism.

6.10 Connectivity of Sensory Tooth Pulp Nerves

The central branches of TG neurons travel via the trigeminal root to the brain stem. Subnucleus caudalis of the spinal trigeminal nucleus is seen as the major nociceptive relay of the trigeminal brain stem complex, since it receives an immense input from pain-transmitting axons that innervate the orofacial region [162, 163]. Morphological investigations using tracing techniques from the tooth have shown that pulpal afferent terminates predominantly in the superficial laminae of subnucleus caudalis, but also in its deep laminae [164–166]. Furthermore, many dental pulp fibers have their central endings more rostrally, especially in the trigeminal subnuclei interpolaris and oralis. When tooth pulps are electrically stimulated, the responses of postsynaptic neurons in all three spinal trigeminal subnuclei correspond to the anatomical findings [162] and largely agree with what would be expected from nociceptors. This is remarkable, since most pulpal sensory afferents have anatomical and electrophysiological characteristics of LTMs and not primary nociceptive neurons. However, since pulpal axons do have the capacity to deliver pain messages to higher brain centers even upon very weak and subtle stimulation, they would have to terminate synaptically on spinal trigeminal nuclei neurons in order to connect into the pain-mediating network.

The fact that pulpal afferents are LTMs whose signals evoke pain rather than touch, due to idiosyncratic connectivity and/or neurotransmitter content, makes them unique among pain-mediating neurons. Since they have very different characteristics from classical nociceptors, we have proposed a novel definition, "algoneurons," for peripheral neurons that, when activated, evoke a sensation of pain. In contrast to the term nociceptor, the term algoneuron focuses on the sensory effect of the afferent's signal and not its response properties. According to this, a majority of trigeminal tooth pulp neurons are low-threshold mechanoalgoneurons [38].

In the thalamus, tooth pulp-driven neurons have been identified in ventral posteromedial (VPM) and mediodorsal (MD) nuclei [167]. Considering even higher CNS levels, functional magnetic resonance imaging (fMRI) has demonstrated that painful electrical tooth pulp stimulation leads to bilateral activation of S1, S2, and the insular region of the cerebral cortex. The cingulate gyrus is also activated, as well as motor and frontal areas including the orbital frontal cortex. Tooth pulp pain involves a cortical network, which in several respects appears to be different from that activated by painful stimulation of a hand [5]. Seemingly specific tooth pulp projections to the somatosensory cortex were also shown with magnetic field recording methods. Here, the latencies clearly indicated that the input came from intradental Aβ fibers [168].

6.11 Aging of Pulpal Nerves

With increasing age odontoblasts shrink, apparently due to changes in autophagy [169]. However, secondary dentin formation continues at a slow rate during the life of the tooth, causing a gradual reduction of the pulpal space. This may be aggravated by irregular dentin formed in response to external stimuli. Concomitant with this, a protracted phase of age-related axonal alterations and axon loss occurs. In parallel, there are changes in pulpal nerve cytochemistry. Some of these likely are responses to wear and/or trauma, since they are typically seen proximal to nerve injuries [170, 171]. Pulpal nerve deterioration in senescence is paralleled by a reduced sensitivity to electrical pulp stimulation in human subjects [172].

6.12 Neurotrophins/Receptors in Pulpal Nerve Plasticity

In addition to their important role in establishing innervation of pulpal tissues during development, the neurotrophins and their respective receptors are critical in maintaining the unique phenotype of pulpal afferents in the mature pulp and are important mediators of neuronal plasticity in response to injury. Nerve growth factor (NGF) is the most studied neurotrophin, and indeed all pulpal neurons are at some point dependent on NGF. The receptors for NGF include the high-affinity tyrosine kinase receptor trkA and the low-affinity neurotrophin receptor p75. The importance of the trkA receptor to pulpal innervation is highlighted by the finding that sensory and sympathetic innervation of the dental pulp is eliminated in trkA knockout mice [25]. In the mature pulp, many afferents lose their dependence on NGF with many of the larger fibers becoming dependent on glial-derived neurotrophic factor (GDNF) by expressing the GDNF receptor GFR-α1 [87, 173]. Neurotrophin and neurotrophin receptor expression is altered by the presence of injury and inflammation in the pulp. For example, an upregulation in NGF is observed in pulpal fibroblasts after dentinal injury and is thought to promote sprouting of pulpal afferents [174]. Importantly, neurotrophin expression at the site of injury affects the transcription of genes encoding neurotransmitters, receptors, and ion channels that are key to pain transduction including CGRP, SP, TRPV1, TRPA1, and $Na_v1.8$ [175, 176]. This plasticity is thought to contribute to the hypersensitivity and spontaneous pain that occur after injury [177].

6.13 Neuroplasticity in the Peripheral and Central Nervous System Subsequent to Pulpal Injury

Both the peripheral and central nervous systems demonstrate remarkable neuroplasticity in response to pulpal injury. In this chapter, we have previously described neurotrophin-dependent changes in neuropeptide and receptor expression that occurs in response to inflammation, as well as sprouting of afferent terminals at the site of injury. Both of these mechanisms are thought to contribute to the development of hypersensitivity in the setting of inflammation. In the trigeminal ganglion, activation of the satellite glial cells surrounding neuronal cell bodies occurs subsequent to pulpal inflammation [178, 179]. Activated satellite cells can contribute to neuronal hyperexcitability via the intraganglionic release of proinflammatory cytokines. Astroglial induction and proliferation in the trigeminal nucleus also contributes to hypersensitivity after dental pulp injury [180]. In fact, significant anatomical and functional changes in activity are observed in the trigeminal nucleus subsequent to pulpal injury [181, 182]. These findings are important because they parallel observations from studies using animal models of neuropathic pain, most of which involve a partial nerve injury that produces persistent mechanical and/or thermal hypersensitivity in the region innervated by the injured nerve. In total, these studies support the existence of neuroplastic mechanisms that occur in response to deafferentation of the dental pulp and have the potential to contribute to persistent pain states subsequent to natural or iatrogenic dental pulp injury.

The possibility of persistent pain after clinical interventions that remove dental pulp, such as root canal treatment, has been recognized for quite some time [183–186]. Although persistent symptoms could be due to ongoing odontogenic causes (e.g., an undetected root fracture or recurrent infection), there are cases when pain persists despite the absence of obvious pathology. Historically such persistent pain was referred to as atypical odontalgia, or phantom tooth pain, or more currently, persistent dentoalveolar pain or peripheral painful traumatic trigeminal neuropathy [187, 188]. Although debates regarding the criteria for classification of this clinical entity are ongoing, it likely represents a very specific type of persistent postsurgical pain. The etiology of non-odontogenic persistent post endodontic therapy pain is unknown, but there is some evidence that neuropathic mechanisms are involved [189–191]. More research is needed to continue to gain knowledge relating to the biological mechanisms contributing to the development of persistent postsurgical pain.

References

1. Hildebrand C, Fried K, Tuisku F, Johansson CS. Teeth and tooth nerves. Prog Neurobiol. 1995;45(3):165–222.
2. Cook SP, Vulchanova L, Hargreaves KM, Elde R. Distinct ATP receptors on pain-sensing and stretch-sensing neurons. Nature. 1997;387(6632):505–8.
3. Northcutt RG, Gans C. The genesis of neural crest and epidermal placodes: a reinterpretation of vertebrate origins. Q Rev Biol. 1983;58(1):1–28.
4. Young GC, Karatajute-Talimaa VN, Smith MM. A possible late Cambrian vertebrate from Australia. Nature. 1996;38(3):810–2.
5. Calloni GW, Le Douarin NM, Dupin E. High frequency of cephalic neural crest cells shows coexistence of neurogenic, melanogenic, and osteogenic differentiation capacities. Proc Natl Acad Sci U S A. 2009;106(22):8947–52.
6. Farahani RM, Simonian M, Hunter N. Blueprint of an ancestral neurosensory organ revealed in glial networks in human dental pulp. J Comp Neurol. 2011;519(16):3306–26.
7. Dong WK, Chudler EH, Martin RF. Physiological properties of intradental mechanoreceptors. Brain Res. 1985;334(2):389–95.
8. Paphangkorakit J, Osborn JW. Discrimination of hardness by human teeth apparently not involving periodontal receptors. Arch Oral Biol. 1998;43(11):833–9.
9. Robertson LT, Levy JH, Petrisor D, Lilly DJ, Dong WK. Vibration perception thresholds of human maxillary and mandibular central incisors. Arch Oral Biol. 2003;48(4):309–16.
10. Boissonade FM, Matthews B. Responses of trigeminal brain stem neurons and the digastric muscle to tooth-pulp stimulation in awake cats. J Neurophysiol. 1993;69(1):174–86.
11. Olgart L, Gazelius B, Sundstrom F. Intradental nerve activity and jaw-opening reflex in response to mechanical deformation of cat teeth. Acta Physiol Scand. 1988;133(3):399–406.
12. Trulsson M. Sensory-motor function of human periodontal mechanoreceptors. J Oral Rehabil. 2006;33(4):262–73.
13. Fried K, Nosrat C, Lillesaar C, Hildebrand C. Molecular signaling and pulpal nerve development. Crit Rev Oral Biol Med. 2000;11(3):318–32.
14. Luukko K, Moe K, Sijaona A, Furmanek T, Hals Kvinnsland I, Midtbø M, Kettunen P. Secondary induction and the development of tooth nerve supply. Ann Anat. 1998;210(4):463–71.
15. Kettunen P, Løes S, Furmanek T, Fjeld K, Kvinnsland IH, Behar O, Yagi T, Fujisawa H, Vainio S, Taniguchi M, Luukko K. Coordination of trigeminal axon navigation and patterning with tooth organ formation: epithelial-mesenchymal interactions, and epithelial Wnt4 and Tgfbeta1 regulate semaphorin 3a expression in the dental mesenchyme. Development. 2005;132(2):323–34.
16. Fried K, Lillesaar C, Sime W, Kaukua N, Patarroyo M. Target finding of pain nerve fibers: neural growth mechanisms in the tooth pulp. Physiol Behav. 2007;92(1–2):40–5.
17. Moe K, Kettunen P, Kvinnsland IH, Luukko K. Development of the pioneer sympathetic innervation into the dental pulp of the mouse mandibular first molar. Arch Oral Biol. 2008;53(9):865–73.
18. Kettunen P, Spencer-Dene B, Furmanek T, Kvinnsland IH, Dickson C, Thesleff I, Luukko K. Fgfr2b mediated epithelial-mesenchymal interactions coordinate tooth morphogenesis and dental trigeminal axon patterning. Mech Dev. 2007;124(11–12):868–83.
19. Lillesaar C, Fried K. Neurites from trigeminal ganglion explants grown in vitro are repelled or attracted by tooth-related tissues depending on developmental stage. Neuroscience. 2004;125(1):149–61.

20. Moe K, Sijaona A, Shrestha A, Kettunen P, Taniguchi M, Luukko K. Semaphorin 3A controls timing and patterning of the dental pulp innervation. Differentiation. 2012;84(5):371–9.
21. Sijaona A, Luukko K, Kvinnsland IH, Kettunen P. Expression patterns of Sema3F, PlexinA4, -A3, Neuropilin1 and -2 in the postnatal mouse molar suggest roles in tooth innervation and organogenesis. Acta Odontol Scand. 2012;70(2):140–8.
22. Luukko K, Arumae U, Karavanov A, Moshnyakov M, Sainio K, Sariola H, Saarma M, Thesleff I. Neurotrophin mRNA expression in the developing tooth suggests multiple roles in innervation and organogenesis. Dev Dyn. 1997;210(2):117–29.
23. Naftel JP, Qian XB, Bernanke JM. Effects of postnatal anti-nerve growth factor serum exposure on development of apical nerves of the rat molar. Brain Res Dev Brain Res. 1994;80(1–2):54–62.
24. Nosrat CA, Fried K, Ebendal T, Olson L. NGF, BDNF, NT3, NT4 and GDNF in tooth development. Eur J Oral Sci. 1998;106 Suppl 1:94–9.
25. Matsuo S, Ichikawa H, Henderson TA, Silos-Santiago I, Barbacid M, Arends JJ, Jacquin MF. trkA modulation of developing somatosensory neurons in orofacial tissues: tooth pulp fibers are absent in trkA knockout mice. Neuroscience. 2001;105(3):747–60.
26. Kvinnsland IH, Luukko K, Fristad I, Kettunen P, Jackson DL, Fjeld K, von Bartheld CS, Byers MR. Glial cell line-derived neurotrophic factor (GDNF) from adult rat tooth serves a distinct population of large-sized trigeminal neurons. Eur J Neurosci. 2004;19(8):2089–98.
27. Luukko K, Suvanto P, Saarma M, Thesleff I. Expression of GDNF and its receptors in developing tooth is developmentally regulated and suggests multiple roles in innervation and organogenesis. Dev Dyn. 1997;210(4):463–71.
28. Lillesaar C, Eriksson C, Fried K. Rat tooth pulp cells elicit neurite growth from rat trigeminal neurones and express mRNA for neurotrophic factors *in vitro*. Neurosci Lett. 2001;308(3):161–4.
29. Fried K, Risling M, Tidcombe H, Gassmann M, Lillesaar C. Expression of ErbB3, ErbB4 and neuregulin-1 mRNA during tooth development. Dev Dyn. 2002;224(3):356–60.
30. Fried K, Sime W, Lillesaar C, Virtanen I, Tryggvasson K, Patarroyo M. Laminins 2 (alpha2beta1gamma1, Lm-211) and 8 (alpha4beta1gamma1, Lm-411) are synthesized and secreted by tooth pulp fibroblasts and differentially promote neurite outgrowth from sensory trigeminal ganglion neurons. Exp Cell Res. 2005;307(2):329–41.
31. Yuasa K, Fukumoto S, Kamasaki Y, Yamada A, Fukumoto E, Kanaoka K, Saito K, Harada H, Arikawa-Hirasawa E, Miyagoe-Suzuki Y, Takeda S, Okamoto K, Kato Y, Fujiwara T. Laminin alpha2 is essential for odontoblast differentiation regulating dentin sialoprotein expression. J Biol Chem. 2004;279(11):10286–92.
32. Maurin JC, Couble ML, Didier-Bazes M, Brisson C, Magloire H, Bleicher F. Expression and localization of reelin in human odontoblasts. Matrix Biol. 2004;23(5):277–85.
33. Gotliv BA, Veis A. Peritubular dentin, a vertebrate apatitic mineralized tissue without collagen: role of a phospholipid-proteolipid complex. Calcif Tissue Int. 2007;81(3):191–205.
34. Byers MR, Suzuki H, Maeda T. Dental neuroplasticity, neuro-pulpal interactions, and nerve regeneration. Microsc Res Tech. 2003;60(5):503–15.
35. Fried K, Aldskogius H, Hildebrand C. Proportion of unmyelinated axons in rat molar and incisor tooth pulps following neonatal capsaicin treatment and/or sympathectomy. Brain Res. 1998;463(1): 118–23.
36. Paik SK, Park KP, Lee SK, Ma SK, Cho YS, Kim YK, Rhyu IJ, Ahn DK, Yoshida A, Bae YC. Light and electron microscopic analysis of the somata and parent axons innervating the rat upper molar and lower incisor pulp. Neuroscience. 2009;162(4): 1279–86.
37. Cadden SW, Lisney SJ, Matthews B. Thresholds to electrical stimulation of nerves in cat canine tooth-pulp with A beta-, A delta- and C-fibre conduction velocities. Brain Res. 1983;261(1):31–41.
38. Fried K, Sessle BJ, Devor M. The paradox of pain from tooth pulp: low-threshold "algoneurons"? Pain. 2011;152(12):2685–9.
39. Henry MA, Luo S, Levinson SR. Unmyelinated nerve fibers in the human dental pulp express markers for myelinated fibers and show sodium channel accumulations. BMC Neurosci. 2012;13:29. doi:10.1186/1471-2202-13-29.
40. Paik SK, Lee DS, Kim JY, Bae JY, Cho YS, Ahn DK, Yoshida A, Bae YC. Quantitative ultrastructural analysis of the neurofilament 200-positive axons in the rat dental pulp. J Endod. 2010;36(10):1638–42.
41. Gibbs JL, Melnyk JL, Basbaum AI. Differential TRPV1 and TRPV2 channel expression in dental pulp. J Dent Res. 2011;90(6):765–70.
42. Olgart L. Neural control of pulpal blood flow. Crit Rev Oral Biol Med. 1996;7(2):159–71.
43. Lewinter RD, Skinner K, Julius D, Basbaum AI. Immunoreactive TRPV-2 (VRL-1), a capsaicin receptor homolog, in the spinal cord of the rat. J Comp Neurol. 2004;470(4):400–8.
44. Byers MR. Terminal arborization of individual sensory axons in dentin and pulp of rat molars. Brain Res. 1985;345(1):181–5.
45. Carda C, Peydró A. Ultrastructural patterns of human dentinal tubules, odontoblasts processes and nerve fibres. Tissue Cell. 2006;38(2):141–50.
46. Holland GR. Morphological features of dentine and pulp related to dentine sensitivity. Arch Oral Biol. 1994;39(Suppl):3S–11.
47. Ibuki T, Kido MA, Kiyoshima T, Terada Y, Tanaka T. An ultrastructural study of the relationship between sensory trigeminal nerves and odontoblasts in rat dentin pulp as demonstrated by anterograde transport of wheat germ agglutinin-horseradish peroxidase. J Dent Res. 1996;75(12):1963–70.

48. Caviedes-Bucheli J, Munoz HR, Azuero-Holguin MM, Ulate E. Neuropeptides in dental pulp: the silent protagonists. J Endod. 2008;34(7):773–88.
49. Fristad I, Bletsa A, Byers MR. Inflammatory nerve responses in the dental pulp. Endod Top. 2010;17:12–41.
50. Nagy JI, Hunt SP. Fluoride-resistant acid phosphatase-containing neurones in dorsal root ganglia are separate from those containing substance P or somatostatin. Neuroscience. 1982;7(1):89–97.
51. Zylka MJ, Rice FL, Anderson DJ. Topographically distinct epidermal nociceptive circuits revealed by axonal tracers targeted to Mrgprd. Neuron. 2005;45(1):17–25.
52. Ivanavicius SP, Blake DR, Chessell IP, Mapp PI. Isolectin B4 binding neurons are not present in the rat knee joint. Neuroscience. 2004;128(3):555–60.
53. Tan LL, Bornstein JC, Anderson CR. Distinct chemical classes of medium-sized transient receptor potential channel vanilloid 1-immunoreactive dorsal root ganglion neurons innervate the adult mouse jejunum and colon. Neuroscience. 2008;156(2):334–43.
54. Cavanaugh DJ, Lee H, Lo L, Shields SD, Zylka MJ, Basbaum AI, Anderson DJ. Distinct subsets of unmyelinated primary sensory fibers mediate behavioral responses to noxious thermal and mechanical stimuli. Proc Natl Acad Sci U S A. 2009;106(22):9075–80.
55. Joseph EK, Levine JD. Hyperalgesic priming is restricted to isolectin B4-positive nociceptors. Neuroscience. 2010;169(1):431–5.
56. Awawdeh L, Lundy FT, Shaw C, Lamey PJ, Linden GJ, Kennedy JG. Quantitative analysis of substance P, neurokinin A and calcitonin gene-related peptide in pulp tissue from painful and healthy human teeth. Int Endod J. 2002;35(1):30–6.
57. Fried K, Arvidsson J, Robertson B, Brodin E, Theodorsson E. Combined retrograde tracing and enzyme/immunohistochemistry of trigeminal ganglion cell bodies innervating tooth pulps in the rat. Neuroscience. 1989;33(1):101–9.
58. Heyeraas KJ, Kvinnsland I, Byers MR, Jacobsen EB. Nerve fibers immunoreactive to protein gene product 9.5, calcitonin gene-related peptide, substance P, and neuropeptide Y in the dental pulp, periodontal ligament, and gingiva in cats. Acta Odontol Scand. 1993;51(4):207–21.
59. Mori H, Ishida-Yamamoto A, Senba E, Ueda Y, Tohyama M. Calcitonin gene-related peptide containing sensory neurons innervating tooth pulp and buccal mucosa of the rat: an immunohistochemical analysis. J Chem Neuroanat. 1990;3(3):155–63.
60. Buck S, Reese K, Hargreaves KM. Pulpal exposure alters neuropeptide levels in inflamed dental pulp and trigeminal ganglia: evaluation of axonal transport. J Endod. 1999;25(11):718–21.
61. Pan Y, Wheeler EF, Bernanke JM, Yang H, Naftel JP. A model experimental system for monitoring changes in sensory neuron phenotype evoked by tooth injury. J Neurosci Methods. 2003;126(1):99–109.
62. Rodd HD, Boissonade FM. Comparative immunohistochemical analysis of the peptidergic innervation of human primary and permanent tooth pulp. Arch Oral Biol. 2002;47(5):375–85.
63. Taylor PE, Byers MR, Redd PE. Sprouting of CGRP nerve fibers in response to dentin injury in rat molars. Brain Res. 1988;461(2):371–6.
64. Kimberly CL, Byers MR. Inflammation of rat molar pulp and periodontium causes increased calcitonin gene-related peptide and axonal sprouting. Anat Rec. 1988;222(3):289–300.
65. Ballica R, Valentijn K, Khachatryan A, Guerder S, Kapadia S, Gundberg C, Gilligan J, Flavell RA, Vignery A. Targeted expression of calcitonin gene-related peptide to osteoblasts increases bone density in mice. J Bone Miner Res. 1999;14(7):1067–74.
66. Fang Z, Yang Q, Xiong W, Li GH, Liao H, Xiao J, Li F. Effect of CGRP-adenoviral vector transduction on the osteoblastic differentiation of rat adipose-derived stem cells. PLoS One. 2013;8(8):e72738.
67. Hukkanen M, Konttinen YT, Santavirta S, Paavolainen P, Gu XH, Terenghi G, Polak JM. Rapid proliferation of calcitonin gene-related peptide-immunoreactive nerves during healing of rat tibial fracture suggests neural involvement in bone growth and remodelling. Neuroscience. 1993;54(4):969–79.
68. Bongenhielm U, Haegerstrand A, Theodorsson E, Fried K. Effects of neuropeptides on growth of cultivated rat molar pulp fibroblasts. Regul Pept. 1995;60(2–3):91–8.
69. Calland JW, Harris SE, Carnes Jr DL. Human pulp cells respond to calcitonin gene-related peptide in vitro. J Endod. 1997;23(8):485–9.
70. Kline LW, Yu DC. Effects of calcitonin, calcitonin gene-related peptide, human recombinant bone morphogenetic protein-2, and parathyroid hormone-related protein on endodontically treated ferret canines. J Endod. 2009;35(6):866–9.
71. Zhang M, Fukuyama H. CGRP immunohistochemistry in wound healing and dentin bridge formation following rat molar pulpotomy. Histochem Cell Biol. 1999;112(5):325–33.
72. Berggreen E, Heyeraas KJ. The role of sensory neuropeptides and nitric oxide on pulpal blood flow and tissue pressure in the ferret. J Dent Res. 1999;78(9):1535–43.
73. Kerezoudis NP, Olgart L, Edwall L. Evans blue extravasation in rat dental pulp and oral tissues induced by electrical stimulation of the inferior alveolar nerve. Arch Oral Biol. 1993;38(10):893–901.
74. Edwall B, Gazelius B, Fazekas A, Theodorsson-Norheim E, Lundberg JM. Neuropeptide Y (NPY) and sympathetic control of blood flow in oral mucosa and dental pulp in the cat. Acta Physiol Scand. 1985;125(2):253–64.
75. Rodd HD, Boissonade FM. Immunocytochemical investigation of neurovascular relationships in human tooth pulp. J Anat. 2003;202(2):195–203.
76. Gazelius B, Edwall B, Olgart L, Lundberg JM, Hokfelt T, Fischer JA. Vasodilatory effects and coexistence of calcitonin gene-related peptide (CGRP) and

substance P in sensory nerves of cat dental pulp. Acta Physiol Scand. 1987;130(1):33–40.
77. Holzmann B. Modulation of immune responses by the neuropeptide CGRP. Amino Acids. 2013;45(1):1–7. doi:10.1007/s00726-011-1161-2.
78. Okiji T, Jontell M, Belichenko P, Dahlgren U, Bergenholtz G, Dahlstrom A. Structural and functional association between substance P- and calcitonin gene-related peptide-immunoreactive nerves and accessory cells in the rat dental pulp. J Dent Res. 1997;76(12):1818–24.
79. Yamaguchi M, Kojima T, Kanekawa M, Aihara N, Nogimura A, Kasai K. Neuropeptides stimulate production of interleukin-1 beta, interleukin-6, and tumor necrosis factor-alpha in human dental pulp cells. Inflamm Res. 2004;53(5):199–204.
80. Chiu IM, Heesters BA, Ghasemlou N, Von Hehn CA, Zhao F, Tran J, Wainger B, Strominger A, Muralidharan S, Horswill AR, Bubeck Wardenburg J, Hwang SW, Carroll MC, Woolf CJ. Bacteria activate sensory neurons that modulate pain and inflammation. Nature. 2013;501(7465):52–7.
81. Byers MR, Taylor PE. Effect of sensory denervation on the response of rat molar pulp to exposure injury. J Dent Res. 1993;72(3):613–8.
82. Awawdeh LA, Lundy FT, Linden GJ, Shaw C, Kennedy JG, Lamey PJ. Quantitative analysis of substance P, neurokinin A and calcitonin gene-related peptide in gingival crevicular fluid associated with painful human teeth. Eur J Oral Sci. 2002;110(3): 185–91.
83. Bowles WR, Withrow JC, Lepinski AM, Hargreaves KM. Tissue levels of immunoreactive substance P are increased in patients with irreversible pulpitis. J Endod. 2003;29(4):265–7.
84. Rodd HD, Boissonade FM. Substance P expression in human tooth pulp in relation to caries and pain experience. Eur J Oral Sci. 2000;108:467–74.
85. Hill R. NK1 (substance P) receptor antagonists – why are they not analgesic in humans? Trends Pharmacol Sci. 2000;21(7):244–6.
86. Ho TW, Edvinsson L, Goadsby PJ. CGRP and its receptors provide new insights into migraine pathophysiology. Nat Rev Neurol. 2010;6(10):573–82.
87. Michot B, Bourgoin S, Viguier F, Hamon M, Kayser V. Differential effects of calcitonin gene-related peptide receptor blockade by olcegepant on mechanical allodynia induced by ligation of the infraorbital nerve vs the sciatic nerve in the rat. Pain. 2012;153(9): 1939–48.
88. Gibbs JL, Flores CM, Hargreaves KM. Attenuation of capsaicin-evoked mechanical allodynia by peripheral neuropeptide Y Y1 receptors. Pain. 2006;124(1–2): 167–74.
89. Solway B, Bose SC, Corder G, Donahue RR, Taylor BK. Tonic inhibition of chronic pain by neuropeptide Y. Proc Natl Acad Sci U S A. 2011;108(17):7224–9.
90. Patapoutian A, Tate S, Woolf CJ. Transient receptor potential channels: targeting pain at the source. Nat Rev Drug Discov. 2009;8(1):55–68.
91. Malin S, Molliver D, Christianson JA, Schwartz ES, Cornuet P, Albers KM, Davis BM. TRPV1 and TRPA1 function and modulation are target tissue dependent. J Neurosci. 2011;31(29):10516–28.
92. Caterina MJ, Schumacher MA, Tominaga M, Rosen TA, Levine JD, Julius D. The capsaicin receptor: a heat-activated ion channel in the pain pathway. Nature. 1997;389(6653):816–24.
93. Szallasi A, Cortright DN, Blum CA, Eid SR. The vanilloid receptor TRPV1: 10 years from channel cloning to antagonist proof-of-concept. Nat Rev Drug Discov. 2007;6(5):357–72.
94. Ichikawa H, Sugimoto T. VR1-immunoreactive primary sensory neurons in the rat trigeminal ganglion. Brain Res. 2001;890(1):184–8.
95. Stenholm E, Bongenhielm U, Ahlquist M, Fried K. VRl- and VRL-l-like immunoreactivity in normal and injured trigeminal dental primary sensory neurons of the rat. Acta Odontol Scand. 2002;60(2):72–9.
96. Petersson K, Soderstrom C, Kiani-Anaraki M, Levy G. Evaluation of the ability of thermal and electrical tests to register pulp vitality. Endod Dent Traumatol. 1999;15(3):127–31.
97. Chung MK, Lee J, Duraes G, Ro JY. Lipopolysaccharide-induced pulpitis up-regulates TRPV1 in trigeminal ganglia. J Dent Res. 2011;90(9):1103–7.
98. Morgan CR, Rodd HD, Clayton N, Davis JB, Boissonade FM. Vanilloid receptor 1 expression in human tooth pulp in relation to caries and pain. J Orofac Pain. 2005;19(3):248–60.
99. Fehrenbacher JC, Sun XX, Locke EE, Henry MA, Hargreaves KM. Capsaicin-evoked iCGRP release from human dental pulp: a model system for the study of peripheral neuropeptide secretion in normal healthy tissue. Pain. 2009;144(3):253–61.
100. Gibbs JL, Hargreaves KM. Neuropeptide Y Y1 receptor effects on pulpal nociceptors. J Dent Res. 2008;87(10):948–52.
101. Lawson JJ, McIlwrath SL, Woodbury CJ, Davis BM, Koerber HR. TRPV1 unlike TRPV2 is restricted to a subset of mechanically insensitive cutaneous nociceptors responding to heat. J Pain. 2008;9(4): 298–308.
102. Park U, Vastani N, Guan Y, Raja SN, Koltzenburg M, Caterina MJ. TRP vanilloid 2 knock-out mice are susceptible to perinatal lethality but display normal thermal and mechanical nociception. J Neurosci. 2011;31(32):11425–36.
103. Newton CW, Hoen MM, Goodis HE, Johnson BR, McClanahan SB. Identify and determine the metrics, hierarchy, and predictive value of all the parameters and/or methods used during endodontic diagnosis. J Endod. 2009;35(12):1635–44.
104. Rees JS, Addy M. A cross-sectional study of dentine hypersensitivity. J Clin Periodontol. 2002;29(11): 997–1003.
105. Levin LG, Law AS, Holland GR, Abbott PV, Roda RS. Identify and define all diagnostic terms for pulpal health and disease states. J Endod. 2009; 35(12):1645–57.

106. McKemy DD, Neuhausser WM, Julius D. Identification of a cold receptor reveals a general role for TRP channels in thermosensation. Nature. 2002;416(6876):52–8.
107. Peier AM, Moqrich A, Hergarden AC, Reeve AJ, Andersson DA, Story GM, Earley TJ, Dragoni I, McIntyre P, Bevan S, Patapoutian A. A TRP channel that senses cold stimuli and menthol. Cell. 2002;108(5):705–15.
108. Alvarado LT, Perry GM, Hargreaves KM, Henry MA. TRPM8 Axonal expression is decreased in painful human teeth with irreversible pulpitis and cold hyperalgesia. J Endod. 2007;33(10):1167–71.
109. Park CK, Kim MS, Fang Z, Li HY, Jung SJ, Choi SY, Lee SJ, Park K, Kim JS, Oh SB. Functional expression of thermo-transient receptor potential channels in dental primary afferent neurons: implication for tooth pain. J Biol Chem. 2006;281(25): 17304–11.
110. Takashima Y, Daniels RL, Knowlton W, Teng J, Liman ER, McKemy DD. Diversity in the neural circuitry of cold sensing revealed by genetic axonal labeling of transient receptor potential melastatin 8 neurons. J Neurosci. 2007;27(51):14147–57.
111. Bautista DM, Movahed P, Hinman A, Axelsson HE, Sterner O, Hogestatt ED, Julius D, Jordt SE, Zygmunt PM. Pungent products from garlic activate the sensory ion channel TRPA1. Proc Natl Acad Sci U S A. 2005;102(34):12248–52.
112. Story GM, Peier AM, Reeve AJ, Eid SR, Mosbacher J, Hricik TR, Earley TJ, Hergarden AC, Andersson DA, Hwang SW, McIntyre P, Jegla T, Bevan S, Patapoutian A. ANKTM1, a TRP-like channel expressed in nociceptive neurons, is activated by cold temperatures. Cell. 2003;112(6):819–29.
113. Bautista DM, Pellegrino M, Tsunozaki M. TRPA1: a gatekeeper for inflammation. Annu Rev Physiol. 2013;75:181–200.
114. Hermanstyne TO, Markowitz K, Fan L, Gold MS. Mechanotransducers in rat pulpal afferents. J Dent Res. 2008;87(9):834–8.
115. Kim YS, Jung HK, Kwon TK, Kim CS, Cho JH, Ahn DK, Bae YC. Expression of transient receptor potential ankyrin 1 in human dental pulp. J Endod. 2012;38(8):1087–92.
116. Dib-Hajj SD, Cummins TR, Black JA, Waxman SG. Sodium channels in normal and pathological pain. Annu Rev Neurosci. 2010;33:325–47.
117. Wood JN, Boorman JP, Okuse K, Baker MD. Voltage-gated sodium channels and pain pathways. J Neurobiol. 2004;61(1):55–71.
118. Akopian AN, Sivilotti L, Wood JN. A tetrodotoxin-resistant voltage-gated sodium channel expressed by sensory neurons. Nature. 1996;379(6562):257–62.
119. Dib-Hajj SD, Tyrrell L, Black JA, Waxman SG. NaN, a novel voltage-gated Na channel, is expressed preferentially in peripheral sensory neurons and down-regulated after axotomy. Proc Natl Acad Sci U S A. 1998;95(15):8963–8.
120. Abrahamsen B, Zhao J, Asante CO, Cendan CM, Marsh S, Martinez-Barbera JP, Nassar MA, Dickenson AH, Wood JN. The cell and molecular basis of mechanical, cold, and inflammatory pain. Science. 2008;321(5889):702–5.
121. Zimmermann K, Leffler A, Babes A, Cendan CM, Carr RW, Kobayashi J, Nau C, Wood JN, Reeh PW. Sensory neuron sodium channel Nav1.8 is essential for pain at low temperatures. Nature. 2007;447(7146):855–8.
122. Coggeshall RE, Tate S, Carlton SM. Differential expression of tetrodotoxin-resistant sodium channels Nav1.8 and Nav1.9 in normal and inflamed rats. Neurosci Lett. 2004;355(1–2):45–8.
123. Renton T, Yiangou Y, Plumpton C, Tate S, Bountra C, Anand P. Sodium channel Nav1.8 immunoreactivity in painful human dental pulp. BMC Oral Health. 2005;5:5. doi:1472-6831-5-5.
124. Suwanchai A, Theerapiboon U, Chattipakorn N, Chattipakorn SC. NaV 1.8, but not NaV 1.9, is upregulated in the inflamed dental pulp tissue of human primary teeth. Int Endod J. 2012;45(4):372–8.
125. Warren CA, Mok L, Gordon S, Fouad AF, Gold MS. Quantification of neural protein in extirpated tooth pulp. J Endod. 2008;34(1):7–10.
126. Dib-Hajj SD, Yang Y, Waxman SG. Genetics and molecular pathophysiology of Na(v)1.7-related pain syndromes. Adv Genet. 2008;63:85–110.
127. Cox JJ, Reimann F, Nicholas AK, Thornton G, Roberts E, Springell K, Karbani G, Jafri H, Mannan J, Raashid Y, Al-Gazali L, Hamamy H, Valente EM, Gorman S, Williams R, McHale DP, Wood JN, Gribble FM, Woods CG. An SCN9A channelopathy causes congenital inability to experience pain. Nature. 2006;444(7121):894–8.
128. Beneng K, Renton T, Yilmaz Z, Yiangou Y, Anand P. Sodium channel Nav1.7 immunoreactivity in painful human dental pulp and burning mouth syndrome. BMC Neurosci. 2010;11:71.
129. Luo S, Perry GM, Levinson SR, Henry MA. Nav1.7 expression is increased in painful human dental pulp. Mol Pain. 2008;4:16.
130. Sasano T, Shoji N, Kuriwada S, Sanjo D, Izumi H, Karita K. Absence of parasympathetic vasodilatation in cat dental pulp. J Dent Res. 1995;74(10):1665–70.
131. Uddman R, Grunditz T, Sundler F. Neuropeptide Y: occurrence and distribution in dental pulps. Acta Odontol Scand. 1984;42(6):361–5.
132. Wakisaka S, Ichikawa H, Akai M. Distribution and origins of peptide- and catecholamine-containing nerve fibres in the feline dental pulp and effects of cavity preparation on these nerve fibres. J Osaka Univ Dent Sch. 1986;26:17–28.
133. Kim S. Regulation of pulpal blood flow. J Dent Res. 1985;64 Spec No:590–6.
134. Pohto P, Antila R. Innervation of blood vessels in the dental pulp. Int Dent J. 1972;22(2):228–39.
135. Qian XB, Naftel JP. The effects of anti-nerve growth factor on retrograde labelling of superior cervical ganglion neurones projecting to the molar pulp in the rat. Arch Oral Biol. 1994;39(12):1041–7.
136. Brumovsky P, Villar MJ, Hökfelt T. Tyrosine hydroxylase is expressed in a subpopulation of small

dorsal root ganglion neurons in the adult mouse. Exp Neurol. 2006;200(1):153–65.
137. Fristad I, Heyeraas KJ, Kvinnsland IH. Neuropeptide Y expression in the trigeminal ganglion and mandibular division of the trigeminal nerve after inferior alveolar nerve axotomy in young rats. Exp Neurol. 1996;142(2):276–86.
138. Rodd HD, Loescher AR, Boissonade FM. Immunocytochemical and electron-microscopic features of tooth pulp innervation in hereditary sensory and autonomic neuropathy. Arch Oral Biol. 1998;43(6):445–54.
139. Elenkov IJ, Wilder RL, Chrousos GP, Vizi ES. The sympathetic nerve – an integrative interface between two supersystems: the brain and the immune system. Pharmacol Rev. 2000;52(4):595–638.
140. Haug SR, Brudvik P, Fristad I, Heyeraas KJ. Sympathectomy causes increased root resorption after orthodontic tooth movement in rats: immunohistochemical study. Cell Tissue Res. 2003;313(2):167–75.
141. Haug SR, Heyeraas KJ. Modulation of dental inflammation by the sympathetic nervous system. J Dent Res. 2006;85(6):488–95.
142. Avery JK, Cox CF, Corpron RE. The effects of combined nerve resection and cavity preparation and restoration on response dentine formation in rabbit incisors. Arch Oral Biol. 1974;19(7):539–48.
143. Dababneh RH, Khouri AT, Addy M. Dentine hypersensitivity – an enigma? A review of terminology, epidemiology, mechanisms, aetiology and management. Br Dent J. 1999;187(11):606–11.
144. Brannstrom M, Astrom A. The hydrodynamics of the dentin; its possible relationship to dentinal pain. Int Dent J. 1972;22(2):219–27.
145. Vongsavan N, Matthews B. The relationship between the discharge of intradental nerves and the rate of fluid flow through dentine in the cat. Arch Oral Biol. 2007;52(7):640–7.
146. Ichikawa H, Fukuda T, Terayama R, Yamaai T, Kuboki T, Sugimoto T. Immunohistochemical localization of gamma and beta subunits of epithelial Na + channel in the rat molar tooth pulp. Brain Res. 2005;1065(1–2):138–41.
147. Chung G, Jung SJ, Oh SB. Cellular and molecular mechanisms of dental nociception. J Dent Res. 2013;92(11):948–55.
148. Chung G, Oh SB. TRP channels in dental pain. Open Pain J. 2013;6:31–6.
149. Son AR, Yang YM, Hong JH, Lee SI, Shibukawa Y, Shin DM. Odontoblast TRP channels and thermo/mechanical transmission. J Dent Res. 2009;88(11):1014–9.
150. Chidchuangcha W, Vongsavan N, Matthews B. Sensory transduction mechanisms responsible for pain caused by cold stimulation of dentine in man. Arch Oral Biol. 2007;52(2):154–60.
151. Lin M, Luo ZY, Bai BF, Xu F, Lu TJ. Fluid mechanics in dentinal microtubules provides mechanistic insights into the difference between hot and cold dental pain. PLoS ONE. 2011;6(3):e18068.
152. El Karim IA, Linden GJ, Curtis TM, About I, McGahon MK, Irwin CR, Lundy FT. Human odontoblasts express functional thermosensitive TRP channels: implications for dentin sensitivity. Pain. 2011;152(10):2211–23.
153. Allard B, Couble ML, Magloire H, Bleicher F. Characterization and gene expression of high conductance calcium-activated potassium channels displaying mechanosensitivity in human odontoblasts. J Biol Chem. 2000;275(33):25556–61.
154. Magloire H, Lesage F, Couble ML, Lazdunski M, Bleicher F. Expression and localization of TREK-1 K + channels in human odontoblasts. J Dent Res. 2003;82(7):542–5.
155. Magloire H, Couble ML, Thivichon-Prince B, Maurin JC, Bleicher F. Odontoblast: a mechano-sensory cell. J Exp Zool B Mol Dev Evol. 2009;312B(5):416–24.
156. Thivichon-Prince B, Couble ML, Giamarchi A, Delmas P, Franco B, Romio L, Struys T, Lambrichts I, Ressnikoff D, Magloire H, Bleicher F. Primary cilia of odontoblasts: possible role in molar morphogenesis. J Dent Res. 2009;88(10):910–5.
157. Allard B, Magloire H, Couble ML, Maurin JC, Bleicher F. Voltage-gated sodium channels confer excitability to human odontoblasts: possible role in tooth pain transmission. J Biol Chem. 2006;281(39):29002–10.
158. Yang SY, Jeon SK, Kang JH, Yoo HI, Kim YS, Moon JS, Kim MS, Koh JT, Oh WM, Kim SH. Synaptic vesicle protein 2b is expressed temporospatially in (pre)odontoblasts in developing molars. Eur J Oral Sci. 2012;120(6):505–12.
159. Alavi AM, Dubyak GR, Burnstock G. Immunohistochemical evidence for ATP receptors in human dental pulp. J Dent Res. 2001;80(2):476–83.
160. Adachi K, Shimizu K, Hu JW, Suzuki I, Sakagami H, Koshikawa N, Sessle BJ, Shinoda M, Miyamoto M, Honda K, Iwata K. Purinergic receptors are involved in tooth-pulp evoked nocifensive behavior and brainstem neuronal activity. Mol Pain. 2010;6:59.
161. Lim JC, Mitchell CH. Inflammation, pain, and pressure – purinergic signaling in oral tissues. J Dent Res. 2012;91(12):1103–9.
162. Sessle BJ. Acute and chronic craniofacial pain: brainstem mechanisms of nociceptive transmission and neuroplasticity, and their clinical correlates. Crit Rev Oral Biol Med. 2000;11(1):57–91.
163. Tal M, Devor M. Anatomy and neurophysiology of orofacial pain. In: Sharav Y, Benouliel R, editors. Orofacial pain and headache. London: Blackwell; 2008. p. 19–44.
164. Arvidsson J, Gobel S. An HRP study of the central projections of primary trigeminal neurons which innervate tooth-pulps in the cat. Brain Res. 1981; 210(1–2):1–16.
165. Marfurt CF, Turner DF. The central projections of tooth pulp afferent neurons in the rat as determined by the transganglionic transport of horseradish peroxidase. J Comp Neurol. 1984;223(4):535–47.

166. Sugimoto T, Fujiyoshi Y, He YF, Xiao C, Ichikawa H. Trigeminal primary projection to the rat brain stem sensory trigeminal nuclear complex and surrounding structures revealed by anterograde transport of cholera toxin B subunit-conjugated and Bandeiraea simplicifolia isolectin horseradish peroxidase. Neurosci Res. 1997;28(4):361–71.
167. Kaneko M, Sunakawa M, Matsui Y, Suda H. Responsiveness of tooth pulp-driven neurons in thalamic ventral posteromedial and mediodorsal nuclei following experimental pulpitis and naloxone administration in rats. J Oral Biosci. 2005;47:135–48.
168. Kubo K, Shibukawa Y, Shintani M, Suzuki T, Ichinohe T, Kaneko Y. Cortical representation area of human dental pulp. J Dent Res. 2008;87(4):358–62.
169. Couve E, Osorio R, Schmachtenberg O. Mitochondrial autophagy and lipofuscin accumulation in aging odontoblasts. J Dent Res. 2012;92(9):765–72.
170. Fried K. Changes in innervation of dentin and pulp with age. Front Oral Physiol. 1987;6:63–84.
171. Swift ML, Byers MR. Effect of ageing on responses of nerve fibres to pulpal inflammation in rat molars analysed by quantitative immunocytochemistry. Arch Oral Biol. 1992;37(11):901–12.
172. Farac RV, Morgental RD, Lima RK, Tiberio D, dos Santos MT. Pulp sensibility test in elderly patients. Gerodontology. 2012;29(2):135–9.
173. Yang H, Bernanke JM, Naftel JP. Immunocytochemical evidence that most sensory neurons of the rat molar pulp express receptors for both glial cell line-derived neurotrophic factor and nerve growth factor. Arch Oral Biol. 2006;51(1):69–78.
174. Byers MR, Wheeler EF, Bothwell M. Altered expression of NGF and P75 NGF-receptor by fibroblasts of injured teeth precedes sensory nerve sprouting. Growth Factors. 1992;6(1):41–52.
175. Diogenes A, Akopian AN, Hargreaves KM. NGF up-regulates TRPA1: implications for orofacial pain. J Dent Res. 2007;86(6):550–5.
176. Schmidt Y, Unger JW, Bartke I, Reiter R. Effect of nerve growth factor on peptide neurons in dorsal root ganglia after taxol or cisplatin treatment and in diabetic (db/db) mice. Exp Neurol. 1995;132(1):16–23.
177. Sah DW, Ossipo MH, Porreca F. Neurotrophic factors as novel therapeutics for neuropathic pain. Nat Rev Drug Discov. 2003;2(6):460–72.
178. Matsuura S, Shimizu K, Shinoda M, Ohara K, Ogiso B, Honda K, Katagiri A, Sessle BJ, Urata K, Iwata K. Mechanisms underlying ectopic persistent tooth-pulp pain following pulpal inflammation. PLoS One. 2013;8(1):e52840.
179. Stephenson JL, Byers MR. GFAP immunoreactivity in trigeminal ganglion satellite cells after tooth injury in rats. Exp Neurol. 1995;131(1):11–22.
180. Tsuboi Y, Iwata K, Dostrovsky JO, Chiang CY, Sessle BJ, Hu JW. Modulation of astroglial glutamine synthetase activity affects nociceptive behaviour and central sensitization of medullary dorsal horn nociceptive neurons in a rat model of chronic pulpitis. Eur J Neurosci. 2011;34(2):292–302.
181. Gobel S, Binck JM. Degenerative changes in primary trigeminal axons and in neurons in nucleus caudalis following tooth pulp extirpations in the cat. Brain Res. 1977;132(2):347–54.
182. Hu JW, Dostrovsky JO, Lenz YE, Ball GJ, Sessle BJ. Tooth pulp deafferentation is associated with functional alterations in the properties of neurons in the trigeminal spinal tract nucleus. J Neurophysiol. 1986;56(6):1650–68.
183. Marbach JJ, Hulbrock J, Hohn C, Segal AG. Incidence of phantom tooth pain: an atypical facial neuralgia. Oral Surg Oral Med Oral Pathol. 1982;53(2):190–3.
184. Nixdorf DR, Moana-Filho EJ, Law AS, McGuire LA, Hodges JS, John MT. Frequency of nonodontogenic pain after endodontic therapy: a systematic review and meta-analysis. J Endod. 2010;36(9):1494–8.
185. Nixdorf DR, Moana-Filho EJ, Law AS, McGuire LA, Hodges JS, John MT. Frequency of persistent tooth pain after root canal therapy: a systematic review and meta-analysis. J Endod. 2010;36(2):224–30.
186. Polycarpou N, Ng YL, Canavan D, Moles DR, Gulabivala K. Prevalence of persistent pain after endodontic treatment and factors affecting its occurrence in cases with complete radiographic healing. Int Endod J. 2005;38(3):169–78.
187. Benoliel R, Zadik Y, Eliav E, Sharav Y. Peripheral painful traumatic trigeminal neuropathy: clinical features in 91 cases and proposal of novel diagnostic criteria. J Orofac Pain. 2012;26(1):49–58.
188. Nixdorf DR, Drangsholt MT, Ettlin DA, Gaul C, De Leeuw R, Svensson P, Zakrzewska JM, De Laat A, Ceusters W. Classifying orofacial pains: a new proposal of taxonomy based on ontology. J Oral Rehabil. 2012;39(3):161–9.
189. Haroutiunian S, Nikolajsen L, Finnerup NB, Jensen TS. The neuropathic component in persistent postsurgical pain: a systematic literature review. Pain. 2013;154(1):95–102.
190. List T, Leijon G, Svensson P. Somatosensory abnormalities in atypical odontalgia: a case-control study. Pain. 2008;139(2):333–41.
191. Oshima K, Ishii T, Ogura Y, Aoyama Y, Katsuumi I. Clinical investigation of patients who develop neuropathic tooth pain after endodontic procedures. J Endod. 2009;35(7):958–61.

Inflammatory Processes in the Dental Pulp

7

Paul R. Cooper and Anthony J. Smith

7.1 Introduction

Innate and adaptive immune mechanisms are prevalent in the dental pulp and are a key feature of its defense capacity to minimize the effects of injurious challenge. Thus, the pulp shows similarities to many of the other connective tissues of the body but perhaps differs due to its noncompliant environment where the rigid covering shell of hard mineralized tissue constrains the pulp tissue swelling. The architecture of the tissue provides further constraints clinically and limits attempts to remove and repair the causes of the injury. Bacterial infection of the dental pulp represents the most common injurious challenge to the tissue due to the effects of dental caries, clinical operative procedures, and trauma. As a consequence, a mixed microbial flora, particularly including gram-negative, anaerobic bacteria, is present in the diseased pulp [1].

Inflammatory processes are important in the host's immune response to injurious challenge and represent a broad array of cellular and molecular events. These processes aim to both recruit circulating immunocompetent cells from the vasculature to eliminate pathogens and necrotic tissue debris and stimulate responses by resident cells in the pulp to minimize tissue damage and initiate reparative and regenerative events. Innate immune responses will particularly trigger local cytokine production and promote an influx of phagocytic leukocytes as part of the proinflammatory response. The relatively noncompliant and non-self-cleansing environment of the pulp may often lead to infections becoming chronic, and adaptive immune responses can also come into play. The latter process leads to T- and B-cell recruitment and activation and adds further complexity to the inflammatory response. Although clearly these various responses are defensive in nature, both the combination of their complexity and the constraints imposed by the structure of the tooth can lead to exacerbation of tissue injury and compromise tooth vitality. There is also increasing evidence of the sequestration of a variety of bioactive molecules within the dentin matrix [2] and their release during carious matrix dissolution will further complicate the cellular signaling taking place in the pulp. Clearly, while it is possible to generalize about the various defense responses occurring in the diseased and infected pulp, each individual case will be unique in terms of the extent of disease activity and the consequent involvement and timing of the various defense responses. Thus, the clinical management of pulpal inflammation can be a significant challenge.

Although there is a good clinical appreciation of the impact of inflammation on disease progression and treatment outcomes (see Chap. 9), the correlations between the biological events of pulp inflammation and the clinical presentation

P.R. Cooper, PhD (✉) • A.J. Smith, PhD
Department of Oral Biology,
School of Dentistry, University of Birmingham,
St Chads Queensway, Birmingham B4 6NN, UK
e-mail: p.r.cooper@bham.ac.uk; a.j.smith@bham.ac.uk

M. Goldberg (ed.), *The Dental Pulp*,
DOI 10.1007/978-3-642-55160-4_7, © Springer-Verlag Berlin Heidelberg 2014

of disease progression are currently poorly understood. This represents a significant challenge to clinical diagnosis and management of pulpal disease [3, 4], particularly as regenerative approaches to therapeutically promote tooth vitality emerge [5–10]. The importance of understanding the cellular and molecular basis of pulpal inflammatory processes is now further emphasized with the recognition that there is considerable cross talk between inflammatory and regenerative events. Traditionally, tissue defense and repair/regeneration have been considered as distinct processes but clearly, these processes should now be considered as working in tandem.

7.2 The Pulpal Environment and Injury Responses

Dental caries represents the most prevalent infectious disease globally and affects the majority of the population. While not a life-threatening disease, caries has significant impact economically, nutritionally, and in terms of pain and quality of life. In health, the architecture of the tooth protects the pulp well from the infectious influences of the oral cavity. Carious infection of the tooth, however, soon exposes the pulp to bacteria and their products. Traumatic injury to the tooth, although physical in nature, provides indirect exposure of the pulp to these bacterial influences since the tooth is constantly bathed in bacteria-containing oral secretions. Increasing identification of tooth wear in the population is also a risk factor for pulpal infection with dentin exposure due to erosion, abrasion, and attrition opening diffusion pathways to the pulp.

The nature of the bacterial challenge will vary depending on disease progression and the extent of existing tooth tissue loss. At earlier stages, relatively small bacterial products may begin to diffuse within the tubules to the pulp, but with increasing disease progression, permeability of the tissues will increase and allow intact bacteria to migrate and colonize the deeper areas of the dentin and pulp (Fig. 7.1a–c). Thus, both the range of bacterial pathogenic challenges and their intensity may show considerable variation during the course of disease. As well as the challenge from intact bacteria, both cellular breakdown products and metabolites will likely contribute. Thus, cell membrane degradation products from gram-positive and gram-negative microbes, such as lipoteichoic acids, lipopolysaccharides, and DNA as well as other cell-derived products, may all contribute to the challenge posed by bacteria. Bacterial metabolites such as the weak organic acids produced during carbohydrate fermentation are well-established as major factors in tooth tissue degradation during caries, and these may also contribute to the insult caused by the bacteria. It is important, however, to recognize that there may also be an indirect bacterial challenge posed through the action of these bacterial metabolites on the dental tissues. Carious demineralization of dentin by bacterial acids will be accompanied by the dissolution of a significant proportion of the noncollagenous extracellular matrix of dentin. These dentin matrix components are now recognized to comprise a diverse range of molecules; a number of which include structural matrix molecules while others include cytokines, growth factors, and inflammatory mediators [2, 11]. Proteomic analysis of dentin already indicates the presence of up to nearly 300 distinct proteins [12, 13], and many of these display bioactive properties capable of signaling a multitude of cellular events in both tissue-resident cells and those recruited to the pulp as a part of the immune defense and wound healing processes. Mineralized tissues provide a unique environment in that expression of bioactive molecules by their formative cells frequently leads to their subsequent sequestration in the extracellular matrix in a fossilized state. This is especially true of dentin, which shows limited remodeling, unlike bone, and these sequestrated molecules may remain with their bioactivity in a protected state until the matrix is demineralized during injurious events, such as caries. While bacteria and their products may initiate defense responses classically associated with many of the other tissues of the body, superimposition of the effects of dentin matrix components released during carious demineralization may significantly

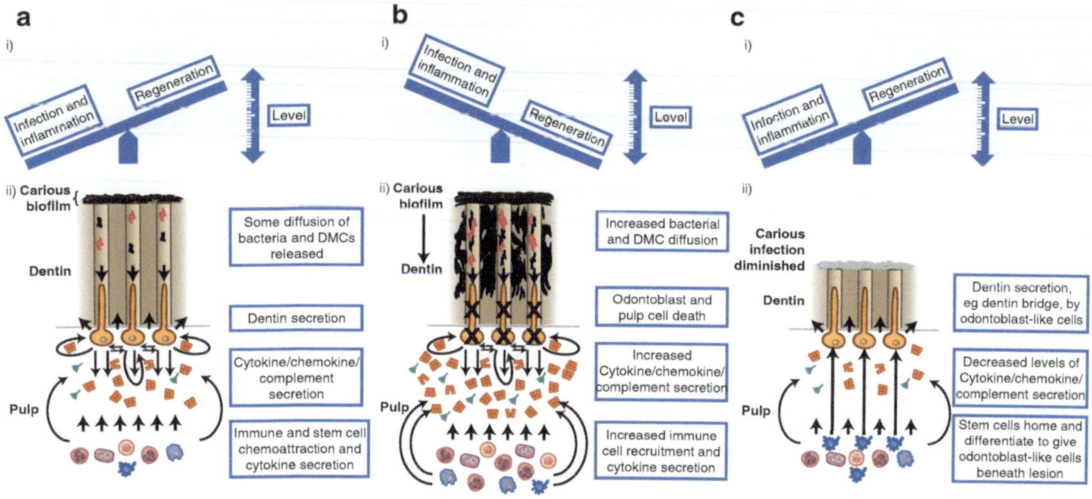

Fig. 7.1 (**a**) Early stage of carious disease with minimal hard tissue involvement. (*i*) Inflammation and infection are at relatively low levels which enable and promote tissue regenerative mechanisms, such as reactionary dentinogenesis. (*ii*) Bacteria and their products, as well as released dentin matrix components (*DMCs*), diffuse within the dentinal tubules where they are detected by odontoblasts, which can then elicit reactionary dentinogenic events and cytokine and complement secretion. Immune and potentially stem cells can be attracted to the site beneath the lesion at relatively low levels where they contribute further to proinflammatory mediator production. (**b**) Chronic and later stages of carious disease with increasing hard tissue involvement. (*i*) Relatively high levels of infection and inflammation lead to the impeding of tissue regenerative events. (*ii*) Increased amounts of bacteria and their products, as well as released DMCs, diffuse down the dentinal tubules where they signal odontoblast death. Relatively high levels of cytokines and immune system cells are present in the infected pulpal tissue. (**c**) Resolution of infection and modulation of inflammation, e.g., following clinical intervention. (*i*) Dental tissue regenerative events, such as reparative dentinogenesis, are enabled as infection and inflammation levels are decreased. (*ii*) Progenitor cells are recruited and differentiate to give a new population of odontoblast-like cells. Potential sources of progenitor cells include dental pulp stem cells (*DPSCs*). Low-level proinflammatory mediators, e.g., complement, cytokines, and reactive oxygen species (*ROS*), may promote signaling of these events. *Arrows* within pulp indicate signaling or secretory activity

modulate the pulpal and immune responses (Fig. 7.1a–c). Indeed, antibacterial activity displayed by some of these dentin matrix components [14] may modify the nature or intensity of the bacterial challenge to the pulp. Some signaling pathways are common to a number of cell types and their processes, which lead to more unpredictable effects of the combined challenges from bacterial and dentin matrix component exposure. For example, p38 mitogen-activated protein kinase (MAPK) signaling has been implicated in the control of odontoblast secretory activity during tertiary dentinogenesis [15]. Exposure to reactive oxygen species (ROS), which are generated during bacterial challenge, can also activate MAPK and NF-κB signaling pathways [16, 17]. These pathways can be initiated through a variety of cellular stresses, such as cytokine and bacterial LPS exposure as well as heat shock [18, 19].

Traditionally, wound generation and healing in the body's tissues follow a distinct chronological pattern with defense mechanisms initiated first, and once clearance of the injurious challenge has been largely achieved, healing processes are invoked (Fig. 7.1a–c). Such a pattern may be less distinct in the dentin-pulp where significant release of pro-regenerative factors at the time of tissue injury may lead to competing influences of defense and regeneration or healing occurring alongside one another. In such circumstances, the relative intensities of these competing influences may direct the outcomes of tissue events, although other factors may also affect outcomes. Tertiary dentinogenesis represents a repair response of the dentin-pulp, ultimately aimed at tissue regeneration if conditions are permissive. This repair process may be further subclassified into reactionary and reparative dentinogenesis depending on whether the formative cells are upregulated surviving postmitotic primary odontoblasts or a new generation of odontoblasts-like cells arising from differentiation of stem/progenitor cells due to local death of the primary odontoblasts (see Chap. 2). Clearly, the complexity of these two processes differs significantly, and in the context of competing tissue defense and healing influences, simple upregulation of secretory activity of an existing population of odontoblasts (reactionary dentinogenesis) may be more easily achieved. During reparative dentinogenesis, involvement of pulp-derived mesenchymal stem cells (MSCs) may influence defense events through their immunomodulatory properties [20–22]. Thus, it is important that inflammation and repair/regeneration are considered as overlapping and interrelated processes.

7.3 Environmental Sensing by Odontoblasts and Pulp Cells

The main role of odontoblasts has long been considered to be that of dentin matrix secretion, and morphologically, these cells are well adapted to this function. However, it is becoming increasingly apparent that odontoblasts have much broader roles in the defense of the tooth and environmental sensing (Fig. 7.1a–c). This is emphasized by the histological structure of the dentin-pulp where the intricate and elaborate permeation of dentin matrix by the odontoblast process and its lateral branches [23] ensures that the cell communicates intimately with its extracellular matrix. Thus, the odontoblasts are well positioned to detect invading bacteria and their products, as well as dentin matrix components released during carious demineralization, at an early stage of the disease process. While odontoblasts are likely to be the first cells of the pulp that come into contact with the bacterial pathogens and their components, other pulpal cells will also subsequently be exposed to these stimuli. Indeed, recent evidence implicates odontoblasts, pulpal fibroblasts, and endothelial cells in the detection of exposure to bacterial pathogens. Consequently, these cells should be regarded as a constitutive part of the pulp's defense response to bacterial pathogens [16].

7.4 Innate and Adaptive Immune Responses

Both innate and adaptive immune responses encompass a complex range of cellular and molecular events. While the earlier responses to bacteria and other injurious challenges generally reflect innate immunity, the transition to adaptive immunity is a gradual one as infections become chronic and will vary in the same way as disease progression in each individual patient varies. Thus, at later stages of disease progression, adaptive responses will likely be superimposed on innate responses. This situation provides significant challenges to the identification of suitable targets for diagnosis or therapeutic intervention. In describing the innate and adaptive immune responses of the pulp, it is difficult to categorize or assign the molecular and cellular changes observed in the tissue as being distinct to either of these responses. Instead, it is probably more helpful for the reader to consider these changes in a chronological order in relation to disease progression. In this way, it is perhaps easier to understand their involvement in the clinical presentation of pulpal inflammation (see Chap. 9).

7.4.1 Bacterial Pathogen Recognition

Pattern recognition receptors (PRRs) are a group of cell membrane- and endosome-bound receptors, which can recognize ligands (pathogen-associated molecular patterns or PAMPs) that are broadly shared by pathogens but which are distinct from host molecules [24]. The Toll-like receptors (TLRs) are a key family of PRRs, which play a central role within the innate immune system in the recognition of their ligands or PAMPs. These ligands predominantly include the surface components of bacteria, including lipopolysaccharides (LPS), lipoteichoic acids (LTS), flagellin, peptidoglycans, and lipoproteins as well as nucleic acid ligands from bacterial or viral pathogens. TLR-1 to TLR-6 and TLR-9 expression has been detected in odontoblasts and pulpal fibroblasts, and binding of these PRRs to their respective ligands initiates an acute inflammatory response, leading to activation of cells and release of proinflammatory mediators [16, 25–31]. These molecules include those associated predominantly with the vascular responses of the pulp including histamine, endothelin, serotonin, and neuropeptides and a broad array of cytokines and chemokines with potent cellular signaling properties (Fig. 7.1a–c). Other PRRs include the cytoplasmic NOD-like receptors (NLRs) and retinoic acid-inducible gene (RIG)-like receptors (RLRs) [32], although minimal information is currently available on their involvement in dentin-pulp-mediated inflammation.

7.4.2 Early Vascular Responses

An early feature of pulpal inflammation is changes to the vascular flow in the pulp with vasodilation and increases in blood flow. These changes are associated with increased fluid and plasma protein exudation and recruitment of leukocytes. Fluid exudation or edema during acute inflammation classically gives rise to swelling in soft tissues, although, as noted previously, such swelling is constrained in the pulp by the covering hard shell of mineralized dentin. Key molecular mediators of these vascular responses may include histamine, endothelin, neuropeptides, and serotonin. Both in vitro [33, 34] and in vivo [35, 36] studies indicate that histamine can produce vasodilation and reductions in blood flow in the pulp. Endothelin-1, a vasoconstrictor, and its receptors are constitutively expressed in odontoblasts and dental papilla of the developing teeth [37], and its application to pulp causes a decrease in blood flow [38]. A number of neuropeptides have been reported in pulp including substance P, calcitonin gene-related peptide, neurokinin A, neurokinin K, neuropeptide K, neuropeptide Y, somatostatin, and vasoactive intestinal peptide [39], and while these are largely associated with neural structures, some neuropeptides have been reported to be expressed in pulp fibroblasts [40, 41]. A number of these neuropeptides are vasodilators, while others are vasoconstrictors. Their involvement in neurogenic inflammation is complex [39], but these neuropeptides may provide novel therapeutic targets for the control of both pain and inflammation simultaneously [42]. Serotonin, a vasoconstrictor, can stimulate prostaglandin E_2 (PGE_2) [43] and prostacyclin (PGI_2) [44] production as well as increase blood flow in the pulp tissue [45]. PGE_2 and PGI_2 are among a broader family of prostaglandins, which have been implicated in pulpal inflammation [46–48]. Interestingly, histamine synergistically activates COX-2 (one of the cyclooxygenase enzymes involved in prostaglandin generation) expression and PGE_2 production in pulp fibroblasts through a TLR2-mediated process [49].

Several molecular cascade systems, based on plasma proteins, act in parallel to these cell-derived mediators to further initiate and propagate the acute inflammatory response. The complement system is activated by bacteria, with the outcome being the lysis of the bacterial membrane. Early reports have provided rather variable evidence for the contribution of complement activation in the pulp [50–54]; however, these data may reflect the difficulties in detecting the rather transient presence of these proteins during the inflammatory process. Many inflammatory mediators have relatively short half-lives, which in part explains why acute inflammation readily subsides in some tissues once the stimulus has been removed [55]. However, in the pulp, the problems of elimination of the infectious agents and their components ensure that the stimulus will often be ongoing. Complement will likely play a role in leukocyte recruitment (see next section) in the pulp as well as potentially progenitor cell recruitment for subsequent regenerative events [56] (Fig. 7.1a–c). The other molecular cascade systems closely associated with inflammation are the clotting and fibrinolytic systems. Dental pulp has long been recognized to show fibrinolytic activity [57] and the fibrinolytic system may contribute to early wound organization during pulp healing. In the inflamed pulp, gene transcript and protein levels of tissue-type plasminogen activator are increased significantly [58] and are upregulated in the presence of proinflammatory cytokines [58, 59]. As well as allowing plasminogen cleavage to plasmin for fibrin clot lysis, the proteolytic action of plasmin may breakdown C3 facilitating initiation of the complement cascade.

Even at the level of the vascular responses to injurious challenges, the complexity and interrelationships of inflammatory mediator involvement in the pulp are apparent, and our understanding of these events is currently limited.

7.4.3 Leukocyte Recruitment

Recruitment of leukocytes to sites of inflammation is an important aspect of pathogen elimination through phagocytosis and degranulation mechanisms. Increases in vascular permeability facilitate their migration through the endothelial lining, and such extravasation is carefully regulated by the action of molecules involved in their adhesion and transmigration. These molecules include integrins, selectins, endothelial adhesion molecules, and the cell adhesion molecules (intercellular adhesion molecules 1–5, ICAM 1–5; vascular cell adhesion molecule 1, VCAM-1; junctional adhesion molecules, JAMs; platelet endothelial cell adhesion molecule 1, PECAM-1; endothelial cell adhesion molecule, ECAM). While some of these adhesion molecules are constitutively expressed in odontoblasts and other pulp cells in health [60, 61] for the maintenance of tissue architecture, expression of others increases significantly during episodes of inflammation. For instance, weak reactivity for E- and P-selectins in the healthy pulp is strongly upregulated following injury [62, 63].

Recruitment of leukocytes and other cells to sites of inflammation involves attraction along gradients of chemotactic molecules (Fig. 7.1a–c). These chemotactic molecules are diverse in their origins, perhaps reflecting to some extent the lack of specificity of their influences on cell type. Bacterial components are chemotactic to neutrophils in the pulp [64], and components of the dentin matrix released during carious demineralization are chemotactic to both inflammatory cells [65–68] and resident pulp cells [69]. Recognition that the composition of the dentin matrix reflects the expression of a diverse range of molecules by odontoblasts in addition to the well-established structural extracellular matrix components [2, 11] explains why these matrix components have immunomodulatory effects. Certain interleukins are basally expressed by odontoblasts [70] leading to their sequestration within the dentin matrix, and a complex cocktail of pro- and anti-inflammatory molecules have been detected in the dentin matrix [71]. Several inflammatory mediators also show chemotactic properties including complement components C3a and C5a, the arachidonic acid metabolites, and the leukotrienes (especially leukotriene B_4 – LTB_4). Growth factors, cytokines, and chemokines also modulate chemotaxis. The chemotactic effects of some of these molecules influence migration of both pulp progenitor and inflammatory cells [56, 69] highlighting the interplay between inflammatory defense and regenerative events in this tissue following injury (Fig. 7.1a–c).

7.4.4 Cytokine and Chemokine Mediation of Inflammatory and Post-injury Events

PAMP recognition by TLRs on odontoblasts and pulpal fibroblasts results in activation of the nuclear factor kappa B (NF-κB) intracellular signaling pathway, which is central to regulation of the molecular inflammatory response in many cell types [16, 25–31]. A range of cytokines and chemokines are produced as a result of activation of NF-κB signaling, and these molecules regulate much of the immune and inflammatory response. These cytokines and chemokines are synthesized by a variety of immune and tissue structural cells in response to infectious and traumatic challenge and have potent cellular signaling properties. Binding of these molecules to specific cell surface receptors further modulates target-cell gene expression and molecular responses via second messenger signaling mechanisms [72, 73]. The actions of these cytokines and chemokines are often synergistic with stimulation of a cascade of release of other related molecules following their initial receptor binding [74]. The immunomodulatory actions of these cytokines and chemokines will impact on both innate and adaptive immune processes, including extravasation, leukocyte recruitment, cell activation and differentiation, and antibody production, as well as regenerative events associated with the wound healing response.

The archetypal proinflammatory regulatory cytokines include interleukin-1α and interleukin-1β (IL-1α,IL-β) and tumor necrosis factor-α (TNF-α), which have been demonstrated to play important roles in pulp's response to bacterial challenge [66, 67, 75–81]. Dentin matrix dissolution during caries may also stimulate expression of both TNF-α and IL-1β by macrophages [65] emphasizing that environmental sensing and defense mechanisms in the pulp may be broad ranging. The cascades of inflammatory mediator released after early PAMP-PRP interaction are well illustrated by the induction of the proinflammatory cytokine, IL-8 (which is central to neutrophil recruitment and activation), by stimulation with bacterial components or due to IL-1β and TNF-α exposure [70]. Upregulation of IL-8 has been reported in carious human pulp tissue [71, 78, 82], and the constitutive expression of this cytokine in odontoblasts [70] also highlights the complexity of the cellular interrelationships taking place in the defense of the pulp. Both gene and antibody array technologies have allowed demonstration of increased levels of several inflammatory cytokines and S100 transcripts and proteins in carious compared with healthy pulpal tissue [71, 78, 83]. These data are corroborated by other reports demonstrating increased interleukin levels in bacterially challenged pulpal tissue, including increases in IL-4 [84], IL-6 [85], and IL-10 [84]. The release of proinflammatory cytokines within the diseased pulp will have wide-ranging effects generally aimed at reinforcing control of the pathogenic challenge and subsequently resolution of inflammatory processes and stimulation of regenerative events. Among these proinflammatory effects will be the development of chemotactic gradients which promote the recruitment and activation of immune system cells [86, 87] to underpin the innate and adaptive immune responses (Fig. 7.1a–c). These chemotactic mechanisms will operate in tandem with those described previously for leukocyte recruitment to sites of inflammation.

7.4.5 Immune Cell Mediation of Innate and Adaptive Immune Responses

T- and B-lymphocytes, plasma cells, neutrophils, and macrophages are observed to infiltrate the pulp in increasing numbers as carious disease progresses [88, 89] (Fig. 7.1a–c). These cells constitute the effectors of the innate and adaptive immune responses; the latter of which will become increasingly superimposed on the former as caries extends deeply and more extensively into the dentin-pulp and the inflammation becomes more chronic in nature. As caries extends, the immune cell infiltrate in the pulp will also increase and will change from being more focal and localized to a much more extensive presentation. Such changes reflect the transition from more acute to chronic inflammation (Fig. 7.1a–c).

A prime role for the neutrophils and macrophages is that of phagocytosis, especially during the earlier acute phase of inflammation when bacterial pathogens are first encountered. Extravasation of natural killer (NK) cells to sites of inflammation in response to cytokines [90] allows their interaction with immature dendritic cells (DCs), which can lead to reciprocal activation and increased cytokine production by these cells [91]. NK cells likely contribute to further cytokine production during caries including that of IFN-γ [84], which can activate macrophages to stimulate phagocytosis as well as promoting T-cell responses [92]. Tissue-resident DCs are found in the pulp and following PAMP recognition; immature DCs will undergo maturation after which they will likely function in antigen presentation to naïve T cells. T cells have been shown to be present in healthy pulp [93] with CD8+ T cells predominating [88, 89, 94]. An immunosurveillance role is generally assumed for these cells. This contrasts with B cells, which appear to be largely absent from the healthy pulp [89, 93] as also are their plasma cell progeny [51]. However, with establishment of deeper infection within the dentin-pulp, the initial inflammatory cell infiltrate of neutrophils and monocytes intensifies with accumulation of helper T (T_H) cells, cytotoxic T (T_C) cells, regulatory T (T_{reg}) cells, B cells, and plasma cells as adaptive immune defenses develop [88]. While these immune cells are recruited to the tissue for defense purposes, their ability to achieve clearance of the infection is frequently insufficient, and tissue destruction will often result collaterally. Such tissue destruction may in part be a direct result of the immune cells' scavenging actions on bacterial pathogens during which release of degradative enzymes and molecules, such as matrix metalloproteinases (MMPs), and reactive oxygen species (ROS) can negatively impact on the host tissue extracellular environment. ROS, which include superoxide anions, hydrogen peroxide, and hydroxyl radicals, can exacerbate tissue injury due to their damaging effects on DNA, proteins, and lipids. Apoptosis can result from cell exposure to ROS through activation of mitogen-activated protein kinase (MAPK) and NF-κB signaling pathways [16, 17]. These pathways can be initiated through a variety of cellular stresses, such as cytokine and bacterial LPS exposure, and heat shock [18, 19] highlighting the many opportunities for their activation during infection and inflammation in the pulp. Triggering of these pathways will further stimulate immune cell activity and contribute to an increasing accumulation of inflammatory mediators. It is abundantly clear that the cellular responses initiated during the innate responses associated with acute inflammation and their increasing complexity as adaptive immune responses come into play during chronic inflammation are still poorly defined in the pulp as disease progresses. While many observational studies have reported the presence of various immune cells and their inflammatory mediators in carious pulpal tissue, few functional studies have provided a clear picture of the specific functional activities of these cells and their complex interrelationships. This potentially reflects the individual variation in disease progression in the pulp, which is controlled by diverse factors. Significant challenges still exist to our understanding of the innate and adaptive immune defenses in the pulp, and this constrains clinical approaches to management of pulpal infection and inflammation.

7.4.6 Anti-inflammatory Activities and Inflammation Resolution

In ideal circumstances, immune defense following injurious challenge to a tissue will lead to elimination of the infecting agent and ultimately provides a conducive environment within which wound healing can occur. Such circumstances are not easily achieved in the dentin-pulp with its noncompliant environment and the exposure of the tooth to the oral cavity with its complex microflora and abundant supply of nutrients. Nevertheless, mechanisms for the regulation of inflammation require both suppression and activation of responses. Identification of several anti-inflammatory and pro-resolving mediators has started to clarify how inflammation may be suppressed after it has achieved its purpose.

The lipoxins are metabolites of arachidonic acid, which negatively regulate the actions of the leukotrienes inhibiting neutrophil chemotaxis and adhesion as well as stimulating apoptotic cell phagocytosis by macrophages and other anti-inflammatory actions [95]. Although there have been no reports of lipoxins in pulp to date, they represent an interesting potential target for control of pulpal inflammation together with the other families of anti-inflammatory mediators described later in this chapter. Three distinct families of anti-inflammatory pro-resolving lipid mediators are now recognized: resolvins, protectins, and maresins [96]. These families target distinct cell populations by interaction with specific receptors and contribute to the overall resolution of inflammation. Resolvins suppress proinflammatory mediator production and regulate neutrophil movement to sites of inflammation. In a rodent model of pulpal infection and inflammation, resolving E1 (RvE1) application led to a decrease in inflammation at 24 and 72 h [97]. Protectins can block T-cell migration and secretion of TNF-α and IFN-γ and promote T-cell apoptosis as well as upregulating the chemokine receptor CCR5 on neutrophils to suppress chemokine signaling. The recently discovered maresins are produced by macrophages and inhibit proinflammatory mediator production by LTA4 hydrolase [98]. The actions of these various specialized pro-resolving anti-inflammatory mediators are only starting to be elucidated, and little information exists on their involvement in pulpal inflammation. Nevertheless, they represent exciting targets for the modulation of inflammatory activity, and the use of analogs may potentially provide novel therapeutic tools for clinical management of inflammation.

7.4.7 Inflammation-Regeneration Cross Talk in Dentin-Pulp After Injury

The specialized environment of the dentin-pulp can lead to competing pathogenic influences during disease in terms of concomitant release of proinflammatory and pro-reparative/regenerative factors as identified earlier in this chapter. Clearly, a balance in favor of tissue repair is the goal of clinical management of pulpal disease, but this may be unrealistic when relatively high levels of infection and inflammation persist. Current evidence suggests that reparative and regenerative processes ensue only after significant control or resolution of infection and inflammation has occurred [99–101] (Fig. 7.1a–c). Many of the potential cellular signaling mediators present in the post-injury tissue milieu demonstrate pleiotropic effects, which can show dose dependency and contribute to the balance of tissue outcomes. Thus, cytokines and growth factors, such as TNF-α and TGF-β as well as released dentin matrix components, can have detrimental effects on pulpal tissue and induce cell death if present at relatively high concentrations during the infectious and inflammatory processes [66, 68, 102, 103] in contrast to their beneficial effects at lower concentrations. While therapeutic targeting of inflammation may be attractive to change the balance of tissue events, infection control may prove more effective since the inflammatory challenge will continue as long as bacterial involvement persists. However, eradication of bacterial infection of the dental tissues, especially while minimizing host cell damage, has long been a challenge to operative dentistry and endodontics. It is therefore important to better understand the interplay between inflammation and repair/regeneration to guide decision making in the management of dental disease.

It has recently been emphasized that inflammation is an important prerequisite to enable repair and regeneration to subsequently ensue [104]. This may reflect both the need for defense processes to create a conducive environment for repair/regeneration and, also, the pleiotropic effects of some of the proinflammatory mediators, which may impact on repair/regenerative events. A number of proinflammatory mediators can promote degradative events in the oral and dental tissues during their defense responses to pathogenic challenge, for example, in bone resorption in periapical lesions [66, 105]. However, some cytokines, such as TNF-α, can also stimulate pro-regenerative/reparative signaling, including via p38 MAPK pathway activation, leading to odontoblast-like differentiation of dental pulp stem cells with increased dentin phosphoprotein (DPP) and dentin sialoprotein (DSP) expression and tertiary dentinogenesis [106]. The importance of this inflammation-regeneration interplay is further emphasized by the correlation of p38 MAPK signaling with initiation of tertiary dentinogenesis [15]. Such molecular switching is fundamental to the upregulation of odontoblast secretory activity and dentin deposition during the wound healing responses in the pulp. Other proinflammatory mediators, including IL-1β, may also contribute to the interplay of events in the pulp post-injury. IL-1β can stimulate mineralized bone matrix formation by osteoblasts while inhibiting proliferation and differentiation of bone marrow mesenchymal stem cells (BMMSCs) [107]. In the liver, IL-1β can induce the normally quiescent hepatocytes to proliferate, thereby contributing to regeneration of this organ [107]. In the context of the dentin-pulp, proinflammatory cytokines may stimulate wound healing in surviving odontoblasts at earlier stages of disease when the pathogenic challenge is relatively less intense. However, during chronic disease, these cytokines may suppress odontoblast-like cell differentiation from stem/progenitor cells until infection is controlled and a more conducive tissue environment for wound healing prevails (Fig. 7.1a–c).

Other inflammatory mediators may also influence post-injury events in the pulp. While immune cell-derived ROS can contribute to tissue damage, at relatively low levels, these molecules can promote stem/progenitor cell differentiation and mineralization [108]. Clearly, there is a complex interplay occurring between the various molecules mediating events in the pulp post-injury, and the relative concentrations of these molecules may be key to the cellular signaling outcomes. Thus, at relatively low concentrations, these molecules may stimulate regenerative/reparative events, including cellular recruitment, differentiation, and matrix secretion. At higher concentrations, however, such events may be impeded through signaling inhibition and tissue degradative processes (Fig. 7.1a–c). Elucidation of these inflammation-regeneration interplay relationships is complicated by the various origins of these mediator molecules. Contributory sources will be tissue resident and immune cells as well as dentin matrix sequestrated pools released during carious dissolution. The relative contributions from these various sources will fluctuate with both the rate of disease progression (and hence, carious matrix dissolution) and the extent of cellular signaling taking place in the tissue (Fig. 7.1a–c). Some of the cytokine receptors responsible for triggering of cellular responses are common to both the immune/inflammatory and stem cells, which may explain the pleiotropic effects of these cytokines. For example, C-X-C chemokine receptor 4 (CXCR4) is expressed on lymphocytes and granulocytes and is also involved in stem cell recruitment [109, 110]. Both stromal cell-derived factor-1 (SDF-1)/CXCL12 and its receptor are expressed in pulp and are upregulated during disease [111, 112]. The sharing of common cytokine receptors between immune and stem cells probably reflects evolutionary conservation of cellular signaling processes. Recruitment of both cell types is required post-injury, and the extent of injury or infection may be the prime determinant of the level of involvement of these different cell types [113]. Further regulatory control of tissue events in the inflammatory milieu may also be triggered through modulation of stem cell cytokine receptor expression. For instance, increased cytokine levels can modulate the surface expression of CXCR4 on stem cells [109]. Such modulation could result in the suppression of regenerative/reparative responses during active inflammation.

In addition, mesenchymal stem cells (MSCs) can display immunomodulatory effects in tissues further demonstrating the dynamic interplay between inflammation and regeneration [114]. Dampening of excessive inflammatory responses by MSCs through modulation of immune cells will likely involve a number of mechanisms. One such process has recently been identified in which MSCs have been demonstrated to inhibit the NOD-like receptor family, pyrin domain containing 3 (NLRP3) inflammasome in monocytes/macrophages by decreasing ROS generation [115]. Although the scope of the immunomodulatory properties of dental pulp stem cells is still being explored [20–22], their contribution to the inflammation-regeneration interplay in pulp is already evident. Notably, TLR binding to stem cells, including those from pulp, can activate the NF-κB proinflammatory signaling cascade resulting in suppression of their differentiation [25, 26]. In addition, LPS binding to TLR4 in stem cells from the apical papilla (SCAPs) induces IL-6, IL-8, and TNF-α production in a time-dependent manner, and this can be suppressed by treatment with the transcription factor nuclear factor I C (NFIC) [116].

The influences of cell death on the tissue environment should also be considered in the context of the interplay between inflammation and regeneration. Pulp-capping agents, such as calcium hydroxide and mineral trioxide aggregate (MTA), have long been used to stimulate reparative dentinogenesis following pulpal disease. Although the precise mechanisms of action of these agents remain controversial, it has been suggested that hydroxyl ion release from the material [117] leads to high pH conditions locally in the tissue resulting in cell necrosis [118, 119]. Chemical irritation of vital pulp tissue beneath the area of necrosis was proposed to stimulate reparative processes. Other possible mechanisms of action, including the local dissolution of growth factors and cytokines from the dentin matrix [120, 121], have also been proposed. It is now known that necrotic cells release low levels of proinflammatory mediators [122–125], and these may promote regenerative/reparative events if the levels of release do not become excessive. Increased cytokine release (IL-1α, IL-1β, IL-2, IL-6, and IL-8) from mineralizing cells has also been reported following exposure to MTA [126–128]. The release of low levels of cytokines from necrotic and mineralizing cells, in addition to dentin matrix dissolution of these molecules, may provide concentrations which favor regeneration/repair rather than promoting significant inflammation during milder disease conditions in the tissues. As the disease stimulus becomes more intense, the increasing levels of these cytokines released may then tip the balance towards more chronic inflammation.

It is clear from this that there are many complex molecular responses and interactions occurring in the diseased pulp. Identification of the involvement of these molecules is important both to our understanding of inflammatory and regenerative/reparative events, the identification of diagnostic markers, and the development of novel clinical therapeutic strategies to maintain pulp vitality. As new technologies become available, opportunities for more sensitive molecular profiling of diseased tissues arise. High-throughput transcriptional profiling and subsequent bioinformatic analyses represent one such technology. The use of this approach to investigate carious pulp tissue identified that inflammation was the predominant ontological tissue response upregulated; however, several other activated processes were also detected [129]. Notably, scrutiny of pro-regenerative/reparative responses indicated that the cytokine adrenomedullin (ADM) was also upregulated in carious pulp tissue. This molecule has wide-ranging effects, including immunomodulatory and antibacterial capabilities, and can stimulate hard tissue cell differentiation and mineralization [130–133]. We have now demonstrated similar effects for ADM in pulp [134]. Interestingly, ADM is a part of the neuropeptide family released during dental neuro-inflammatory events [135], and as discussed previously, these neuropeptides can exhibit anti-inflammatory actions [136, 137], and therefore ADM may contribute to the inflammation-regeneration interplay in diseased pulp. Exploitation of other new tissue-profiling technologies will undoubtedly further contribute to our understanding of the interplay of tissue events in the diseased pulp at the molecular level and identify further diagnostic and therapeutic molecular targets.

7.5 Future Directions

The environmental sensing and defense roles of odontoblasts and pulpal fibroblasts provide an exquisite system to detect pathogenic challenges which subsequently lead to release of a wide variety of regulatory cytokines and chemokines [138] that signal subsequent inflammatory, immune, and regenerative responses in pulp. The sequestration of many bioactive molecules and proinflammatory mediators within the dentin matrix in a fossilized state provides a further level of modulation once these molecules are released during carious dissolution of the dental tissues. It has become clear that a complex interplay can take place between all of these molecules and that inflammation and regeneration/repair are not distinct, but are intimately interlinked. The cross talk and balance between inflammation and regeneration/repair are dependent on both the presence and concentrations of the various signaling mediators. Thus, in a slowly progressing carious lesion tissue, conditions may be conducive to regenerative/reparative events, which become suppressed as the injurious challenge increases with advancing disease (Fig. 7.1a–c). While there is still considerable scope to better understand the involvement of many of the signaling mediators in both inflammation and regeneration/repair, there is also now a significant opportunity to therapeutically target some of the inflammatory mediators to dampen their effects. For instance, the use of antioxidants, such as N-acetyl cysteine (NAC), in conjunction with dental restorative materials may limit the activation of key proinflammatory signaling pathways, including NK-κB activation, with subsequent effects on cytokine release which may then favor regenerative events within the dentin-pulp [139]. In addition, the anti-inflammatory actions of TGF-β1 [30, 140] may complement its stimulatory effects in reparative dentinogenesis. Furthermore, targeted upregulation of the transcription factor NFIC, which is important in tooth root development, can suppress cytokine production by pulp stem cells [116]. These and many other therapeutic routes offer exciting possibilities to modulate the inflammatory responses in the diseased pulp and to tip the balance of tissue responses towards preservation of pulp vitality and tissue repair.

References

1. Siqueira JF. Pulpal infections including caries. In: Hargreaves KM, Goodis HE, Tay FR, editors. Seltzer & Bender's dental pulp. 2nd ed. Chicago: Quintessence Books; 2012. p. 205–39.
2. Smith AJ, Scheven BA, Takahashi Y, Ferracane J, Shelton RM, Cooper PR. Dentine as a bioactive extracellular matrix. Arch Oral Biol. 2012;57:109–21.
3. Petersson A, Axelsson S, Davidson T, Frisk F, Hakeberg M, Kvist T, et al. Radiological diagnosis of periapical bone tissue lesions in endodontics: a systematic review. Int Endod J. 2012;45:783–801.
4. Mejàre IA, Axelsson S, Davidson T, Frisk F, Hakeberg M, Kvist T, et al. Diagnosis of the condition of the dental pulp: a systematic review. Int Endod J. 2012; 45:597–613.
5. Trope M. Regenerative potential of dental pulp. J Endod. 2008;34(Suppl):S13–7.
6. Hargreaves KM, Diogenes A, Teixeira FB. Treatment options: biological basis of regenerative endodontic procedures. Pediatr Dent. 2013;35:129–40.
7. Law AS. Considerations for regeneration procedures. J Endod. 2013;39(Suppl):S44–56.
8. Mao JJ, Kim SG, Zhou J, Ye L, Cho S, Suzuki T, et al. Regenerative endodontics: barriers and strategies for clinical translation. Dent Clin North Am. 2012;56: 639–49.
9. Smith AJ, Smith JG, Shelton RM, Cooper PR. Harnessing the natural regenerative potential of the dental pulp. Dent Clin North Am. 2012;56: 589–601.
10. Kinaia BM, Chogle SM, Kinaia AM, Goodis HE. Regenerative therapy: a periodontal-endodontic perspective. Dent Clin North Am. 2012;56:537–47.
11. Goldberg M, Smith AJ. Cells and extracellular matrices of dentin and pulp: biological strategies for repair and tissue engineering. Crit Rev Oral Biol Med. 2004;15:4–12.
12. Park ES, Cho HS, Kwon TG, Jang SN, Lee SH, An CH, et al. Proteomics analysis of human dentin reveals distinct protein expression profiles. J Proteome Res. 2009;8:1338–46.
13. Jágr M, Eckhardt A, Pataridis S, Mikšík I. Comprehensive proteomic analysis of human dentin. Eur J Oral Sci. 2012;120:259–68.
14. Smith JS, Smith AJ, Shelton RM, Cooper PR. Antibacterial activity of dentine and pulp extracellular matrix extracts. Int Endod J. 2012;45:749–55.
15. Simon S, Smith AJ, Berdal A, Lumley PJ, Cooper PR. The MAPK pathway is involved in odontoblast stimulation via p38 phosphorylation. J Endod. 2010;3: 256–9.
16. Veerayutthwilai O, Byers MR, Pham TT, Darveau RP, Dale BA. Differential regulation of immune responses by odontoblasts. Oral Microbiol Immunol. 2007;22: 5–13.
17. Fiers W, Beyaert R, Declercq W, Vandenabeele P. More than one way to die: apoptosis, necrosis and reactive oxygen damage. Oncogene. 1999;18:7719–30.
18. Guha M, Mackman N. LPS induction of gene expression in human monocytes. Cell Signal. 2001;13:85–94.
19. Hagemann C, Blank JL. The ups and downs of MEK kinase interactions. Cell Signal. 2001;13:863–75.
20. Wada N, Menicanin D, Shi S, Bartold PM, Gronthos S. Immunomodulatory properties of human periodontal ligament stem cells. J Cell Physiol. 2009;219:667–76.
21. Leprince JG, Zeitlin BD, Tolar M, Peters OA. Interactions between immune system and mesenchymal stem cells in dental pulp and periapical tissues. Int Endod J. 2012;45:689–701.
22. Li Z, Jiang CM, An S, Cheng Q, Huang YF, Wang YT, et al. Immunomodulatory properties of dental tissue-derived mesenchymal stem cells. Oral Dis. 2014;20: 25–34.
23. Lu Y, Xie Y, Zhang S, Dusevich V, Bonewald LF, Feng JQ. DMP1-targeted Cre expression in odontoblasts and osteocytes. J Dent Res. 2007;86:320–5.
24. Medzhitov R. Toll-like receptors and innate immunity. Nat Rev Immunol. 2001;1:135–45.
25. Pevsner-Fischer M, Morad V, Cohen-Sfady M, Rousso-Noori L, Zanin-Zhorov A, Cohen S, et al. Toll-like receptors and their ligands control mesenchymal stem cell functions. Blood. 2007;109: 1422–32.

26. Chang J, Zhang C, Tani-Ishii N, Shi S, Wang CY. NF-kappaB activation in human dental pulp stem cells by TNF and LPS. J Dent Res. 2005;84:994–8.
27. Zampetaki A, Xiao Q, Zeng L, Hu Y, Xu Q. TLR4 expression in mouse embryonic stem cells and in stem cell-derived vascular cells is regulated by epigenetic modifications. Biochem Biophys Res Commun. 2006;347:89–99.
28. Farges JC, Keller JF, Carrouel F, Durand SH, Romeas A, Bleicher F, et al. Odontoblasts in the dental pulp immune response. J Exper Zool B Mol Dev Evol. 2009;312B:425–36.
29. Hirao K, Yumoto H, Takahashi K, Mukai K, Nakanishi T, Matsuo T. Roles of TLR2, TLR4, NOD2, and NOD1 in pulp fibroblasts. J Dent Res. 2009;88:762–7.
30. Horst OV, Tompkins KA, Coats SR, Braham PH, Darveau RP, Dale BA. TGF-β1 inhibits TLR-mediated odontoblast responses to oral bacteria. J Dent Res. 2009;88:333–8.
31. Botero TM, Son SJ, Vodopyanov D, Hasegawa M, Shelburne CE, Nor JE. MAPK signaling is required for LPS-induced VEGF in pulp stem cells. J Dent Res. 2010;89:264–9.
32. Creagh EM, O'Neill LA. TLRs, NLRs and RLRs: a trinity of pathogen sensors that co-operate in innate immunity. Trends Immunol. 2006;27:352–7.
33. Yu CY, Boyd NM, Cringle SJ, Su EN, Alder VA, Yu DY. Agonist-induced vasoactive responses in isolated perfused porcine dental pulpal arterioles. Arch Oral Biol. 2002;47:99–107.
34. Yu CY, Boyd NM, Cringle SJ, Su EN, Alder VA, Yu DY. An in vivo and in vitro comparison of the effects of vasoactive mediators on pulpal blood vessels in rat incisors. Arch Oral Biol. 2002;47:723–32.
35. Edwall L, Olgart L, Haegerstam G. Influence of vasodilator substances on pulpal blood flow in the cat. Acta Odontol Scand. 1973;31:289–96.
36. Kim S, Dorscher-Kim J. Hemodynamic regulation of the dental pulp in a low compliance environment. J Endod. 1989;15:404–8.
37. Neuhaus SJ, Byers MR. Endothelin receptors and endothelin-1 in developing rat teeth. Arch Oral Biol. 2007;52:655–62.
38. Yu CY, Boyd NM, Cringle SJ, Su EN, Yu DY. Vasoactive response of isolated pulpal arterioles to endothelin-1. J Endod. 2004;30:149–53.
39. Caviedes-Bucheli J, Munoz HR, Azuero-Holguin MM, Ulate E. Neuropeptides in dental pulp. J Endod. 2008;4:773–88.
40. Killough SA, Lundy FT, Irwin CR. Substance P expression by human dental pulp fibroblasts: a potential role in neurogenic inflammation. J Endod. 2009;35:73–7.
41. Killough SA, Lundy FT, Irwin CR. Dental pulp fibroblasts express neuropeptide Y Y1 receptor but not neuropeptide Y. Int Endod J. 2010;43:835–42.
42. Bowler KE, Worsley MA, Broad L, Sher E, Benschop R, Johnson K, et al. Evidence for anti-inflammatory and putative analgesic effects of a monoclonal antibody to calcitonin gene-related peptide. Neuroscience. 2013;228:271–82.
43. Hirafuji M, Terashima K, Satoh S, Ogura Y. Stimulation of prostaglandin E_2 biosynthesis in rat dental pulp explants in vitro by 5-hysroxytryptamine. Arch Oral Biol. 1982;27:961–4.
44. Hirafuji M, Ogura Y. 5-hydroxytryptamine stimulates the release of prostacyclin but not thromboxane A_2 from isolated rat dental pulp. Eur J Pharmacol. 1987;136:433–6.
45. Liu M, Kim S, Park DS, Markowitz K, Bilotto G, Dorscher-Kim J. Comparison of the effects of intra-arterial and locally applied vasoactive agents on pulpal blood flow in dog canine teeth determined by laser Doppler velocimetry. Arch Oral Biol. 1990;35:405–10.
46. Okiji T, Morita I, Kobayashi C, Sunada I, Murota S. Arachidonic-acid metabolism in normal and experimentally-inflamed rat dental pulp. Arch Oral Biol. 1987;32:723–7.
47. Miyauchi M, Takata T, Ito H, Ogawa I, Kobayashi J, Nikai H, et al. Immunohistochemical demonstration of prostaglandins E2, F2 alpha, and 6-keto-prostaglandin F1 alpha in rat dental pulp with experimentally induced inflammation. J Endod. 1996;22:600–2.
48. Okiji T, Morita I, Sunada I, Murota S. Involvement of arachidonic acid metabolites in increases in vascular permeability in experimental dental pulpal inflammation in the rat. Arch Oral Biol. 1989;34:523–8.
49. Park C, Lee SY, Kim HJ, Park K, Kim JS, Lee SJ. Synergy of TLR2 and H1R on Cox-2 Activation in Pulpal Cells. J Dent Res. 2010;89:180–5.
50. Speer ML, Madonia JV, Heuer MA. Quantitative evaluation of the immunocompetence of the dental pulp. J Endod. 1977;3:418–23.
51. Pulver WH, Taubman MA, Smith DJ. Immune components in normal and inflamed human dental pulp. Arch Oral Biol. 1977;22:103–11.
52. Peković DD, Fillery ED. Identification of bacteria in immunopathologic mechanisms of human dental pulp. Oral Surg Oral Med Oral Pathol. 1984;57:652–61.
53. Okamura K, Maeda M, Nishikawa T, Tsutsui M. Dentinal response against carious invasion: localization of antibodies in odontoblastic body and process. J Dent Res. 1980;59:1368–73.
54. Pekovic DD, Adamkiewicz VW, Shapiro A, Gornitsky M. Identification of bacteria in association with immune components in human carious dentin. J Oral Pathol. 1987;16:223–33.
55. Cotran RS, Kumar V, Collins T, editors. Robbins pathologic basis of disease. 6th ed. Philadelphia: WB Saunders Company; 1998.
56. Chmilewsky F, Jeanneau C, Laurent P, Kirschfink M, About I. Pulp progenitor cell recruitment is selectively guided by a C5a gradient. J Dent Res. 2013;92:532–9.
57. Southam JC, Moody GH. The fibrinolytic activity of human and rat dental pulps. Arch Oral Biol. 1975;20:783–6.
58. Huang FM, Tsai CH, Chen YJ, Liu CM, Chou MY, Chang YC. Upregulation of tissue-type plasminogen

activator in inflamed human dental pulps. Int Endod J. 2005;38:328–33.
59. Huang FM, Tsai CH, Chen YJ, Chou MY, Chang YC. Examination of the signal transduction pathways leading to upregulation of tissue type plasminogen activator by interleukin-1alpha in human pulp cells. J Endod. 2006;32:30–3.
60. Lucchini M, Couble ML, Romeas A, Staquet MJ, Bleicher F, Magloire H, et al. Alpha v beta 3 integrin expression in human odontoblasts and co-localization with osteoadherin. J Dent Res. 2004;83:552–6.
61. Staquet MJ, Couble ML, Roméas A, Connolly M, Magloire H, Hynes RO, et al. Expression and localisation of alpha integrins in human odontoblasts. Cell Tissue Res. 2006;323:457–63.
62. Bagis B, Atilla P, Cakar N, Hasanreisoglu U. Immunohistochemical evaluation of endothelial cell adhesion molecules in human dental pulp: effects of tooth preparation and adhesive application. Arch Oral Biol. 2007;52:705–11.
63. Bagis B, Atilla P, Cakar N, Hasanreisoglu U. An immunohistochemical evaluation of cell adhesion molecules in human dental pulp after tooth preparation and application of temporary luting cements. Oral Surg Oral Med Oral Pathol Oral Radiol Endod. 2009;107:137–44.
64. Warfvinge J, Dahlén G, Bergenholtz G. Dental pulp response to bacterial cell wall material. J Dent Res. 1985;64:1046–50.
65. Tani-Ishii N, Wang CY, Stashenko P. Immunolocalization of bone-resorptive cytokines in rat pulp and periapical lesions following surgical pulp exposure. Oral Microbiol Immunol. 1995;10:213–9.
66. Lara VS, Figueiredo F, da Silva TA, Cunha FQ. Dentin-induced in vivo inflammatory response and in vitro activation of murine macrophages. J Dent Res. 2003;82:460–5.
67. Silva TA, Lara VS, Silva JS, Garlet GP, Butler WT, Cunha FQ. Dentin sialoprotein and phosphoprotein induce neutrophil recruitment: a mechanism dependent on IL-1beta, TNF-beta, and CXC chemokines. Calcif Tissue Int. 2004;74:532–41.
68. Silva TA, Lara VS, Silva JS, Oliveira SH, Butler WT, Cunha FQ. Macrophages and mast cells control the neutrophil migration induced by dentin proteins. J Dent Res. 2005;84:79–83.
69. Smith JG, Smith AJ, Shelton RM, Cooper PR. Recruitment of dental pulp cells by dentine and pulp extracellular matrix components. Exp Cell Res. 2012;318:2397–406.
70. Levin LG, Rudd A, Bletsa A, Reisner H. Expression of IL-8 by cells of the odontoblast layer in vitro. Eur J Oral Sci. 1999;107:131–7.
71. Cooper PR, Takahashi Y, Graham LW, Simon S, Imazato S, Smith AJ. Inflammation-regeneration interplay in the dentine-pulp complex. J Dent. 2010;38:687–97.
72. Kupper TS, Horowitz M, Birchall N, Mizutani H, Coleman D, McGuire J, et al. Hematopoietic, lymphopoietic, and pro-inflammatory cytokines produced by human and murine keratinocytes. Ann N Y Acad Sci. 1998;548:262–70.
73. Akira S, Hirano T, Taga T, Kishimoto T. Biology of multifunctional cytokines: IL 6 and related molecules (IL 1 and TNF). FASEB J. 1990;4:2860–7.
74. Taub DD, Oppenheim JJ. Chemokines, inflammation and the immune system. Ther Immunol. 1994;1:229–46.
75. Hosoya S, Matsushima K, Ohbayashi E, Yamazaki M, Shibata Y, Abiko Y. Stimulation of interleukin-1beta-independent interleukin-6 production in human dental pulp cells by lipopolysaccharide. Biochem Mol Med. 1996;59:138–43.
76. Matsuo T, Ebisu S, Nakanishi T, Yonemura K, Harada Y, Okada H. Interleukin-1 alpha and interleukin-1 beta periapical exudates of infected root canals: correlations with the clinical findings of the involved teeth. J Endod. 1994;20:432–5.
77. Pezelj-Ribaric S, Anic I, Brekalo I, Miletic I, Hasan M, Simunovic-Soskic M. Detection of tumor necrosis factor alpha in normal and inflamed human dental pulps. Arch Med Res. 2002;33:482–4.
78. McLachlan JL, Sloan AJ, Smith AJ, Landini G, Cooper PR. S100 and cytokine expression in caries. Infect Immun. 2004;72:4102–8.
79. de Brito LC, Teles FR, Teles RP, Totola AH, Vieira LQ, Sobrinho AP. T-lymphocyte and cytokine expression in human inflammatory periapical lesions. J Endod. 2012;38:481–5.
80. Dinarello CA. Interleukin-1. Rev Infect Dis. 1984;6:51–95.
81. Smith KA, Lachman LB, Oppenheim JJ, Favata MF. The functional relationship of the interleukins. J Exp Med. 1980;151:1551–6.
82. Guo X, Niu Z, Xiao M, Yue L, Lu H. Detection of interleukin-8 in exudates from normal and inflamed human dental pulp tissues. Chin J Dent Res. 2000; 3:63–6.
83. Graham LW, Smith AJ, Sloan AJ, Cooper PR. Cytokine release from human dentine. J Dent Res 2007;86(Spec Iss B):abstract number 0222 (BSDR). www.dentalresearch.org.
84. Hahn CL, Best AM, Tew JG. Cytokine induced by Streptococcus mutans and pulpal pathogenesis. Infect Immun. 2000;68:6785–9.
85. Barkhordar RA, Hayashi C, Hussain MZ. Detection of interleukin-6 in human pulp and periapical lesions. Endod Dent Traumatol. 1999;15:26–7.
86. Brennan EP, Tang XH, Stewart-Akers AM, Gudas LJ, Badylak SF. Chemoattractant activity of degradation products of fetal and adult skin extracellular matrix for keratinocyte progenitor cells. J Tissue Eng Regen Med. 2008;2:491–8.
87. Reing JE, Zhang L, Myers-Irvin J, Cordero KE, Freytes DO, Heber-Katz E, Bedelbaeva K, McIntosh D, Dewilde A, Braunhut SJ, Badylak SF. Degradation products of extracellular matrix affect cell migration and proliferation. Tissue Eng Part A. 2009;15:605–14.
88. Izumi T, Kobayashi I, Okamura K, Sakai H. Immunohistochemical study on the immunocompetent cells of the pulp in human non-carious and carious teeth. Arch Oral Biol. 1995;40: 609–14.

89. Hahn CL, Falkler Jr WA, Siegel MA. A study of T and B cells in pulpal pathosis. J Endod. 1989;15:20–6.
90. Maghazachi AA. Compartmentalization of human natural killer cells. Mol Immunol. 2005;42:523–9.
91. Kikuchi T, Hahn CL, Tanaka S, Barbour SE, Schenkein HA, Tew JG. Dendritic cells stimulated with Actinobacillus actinomycetemcomitans elicit rapid gamma interferon responses by natural killer cells. Infect Immun. 2004;72:5089–96.
92. Trinchieri G. Interleukin-12: a proinflammatory cytokine with immunoregulatory functions that bridge innate resistance and antigen-specific adaptive immunity. Annu Rev Immunol. 1995;13:251–76.
93. Jontell M, Gunraj MN, Bergenholtz G. Immunocompetent cells in the normal dental pulp. J Dent Res. 1987;66:1149–53.
94. Sakurai K, Okiji T, Suda H. Co-increase of nerve fibers and HLA-DR- and/or factor-XIIIa-expressing dendritic cells in dentinal caries-affected regions of the human dental pulp: an immunohistochemical study. J Dent Res. 1999;78:1596–608.
95. Maderna P, Godson C. Lipoxins: resolutionary road. Br J Pharmacol. 2009;158:947–59.
96. Serhan CN. Systems approach to inflammation resolution: identification of novel anti-inflammatory and pro-resolving mediators. J Thromb Haemost. 2009;7 suppl 1:44–8.
97. Dondoni L, Scarparo RK, Kantarci A, Van Dyke TE, Figueiredo JA, Batista EL Jr. Effect of the pro-resolution lipid mediator Resolvin E1 (RvE1) on pulp tissues exposed to the oral environment. Int Endod J. 2013 Dec 2. doi: 10.1111/iej.12224. [Epub ahead of print].
98. Dalli J, Zhu M, Vlasenko NA, Deng B, Haeggström JZ, Petasis NA, Serhan CN. The novel 13S, 14S-epoxy-maresin is converted by human macrophages to maresin 1 (MaR1), inhibits leukotriene A4 hydrolase (LTA4H), and shifts macrophage phenotype. FASEB J. 2013;27:2573–83.
99. Bergenholtz G. Inflammatory response of the dental pulp to bacterial irritation. J Endod. 1981;7:100–4.
100. Rutherford RB, Gu K. Treatment of inflamed ferret dental pulps with recombinant bone morphogenetic protein-7. Eur J Oral Sci. 2000;108:202–6.
101. Baumgardner KR, Sulfaro MA. The anti-inflammatory effects of human recombinant copper-zinc superoxide dismutase on pulp inflammation. J Endod. 2001;27:190–5.
102. Smith AJ, Patel M, Graham L, Sloan AJ, Cooper PR. Dentine regeneration: key roles for stem cells and molecular signalling. Oral Biosci Med. 2005;2:127–32.
103. He WX, Niu ZY, Zhao SL, Smith AJ. Smad protein mediated transforming growth factor beta1 induction of apoptosis in the MDPC-23 odontoblast-like cell line. Arch Oral Biol. 2005;50:929–36.
104. Goldberg M, Farges JC, Lacerda-Pinheiro S, Six N, Jegat N, Decup F, et al. Inflammatory and immunological aspects of dental pulp repair. Pharmacol Res. 2008;58:137–47.
105. Kjeldsen M, Holmstrup P, Bendtzen K. Marginal periodontitis and cytokines: a review of the literature. J Periodontol. 1993;64:1013–22.
106. Paula-Silva FW, Ghosh A, Silva LA, Kapila YL. TNF-alpha promotes an odontoblastic phenotype in dental pulp cells. J Dent Res. 2009;88:339–44.
107. Lange J, Sapozhnikova A, Lu C, Hu D, Li X, Miclau 3rd T, Marcucio RS. Action of IL-1beta during fracture healing. J Orthop Res. 2010;28:778–84.
108. Lee DH, Lim BS, Lee YK, Yang HC. Effects of hydrogen peroxide (H2O2) on alkaline phosphatase activity and matrix mineralization of odontoblast and osteoblast cell lines. Cell Biol Toxicol. 2006;22:39–46.
109. Murdoch C. CXCR4: chemokine receptor extraordinaire. Immunol Rev. 2000;177:175–84.
110. Miller RJ, Banisadr G, Bhattacharyya BJ. CXCR4 signaling in the regulation of stem cell migration and development. J Neuroimmunol. 2008;198:31–8.
111. Jiang HW, Ling JQ, Gong QM. The expression of stromal cell-derived factor 1 (SDF-1) in inflamed human dental pulp. J Endod. 2008;34:1351–4.
112. Jiang L, Zhu YQ, Du R, Gu YX, Xia L, Qin F, et al. The expression and role of stromal cell-derived factor-1alpha-CXCR4 axis in human dental pulp. J Endod. 2008;34:939–44.
113. About I, Mitsiadis TA. Molecular aspects of tooth pathogenesis and repair: in vivo and in vitro models. Adv Dent Res. 2001;15:59–62.
114. Bernardo ME, Fibbe WE. Mesenchymal stromal cells: sensors and switchers of inflammation. Cell Stem Cell. 2013;13:392–402.
115. Oh JY, Ko JH, Lee HJ, Yu JM, Choi H, Kim MK, et al. Mesenchymal stem/stromal cells inhibit the NLRP3 inflammasome by decreasing mitochondrial reactive oxygen species. Stem Cells. 2014;32:1553–63.
116. Zhang J, Zhang Y, Lv H, Yu Q, Zhou Z, Zhu Q, et al. Human stem cells from the apical papilla response to bacterial lipopolysaccharide exposure and anti-inflammatory effects of nuclear factor I C. J Endod. 2013;39:1416–22.
117. Kardos TB, Hunter AR, Hanlin SM, Kirk EE. Odontoblast differentiation: a response to environmental calcium? Endod Dent Traumatol. 1998;14:105–11.
118. Schröder U, Granath LE. Early reaction of intact human teeth to calcium hydroxide following experimental pulpotomy and its significance to the development of hard tissue barrier. Odontol Revy. 1971;22:379–95.
119. Stanley H. Calcium hydroxide and vital pulp therapy. In: Hargreaves KM, Goodis HE, editors. Seltzer and Bender's dental pulp. 1st ed. Chicago: Quintessence Books; 2002. p. 309–24.
120. Graham L, Cooper PR, Cassidy N, Nor JE, Sloan AJ, Smith AJ. The effect of calcium hydroxide on solubilisation of bio-active dentine matrix. Biomaterials. 2006;27:2865–73.
121. Tomson PL, Grover LM, Lumley PJ, Sloan AJ, Smith AJ, Cooper PR. Dissolution of bio-active den-

tine matrix components by mineral trioxide aggregate. J Dent. 2007;35:636–42.
122. Brentano F, Schorr O, Gay RE, Gay S, Kyburz D. RNA released from necrotic synovial fluid cells activates rheumatoid arthritis synovial fibroblasts via Toll-like receptor 3. Arthritis Rheum. 2005;52:2656–65.
123. Luheshi NM, McColl BW, Brough D. Nuclear retention of IL-1alpha by necrotic cells: a mechanism to dampen sterile inflammation. Eur J Immunol. 2009;39:2973–80.
124. Acosta-Pérez G, Maximina Bertha Moreno-Altamirano M, Rodríguez-Luna G, Javier Sánchez-Garcia F. Differential dependence of the ingestion of necrotic cells and TNF-alpha/ IL-1beta production by murine macrophages on lipid rafts. Scand J Immunol. 2008;68:423–9.
125. Magalhães-Santos IF, Andrade SG. Participation of cytokines in the necrotic-inflammatory lesions in the heart and skeletal muscles of Calomys callosus infected with Trypanosoma cruzi. Mem Inst Oswaldo Cruz. 2005;100:555–61.
126. Huang TH, Yang CC, Ding SJ, Yeng M, Kao CT, Chou MY. Inflammatory cytokines reaction elicited by root-end filling materials. J Biomed Mater Res B Appl Biomater. 2005;73:123–8.
127. Mitchell PJ, Pitt Ford TR, Torabinejad M, McDonald F. Osteoblast biocompatibility of mineral trioxide aggregate. Biomaterials. 1999;20:167–73.
128. Koh ET, McDonald F, Pitt Ford TR, Torabinejad M. Cellular response to mineral trioxide aggregate. J Endod. 1998;24:543–7.
129. McLachlan JL, Smith AJ, Bujalska IJ, Cooper PR. Gene expression profiling of pulpal tissue reveals the molecular complexity of dental caries. Biochim Biophys Acta. 2005;1741:271–81.
130. Zudaire E, Portal-Núñez S, Cuttitta F. The central role of adrenomedullin in host defense. J Leukoc Biol. 2006;80:237–44.
131. Montuenga LM, Martínez A, Miller MJ, Unsworth EJ, Cuttitta F. Expression of adrenomedullin and its receptor during embryogenesis suggests autocrine or paracrine modes of action. Endocrinology. 1997;138:440–51.
132. Ishii M, Koike C, Igarashi A, Yamanaka K, Pan H, Higashi Y, et al. Molecular markers distinguish bone marrow mesenchymal stem cells from fibroblasts. Biochem Biophys Res Commun. 2005;332:297–303.
133. Cornish J, Callon KE, Coy DH, Jiang NY, Xiao L, Cooper GJ, et al. Adrenomedullin is a potent stimulator of osteoblastic activity in vitro and in vivo. Am J Physiol. 1997;273(6 Pt 1):E1113–20.
134. Musson DS, McLachlan JL, Sloan AJ, Smith AJ, Cooper PR. Adrenomedullin is expressed during rodent dental tissue development and promotes cell growth and mineralisation. Biol Cell. 2010;102:145–57.
135. Delgado M, Ganea D. Anti-inflammatory neuropeptides: a new class of endogenous immunoregulatory agents. Brain Behav Immun. 2008;22:1146–51.
136. Fristad I, Bletsa A, Byers M. Inflammatory nerve responses in the dental pulp. Endod Topics. 2010;17:12–41.
137. Haug SR, Heyeraas KJ. Modulation of dental inflammation by the sympathetic nervous system. J Dent Res. 2006;85:488–95.
138. Staquet MJ, Durand SH, Colomb E, Roméas A, Vincent C, Bleicher F, et al. Different roles of odontoblasts and fibroblasts in immunity. J Dent Res. 2008;87:256–61.
139. Yamada M, Kojima N, Paranjpe A, Att W, Aita H, Jewett A, et al. N-acetyl cysteine (NAC)-assisted detoxification of PMMA resin. J Dent Res. 2008;87:372–7.
140. Liu H, Kemeny DM, Heng BC, Ouyang HW, Melendez AJ, Cao T. The immunogenicity and immunomodulatory function of osteogenic cells differentiated from mesenchymal stem cells. J Immunol. 2006;176:2864–71.

8 Pulp Aging: Fibrosis and Calcospherites

Michel Goldberg

8.1 Pulp Anatomy During the Aging Process

Dentinogenesis is a constant process as long as a tooth is alive. During the early stages, the daily dentin formation reported is about 10 μm/day, whereas at later stage of crown formation, 4 μm is formed daily. The rate of dentinogenesis decreases slowly, but the phenomenon is still occurring even in the elderly.

Primary dentin formation takes place at early stages, between the beginning of dentinogenesis, when pre-polarized odontoblasts are facing a basement membrane and when the erupted teeth come into contact with their antagonists, when it ends. The initial formation of mantle dentin leads to an atubular dentin, containing high amount of proteoglycans. In the root, two superficial layers have been reported: the granular Tome's layer followed by the hyaline Hopewell-Smith layer. They are characterized by the presence of bent tubules, with minute diameters, large globules, and interglobular spaces.

Secondary dentin formation follows circadian rhythms, characterized by repetitive von Ebner lines separating 4 μm thick dentin layers. At each 4–5 von Ebner lines, at about 20 μm intervals, a more accentuated Owen's line is found, which is less mineralized and displays a higher content of organic matrix. There is no clear-cut explanation for this other periodicity. Because it is centripetal, dentin is formed to the detriment of the close space occupied by the pulp. As long as a tooth is alive, dentin will be produced at a decreasing speed and gradually reduce the space occupied by the dental pulp. Pathologic dentinogenesis, such as *dentinogenesis imperfecta* or *dentin dysplasia*, contributes to reduce dramatically the width of the pulp, especially in the root part, where apparent pulp closure may occur, whereas some remnants are maintained in the coronal pulp.

Tertiary dentin is formed in reaction to a carious lesion or in case of rapid abrasion. The release of cytotoxic molecules of the monomer of restorative resins may also induce the formation of reparative dentin, another name given to tertiary dentin. Facing pulp exposure due to the rapid progress of the carious lesion, reparative dentin formation may protect the pulp from invading bacteria.

Depending on the species, it has been reported that the dentin layer located in the upper part of the pulp chamber increases in thickness more rapidly than the floor of the pulp chamber. This is the case for rat's molar, but the reverse is also seen in the human situation [1]. Dentin deposition along the lateral walls contributes also to reduce the pulp chamber volume, but at a slower rate

M. Goldberg, DDS, PhD
Department of Oral Biology, Institut National de la Santé et de la Recherche Médicale,
Université Paris Descartes,
45 Rue des Saints Pères, Paris 75006, France
e-mail: mgoldod@gmail.com,
michel.goldberg@parisdescartes.fr

M. Goldberg (ed.), *The Dental Pulp*,
DOI 10.1007/978-3-642-55160-4_8, © Springer-Verlag Berlin Heidelberg 2014

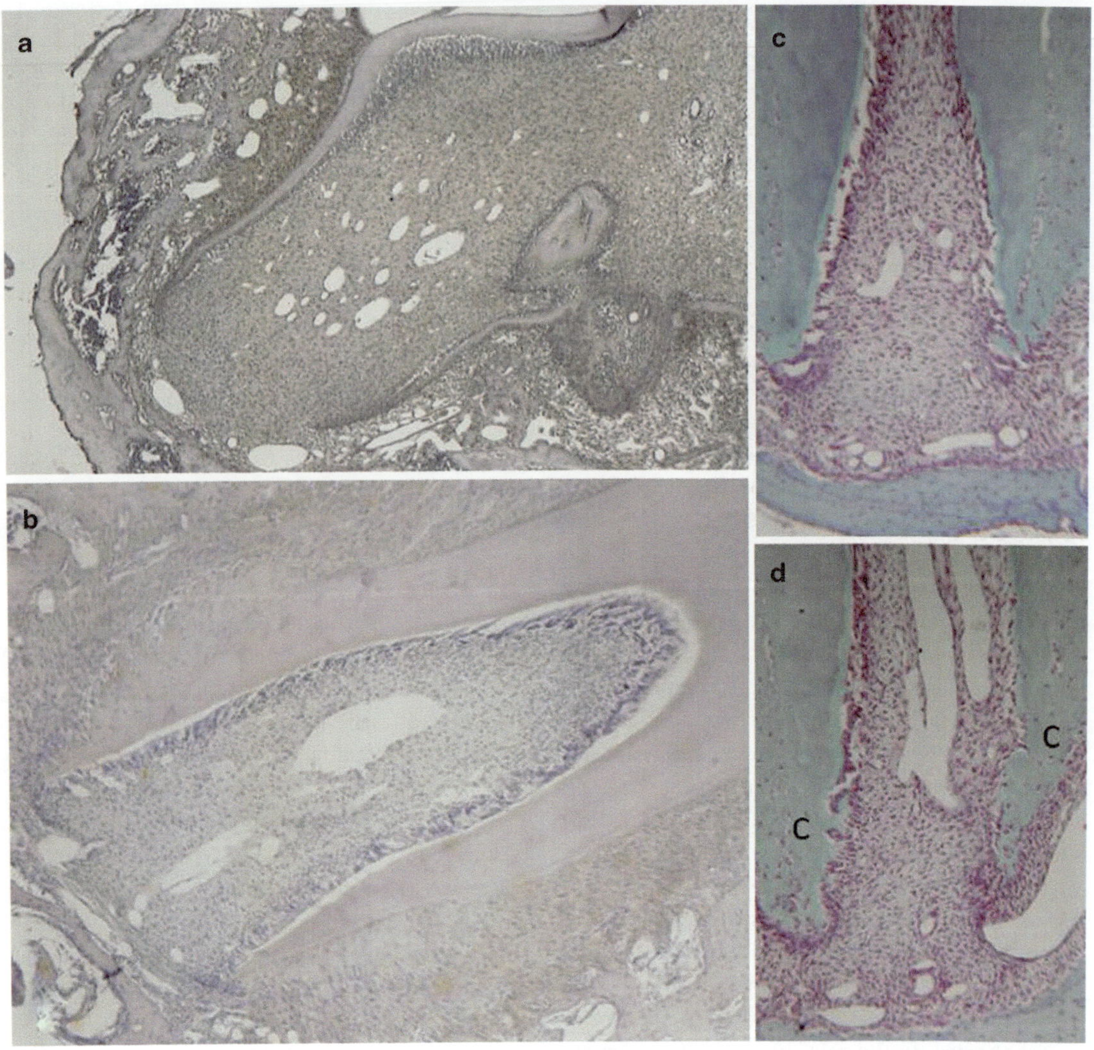

Fig. 8.1 During the formation of the root, the apical part is widely open at the onset of root formation (**a**) and gradually restrained in the three successive stages (**b–d**). Cementum (*C*) deposition contributes to apexification

(Figs. 8.1a–d, 8.2, and 8.3). However, it should be noted that pulp horn stays non-mineralized. Other investigations concluded with the lack of difference between the increased formations of dentin in the horn region compared with the floor. The mesiodistal diameter decreased with age, faster between patients age 20 and 40 and slower afterward [2].

In the root, tubular orthodentin formation is seen in the labial and lingual parts, whereas fibrodentin is added in the mesiodistal surfaces of the root canal lumen. This age-related variation gave the root canal lumen an oval profile. The dentin heterogeneity leads to a smaller pulp volume in the elderly (Fig. 8.4).

During aging, the dental pulp is enriched gradually by fibrous bundles of collagen. Noncollagenous proteins located in the so-called ground substance are somehow restricted, whereas the pulp is enriched in lipidic inclusions and pulp stones. The pulp response time is increased in older people, whereas pain intensity decreases.

With age *odontometric changes* have been noted with respect to pulp cell density, pulp

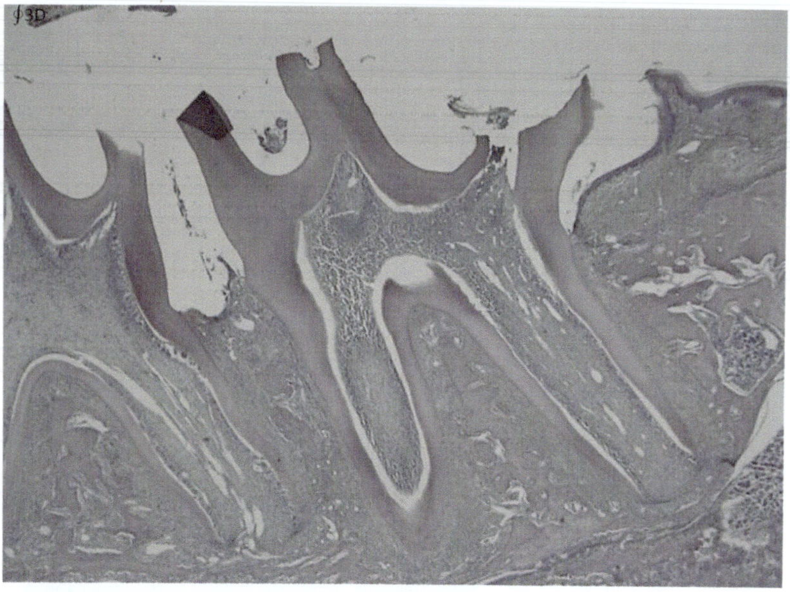

Fig. 8.2 At date 21 the full length of the root is achieved, but there is still a gradual reduction in width of the pulp, together with an increased dentin formation in the root

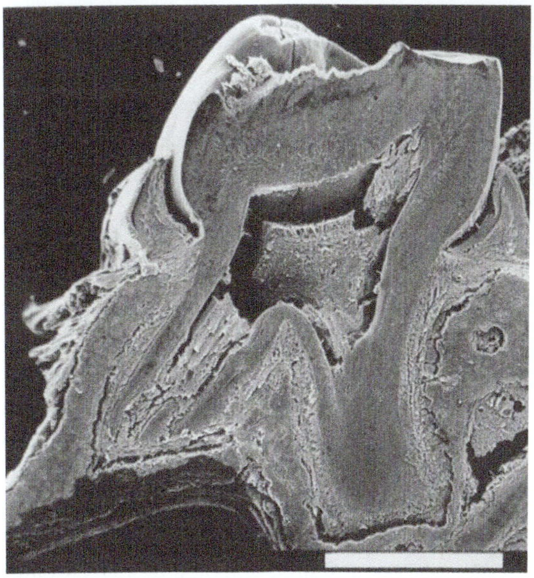

Fig. 8.3 Scanning electron microscopy (SEM). Longitudinal section of a mandibular molar during the formation of roots. Bar, 1,000 μm

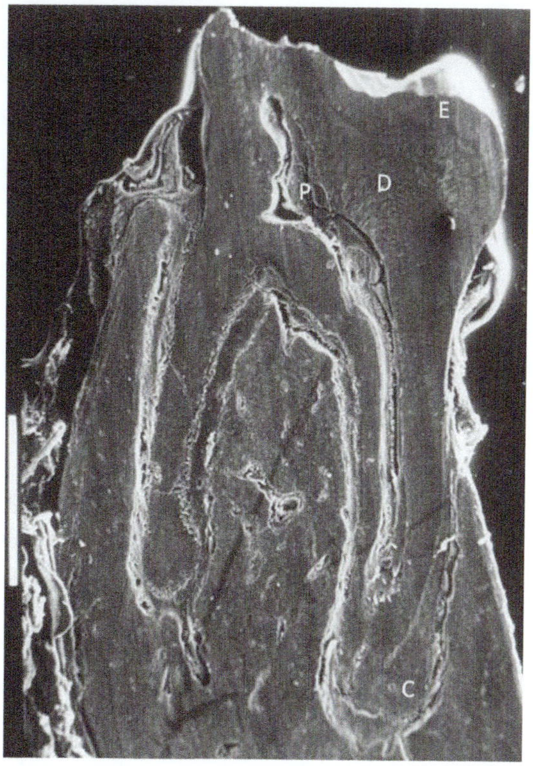

Fig. 8.4 SEM. Longitudinal section of a mandibular molar. The roots are totally formed. The dentin (*D*) thickness is maximal, and the pulp (*P*) volume is reduced. Cementum (*C*) extends and is observed in the apical part of the roots

area, and dentinal thickness. Cell density of odontoblasts, sub-odontoblasts, and pulp fibroblasts is decreasing. There is also a decrease of age-related changes in the root, which is more pronounced than in the crown. Cell density in

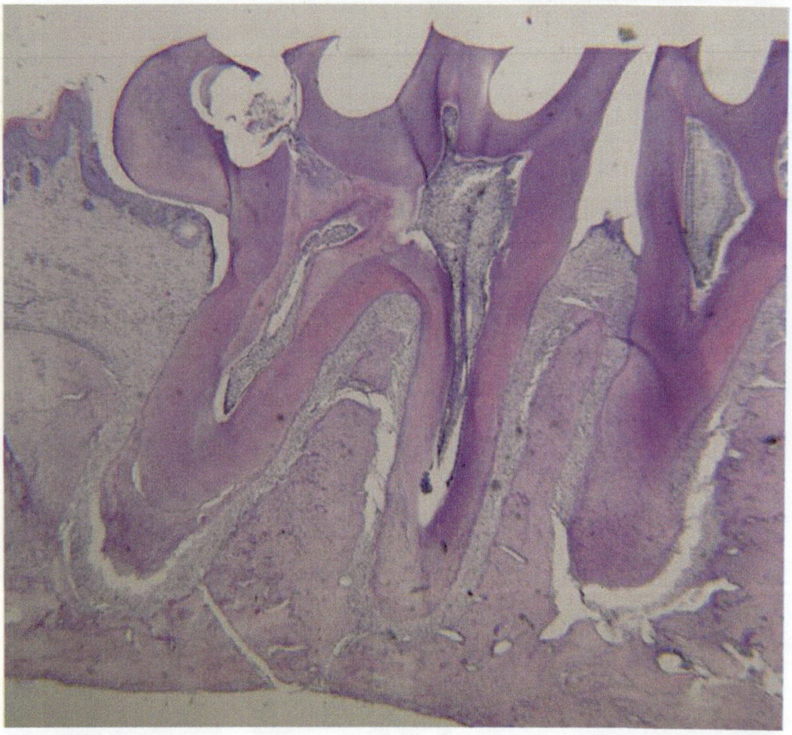

Fig. 8.5 Three-month-old rat. The apical part of the root is closed

the crown was greater than in the root (Fig. 8.5). Dentinal deposition is greater in the root compared to the crown [3].

8.2 Age Determination

Gustafson's method for age determination from the teeth is based on 6 age-associated parameters evaluated on ground sections [4]. The transparency of radicular dentin and secondary dentin deposition constitute the two major criteria. Cementum apposition, periodontal and root recession, and attrition should also be taken into account.

The age-dependent nonenzymatic changes of D- and L-forms of aspartic acid constitute a reliable and accurate method, tooth dentin being considered as one of the best target tissues. Analysis of osteocalcin and elastin also provides accurate results [5]. A strong correlation was found between the population ages and translucency of dentin. Staining abraded sections with 1 % methylene blue stains the "normal" tubular dentin, whereas sclerotic dentin remains unstained. This is probably due to the precipitation of non-apatitic calcium phosphate within the lumen of the tubules. Intratubular mineralization remains unstained by the dye. The cementum took a dark blue color.

Aging can be distinguished from senescence, defined as an essential irreversible arrest of cell division. Senescent cells are metabolically active but no longer capable of dividing. Replicative senescence is an irreversible loss of division capacity of human cells in vitro after a reproducible number of population doublings. Senescence is the only one possible outcome of a DNA damage response, the other possibilities leading to DNA repair or apoptosis [6]. Senescent cells remain alive and differ from apoptotic cells that are enriched in transglutaminase. They display rigidified thicker cytoskeletal proteins and plasma membrane. After the apoptotic disintegration, apoptotic bodies are engulfed by tubulovesicular endocytic vesicles and degraded inside macrophage lysosomal structures. It differs also from cell cycle arrest.

Aging is characterized by the reduction in length of chromosomes [7]. Telomeres form the end of human chromosomes and senescence-associated distension of satellites (SADS). They shorten with each round of cell division and 50–100 bp are eliminated at each cell division. This mechanism limits the proliferation to a finite number of cell divisions. Telomere extends up to a certain length, and then cells stop dividing. The cells enter in senescence. Telomere shortening limits stem cell function, regeneration, and organ maintenance during aging. It is due to an "end replication problem" of DNA polymerase. In addition, processing of telomeres during the cell cycle and reactive oxygen species may contribute to telomere shortening. The telomerase RNA serves as template for telomere sequence synthesis. The telomerase reverse transcriptase is the catalytic subunit of the enzyme. Active during embryogenesis, telomerase is suppressed postnatally. The gradual loss of telomeres is a regulator for cell life span. Telomerase activity counteracts the gradual loss of telomeres by de novo synthesis of telomere repeats. Telomerase declines by age [8].

Generation of induced pluripotent stem (iPS) cells was obtained by introducing genes encoding pluripotent transcription factors into fibroblasts. Reprogramming of somatic cells of old patients is associated with full re-elongation of telomeres to size comparable with embryonic cells [9].

Beta-galactosidase activity seems to be associated with senescence and with lysosomal dysfunction. As marker, beta-GAL is questionable [10]. However, it is recognized that the senescence β-galactosidase staining was higher in senescent pulp cells than in young cells. They contain greater expression of autophagic proteins (microtubule-associated protein light chain 3 and Beclin 1) than young cells [11].

This is also established for human dental pulp stem cells that STRO-1, nestin, CXCR4, Sox2, nucleostemin, CD90, and CD166, which are considered as well-identified markers, play a role in pulp cell senescence. *Ink4a/Arf* expression is a robust biomarker of aging. The gene locus encodes two tumor suppressor molecules, $p16^{INK4a}$ and ARF, which are principal mediators of cellular senescence [12]. Generation of $p53^{+/m}$ mice and other p53 mutants suggests that this gene has a role in regulating organismal aging [13].

Using microarray and RT-PCR, age-related changes in the expression and composition of third molar pulps of young subjects (age group 18–20 years) were compared to older patients (age group 57–60 years).

In young dental pulp, growth factors such as bone morphogenetic protein, TGF, growth factor differentiation family, platelet-derived growth factor α, vascular endothelial growth factor α, and FGF family were highly expressed, suggesting that they regulate the genes controlling cell proliferation and cell differentiation, and were responsible for signaling of many key events in tooth morphogenesis and differentiation.

In the aging pulp, vascular, lymphatic, and nerve supplies decline. Fibroblasts decrease in size and number. A reduction of 15.6 % was scored for crown odontoblasts and 40.6 % in root odontoblasts. The secretory activity was decreased, suggesting that the reparative capacity was compromised. Furthermore, age-related changes include a higher number of collagen cross-linkages, more collagen fibers, lipid infiltration, and calcifications. In the mature tooth, recruitment of dental pulp stem cells allowing their differentiation toward odontoblast cells for dentin bridge formation occurs. The deposition of secondary dentin increases, and the blood, lymphatic, and nerve supply undergo arteriosclerotic changes. This evidences age-related degenerative changes, together with a progressive mineralization. In older dental pulp there was an upregulation of proapoptotic genes like AIFM1, MOAP1, PDCD5, and PDCD7, confirming a possible correlation between apoptosis and the reduction of dental pulp volume. However, the expression in older dental pulp of growth factors like CTGF, FGF1, FGF5, and TGFB1, and genes implicated in the synthesis of collagenous proteins, confirms that, even reduced, the reparative processes were continuous during the entire tooth life [14].

Senescent fibroblasts are flat and display heterogeneous cell shapes. With respect to intracellular proteins, connexin 43 mRNA was abundantly expressed in young adults (about tenfold higher in young adult) and decreased in aged human dental pulp [15]. Vimentin filaments, parallel with the long axis of the cells, are overproduced by senescent fibroblasts [16]. Expression of Cbfa-1 mRNA, VEGF, and HS27 mRNAs was higher in the adult first molar compared with the young animals [17].

With respect to ECM proteins, osteocalcin expression is reduced in the dental pulp of aged human. Osteocalcin mRNA was decreased in aged human dental pulp [18, 19]. According to some reports, apparently no difference was detectable by immunohistochemistry for type I collagen, osteonectin, and BSP in relationship to the degree of maturation. In contrast, collagen concentration increased as the pulp matured. The ratio of type III to type I similarly increased from 13 % at early stage to 32 % at late stage. The major cross-link dihydroxylsinonorleucine (DHLNL) decreased with age. Hydroxylsinonorleucine (HLNL) and lysinonorleucine (LNL) appeared in insignificant amounts [20]. Collagenase and collagenolytic neutral peptidases showed significantly high activity [21]. Mehrazarin et al. [22] reported a reduced expression of Bmi-1, OC, DSPP, BSP, and DMP-1 compared with replicative senescence, whereas $p16^{INK4A}$ level was increased.

NaF effects are dose dependent. NaF produces large DNA fragments. Higher concentrations reduce the number of viable senescent cells compared with young cells. This suggests that cells become resistant to cytotoxicity of NaF with in vitro aging [23].

8.3 Aging of Mesenchymal Stem Cells (MSCs)

Mesenchymal stem cells include both blood and connective tissue cells. Postembryonic, non-hematopoietic bone marrow-derived cells are efficient to undergo multipotent differentiation into osteoblasts, adipocytes, myoblasts, and early progenitors of neural cells as well. They can be isolated from many tissues including the bone marrow, pericytes, and dental tissues. MSCs divide with a donor-dependent average doubling time of 12–24 h dependent. A significant decrease in the growth rate of MSC is observed for aged donors. In some cultures obtained from old mice and measured by 3H-thymidine uptake, the proliferation was more than three times (and even tenfold) what was observed in culture from young animals.

The age-related changes may be due to intrinsic factors or induced by the somatic environment. Dlx3 and Dlx5 are regulators of odontoblastic differentiation. Overexpression of these homeobox domains stimulates osteoblast differentiation while inhibiting adipogenic differentiation of human dental pulp stem cells by suppressing adipogenic marker genes such as C/EBPα, PPARγ, and aP2 levels in preadipocytes [24].

The pulp response to cavity preparation in aged rat molars was evaluated by immunohistochemistry. Heat shock protein (HSP)-25 and nestin were found in odontoblasts, whereas class II MHC-positive cells were densely distributed at the periphery of the pulp, along the pulp-dentin border. They subsequently disappear by 12 h after the preparation of the cavity [25]. Replicative senescence and stress-induced premature senescence (SIPS) intervene in the expression of dentin sialophosphoprotein and dentin matrix-1 and osteogenic markers such as bone morphogenetic protein-2 and protein-7, runt-related transcription factor-2, osteopontin, alkaline phosphatase activity, and mineralized nodule formation [26].

8.4 Apoptosis

A number of gene effects alter apoptotic receptor levels, extrinsic apoptotic pathways through caspases, cytokine effects on apoptotic events, Ca2 +-induced dead signaling, cell cycle checkpoints, and potential effects of surviving factors. Apoptotic potential is decreased in older compared with younger animals.

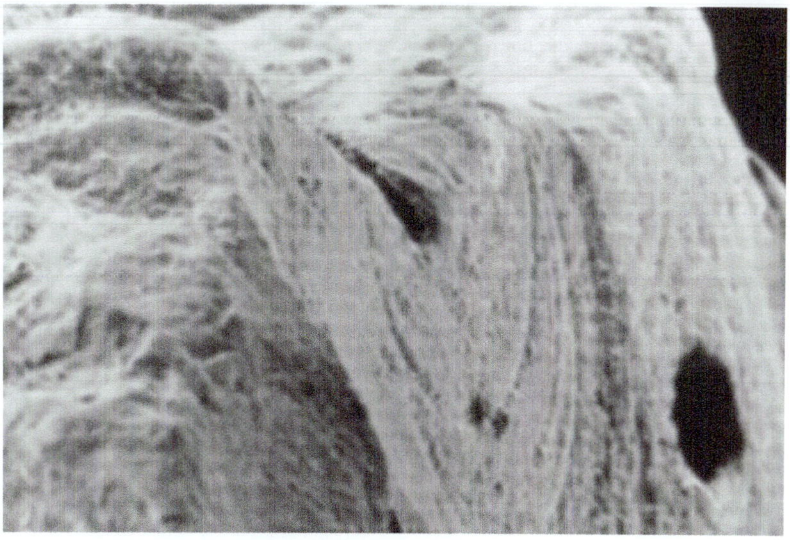

Fig. 8.6 SEM image of a pulp stone after sawing. The spherically mineralized rings are centered on a vessel lumen

8.5 Pulp Stones or Pulp Calcification

These result from discrete calcifications or they are attached to or embedded in dentin, implicated in more diffuse dystrophic calcifications (Fig. 8.6). They appear either as denticles with a central cavity, probably a vascular vessel, or as compact degenerative calcified tissues occupying the whole dental pulp (Fig. 8.7).

Structurally, they appear as either true or false pulp stone. They may include partially fused calcospherites or almost or completely merged pulp mineralization. They may appear network-like or ridgelike or look like spherically mineralized structures. Amorphous appearances were also detected. True pulp stone appears formed by dentin-like structure, lined by odontoblast-like cells. False pulp stone are formed from degenerating cells, which mineralize, or are surrounded by soft tissue. Stones may be free, or adherent, or eventually embedded within the canal wall (Fig. 8.8a, b). Denticle, formed by epithelial remnants surrounded by odontoblasts, fibrodentin, and dystrophic calcifications, was also found.

In a single pulp, 1–12 stones may occlude the pulp space. They are occurring more often in the

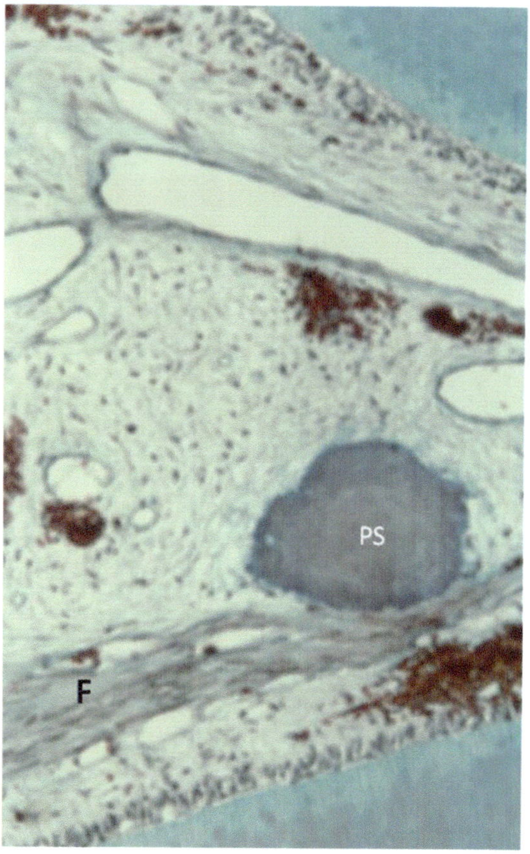

Fig. 8.7 Free mineralized pulp stones (*PS*) form in the central part of the fibrous (*F*) collagen-rich pulp

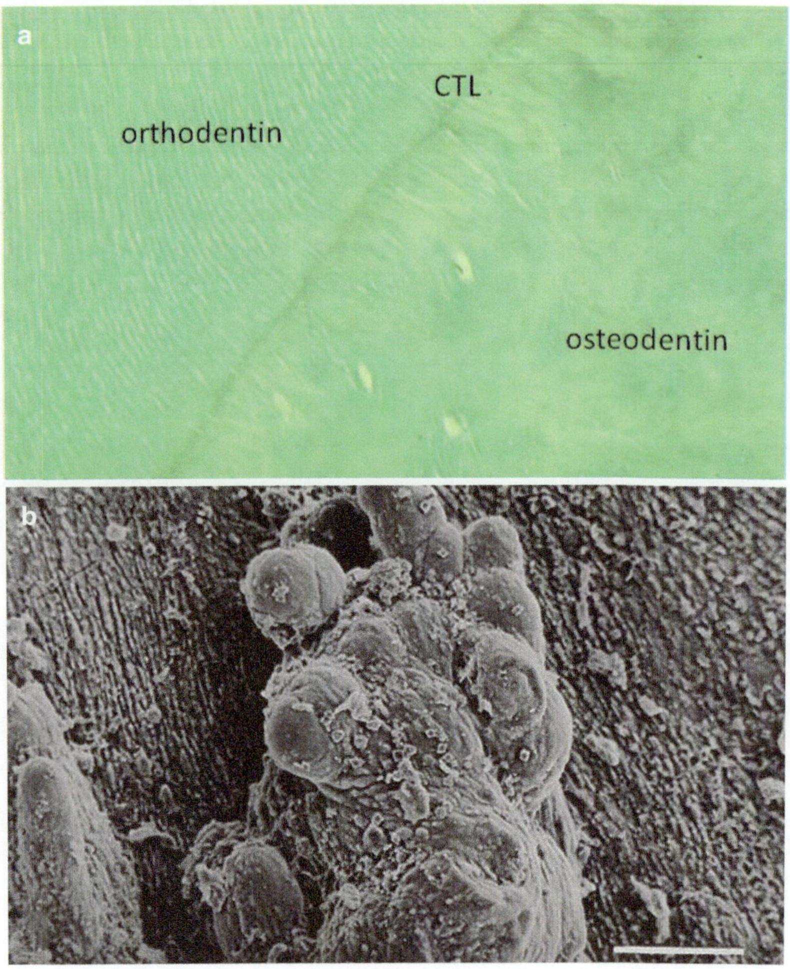

Fig. 8.8 (**a**) A calciotraumatic line (*CTL*) separates the orthodentin (*upper left*) from the osteodentin structure (*lower right*). (**b**) During aging, attached dystrophic calcifications are seen protruding inside the pulp cavity

crown, but they may also be present in the root. Total coronal pulp occlusion is found in *dentin dysplasia* and *dentinogenesis imperfecta*. Ninety percent of teeth in those over 40 years of age display pulp calcifications. Many involve apical blood vessels. The gradual calcifying process becomes circumferential in the endoneurium and/or perineurium. Collagenous bundles are associated with the connective tissue surrounding blood vessels and nerves. Cell density is decreasing by half from 20 to 70 years. Fibrous degeneration or pulp atrophy occurs together with fat deposits which are Sudan black positive (lipidic inclusions or lipofuscin-rich structures) [27].

References

1. Bernick S. Age changes to the dental pulp. Oral morphological changes in older subjects. Front Oral Physiol. 1987;6:7–30.
2. Oi T, Saka H, Ide Y. Three-dimensional observation of pulp cavities in the maxillary first premolar tooth using micro-CT. Int Endod J. 2004;32:46–51.
3. Murray PE, Stanley HR, Matthews JB, Sloan AJ, Smith AJ. Age-related odontometric changes of human teeth. Oral Surg Oral Med Oral Pathol Oral Radiol Endod. 2002;93:474–82.
4. Gustafson G. Age determination on teeth. J Am Dent Assoc. 1950;41:45–54.
5. Ohtani S, Yamamoto T. Strategy for the estimation of chronological age using the aspartic acid racemization method with special reference to coefficient of

correlation between D/L ratios and ages. J Forensic Sci. 2005;50:1020–7.
6. Wang C, Jurk D, Maddick M, Nelson G, Martin-Ruiz C, Von Zglinicki T. DNA damage response and cellular senescence in tissues of aging mice. Aging Cell. 2009;8:311–23.
7. Swanson EC, Manning B, Zhang H, Lawrence JB. Higher-order unfolding of satellite heterochromatin is a consistent and early event in cell senescence. J Cell Biol. 2013;203(6):929–42.
8. Jiang H, Ju Z, Rudolph KL. Telomere shortening and ageing. Z Gerontol Geriat. 2007;40:314–24.
9. Mokry J, Soukup T, Micuda S, Karbanova J, Visek B, Brcakova E, Suchanek J, Bouchal J, Vokurkova D, Ivancakova R. Telomere attrition occurs during ex vivo expansion of human dental pulp stem cells. J Biomed Biotechol. 2010;2010:673513.
10. Sethe S, Scutt A, Stolzing A. Aging of mesenchymal stem cells. Ageing Res Rev. 2006;5:91–116.
11. Li L, Zhu Y-Q, Jiang L, Peng W. Increased autophagic activity in senescent human dental pulp cells. Int Endod J. 2012;45:1074–9.
12. Krishnamurthy J, Torrice C, Ramsey MR, Kovalev GI, Al-Regaiey K, Su L, Sharpless NE. *Ink4a/Arf* expression is a biomarker of aging. J Clin Invest. 2004;114:1299–307.
13. Tyner SD, Venkatachalam S, Choi J, Jones S, Ghebranious N, Igelmann H, Lu X, Soron G, Cooper B, Brayton C, Park SH, Thompson T, Karsenty G, Bradley A, Donehower LA. P53 mutant mice that display early ageing-associated phenotypes. Nature. 2002;415:45–53.
14. Tranasi M, Sberna M-T, Zizzarri V, D'Apolito G, Mastrangelo F, Salini L, Stuppia L, Tetè S. Microarray evaluation of age-related changes in human dental pulp. J Endod. 2009;35:1211–7.
15. Muramatsu T, Hamano H, Ogami K, Ohta K, Inoue T, Shimono M. Reduction of connexin 43 expression in aged human dental pulp. Int Endod J. 2004;37:814–8.
16. Niishio K, Inoue A, Qiao S, Kondo H, Mimura A. Senescence and cytoskeleton: overproduction of vimentin induces senescent-like morphology in human fibroblasts. Histochem Cell Biol. 2001;116:321–7.
17. Matsuzaka K, Muramatsu T, Katakura A, Ishihara K, Hashimoto S, Yoshinari M, Endo T, Tazaki M, Shintani M, Sato Y, Inoue T. Changes in the homeostatic mechanism of dental pulp with age: expression of the core-binding factor alpha-1, dentin sialoprotein, vascular endothelial growth factor, and heat shock protein 27 messenger RNAs. J Endod. 2008;34:818–21.
18. Ranly DM, Thomas HF, Chen J, MacDougall M. Osteocalcin expression in young and aged dental pulps as determined by RT-PCR. J Endod. 1997;23: 374–7.
19. Muramatsu T, Hamano H, Ogami K, Ohta K, Inoue T, Shimono M. Reduction of osteocalcin expression in aged human dental pulp. Int Endod J. 2005;38:817–21.
20. Nielsen CJ, Bentley JP, Marshall FJ. Age-related changes in reducible crosslinks of human dental pulp collagen. Arch Oral Biol. 1983;28:759–64.
21. Hayakawa T, Iijima K, Hashimoto Y, Myokei Y, Takei T, Matsui T. Developmental changes in the collagens and some collagenolytic activities in bovine dental pulps. Arch Oral Biol. 1981;26:1057–62.
22. Mehrazarin S, Oh JE, Chung CL, Chen W, Kim RH, Shi S, Park N-H, Kang MK. Impaired odontogenic differentiation of senescent dental mesenchymal stem cells is associated with loss of Bmi-1 expression. J Endod. 2011;37:662–6.
23. Satoh R, Kishino K, Morshed SR, Takayama F, Otsuki S, Suzuki F, Hashimoto K, Kikuchi H, Nishikawa H, Yasui T, Sakagami H. Changes in fluoride sensitivity during in vitro senescence of normal human oral cells. Anticancer Res. 2005;25:2085–90.
24. Lee H-L, Nam H, Lee G, Baek J-H. Dlx3 and Dlx5 inhibit adipogenic differentiation of human dental pulp stem cells. Int J Oral Biol. 2012;37:31–6.
25. Kawagishi E, Nakakura-Ohshima K, Nomura S, Ohshima H. Pulpal responses to cavity preparation in aged rat molars. Cell Tissue Res. 2006;326:111–22.
26. Lee YH, Kim GE, Cho HJ, Yu MK, Bhattarai G, Lee NH, Yi HK. Aging of in vitro pulp illustrates change of inflammation and dentinogenesis. J Endod. 2013;39:340–5.
27. Goga R, Chandler NP, Oginni AO. Pulp stones: a review. Int Endod J. 2008;41:457–68.

Part II

Pulp Pathology

9 Pulp Inflammation: From the Reversible Pulpitis to Pulp Necrosis During Caries Progression

Lars Bjørndal and Domenico Ricucci

9.1 Setting the Stage

This chapter aims to present an overview of the pulpal events taking place in relation to caries progression and in different stages of lesion activity, from the very first cellular pulp reactions to the non-cavitated enamel caries toward progressive stages of pulpal inflammation including necrosis. Where is the border between a beneficial/reversible inflammation as opposed to unwanted/irreversible inflammation? The former is a prerequisite of repair, from which treatment can be successfully carried out with or without exposure of the pulp, whereas the latter leads to apoptosis and necrosis, and, if left untreated, to further bacterial invasion in the pulp cavity.

To improve the connection between the science of pulp biology in the laboratory and actual clinical treatment concepts, it is important to maintain a link between the visible signs of disease at both a macroscopic and histological level of examination. For example, what do caries look like in progressive stages of caries lesion formation? For obvious reasons, it is not possible to repeat histological data from the same carious lesion over time. Therefore, when progressive cellular events are described, it must be based on different carious lesions in progressive stages of lesion development.

In previous papers [1, 2] and textbooks [3–5], it has been stated that the signs and symptoms do not allow to consistently diagnose the histological status of the pulp and consequently the reversibility or irreversibility of pulp inflammation. However, is it possible to clarify at what lesion stages these problems of interpreting do most often occur? In this chapter we will specify the lesions by using information on progression stage, lesion activity and estimated length of progression time (patient age), including recent clinical evidence from treatment of deep caries lesions. Finally, from a histological viewpoint, it is clear that the point of no return for unwanted inflammation in the pulp can be defined and will be clarified in this chapter because confusion exists among clinicians and researchers.

9.2 The Dilemma of the Difference Between Histology and Clinical Pulp Diagnosis

Although the pulp is the vital tissue connecting the tooth with the body, it constitutes an entity that in the clinical setting is very difficult to monitor in terms of stage of inflammation. No device

L. Bjørndal, DDS, PhD (✉)
Department of Odontology, Section of Cariology and Endodontics/Pediatric and Clinical Genetics, University of Copenhagen, Nørre Allé 20, Copenhagen-N DK-2200, Denmark
e-mail: labj@sund.ku.dk

D. Ricucci, MD, DDS
Private Practice, Piazza Calvario 7, Cetraro 87022, Italy
e-mail: dricucci@libero.it

M. Goldberg (ed.), *The Dental Pulp*,
DOI 10.1007/978-3-642-55160-4_9,

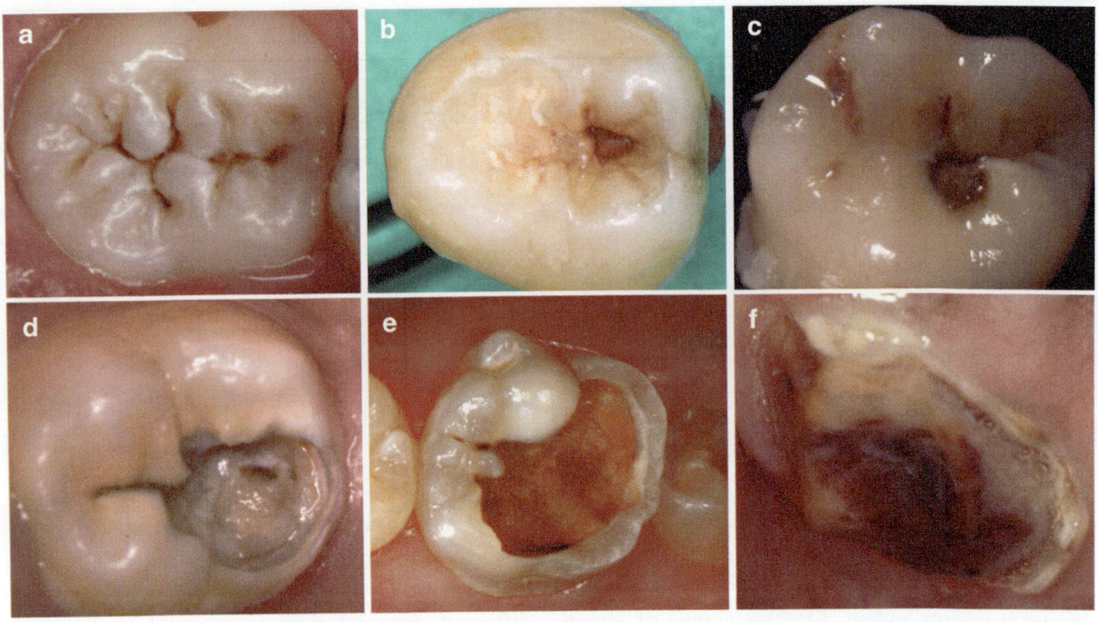

Fig. 9.1 Macroscopical views of progressive stages of caries in different molars. (**a**) Occlusal lesion without clinical dentine exposure in mandibular molar. (**b**) Occlusal lesion with initial dentine exposure reflecting a closed lesion environment in a mandibular molar. (**c**) Occlusal lesion with established dentine exposure. Note that the undermined enamel is reflected as change in enamel translucency surrounding the clinical cavity, also described as retrograde demineralization along the enamel-dentine junction. A macroscopical cutting plane of this lesion is shown in Fig. 9.2b. (**d**) A gradual breakdown is taking place in a mandibular molar; note the milky appearance of the cavity border reflecting the retrograde demineralization. (**e**) In principal, a more advanced stage of enamel breakdown from a maxillary molar. The ecosystem has started to become more open, and the change in microbial growth condition is clearly reflected by a less pronounced biofilm. (**f**) The occlusal enamel here is completely broken away and a "new" occlusal dentine surface has emerged. The clinical appearance of the carious dentine is dark at the central area. This case represent a clinical example of an open lesion environment, with a temporarily arrest at the occlusal site, whereas along the peripheral borders the cariogenic biofilm is clearly present, maintaining a high lesion activity. The case reflects the relative importance of a marked change within a local ecosystem

is available for a noninvasive estimation of pulp inflammation in a routine clinical setting [5, 6], for providing an objective answer to whether the pulp can be saved or not during treatment of deep caries approximating to the pulp. Therefore, the clinicians are forced to apply indirect methods for their clinical diagnoses [7, 8]. Consequently, it is important to underline that when the clinical diagnoses are suggested such as *pulpitis irreversibilis* or *pulpitis reversibilis*, they are *not* based on a true histopathological platform.

Previous histological studies have not always specified the clinical data about the origin of the teeth, age of patients, as well as information about the specific lesion environment being active or arrested [1, 2]. This chapter attempts to present the conditions of the pulp based on different carious lesions in progressive stages of lesion development (Fig. 9.1a–f). This is not optimal as the history of the natural caries lesion may vary, but it will indicate in which lesion types the dilemma of pulp inflammation occurs.

The diagnosis of the condition of the pulp has been systematically reviewed [6] and with only a few papers, accepted for inclusion. Of these, the scientific level of evidence was most often low due to methodological shortcomings. However, the following features were found:

- No obvious association between cold, heat, electric pulp test, and percussion in deep asymptomatic caries and status of pulp inflammation [9].

- Irrespective of the degree of inflammation, the majority of patients may respond to a percussion test, even though the teeth have minimal or no pulp inflammation.
- It is important to stress the unreliability of pain as a parameter for the clinical assessment of reversibility/irreversibility of pulpal inflammation. Severe inflammatory reactions can be observed in teeth with no history of pain.
- Despite the presence of spontaneous pain, indicative of irreversible pulp inflammation, histologically it is possible to find no evidence of pulp exposure or necrosis in the pulp tissue, indicating that the point of irreversibility had not been reached.
- Absence of pain does not exclude the presence of inflammation.

It has recently been addressed [10] that some of the study shortcomings may relate to the fact that the term "deep lesion" has been used without a more detailed definition of the actual depth of the lesion per se. Was it deep dentine involvement? Is it based on x-rays? Is caries in contact with the pulp clinically, radiographically, or by means of subjective symptoms indicating pain, or is it measured in terms of histology? If this information is missing, it is difficult from a clinical aspect to interpret anything about the pulp.

9.3 Previous Histological Shortcomings Have Simplified the Understanding of Caries Pathology

It should not be underestimated that almost all previous histological studies of the pulp have been based on demineralized histological sections in order to be able to cut thin sections. However, during the decalcification procedure of the tooth specimen, not only the mineral content of the dentine but also the entire enamel is removed. Consequently, important information about the lesion environment has therefore been lost. This may have led to several examples of oversimplification in the understanding of caries pathology:

- Studies attempting to correlate enamel caries with pulpal inflammation have been described as being speculative [11].
- The early spread of caries into the dentine was believed to be the same as seen in advanced stages of lesion progress with signs of huge dentine exposure and undermining enamel [12].
- Teeth with rapidly progressing caries have dominated the materials investigated, presumably because they were the ones available for extraction [6].

Consequently, it was taught that even a small enamel lesion without histological contact to the enamel-dentine junction could induce both the translucent zone and the demineralized zone in the dentine [13] and in particular the non-cavitated enamel lesion could hide bacterial invasion [14] as well as undermine sound enamel from the very beginning, even without visible dentine exposure [12–17]. Therefore, many dentists of the last century were trained to both remove the early enamel lesion by operative intervention, as well as to be radical when performing dentine excavation, because carious dentine left over was believed to maintain the pulp inflammation and eventually lead to pulp degradation. The description of the changes in the pulp has often been closely related to scenarios of "points of no returns" in terms of inflammation. Clinically, the so-called deep caries progression should therefore *not* be treated with an indirect pulp capping [4, 11], as the retained carious tissue would maintain further development of pulp inflammation. In contrast, complete excavation was recommended even if it led to exposure of the pulp, because the pulp was judged to be irreversibly inflamed anyway, even though the patient was not in pain. Moreover, if already an enamel lesion is able to induce inflammation, it would be easy to imagine that a deep lesion would be associated with unwanted inflammation.

However, the advances of pulp biology [18–23] combined with at detailed description of the lesion environment, as well as the clinical evidence of treating deep lesions [24–28], have started to modify the traditional view of pulp inflammation as being "the nonstop train" toward apoptosis and pulp necrosis.

9.4 Progression of the Non-cavitated and Cavitated Enamel-Dentinal Lesion Complex

A brief update of principal caries pathology is presented. Detailed histomorphological studies have revealed that the carious enamel-dentine lesion complex is a closed entity as long as there is no destruction and disintegration of the demineralized enamel [8]. The cariogenic biomass is located at the enamel surface only. The enamel lesion contact with the enamel-dentine junction is strictly related to the extent of the demineralized dentine, and no early spread along the enamel-dentine junction can be monitored (Fig. 9.2a). When the established caries remain untreated, they advance in width and depth. The dentine is clinically exposed (Fig. 9.1c, d) and the biofilm is heavily involving the cavity. The microbial ecosystem can be described as being a "closed" environment. At this stage the spread along the enamel-dentine junction is a characteristic feature undermining sound enamel [29] and the lesion becomes wedge shaped, with the base directed toward the surface (Fig. 9.2b). As the deep lesion further progresses, the undermined enamel often breaks off (Fig. 9.1d–f), due to masticatory forces, converting the lesion environment from a "closed" toward a more "open" lesion environment [30]. Clinically and macroscopically the color of the demineralized dentine becomes darker (Fig. 9.2c), and at the central area the cariogenic biofilms overlying the lesion are markedly reduced. Although the lesion has increased in size, the actual activity has temporarily decreased. Even at the macroscopic level, progressive alterations of pulp inflammation can be observed in terms of increasing visualization of the vascular architecture in the pulp (Fig. 9.2a–c).

9.5 Carious Enamel-Dentine Lesion Complex and Activity: Understanding the Initial Pulp Response Subjacent to Enamel Lesion Without Cavitation

Studies have shown that the cytoplasm/nucleus ratio of the mature odontoblast cells, subjacent the superficial enamel lesion, is more reduced than unaffected control sites, even before visible alteration in dentine mineralization. Moreover,

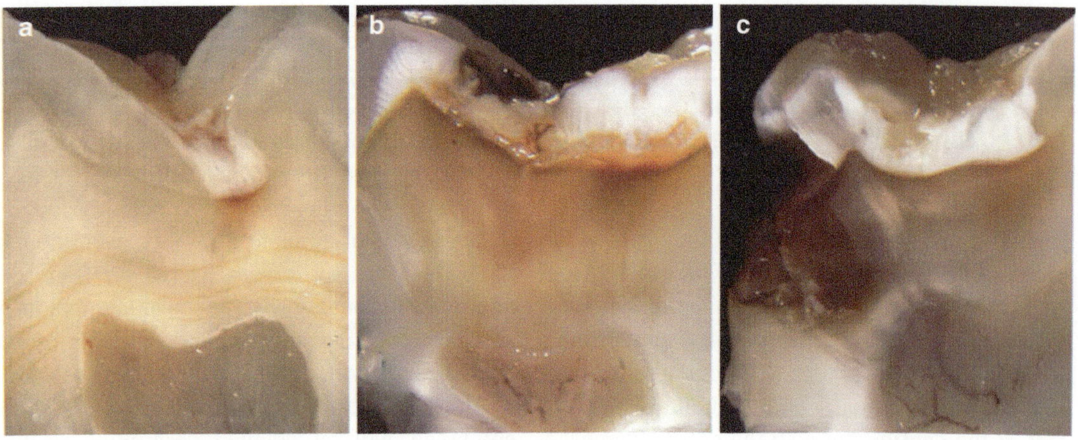

Fig. 9.2 Macroscopical cutting profiles in relation to progressive stages of caries. (**a**) An occlusal lesion without dentine exposure. A close interrelation is noted between the enamel lesion contact and the extent of the affected dentine; no visible alterations in the pulp. (**b**) Macroscopical cutting profile of lesion shown in Fig. 9.1c. Note the retrograde demineralization along the enamel-dentine junction as well as the initial appearance of the pulpal vascular architecture. (**c**) The cutting profile of a mixed lesion environment. Note a more marked appearance of the vascular architecture. From [10] with permission from Elsevier

the subodontoblastic or Höehl's layer appears indistinct [26]. Enhanced mineralization of the dentine is noted in the central and oldest part of the involved lesion area, as the enamel lesion progresses toward the enamel-dentine junction. At the pulpal site substantial growth of the predentine matrix is noted with bundles of collagen fibers aligned with odontoblasts reduced in size [8]. These first findings of cellular alterations and enhanced mineralization of the dentine are probably not associated with antigen-related pulp reactions as the bacterial-induced dentine demineralization is not yet established.

9.6 The Trigger of Pulpal Immunity in Progressive Stages of Carious Dentine

It is well known that the dentine comprises bioactive extracellular matrix [31–35], and during carious demineralization of the dentine, there is a release of bioactive molecules [22, 36, 37], which can trigger the odontoblasts, members of the first line of host defense (see Chap. 7). However, it is unclear whether this may take place even before bacteria have invaded the demineralized dentine (Fig. 9.2a), also defined as affected dentine [38]. In non-cavitated enamel lesions with subjacent dentine demineralization in humans, observations of a reduced odontoblast layer, as well as the alterations in the subodontoblastic region, may qualitatively reflect an early odontoblast-triggered innate immune response [18]. Moreover, a different pulp response is noted subjacent an active versus a slowly progressing lesion. The cytoplasm/nucleus ratio of the odontoblasts appears markedly reduced in both scenarios, but a maintained indistinct subodontoblastic region is noted only in active sites [31].

In a series of *in vitro* studies, the odontoblast-like cells were shown to express Toll-like receptors [39–42] making the cells capable to induce mediators known to influence positively or negatively both the inflammatory as well as the immune responses in pathogen-challenged pulp-like tissue. As an example lipoteichoic acid, which is a by-product from Gram-positive bacteria, was able to elicit proinflammatory cytokines by further promoting immature dendritic cell recruitment [42].

The Gram-positive bacteria represent the first members of invading bacteria in demineralized dentine with a clinical dentine exposure (Fig. 9.2a), eventually accounting for 70 % of the cultivable microbiota in lesion sizes involving two-thirds of the dentine and more [43]. During the development of such a "closed" ecosystem, a remarkable homogeneous lactobacilli microflora can be detected. Gram-negative bacteria take over [44] as the lesion advances, and, because of their content of lipopolysaccharide, they are capable of inducing lipopolysaccharide–binding protein which has an even more complex role in terms of triggering the odontoblast-like cells. Recently, it was found that this protein may interact with lipoteichoic acid, hence reducing the receptor-dependent production of inflammatory cytokines by odontoblast-like cells and in this way modulate host defense in human dental pulp [45]. The various laboratory setups [18–22, 32–37, 39–42] with, for example, the odontoblast-like cells have helped gain more and more complex information, eventually leading to an improved understanding of the host defense as well as of tooth repair in the future. However, we are not there yet, in terms of the application of knowledge transformed into treatment modalities in humans.

9.7 The Dentinal Lesion with Clinical Exposure: The Infected Dentine

The established caries lesion with a visible dentine exposure (Fig. 9.1b, c) will comprise all the classical zones of carious dentine: the zone of sclerotic dentine (i.e., increased intratubular mineralization), the dentine demineralization, and, finally, the zone of bacterial invasion and dentine degradation [8]. When the bacteria invade demineralized dentine and it becomes infected [38], the vast majority of the members of the biofilm are Gram-positive bacteria in primary dentine and the innate immune systems are progressively activated. It is not known how the

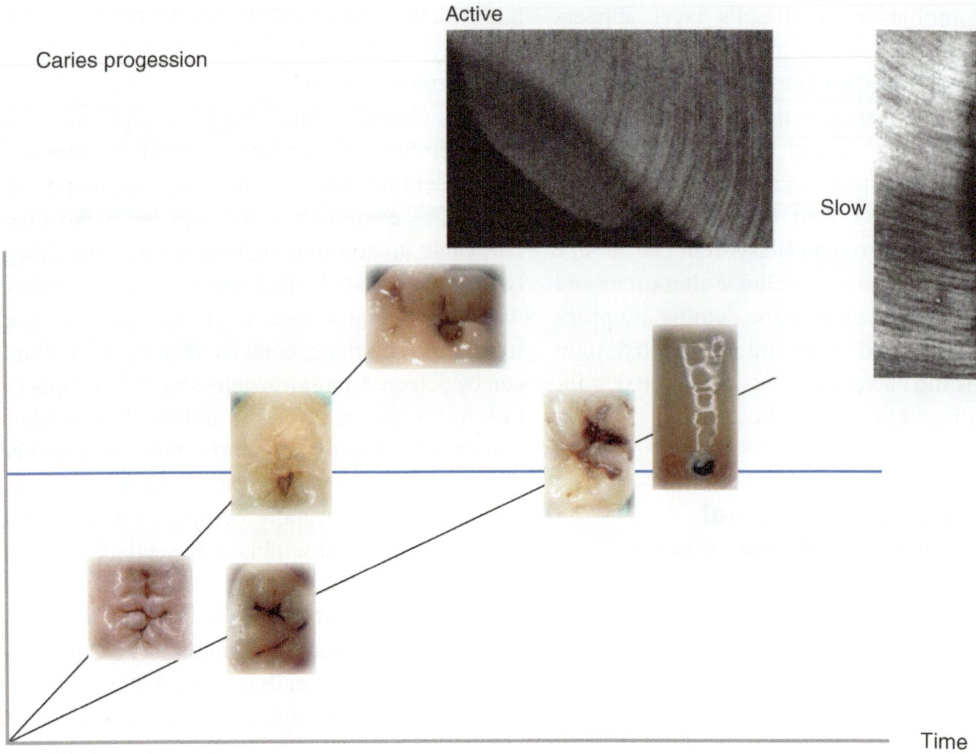

Fig. 9.3 A principal demonstration is shown of an active versus a slow lesion progression. Typically the texture of the tertiary dentine is different within the two scenarios as shown by inserts of two microradiographs of undemineralized thin sections. Along the *x*-axis, the principal caries progression is shown by inserts of clinical illustrations of progressing lesions. The *y*-axis represents the time line. The different slope of the two scenarios reflects the different speed of progression

triggered dental pulp immunity is operating during a slow lesion progression. As tested during stepwise excavation of dentin caries, the cultivable microflora becomes markedly reduced in numbers [46], consequently the acidogenic pH levels decrease, which presumably also leads to decreased production of, for example, lipoteichoic acid. Finally, this may temporarily stop the sequence of events leading to a further unwanted inflammatory response. In confirming this, the tertiary dentine appears like a continuous formation of secondary dentine laid down along a slow lesion progression, whereas the tertiary dentine appears partly atubular during ongoing stages of active lesion progress (Fig. 9.3).

Taken together, the odontoblasts are members of the first line of defense responding to various carious stages and activities. How this modulates the role of the odontoblasts during early tertiary dentine formation in humans remains unclear, but it is detailed in Chap. 10.

9.8 The Definition of Deep and Extreme Deep Carious Lesions

A definition of deep caries has to be made by the x-ray, because it is within the clinical setting that the general dental practitioners will end up using the findings of laboratory research, and therefore it is of paramount importance to have a reference that can be used in a clinical setting. As the depth of the caries lesion represents a diagnostic problem, it may be relevant to further classify it beyond previous attempts. The deep lesion has

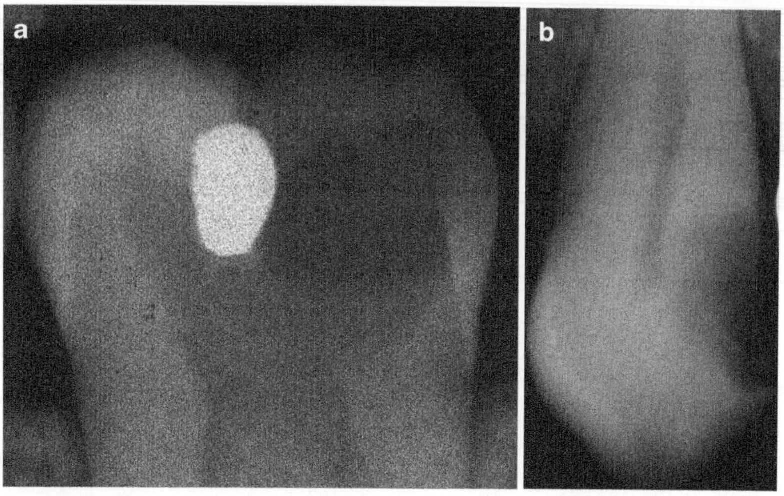

Fig. 9.4 The radiographic presentation of an extreme deep caries lesion versus a deep lesion. (**a**) The entire primary/secondary dentine is penetrated either with no radiodense zone separating the demineralized dentine from the pulp or with a radiodense zone located within the pulp chamber indicative of tertiary dentine. (**b**) The deep lesion involves the pulpal quarter with a radiodense zone separating the translucent zone from the pulp

previously been defined radiographically as being within the pulpal quarter toward actual contact with the pulp [47]. However, we suggest separating the deep lesion in two scenarios. A deep caries lesion is defined as involving the pulpal quarter of the primary/secondary dentine (Figs. 9.4b. 9.5a, and 9.6a) but still with a radiodense zone separating the demineralized dentine [19]. The extreme deep lesion involves the entire primary/secondary dentine either with no radiodense zone separating the demineralized dentine from the pulp or with a radiodense zone located within the pulp chamber indicative of tertiary dentine (Figs. 9.4a, 9.7a, and 9.8a).

These lesions (deep and extreme) may induce initial periapical disturbances/apical periodontitis lesions, which may (Fig. 9.6a) or may not be visible radiographically, but still with the pulp being vital!

9.9 The Further Progression of Deep Caries

Following this radiographically based definition, a deep lesion involves the pulpal quarter of the dentine but still has a radiodense zone separating the pulp and the carious dentine. From a clinical point of view, when bacteria are close to the pulp but are still confined to primary/secondary dentine, pulp inflammatory reactions, even severe, may regress if treatment completely removes the infected and degraded dentinal tissue (Fig. 9.5b, c). There is, to a certain point, a degree of reversibility of pulp inflammation, with tertiary dentin remaining as a permanent "scar" of previous inflammation. However, diagnosing this "borderline-histological-condition" (reversible/irreversible border) by clinical means is very difficult and often misleading. However, it has been known for some time [47] that heavy concentrations of chronic inflammatory cells, macrophages, and a few polymorphonuclear neutrophils (PMNs) represent a characteristic feature beneath the affected dentine tubules in deep lesions. These cells almost obliterate the usual pulp morphology, but liquefaction or coagulation necrosis cannot be found in the pulp (Fig. 9.5c). The inflammatory process is usually confined to the coronal pulp and bacterial cells have advanced to the point of near-exposure. The odontoblasts beneath the affected tubules are very sparse, with no palisading. The coronal blood vessels are engorged. This picture of a long-lasting chronic inflammation has been explained by the fact that the cells in particular macrophages [19] are "frustrated" as they permanently produce cytokines, but without reaching the actual bacteria, because they are still hidden within the carious dentine [8].

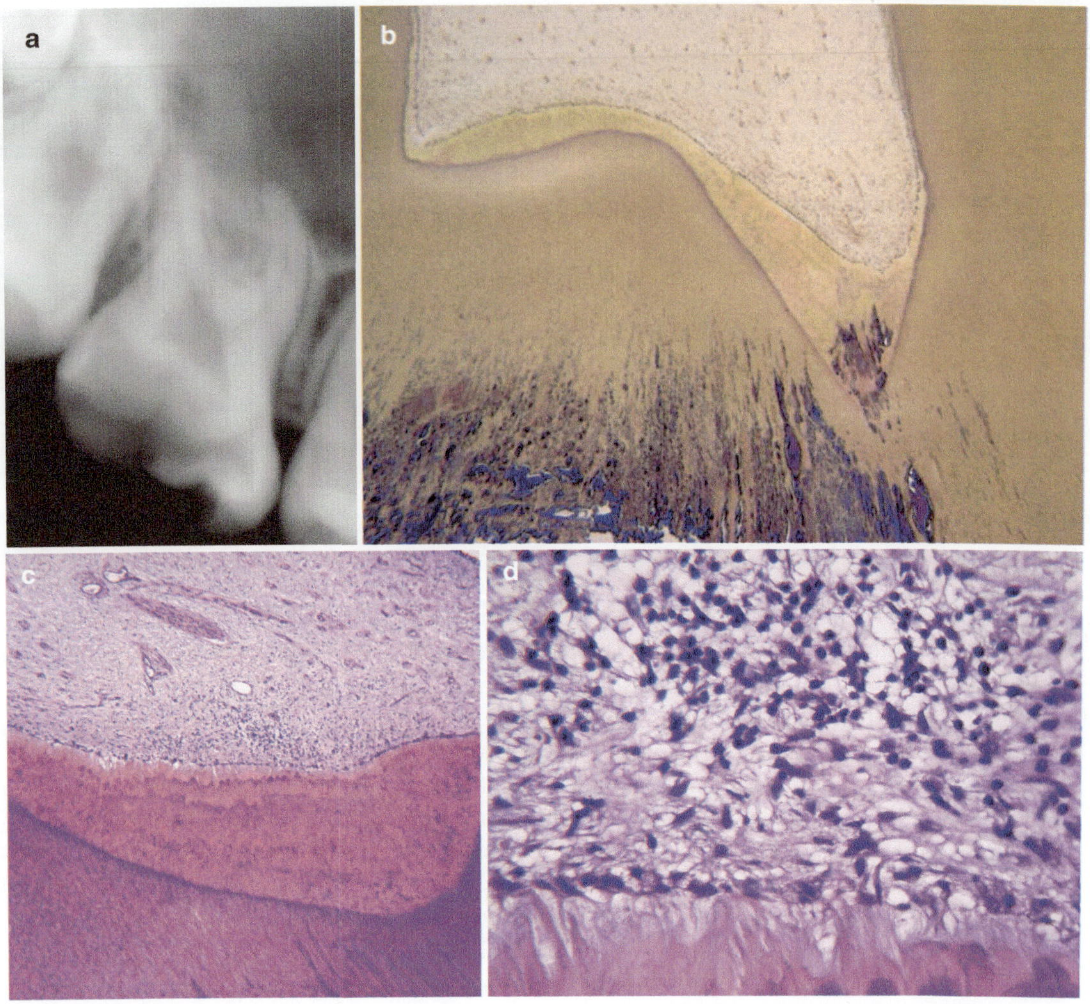

Fig. 9.5 Deep caries is shown with reversible pulp inflammation. (**a**) The caries lesion shown in Fig. 9.1e with a more advanced stage of enamel breakdown due to the undermining nature of caries progression. The patient complained of pain to thermal stimuli and chewing. Sensitivity tests gave exaggerated responses. There is no apical periodontitis lesion. (**b**) Overview encompassing the caries lesion, tertiary dentine and pulp. Note how the tertiary dentine is heavily infiltrated by bacteria in the pulp horn area. No necrosis could be observed in this and in any of the serial sections. The histological diagnosis is "reversible pulp inflammation" (Taylor's modified Brown and Brenn, orig. mag. ×25). (**c**) Considerable amount of tertiary dentine is formed on the pulpal side (H&E, orig. mag. ×50). (**d**) The tertiary dentine is covered by an incomplete layer of flattened odontoblasts. Subjacent to it, a severe concentration of chronic inflammatory cells is present (orig. mag. ×400)

An important aspect related to pulp response is the aforementioned tertiary dentine, which is deposited over the pulpal end of the dentinal tubules centrally affected by caries. During the very active stage of deep lesion progress, where the undermined enamel maintains a "closed" ecosystem, the formation of tertiary dentine may be less pronounced or even absent (Fig. 9.6a, b), and bacterial penetration may be evident in the pulp, even though the caries is not reaching the pulp radiographically. This may be due to the fact that during proximal lesion progress, the cariogenic environment may maintain a high acidogenic environment for a longer period than occlusally,

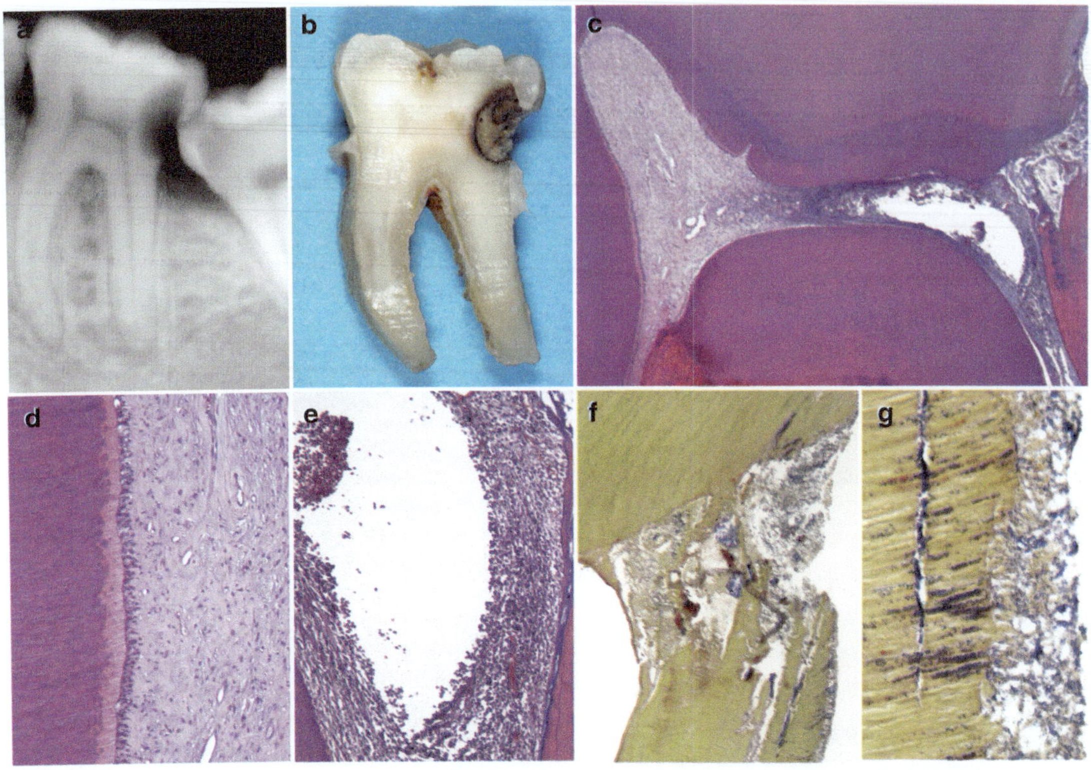

Fig. 9.6 Deep caries is shown with irreversible pulp inflammation. (**a**) Mandibular second molar in a 22-year-old man. Severe spontaneous pain. The radiograph shows a deep distal caries proximal to the pulp. At the distal root there is an indication of an apical periodontitis lesion. The patient did not accept any treatment aimed at conservation of the tooth and requested extraction. (**b**) Photograph taken after grinding the tooth on a mesiodistal plane until the pulp was seen. (**c**) Overview of the pulp chamber. A proximal deep lesion is noted distally. The distal half of the pulp chamber tissue is necrotic, while the mesial portion exhibits a relative normality (H&E, orig. mag. ×16). (**d**) Detail of the left wall of the mesial pulp horn in **c**. The pulp shows normal histological features (orig. mag. ×100). (**e**) Detail of the distal root canal orifice area in **c**. Liquefaction necrosis surrounded by severe concentration of inflammatory cells (orig. mag. ×100). (**f**) Bacterial penetration in the distal pulp horn area (Taylor's modified B&B, orig. mag. ×50). (**g**) High power view of the caries lesion. Dentine is heavily colonized by bacteria. A distinct bacterial biofilm is present on the cavity floor (orig. mag. ×400)

before the undermined enamel gradually breaks down. In addition, the distance toward the pulp may also be shorter.

An active and "closed" ecosystem within the proximal lesion environment is shown in a permanent molar tooth (Fig. 9.6a–g). In this case apical radiolucency is apparent as well, indicative of an apical periodontitis lesion (Fig. 9.6a). A histological study of primary teeth has shown a similar tendency that in subjacent proximal caries lesions, the pulp showed more severe signs of inflammation than compared with occlusal lesions [48].

These observations may also explain why some cases of direct pulp capping fail. If, in the absence of symptoms, during excavation of deep carious lesions, bleeding occurs from a pulp horn, the clinician may decide to perform pulp capping, but the bacteria are in fact "sealed" in the pulp horn. Within recent clinical trials investigating the treatment of well-defined deep caries in adults (using the present definition), it was found very important that excavation did not lead to exposure of the pulp. In case of exposure a pulp capping was performed and the outcome was markedly reduced [24].

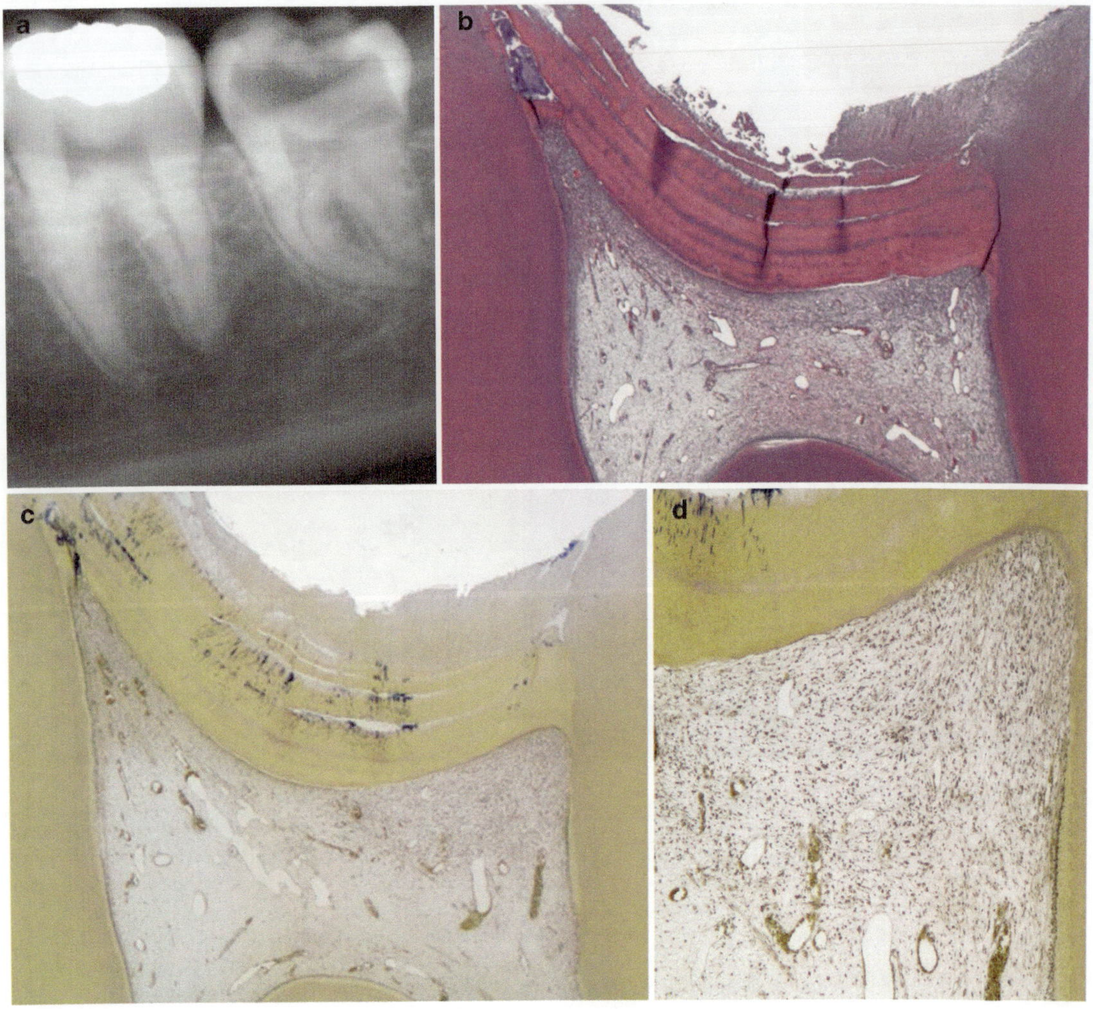

Fig. 9.7 Deep caries and irreversible pulp inflammation are shown. (**a**) Mandibular third molar in a 23-year-old woman. Spontaneous pain. The tooth was extracted. Note that it is classified as an extreme deep lesion. (**b**) Overview of the pulp chamber. Its roof is constituted only by tertiary dentine. No necrosis can be seen in this section (H&E, orig. mag. ×16). (**c**) Section proximal to that in **b** (Taylor's modified B&B, orig. mag. ×16). (**d**) Detail of the distal pulp horn. Severe accumulation of inflammatory cells obscuring the normal pulp architecture (orig. mag. ×50)

9.10 Extreme Deep Caries

During extreme deep caries (Figs. 9.7a–d and 9.8a–g), the bacteria are involved histologically in both peritubular and intertubular dentine. A massive formation of tertiary dentine may be noted, representing the last barrier against bacterial penetration, after the primary/secondary dentine has been completely destroyed (Fig. 9.7b, c). Obvious neutrophil accumulation is occurring (Fig. 9.7c) as bacteria are in direct contact with the pulp tissue reflecting the adaptive immune response [19]. Bacterial invasion of the tertiary dentine per se indicates as well irreversible or unwanted inflammation [49]. In terms of clinical treatment and with a specified treatment protocol which includes the use of magnification, observational studies have actually shown positive outcomes of pulp capping even in extreme deep caries [50].

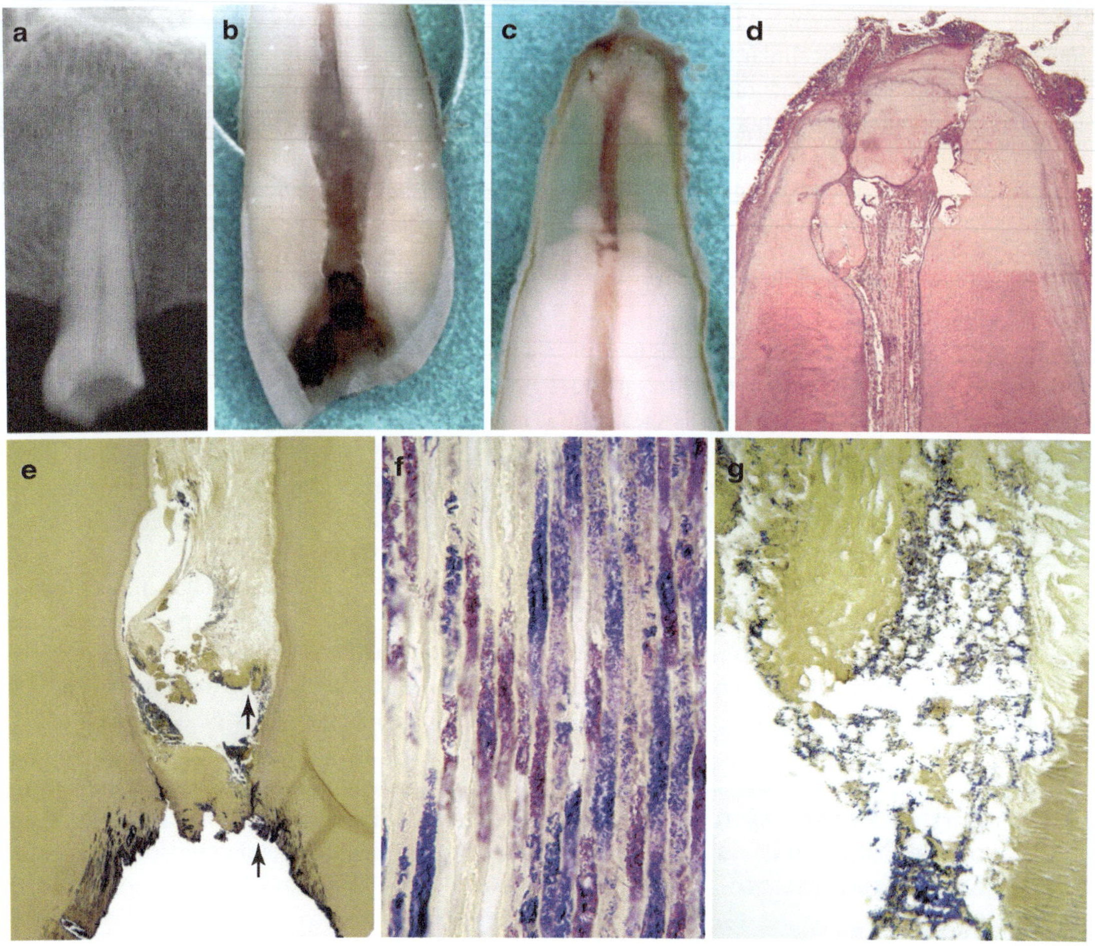

Fig. 9.8 Progression of pulp necrosis is observed. (**a**) Maxillary canine with caries penetrating the pulp chamber in a 61-year-old woman. Spontaneous pain. The tooth was sensitive to percussion. The radiograph shows an absence of the pulp chamber roof and widening of the periodontal ligament space. The patient required extraction of the tooth. The lesion is classified as an extreme caries lesion. (**b**) Photograph taken after grinding the tooth on a buccolingual direction in order to allow proper fixation of the pulp tissue. The pulp tissue in the pulp chamber and in the coronal third appears *dark*, indicating possible pulp necrosis. (**c**) Photograph of the middle and apical third. The pulp tissue appears *reddish* in these areas, indicating the presence of blood and possibly vital pulp tissue. Note the apical ramifications. (**d**) Apical third. The histological section confirms the presence of vital tissue in the apical canal and in the numerous apical ramifications (H&E, orig. mag. ×25). (**e**) Pulp chamber with empty spaces indicating necrosis, the *arrows* are detailed in **f** and **g**. (Taylor's modified B&B, orig. mag. ×25). (**f**) High power view from the carious dentin. Dentinal tubules heavily invaded by Gram-positive and Gram-negative bacteria indicated by the *lower arrow* in **e** (orig. mag. ×1,000). (**g**) Magnification of the area of the right pulp chamber wall indicated by the *upper arrow* in **e**. Dense concentration of bacteria in necrotic tissue (orig. mag. ×400)

9.11 Irreversibility of Pulpal Inflammation

The pulp may be directly exposed by caries, and bacteria have invaded the pulp tissue and a focus of necrosis with limited extension is typically formed in the pulp tissue. It should be emphasized that despite the presence of bacteria in the pulp horn (Fig. 9.8e, f) and a severe acute inflammation in the surrounding areas (Fig. 9.8e, g), the rest of the pulp chamber as well as the radicular pulp may show no signs of inflammation (Fig. 9.8d).

The necrotic area is surrounded by a dense accumulation of PMNs and acellular tissue remnants, indicative of partial liquefaction. Further away from the center of the destruction, there is a typical chronic inflammatory response with a large number of plasma cells, small and large lymphocytes, macrophages, fibroblasts, mast cells, and foam cells.

From a histological point of view, the occurrence of an area of necrosis with bacteria, although of limited extension, constitutes the point of transition from a reversible to an irreversible inflammatory state [51]. It is clear that the treatment of such condition cannot include measures aiming at conservation of the diseased pulp and a more aggressive approach has to be adopted, i.e., full or partial pulpotomy as well as pulpectomy. Once again, from a clinical point of view, the problem lies in our inability to diagnose by clinical means the actual histological condition of the pulp, i.e., the presence of bacteria within the pulp cavity.

9.12 Further Progression of the Pulp Necrosis and Degeneration

The initial area of necrosis slowly expands to involve increasing areas of the coronal pulp. It has to be emphasized that the necrotic tissue colonized by bacteria is clearly demarcated from the adjacent tissue, which continues to be vital and relatively normal (Fig. 9.8a–g). It is important to note that pulp necrosis and infection are slow processes that gradually move in apical direction [51].

9.13 Treatment of Deep Caries and Status of Pulp Inflammation

Know-how on treatment outcomes has improved concerning the treatment of deep caries, because studies and reviews reporting higher evidence have started to emerge [24–28, 52], including longtime data [53, 54], but we are still far from having the best clinical evidence for the objective treatment of actual pulp inflammation.

If a deep or extreme deep lesion progression is present even before the root complex is fully completed, the speed of caries progression must have been quite fast. In a global context, it is known today that the speed of caries progression has been markedly reduced during the past decades [55]. On this basis it could be speculated that the degree of inflammation in "cariously exposed pulps" going back 40–50 years was more often in a severe and irreversible stage of inflammation than in the population of "current pulps." In other words, the rationale and the magnitude of using various treatment modalities of deep caries intervention have started to be modified [56–58]. As examples, the use of less invasive caries excavation approaches in deep caries scenarios has shown that a stepwise excavation also defined as two-step caries removal may be more convenient as opposed to complete excavation in well-defined stages of deep caries [24]. In addition, pulp-capping procedures as well as "full" pulpotomy treatments have started to be reevaluated, and it may become a broader alternative to the vital pulpectomy [59, 60]. Yet, clinical evidence of high quality seems a mandatory prerequisite.

9.14 Summary and Conclusions

In this chapter we have focused on caries as the cause of inflammation, mainly because caries seem to be the most frequent reason for performing a root canal treatment [61]. Other reasons will cause inflammation in the pulp, such as trauma or iatrogenic injuries in general, but the description of these is beyond the scope of this chapter.

The research area of pulp inflammation, repair, as well as regeneration is currently running along a fast track. Histologically, it has long been possible to show the point of irreversibility of pulp inflammation during caries development. How do we apply this information in a clinical setting before it happens and would it be possible to cure unwanted inflammation without removing the entire pulp? In this context it is important to know:

- The stage of lesion progress (how deep has caries progressed).
- An estimated clinical assessment of the caries activity.
- Estimated age of the caries lesion (approximation of patient age).
- The value of classifying the depth of caries into "extreme deep" as well as "deep" (Fig. 9.4a, b).
- As acidogenic by-products of cariogenic microorganisms are important for the odontoblast triggering of the innate immune response, the simple act of arresting the cariogenic environment does influence the condition of the pulp.
- That treatment of deep lesion (Fig. 9.4b) using a noninvasive excavation approach does not seem to lead to unwanted inflammation, in particular, if the pulp remains unexposed.
- The turn-on of the adapted immune response histologically reflected by a huge accumulation of inflammatory cells demonstrates bacteria either advancing to the point of near-exposure or actual pulp invasion.
- When bacteria have invaded the tertiary dentine in histological terms, the subjacent pulp has become irreversibly inflamed, eventually leading to further apoptosis and pulp necrosis.
- When the radiograph indicates extreme caries (Fig. 9.4a), the bacteria have almost certainly invaded tertiary dentine.
- That extreme deep caries lesion needs a pulp invasive treatment concept.
- Although being at the stage of an extreme deep lesion with focal areas of necrosis and infection, the radicular pulp may be vital.
- The clinical estimation of pulp inflammation remains a challenge.

An obvious goal for pulp inflammation research and for the clinical application is as follows: when would we be able to actively interfere with the pulp host response, by introducing biomolecules which could be used to mitigate the deleterious effects of the cariogenic biofilm? At this time, the simple act of clinically arresting the deep lesion environment by less invasive excavation modalities as well as the treatment of pulp cappings using magnification combined with various applications of basc matcrials comprising bioactivity seems to be efficient approaches (see Chaps. 17 and 19). However, if the bacterial invasion cannot be controlled, it seems impossible to see advances in pulp inflammation and repair research being successfully integrated into a general dental practitioner environment.

References

1. Seltzer S, Bender IB, Ziontz M. The dynamics of pulp inflammation: correlations between diagnostic data and actual histologic findings in the pulp. Oral Surg Oral Med Oral Pathol. 1963;16:846–71, 969–77.
2. Dummer PM, Hicks R, Huws D. Clinical signs and symptoms in pulp disease. Int Endod J. 1980;13:27–35.
3. Trowbridge HO. Histology of pulpal inflammation. In: Hargreaves KM, Goodis HE, editors. Seltzer and Bender's dental pulp. Chicago: Quintessence Publish. Co, Inc; 2002. p. 227–45.
4. Tronstad L. Clinical endodontics. 2nd ed. Stuttgard: Thieme; 2003. p. 76.
5. Reit C, Petersson K. Diagnostic considerations and clinical decision making. In: Bergenholtz G, Hørsted-Bindslev P, Reit C, editors. Textbook of endodontology. 2nd ed. Oxford: Wiley-Blackwell; 2010. p. 235–54.
6. Mejáre IA, Axelsson S, Davidson T, et al. Diagnosis of the condition of the dental pulp: a systematic review. Int Endod J. 2012;45:597–613.
7. Reit C. Värdering av information. Tandläkertidningen. 1995;87:67–77.
8. Bjørndal L, Mjör IA. Dental caries: characteristics of lesions and pulpal reactions. In: Mjör IA, editor. Pulp-dentin biology in restorative dentistry. Chicago: Quintessence; 2002. p. 55–75.
9. Hasler JE, Mitchell DF. Painless pulpitis. JADA. 1970;81:671–7.
10. Bjørndal L, Demant S, Dabelsteen S. Depth and activity of carious lesions as indicators for the regenerative potential of dental pulp after intervention. J Endod. 2014;40:S76–81.
11. Langeland K. Tissue response to dental caries. Endod Dent Traumatol. 1987;3:149–71.
12. Silverstone LM, Hicks MJ. The structure and ultrastructure of the carious lesion in human dentin. Gerodontics. 1985;1:185–93.
13. Fejerskov O, Thylstrup A. Pathology of dental caries. In: Thylstrup A, Fejerskov O, editors. Textbook of cariology. Copenhagen: Munksgaard; 1986. p. 204–30.
14. Driessens FCM, Wöltgens IHM. Tooth development and caries. 2nd ed. Florida: CRC-Press; 1986.

15. Amprino R, Camanni F. Historadiographics and autoradiographic researches on dental tissue. Acta Anat. 1956;28:217–58.
16. Silverstone LM, Johnson NW, Hardie IM. Dental caries, aetiology, pathology and prevention. London: The Macmillan press; 1981.
17. Newbrun E. Cariology. 2nd ed. Baltimore: Williams & Wilkens; 1983.
18. Hahn CL, Liewehr FR. Innate immune responses of the dental pulp to caries. J Endod. 2007;33:643–51.
19. Hahn CL, Liewehr FR. Update on the adaptive immune responses of the dental pulp. J Endod. 2007; 33:773–81.
20. Goldberg M, Farges J-C, Lacerda-Pinheiro S, Six N, Jegat N, Decup F, Septier D, Carrouel F, Durand S, Chaussain-Miller C, DenBesten P, Veis A, Poliard A. Inflammatory and immunological aspects of dental pulp repair. Pharmacol Res. 2008;58:137–47.
21. Cooper PR, Takahashi Y, Graham LW, Simon S, Imazato S, Smith AJ. Inflammation-regeneration interplay in the dentine-pulp complex. J Dent. 2010; 38:687–97.
22. Smith AJ, Scheven BA, Takahashi Y, Ferracane JL, Shelton RM, Cooper PR. Dentine as a bioactive extracellular matrix. Arch Oral Biol. 2012;57:109–21.
23. Couve E, Osorio R, Schmachtenberg O. The amazing odontoblast: activity, autophagy, and aging. J Dent Res. 2013;92:765–72.
24. Bjørndal L, Reit C, Bruun G, Markvart M, Kjäldgaard M, Näsman P, Thordrup M, Dige I, Nyvad B, Fransson H, Lager A, Ericson D, Petersson K, Olsson J, Santimano EM, Wennström A, Winkel P, Gluud C. Treatment of deep caries lesions in adults: randomized clinical trials comparing stepwise vs. direct complete excavation vs. partial pulpotomy. Eur J Oral Sci. 2010;118:290–7.
25. Maltz M, Garcia R, Jardim JJ, de Paula LM, Yamaguti PM, Moura MS, Garcia F, Nascimento C, Oliveira A, Mestrinho HD. Randomized trial of partial vs. stepwise caries removal: 3-year follow-up. J Dent Res. 2012;91:1026–31.
26. Phonghanyudh A, Phantumvanit P, Songpaisan Y, Petersen PE. Clinical evaluation of three caries removal approaches in primary teeth: a randomised controlled trial. Community Dent Health. 2012;29:173–8.
27. Orhan AI, Oz FT, Orhan K. Pulp exposure occurrence and outcomes after 1- or 2-visit indirect pulp therapy vs complete caries removal in primary and permanent molars. Pediatr Dent. 2010;32:347–55.
28. Ricketts D, Lamont T, Innes NP, Kidd E, Clarkson JE. Operative caries management in adults and children. Cochrane Database Syst Rev. 2013;(3):CD003808.
29. Bjørndal L, Kidd EA. The treatment of deep dentine caries lesions. Dent Update. 2005;32:402–4, 407–10, 413.
30. Bjørndal L, Darvann T. A light microscopic study of odontoblastic and non-odontoblastic cells involved in the tertiary dentinogenesis in well-defined cavitated carious lesions. Caries Res. 1999;33:50–60.
31. Bjørndal L, Darvann T, Thylstrup A. A quantitative light microscopic study of the odontoblastic and subodontoblastic reactions to active and arrested enamel caries without cavitation. Caries Res. 1998;32:59–69.
32. Bègue-Kirn C, Smith AJ, Ruch JV, Wozney JM, Purchio A, Hartmann D, Lesot H. Effects of dentin proteins, transforming growth factor β1 (TGFβ1) and bone morphogenetic protein 2 (BMP2) on the differentiation of odontoblasts in vitro. Int J Dev Biol. 1992;36:491–503.
33. Bègue-Kirn C, Smith AJ, Loriot M, Kupferle C, Ruch JV, Lesot H. Comparative analysis of TGF beta s, BMPs, IGF1, msxs, fibronectin, osteonectin and bonesialoprotein gene expression during normal and in vitro-induced odontoblast differentiation. Int J Dev Biol. 1994;38:405–20.
34. Smith AJ, Tobias RS, Plant CG, Browne RM, Lesot H, Ruch JV. In vivo morphogenetic activity of dentine matrix proteins. J Biol Buccale. 1990;18:123–9.
35. Sloan AJ, Smith AJ. Stimulation of the dentine–pulp complex of rat incisor teeth by transforming growth factor-β isoforms 1–3 in vitro. Arch Oral Biol. 1999;44:149–56.
36. Graham L, Cooper PR, Cassidy N, Nor JE, Sloan AJ, Smith AJ. The effect of calcium hydroxide on solubilisation of bio-active dentine matrix. Biomaterials. 2006;27:2865–73.
37. Tomson PL, Grover LM, Lumley PJ, Sloan AJ, Smith AJ, Cooper PR. Dissolution of bio-active dentine matrix components by mineral trioxide aggregate. J Dent. 2007;35:636–42.
38. Massler M, Pawlak J. The affected and infected pulp. Oral Surg. 1977;43:929–47.
39. Durand SH, Flacher V, Roméas A, Carrouel F, Colomb E, Vincent C, Magloire H, Couble M-L, Bleicher F, Staquet M-J, Lebecque S, Farges J-C. Lipoteichoic acid increases TLR and functional chemokine expression while reducing dentin formation in in vitro differentiated human odontoblasts. J Immunol. 2006;176:2880–7.
40. Staquet M-J, Durand SH, Colomb E, Roméas A, Vincent C, Bleicher F, Lebecque S, Farges J-C. Different roles of odontoblasts and fibroblasts in immunity. J Dent Res. 2008;87:256–61.
41. Keller JF, Carrouel F, Colomb E, Durand SH, Baudouin C, Msika P, Bleicher F, Vincent C, Staquet M-J, Farges J-C. Toll-like receptor 2 activation by lipoteichoic acid induces differential production of pro-inflammatory cytokines in human odontoblasts, dental pulp fibroblasts and immature dendritic cells. Immunobiology. 2010;215:53–9.
42. Farges J-C, Carrouel F, Keller JF, Baudouin C, Msika P, Bleicher F, Staquet M-J. Cytokine production by human odontoblast-like cells upon Toll-like receptor-2 engagement. Immunobiology. 2011;216:513–7.
43. Bjørndal L, Larsen T. Changes in the cultivable flora in deep carious lesions following a stepwise excavation procedure. Caries Res. 2000;34:502–8.
44. Hamilton IR. Ecological basis for dental caries. In: Kuramitsu HK, Ellen RP, editors. Oral bacterial ecology: the molecular basis. Wymondham: Horizon Scientific Press; 2000. p. 219–74.

45. Carrouel F, Staquet M-J, Keller JF, Baudouin C, Msika P, Bleicher F, Alliot-Licht B, Farges J-C. Lipopolysaccharide-binding protein inhibits toll-like receptor 2 activation by lipoteichoic acid in human odontoblast-like cells. J Endod. 2013;39:1008–14.
46. Bjørndal L, Thylstrup A, Larsen T. A clinical and microbiological study of deep carious lesions during stepwise excavation using long treatment intervals. Caries Res. 1997;31:411–7.
47. Anderson DM, Langeland K, Clark GE, Galich JW. Diagnostic criteria for the treatment of caries-induced pulpitis. NDRI-PR 1981;81–03.
48. Kassa D, Day P, High A, Duggal M. Histological comparison of pulpal inflammation in primary teeth with occlusal or proximal caries. Int J Paediatr Dent. 2009;19:26–33.
49. Reeves R, Stanley HR. The relationship of bacterial penetration and pulpal pathosis in carious teeth. Oral Surg Oral Med Oral Pathol. 1966;22:59–65.
50. Bogen G, Kim JS, Bakland LK. Direct pulp capping with mineral trioxide aggregate: an observational study. J Am Dent Assoc. 2008;139:305–15.
51. Ricucci D, Siqueira Jr JF. Endodontology. An integrated biological and clinical view. Berlin: Quintessence Publishing Co. Ltd; 2013. p. 31.
52. Schwendicke F, Meyer-Lueckel H, Dörfer C, Paris S. Failure of incompletely excavated teeth – a systematic review. J Dent. 2013;41:569–80.
53. Gruythuysen R, van Strijp G, Wu M-K. Long-term survival of indirect pulp treatment performed in primary and permanent teeth with clinically diagnosed deep carious lesions. J Endod. 2010;36:1490–3.
54. Fagundes TC, Barata TJE, Prakki A, Bresciani E, Pereira JC. Indirect pulp treatment in a permanent molar: case report of 4-year follow-up. J Appl Oral Sci. 2009;17:70–4.
55. Marthaler TM. Changes in dental caries 1953–2003. Caries Res. 2004;38:173–81.
56. Dumsha T, Hovland E. Considerations and treatment of direct and indirect pulp-capping. Dent Clin N Am. 1985;29:251–9.
57. Seale NS. Indirect pulp therapy: an alternative to pulpotomy in primary teeth. Tex Dent J. 2010;127:1175–83.
58. Seal NS, Glickman GN. Contemporary perspective on vital pulp therapy: views from the endodontists and pediatric dentists. J Endod. 2008;34:S57–61.
59. Simon S, Perard M, Zanini M, Smith AJ, Charpinter E, Djole SX, Lumley PJ. Should pulp champer pulpotomy be seen as a permanent treatment? Some preliminary thoughts. Int Endod J. 2013;46:79–87.
60. Asgary S, Eghbal MJ. Treatment outcomes of pulpotomy in permanent molars with irreversible pulpitis using biomaterials: a multi-center randomized controlled trial. Acta Odontol Scand. 2013;71:130–6.
61. Bjørndal L, Laustsen MH, Reit C. Root canal treatment in Denmark is most often carried out in carious vital molar teeth and retreatments are rare. Int Endod J. 2006;39:785–90.

10 Reactionary and Reparative Dentin-Like Structures

Michel Goldberg

10.1 Introduction

In reaction to a slow carious decay, continuous abrasion or noxious effects of monomers released by resin composites, odontoblasts may be wounded but are still alive (Figs. 10.1, 10.2, and 10.3a–c). They synthesize and secrete a true extracellular matrix which contributes to the formation of a mineralized *reactionary tubular dentin-like or a bone-like structure*. A "calciotraumatic" or a "reverse" line separates the newly formed reactionary dentin from the primary or secondary dentin already formed before any development of the traumatic decay (Figs. 10.4a, b and 10.5a–d). It happens that odontoblasts wounded by the release of carious toxins are unable to survive. During dentinogenesis, odontoblast cell bodies constitute a thick layer at the pulp periphery, and four rows of cells may be scored during early dentin formation. Aging influences the thickness of this layer, which is gradually reduced. Although there is no direct evidence that the cells from the Hoehl's layer may be reactivated, differentiated, and become secondary odontoblasts, this is probably what happens. Even if some differences are well identified between a true orthodentin and reactionary dentin-like structure, odontoblasts and/or Hoehl's cells are responsible for the formation of reactionary dentin.

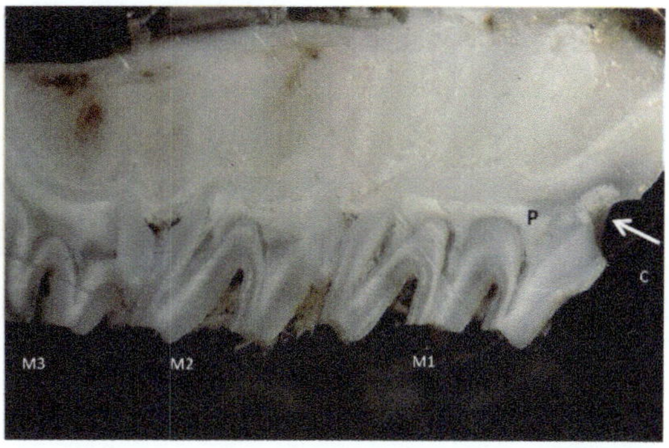

Fig. 10.1 Longitudinal section of the three mandibular molars (M1, M2, and M3). A cavity was prepared in the mesial part of M1. *P* pulp, *C* cavity

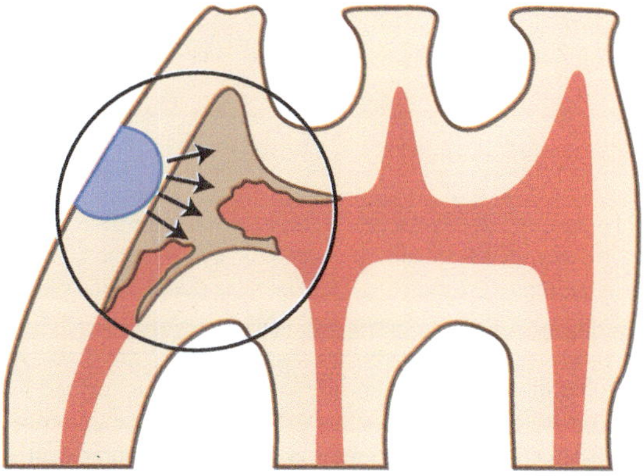

Fig. 10.2 Schematic drawing showing the formation of reactionary dentin beneath a cavity prepared in the mesial aspect of the first molar

M. Goldberg, DDS, PhD
Department of Oral Biology, Institut National de la Santé et de la Recherche Médicale, Université Paris Descartes, 45 Rue des Saints Pères, Paris 75006, France
e-mail: mgoldod@gmail.com, michel.goldberg@parisdescartes.fr

M. Goldberg (ed.), *The Dental Pulp*,
DOI 10.1007/978-3-642-55160-4_10,

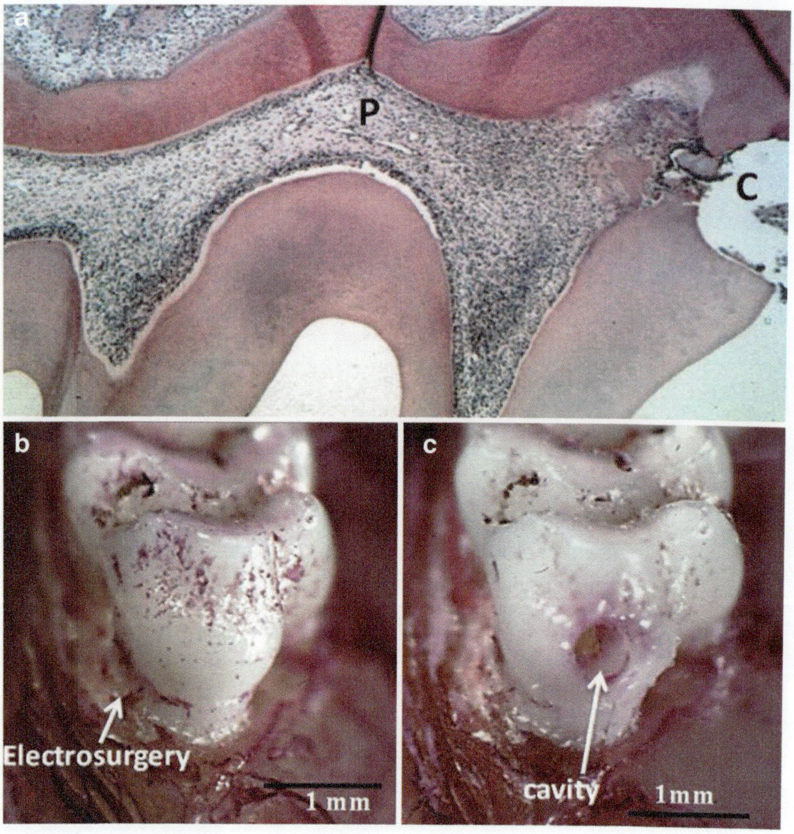

Fig. 10.3 (**a**) Histological section of the demineralized first mandibular molar. After the preparation of a cavity in the mesial aspect (*c*), a slight reaction occurs in the pulp (*P*). In (**b**), the tooth is seen after a cervical electrosurgery, and in (**c**), a cavity is drilled in the mesial aspect

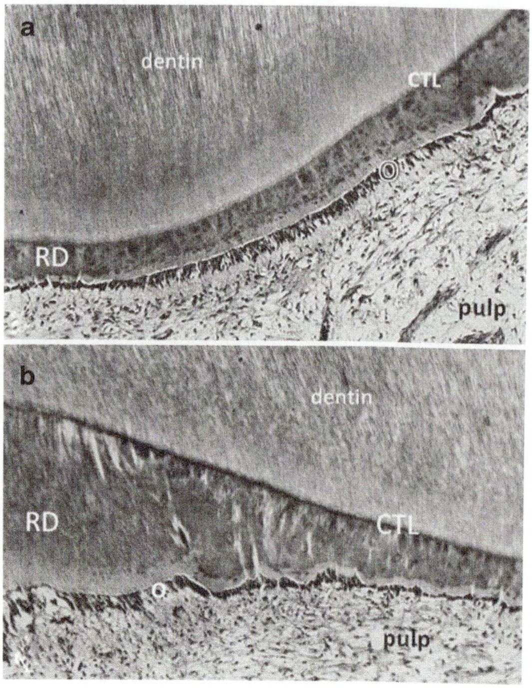

Fig. 10.4 Atubular reactionary dentin (*RD*) is formed by the odontoblasts and Hoehl's cells, beneath a calciotraumatic line (*CTL*) at the junction between a tubular dentin and odontoblast layer (*O*). The dental pulp is fibrous but still alive. (**a**) Thin reactionary dentin formation; (**b**) thicker reparative dentin formation, produced in the same experimental conditions

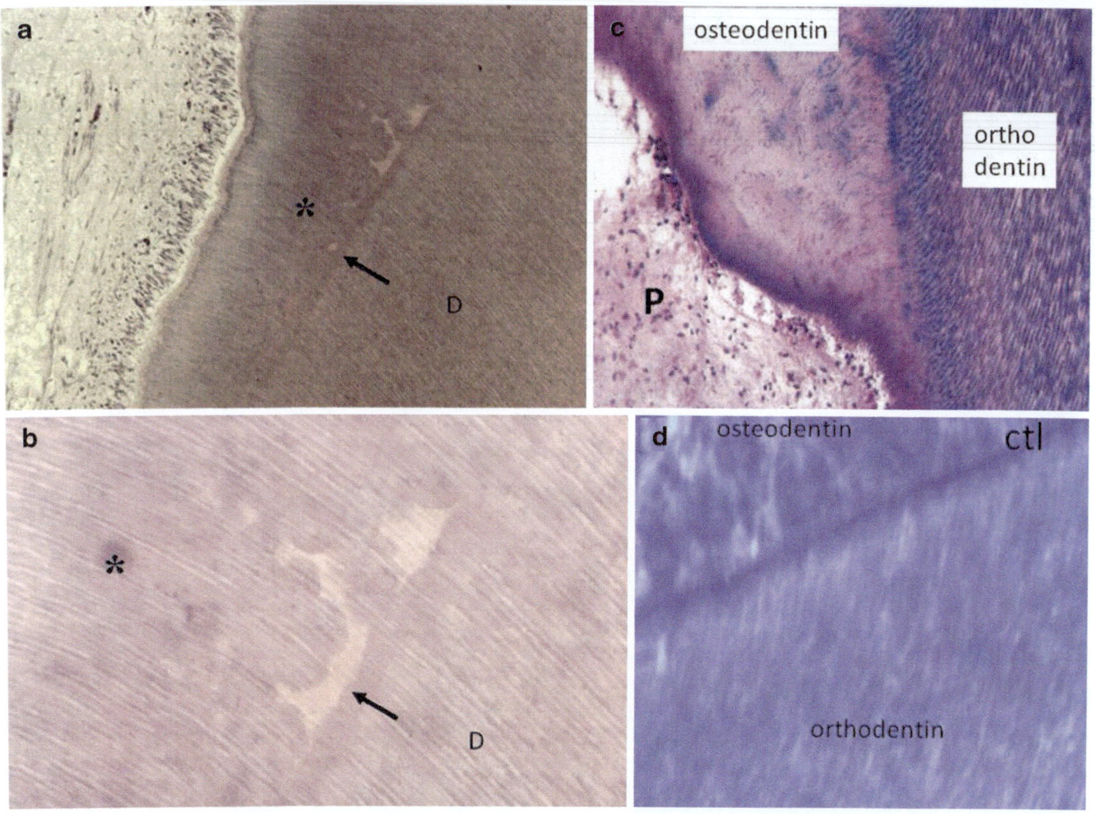

Fig. 10.5 (**a**, **b**) The preparation of a cavity induces the interruption of dentinogenesis. Beneath the tubular dentin formation, the *dark arrows* show interrupted physiological dentin deposit. The effects of the trauma lead to calcospherite formation (globular structures). The newly formed dentin (*asterisks*) comprises globular structures and interglobular spaces. Odontoblasts and pulp are "normal." (**c**) After a more severe decay, tubular orthodentin (stained in purple by the "stains all" method) is interrupted; and the new osteodentin formed in reaction to the trauma, is atubular, and contains a few osteoblast lacunas. The dental pulp is apparently sound. (**d**) Reactionary osteodentin is separated from orthodentin by a calciotraumatic line (*CTL*)

If the carious lesion is progressing quickly, the lesion may destroy the residual dentin and penetrate the dental pulp. The pulp exposure favors the diffusion of bacteria from the contaminated dentin within the dental pulp. Pulp inflammatory cells control the infection and immune cells tend to slow down the progression of the lesion.

The *reparative dentin-like* structure that is formed constitutes an attempt to close the pulp exposure. In this case, odontoblasts and Hoehl's cells are irreversibly wounded. Pulp progenitors or stem cells are implicated in the formation of a reparative dentin bridge or in a "bone-like" structure, also named osteodentin (Fig. 10.6a–d). Pulp cells are embedded in a bone-like structure similar to osteocyte-like cell being. These cells are located inside a lacuna. Reparative dentin-like fills partially or totally the mesial part of the pulp chamber of a rat's molar. The dentin bridge seen at early stages of pulp capping expands and seals the mesial pulp chamber. The mineralization process expands up to the isthmus between the mesial and central parts. In humans, the reparative process is initially located beneath the pulp exposure. Then an osteodentin bridge gradually loads the pulp chamber. The mineralized structure is either in continuity with the newly formed reactionary dentin or appears as free pulp stone (calcospherites), developing around endothelial cells of capillaries.

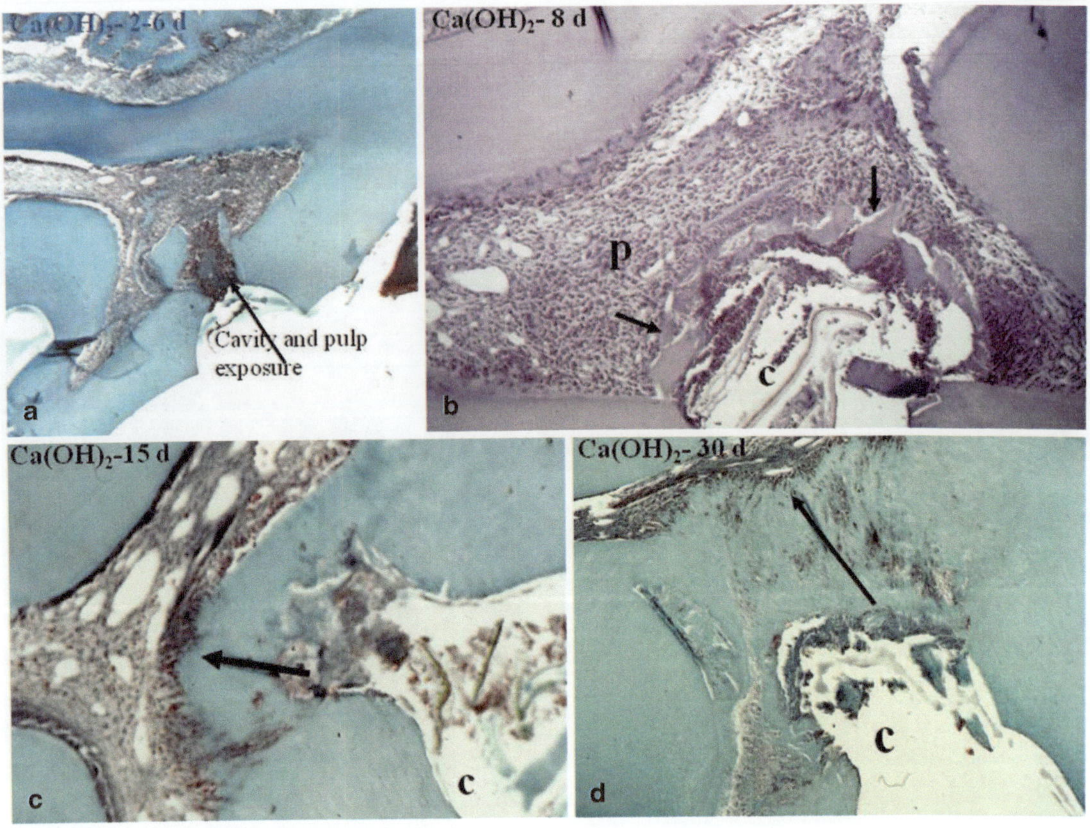

Fig. 10.6 In (**a**), 2–6 days after pulp capping a pulp exposure with $Ca(OH)_2$, some dentin debris are pushed within the pulp. A pulp inflammatory process is limited to the mesial part of the pulp chamber. In (**b**), after 8 days, the calcium hydroxide induces the early formation of a dentin bridge, in close association with dentin debris (*arrows*). In (**c**), after 15 days, the dentin bridge is thicker and more homogeneous. (**d**) After 30 days, the dentin bridge occludes totally the pulp exposure. However, some tunnellike structures or cell remnants persist, allowing bacterial penetration toward the dental pulp. *P* pulp, *C* cavity

In many publications, some confusion occurs between reactionary and reparative dentin-like structures. To be precise, a clear-cut difference should be made between a tubular or atubular reactionary dentin, formed under the control of odontoblasts and Hoehl's cells, and reparative dentin, appearing mostly as osteodentin and produced by pulp stem cells or progenitors.

A series of questions result from the available valid data. Wound healing results from repair or regeneration processes. Many major questions arise, including can we restore the original architecture and the biological function of the injured pulp tissue? Complete regeneration occurs during the fetal period, within the 24 weeks of gestation. However, in a clinical setting, we are treating young patients with already erupted teeth and/or adult patients. This implies that the postnatal wound healing combines a cascade of events leading to pulp repair and/or regeneration. The recruitment and differentiation of potential progenitors/stem cells are prerequisites. This is followed by pulp reconstruction.

10.2 In Vivo Approach

10.2.1 Preparation of a Cavity Without Pulp Exposure

In response to the preparation of a cavity in the mesial aspect of the first maxillary molar, odontoblasts, Hoehl's cells, and pulp tissue reflect the reaction to the preparation of a cavity. Drilling a cavity caused the displacement of some odontoblasts and their penetration into dentin tubules. Inflammatory exudation was seen soon after drilling. The endothelium of capillaries showed an increase of pinocytic vesicles, an event associated with the formation of an exudative lesion. After 1 day, many damaged odontoblasts degenerate. Cells with a high nucleus/cytoplasm ratio and prominent nucleoli accumulate around the subodontoblastic capillaries. Newly differentiating odontoblasts received nutritional supply from the capillaries. After 3 days, differentiating odontoblasts increased in number. They start to produce reactionary dentin by 5 days after cavity preparation. Differences were seen between the original dentin layer formed before the drilling of the cavity and reactionary dentin-like formation [1].

10.2.2 Preparation of a Cavity Followed by a Pulp Exposure

Different repair responses were recorded between the coronal and radicular areas after the implantation of bone morphogenetic protein-7 (BMP-7) in a pulp exposure. Used as a capping agent, BMP-7 induces after 8 days the formation of a reparative osteodentin bridge, formed by globular mineralized areas at the exposure site and leaving unmineralized interglobular areas containing pulp remnants. A heterogeneous mineralization was seen in the coronal part of the pulp. Complete filling of the radicular pulp by a homogeneous mineralization was seen in the root, behind a calciotraumatic line. These results emphasize the biological differences between the coronal and radicular parts of the pulp [2]. Indeed, the crown is formed under the control of enamel organ, whereas the extracellular matrix (ECM) secreted by the epithelial Hertwig's root sheath influences the root construction.

10.3 Reactionary Dentin (Tertiary Dentin)

Reviewing data on reactionary dentinogenesis, Smith et al. [3] differentiate between tertiary dentin and reparative dentinogenesis (Figs. 10.6a–d and 10.7a–d). It is indeed difficult to discriminate between the dentin secreted by postmitotic odontoblasts and a mineralized structure formed by a new population of pulp-derived, odontoblast-like cells. The tertiary dentin beneath a carious decay may comprise both reactionary and reparative dentins. Hence, a clear-cut definition for each of the tissues is needed and is difficult to obtain.

Reactionary dentin is formed in response to a carious decay, to excessive abrasion, or to the cytotoxic effects of monomers released by a restorative material. This is how the pulp limits undesirable noxious effects. Trans-dentin stimulation is mediated by bioactive extracellular matrix components during a cavity preparation, the dental pulp being nonexposed. As an example, OP-1, used as a cavity liner, stimulates the formation of reactionary dentin [4]. Odontoblasts respond to the stimulating presence of ECM. Diffusion of ECM proteins, as determined by the residual dentin thickness (RDT) after preparation of a cavity, influences the signaling process [5]. They synthesize and secrete the ECM of a tubular or atubular dentin-like structure beneath a calciotraumatic line [6]. This line is similar to the reverse line seen in the bone. Sometimes this dentin-like structure is globular, but very often appears as lamellar or amorphous. Reactionary dentin also named tertiary dentin differs from the orthodentin formed prior to the lesion. The matrix of reactionary dentin displays reduced hardness and lowers elastic modulus. The residual dentin thickness (RDT) interferes with the pulp activity. The maximal dentin deposition of reactionary dentin appears beneath cavities with an RDT varying between 0.5 and 0.25 mm. Odontoblasts are reducing in number in cavities closer to the indirectly injured pulp. The restoration material influences also the odontoblast activities [7].

10.3.1 Molecules Expressed or Influencing Reactionary Dentin Formation

Differences were observed between reactionary dentin (RD) and primary or secondary orthodentin with respect to the distribution of *five SIBLINGs* (*BSP*, *OPN*, *DMP-1*, *DSPP*, and *DSP* (a fragment of DSPP)). BSP and OPN were

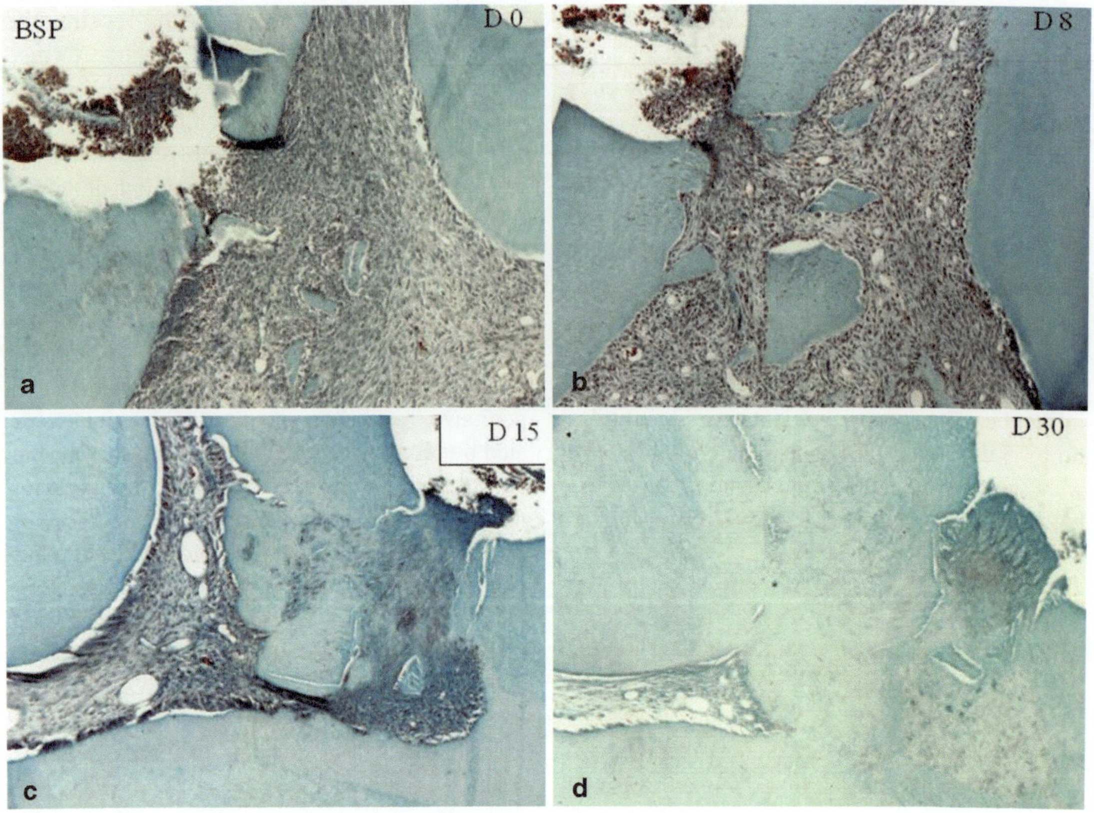

Fig. 10.7 Capping effects of bone sialoprotein (*BSP*). At day 0 (**a**) and at day 8 (**b**). Dentin debris are pushed within the pulp. (**c**) At day 15, beginning of the formation of a reparative dentin bridge. (**d**) 30 days after pulp capping with BSP, a thick and homogeneous dentin bridge closes totally the pulp exposure

observed in RD but not in predentin, whereas the expression of DMP-1 and DSP was lower in RD compared with predentin [8]. DSP was increased in density and spatial resolution. This molecule contributes to putative HAp nucleation on collagen scaffold. DSP antibodies showed weak staining in RD, whereas osteopontin (OPN) was extensively positive in RD [9]. OPN was not detected in physiological and reactionary dentin, but seen to be strongly immunoreactive in reparative dentin alone [10] (Fig. 10.7a–d).

10.3.2 Other ECM Molecules

- *Metalloproteinase-2* (*MMP-2*) is increased. Parallel upregulation occurs for TIMP-2 and for the membrane type 1-matrix metalloproteinase (MT1-MMP). Furthermore, the genes encoding components of the TGF-β signaling pathway, namely, *SMAD-2* and *SMAD-4*, may explain the increased synthesis of collagen [11, 12].
- *Odontogenic ameloblast-associated protein* (*ODAM*) is expressed by ameloblasts and odontoblasts. The molecule plays a role in enamel mineralization, possibly through the regulation of MMP-20. Experimental results show that rODAM accelerates reactionary dentin formation near the pulp exposure area, preserving normal odontoblasts in the remaining pulp [13].
- Finally, the expression of *Toll-like receptors* was altered in response to caries.

10.4 Reparative Dentin-Like: From the Initial Formation of a dentin Bridge to Pulp Mineralization

10.4.1 Cells Implicated in Reparative Dentin-Like Formation

When odontoblasts are injured and destroyed, replacement cells express types I and III collagen gene-specific riboprobes. The cells forming reparative dentin synthesize type I collagen but not type III. Antibodies raised against DSP positively stain the cells. Therefore, they are odontoblast-like cells [14].

Adult human dental pulp stem cells (hDPSCs) have been isolated from the pulp as precursor cells. They differentiate into cells implicated in the formation of reparative dentin. Lysyl oxidase-like 2 (LOXL2) was downregulated when hDPSCs differentiate into odontoblast-like cells. Therefore, LOXL2 has negative effect on the differentiation of pulp cells into odontoblasts [15]. In contrast, the other LOX family members including LOX, LOXL1, LOXL3, and LOXL4 are increased.

Cells emerging from dental pulp explant were studied to elucidate the origin of precursor cells implicated in the formation of reparative dentin. Early outgrowing cells emerging from cultured explant were round or elongated, with thin spinous processes. They were highly mobile and contain numerous lipid vesicles. Radioautography suggested that these lipids resulted from micropinocytosis. After 10–20 days, the cells started to be converted into fibroblast-like cells, less mobile and lacking lipid vesicles. It was concluded that they might be mononuclear phagocytic/histiocytic mesenchymal cells [16].

In vivo, 10 days after $Ca(OH)^2$ pulp capping, focal calcifications were seen within the collagen-rich matrix. Numerous extracellular matrix vesicles display electron-dense material composed of hydroxyapatite crystals [17]. At this early stage, similarities were detected between chondrocyte and the mineralization of cartilage septa and the early formation of reparative dentin.

Frozoni et al. [18] have used the 3.6-green fluorescent protein (GFP) transgenic mice to in vivo the biological sequence of events during pulp healing and reparative dentinogenesis. After pulp exposure and capping, followed by chemical fixation and processing for histological and epifluorescence analysis, immediately after pulp exposure no 3.6-GFP-labeled odontoblast was detected at the injury site. Four weeks after the surgery, reparative dentin started to be formed, and this was even more obvious at 8 weeks. The cells expressing 3.6-GFP lined an atubular dentin bridge. A few fluorescent cells were embedded within the atubular matrix, hence appeared as a bone-like structure (Figs. 10.8a–c and 10.9a–d).

Dentin regeneration may be obtained by using porcine deciduous pulp stem cells mixed with β-tricalcium phosphate. Four months after transplantation, regeneration of a dentin-like structure was completely completed, closing the roof defects [19] (Fig. 10.10a, b).

Canonical Wnt/β-catenin regulates odontoblast-like differentiation, improves moderately the expression of type I collagen, and enhances strongly the expression of osteopontin. Wnt-1 inhibited alkaline phosphatase and the formation of mineralized nodules. Scheller et al. [20] concluded that the canonical Wnt signaling regulates negatively the differentiation and mineralization of dental pulp stem cells.

10.4.2 Cell Mediators Implicated in Reparative Dentin-Like Formation

Bone morphogenetic proteins (BMPs) are multifunctional proteins structurally related to the *transforming growth factor-beta* (TGF-beta) and *activin*, which can induce cartilage and bone growth in vivo. BMPs are members of the TGF-beta superfamily. Their effects on pulp cells and reparative dentin formation have been reported.

The *TGF-beta1* superfamily influences the expression of BSP, DSP, TGF-beta1 receptor I, and Smad 2/3 proteins during reparative dentinogenesis. In vitro, TGF-beta2 had minimal effect on cultured tissues, whereas TGF-beta1 and

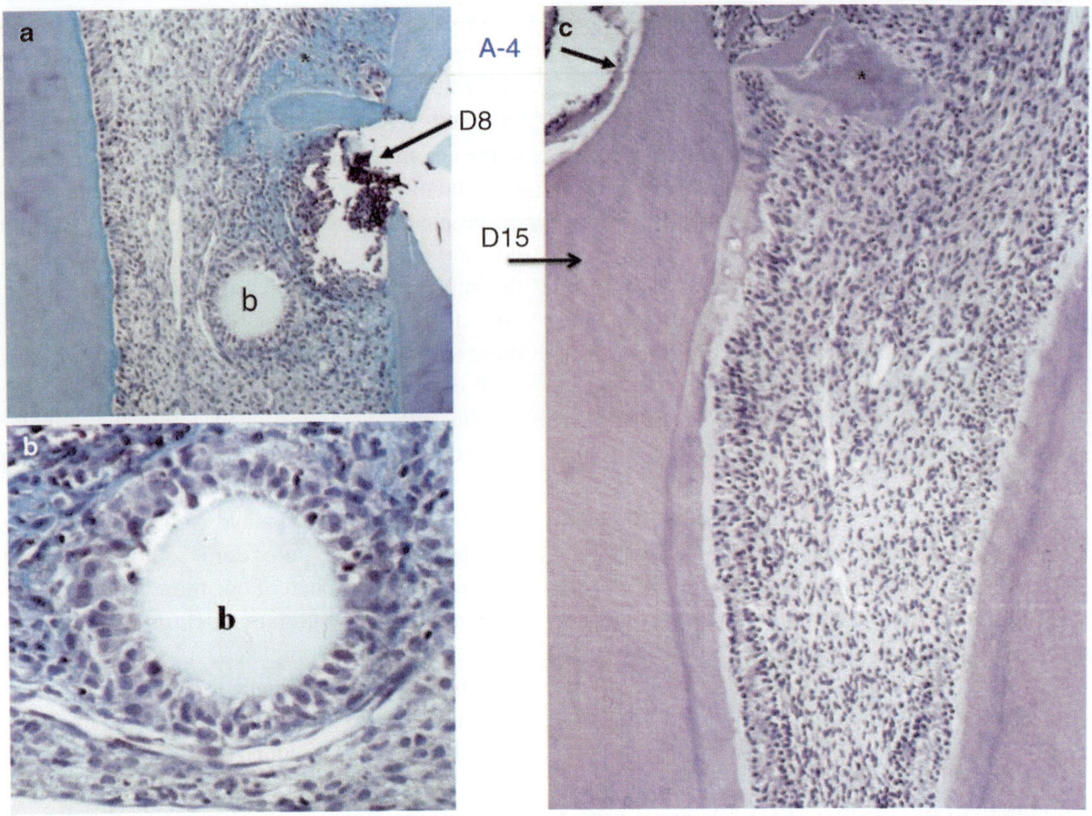

Fig. 10.8 (**a**) Implantation within the dental pulp of beads loaded with an amelogenin gene splice product A-4 after 8 days. (**b**) Committed cells are recruited and form a ring around the agarose bead (*b*). (**c**) At 15 days, reactionary dentin is formed uniformly along the root canal, behind a calciotraumatic line. In the crown, a mineralized area starts to be formed (*asterisk*)

TGF-beta3 stimulate the secretion of ECM by odontoblasts. They are mitogenic for pulp cells and might be important during reparative processes [21]. The level of BSP is increased, but DSPP is decreased. Smad 2/3 level was higher in the reparative dentin than in the normal dentin. Hwang et al. [22] concluded that both dentinogenetic and osteogenetic characteristics are mediated by TGF-beta1.

10.4.2.1 Bone Morphogenetic Proteins

The nuclear proto-oncogenes c-jun and jun-B were induced by growth factors identified as *bone morphogenetic proteins* (*BMPs*). The gene products enhance the expression of osteocalcin and collagen types. In tooth germs, c-jun and jun-B were co-expressed in the odontoblast lineage. In rat adult molars, c-jun was expressed in the odontoblast layer, in contrast with jun-B, which was absent from the cells. After cavity preparation, c-jun and jun-B were expressed only in the pulp cells lining the irregular surface of the thick reparative dentin. The limited expression of jun-B suggests that this gene is involved in active formation of reparative dentin, but is not a major actor [23].

Reparative dentin was formed by recombinant *human osteogenic protein-1* (*OP-1 or BMP-7*) in a time and dose dependences [24]. hOP-1 implanted in vivo in extra- and intra-skeletal sites induces cartilage and bone. The reparative dentin was not completely mineralized after 6 weeks of healing. Radicular pulp vitality was maintained and reparative dentin formed, and mineralization

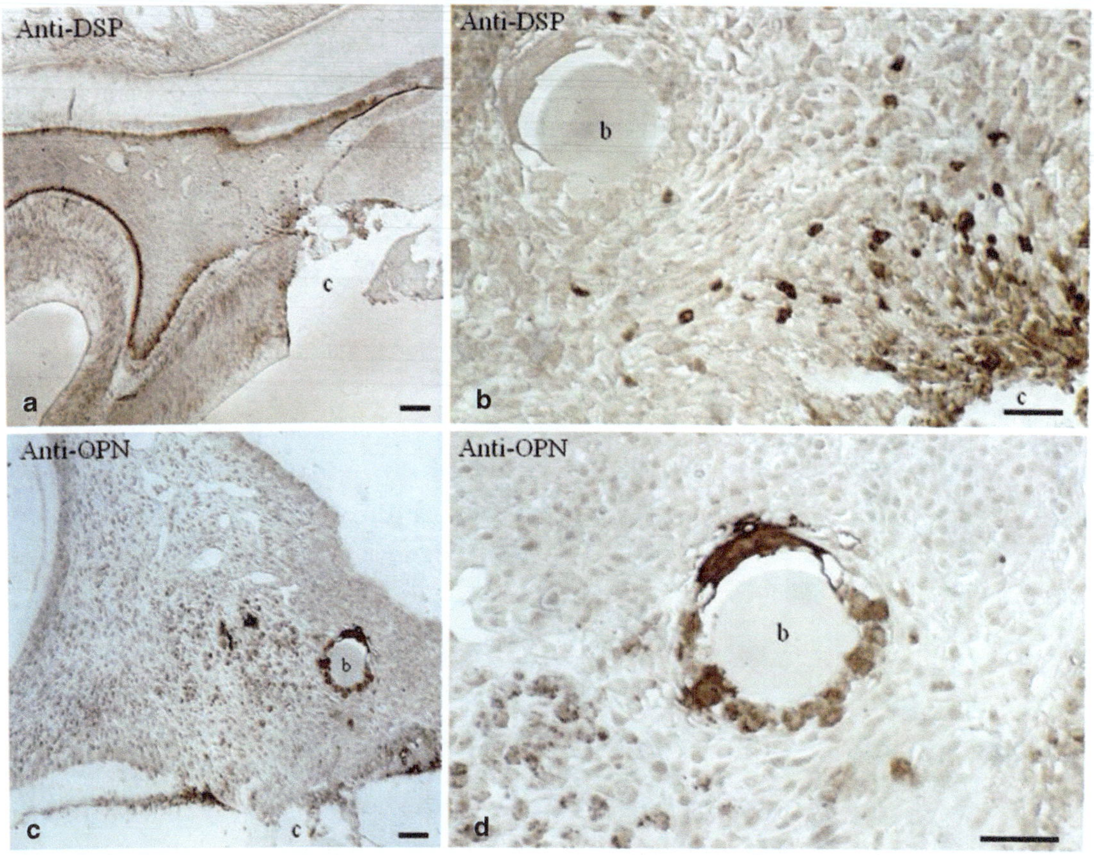

Fig. 10.9 (**a**) After a pulp exposure and dentin sialoprotein (*DSP*) immunolabeling, an odontoblast layer is densely stained. (**b**) No staining is detected around the beads (*b*), but near the pulp exposure (*c* cavity), odontoblast-like cells are well stained by the antibody. (**c**) In contrast, osteopontin immunolabeling is dense around the carrier bead (*b*) and in close proximity. (**d**) The cells located around the bead are densely immunostained by the anti-osteopontin. The two immunostainings reveal that the nature of cells nearby the pulp exposure (*DSP* dentin ECM protein) is very different from the cells implicated in pulp inflammation and bone formation (osteopontin)

was nearly 75 % complete after 1 month and more than 95 % after 4 months [25]. However, implantation of BMP-7 in inflamed dental pulp did not produced reparative dentin in LPS-treated pulp [26]. Indeed, BMP-7 gene induces reparative dentin formation, except when the dental pulp is inflamed due to a bacterial infection [27].

Pellets of *bone morphogenetic protein-2* (*BMP-2*) implanted into amputated pulps stimulate direct progenitor/stem cell differentiation into odontoblasts and reparative dentin formation [28]. The BMP signal is probably mediated by interaction of types I and II BMP receptors. RT-PCR suggests that resident pulp cells are able to respond to BMPs to initiate tissue formation [29].

10.4.2.2 Other Growth Factors

- The *lymphoid enhancer-binding factor 1* (*Lef1*) is a transcription factor that mediates Wnt signaling and regulates DSPP expression [30].
- The in vivo transfer of *growth/differentiation factor 11* (*Gdf11*) by electroporation stimulates the reparative dentin formation [31].
- The proinflammatory cytokine *tumor necrosis factor-α* (*TNF-α*) may be a mediator involved in the differentiation of pulp cells toward the odontoblast phenotype. TNF-α stimulates the differentiation of dental pulp cells toward an odontoblast phenotype via p38, while negatively regulating MMP-1 expression. Extracellular DPP and DSP were detected

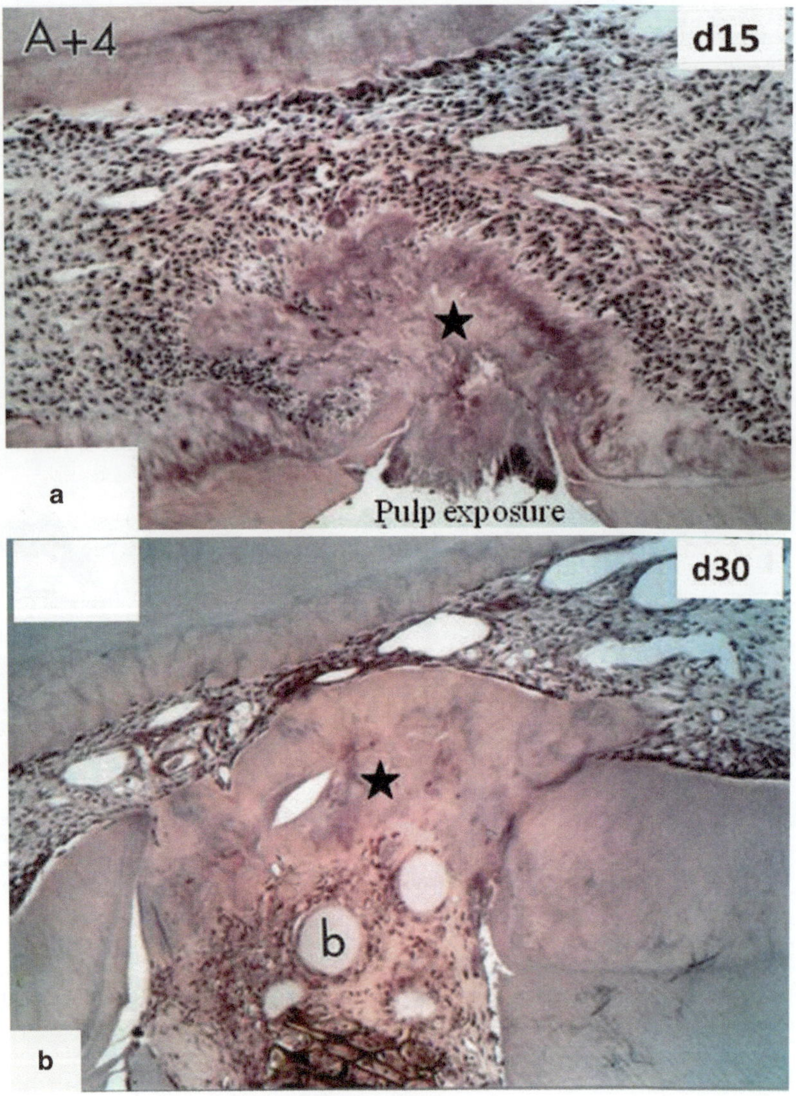

Fig. 10.10 In (**a**), A+4 was implanted in the dental pulp 15 days earlier. A forming homogeneous reparative dentin bridge (*asterisk*) is closing the pulp exposure. (**b**) After 30 days, the reparative dentin bridge is much thicker (*asterisk*) and many beads (*b*) are embedded within dentin

in higher amount in conditioned media from TNF-α-treated pulp cells [32].

- The *connective tissue growth factor/CCN family 2* (*CTGF/CCN2*) is involved in reparative dentinogenesis through the formation of mineralized tissues in human carious teeth [33].
- Recombinant human *insulin-like growth factor-1* (*rhIGF-1*) used for direct pulp capping was seen to induce after 28 days a whole dentin bridge where tubular dentin formation was observed [34].
- *Glypican-1* (*GPC-1*), a cell surface heparin sulfate proteoglycan, is acting as a coreceptor for heparin-binding growth factor and member of the TGF-β. It is related to reparative dentin formation. Downregulation of expression resulted in a 3.9-fold increase of *TGF-β1* expression in the pulp cells and 0.3-fold decrease in *DSPP* expression compared with control. The two molecules are necessary at the onset of differentiation, but should be downregulated before other molecules could be in formation [35].
- *Fibroblast growth factor 23* (*FGF23*) increased the predentin volume and expression of biglycan in dentin. FGF23 overexpression plays a negative role on dentinogenesis [36].

10.4.2.3 ECM

Collagens

Absent in normal dentin, type III collagen is present in reparative dentin. Type III collagen is more frequently observed in the root than in the crown. They are mostly located in the peritubular dentin. Collagen fibrils showed a clear cross-banding and unusual collagen aggregations, segment, and fibrous long-spacing-like structures, intensely stained for type I collagen, but weakly for type III collagen. The type III collagen-positive fibers often extended toward the pulp beyond the odontoblast layer. They may be produced at least partially by the pulp cells [37].

10.4.2.4 SIBLINGs

- *Dentin phosphophoryn* (*DPP*), *either phosphorylated or dephosphorylated*, promotes cell migration in a concentration-dependent manner, but has no effect on cell proliferation. The addition of αvβ3 integrin antibody to the medium suppressed the cell migration. Porcine DPP-derived RGD peptide significantly promotes the cell migration of human dental pulp cells [38].
- *Bone sialoprotein* (*BSP*) implanted in the pulp of the first maxillary molar induced a slight inflammation 8 days after implantation. After 15 days, a dentin bridge starts to be formed. After 30 days osteodentin formation filled the mesial part of the pulp chamber, which occluded totally the pulp exposure [39].
- BSP induced a homogeneous and well-mineralized reparative dentin. BMP-7 gave reparative dentin of the osteodentin type in the coronal part of the tooth and generated the formation of a homogeneous mineralized structure in the root canal [40].
- *Dentin matrix protein-1* (*DMP-1*) induces cytodifferentiation of dental pulp stem cells into odontoblasts [41]. After pulp exposure and implantation of collagen matrix impregnated with DMP-1, it was seen that DMP-1 acts as a morphogen on undifferentiated cells that have the capacity to regenerate a collagen-rich dentin-like tissue.
- *Dentonin*, a peptide derived from the phosphorylated extracellular matrix protein (*MEPE*), promotes the proliferation of pulp cells at day 8 after implantation in the pulp exposure. Osteopontin, weakly labeled at day 8, was increased at 15 days. DSP was undetectable at any time. Dentonin affects primarily the initial cascade of events leading to pulp healing [42].
- The layer of *fibronectin* is denser in fibrodentin after 3 days after MTA pulp capping, more than after Ca(OH)2. Fibronectin is involved in the initial stages of odontoblast differentiation. Dystrophic calcifications in association with cell debris and irregular fibrous matrix were fibronectin positive, supporting that fibronectin plays a mediating role during reparative dentinogenesis [43].
- *Ameloblastin* (also named amelin or sheathlin) influences reparative dentin formation [44], whereas the *amelogenin* gene splice products A+4 and A-4 determine the reorientation of CD45-positive cells to an osteochondrogenic lineage as shown by the positive markers RP59, Sox9, and BSP [45]. A peptide including two amelogenin exons (exons 8 and 9) enhances leucine-rich amelogenin peptide (LRAP)-mediated dental pulp repair [46].
- *MMP-3 and MMP-9* are upregulated at 24 h and 12 h, respectively, after the pulp wound, whereas MMP-2 and MMP-14 are unchanged. MMP-3 induced angiogenesis, pulp healing, and reparative dentin formation [47].

10.5 Summary

In response to a dentin lesion (caries or abrasion) or to the diffusion of resin monomers or to the release of bacterial toxins, odontoblasts and/or Hoehl's cells produce a reactionary dentin-like layer that increases the residual dentin thickness. This leads to the formation of a *reactionary dentin-like* layer (Fig. 10.11a, b). In such a case, dentin adsorbs the noxious molecules and contributes to the closure of dentin lumens by mineral precipitation, and the protection of the pulp is increased.

After a pulp exposure, pulp cells are recruited, differentiate into odontoblast-like cells, and contribute to the formation of a *dentin bridge*.

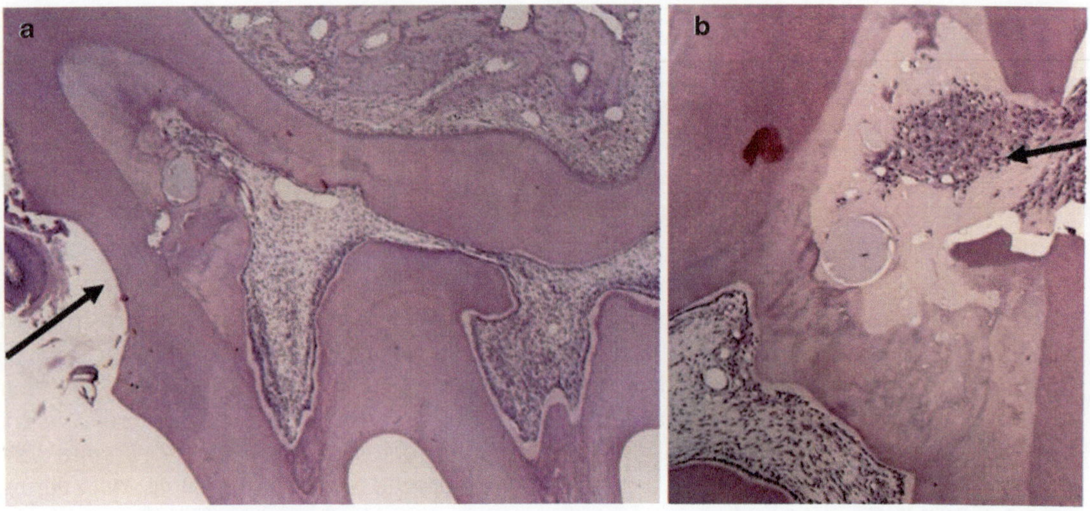

Fig. 10.11 Ninety days after A-4 implantation, reactionary dentin is formed in the pulp chamber and in the root canal (**a**). Reparative dentin fills the pulp exposure (**b**). Here an agarose bead and cell remnants are seen, embedded in the dentin bridge

For a long time, pulp capping has been a therapy, but defects in the dentin bridge lead to failures of the capping method. New therapies arise from the implantation of ECM molecules and the massive formation of *reparative dentin-like* structures. Either as appended dentin layers or as diffuse mineralization within the pulp, the remaining vital tissue is alive.

Comprehensive studies on the formation of the two different dentin-like layers will lead to a better understanding of the mechanisms that are involved. They may be used in the framework of new regenerative therapies.

References

1. Ohshima H. Ultrastructural changes in odontoblasts and pulp capillaries following cavity preparation in rat molars. Arch Histol Cytol. 1990;53:423–38.
2. Six N, Lasfargues JJ, Goldberg M. Differential repair responses in the coronal and radicular areas of the exposed rat molar pulp induced by recombinant human bone morphogenetic protein 7 (osteogenic protein 1). Arch Oral Biol. 2002;47:177–87.
3. Smith AJ, Cassidy N, Perry H, Bégue-Kirn C, Ruch J-V, Lesot H. Reactionary dentinogenesis. Int J Dev Biol. 1995;39:273–80.
4. Rutherford B, Spångberg L, Tucker M, Charette M. Transdentinal stimulation of reparative dentine formation by osteogenic protein-1 in monkeys. Arch Oral Biol. 1995;40:681–3.
5. Smith AJ, Tobias RS, Murray PE. Transdentinal stimulation of reactionary dentinogenesis in ferrets by dentine matrix components. J Dent. 2001;29:341–6.
6. Smith AJ, Murray PE, Sloan AJ, Matthews JB, Zhao S. Trans-dentinal stimulation of tertiary dentinogenesis. Adv Dent Res. 2001;15:51–4.
7. Murray PE, About I, Lumley PJ, Franquin JC, Remusat M, Smith AJ. Cavity remaining dentin thickness and pulpal activity. Am J Dent. 2002;15:41–6.
8. Moses KD, Butler WT, Qin C. Immunohistochemical study of small integrin-binding ligand, N-linked glycoproteins in reactionary dentin of rat molars at different ages. Eur J Oral Sci. 2006;114:216–22.
9. Yuan G, Yang G, Song G, Chen Z, Chen S. Immunohistochemical localization of the NH(2)-terminal and COOH-terminal fragments of dentin sialoprotein in mouse teeth. Cell Tissue Res. 2012; 349:605–14.
10. Aguiar MC, Arana-Chavez VE. Ultrastructural and immunocytochemical analyses of osteopontin in reactionary and reparative dentine formed after extrusion of upper rat incisors. J Anat. 2007;210:418–27.
11. Charadram N, Farahani RM, Harty D, Rathsam C, Swain MV, Hunter N. Regulation of reactionary dentin formation by odontoblasts in response to polymicrobial invasion of dentin matrix. Bone. 2012;50:265–75.
12. Charadram N, Austin C, Trimby P, Simonian M, Swain MV, Hunter N. Structural analysis of reactionary dentin formed in response to polymicrobial invasion. J Struct Biol. 2013;181:207–22.
13. Yang IS, Lee DS, Park JT, Kim HJ, Son HH, Park JC. Tertiary dentin formation after direct pulp capping

with odontogenic ameloblast-associated protein in rat teeth. J Endod. 2010;36:1956–62.

14. D'Souza RN, Bachman T, Baumgardner KR, Butler WT, Litz M. Characterization of cellular responses involved in reparative dentinogenesis in rat molars. J Dent Res. 1995;74:702–9.
15. Kim JH, Lee EH, Park HJ, Park EK, Kwon TG, Shin HI, Cho JY. The role of lysyl oxidase-like 2 in the odontogenic differentiation of human dental pulp stem cells. Mol Cells. 2013;35:543–9.
16. Stanislawski L, Carreau JP, Pouchelet M, Chen ZH, Goldberg M. In vitro culture of human dental pulp cells: some aspects of cells emerging early from the explant. Clin Oral Investig. 1997;1:131–40.
17. Sela J, Tamari I, Hirschfeld Z, Bab I. Transmission electron microscopy of reparative dentin in rat molar pulps. Primary mineralization via extracellular matrix vesicles. Acta Anat (Basel). 1981;109:247–51.
18. Frozoni M, Balic A, Sagomonyants K, Zaia AA, Line SR, Mina M. A feasibility study for the analysis of reparative dentinogenesis in pOBCol3.6GFPtpz transgenic mice. Int Endod J. 2012;45:907–14.
19. Zheng Y, Wang XY, Wang YM, Liu XY, Zhang CM, Hou BX, Wang SL. Dentin regeneration using deciduous pulp stem/progenitor cells. J Dent Res. 2012; 91:676–82.
20. Scheller EL, Chang J, Wang CY. Wnt/β-catenin inhibits dental pulp stem cell differentiation. J Dent Res. 2008;87:126–30.
21. Sloan AJ, Smith AJ. Stimulation of the dentine-pulp complex of rat incisor teeth by transforming growth factor-beta isoforms 1–3 in vitro. Arch Oral Biol. 1999;44:149–56.
22. Hwang YC, Hwang IN, Oh WM, Park JC, Lee DS, Son HH. Influence of TGF-beta1 on the expression of BSP, DSP, TGF-beta1 receptor I and Smad proteins during reparative dentinogenesis. J Mol Histol. 2008; 39:153–60.
23. Kitamura C, Kimura K, Nakayama T, Terashita M. Temporal and spatial expression of c-jun and jun-B proto-oncogenes in pulp cells involved with reparative dentinogenesis after cavity preparation of rat molars. J Dent Res. 1999;78:673–80.
24. Rutherford RB, Wahle J, Tucker M, Rueger D, Charette M. Induction of reparative dentine formation in monkeys by recombinant human osteogenic protein-1. Arch Oral Biol. 1993;38:571–6.
25. Rutherford RB, Spångberg L, Tucker M, Rueger D, Charette M. The time-course of the induction of reparative dentine formation in monkeys by recombinant human osteogenic protein-1. Arch Oral Biol. 1994;39:833–8.
26. Rutherford RB, Gu K. Treatment of inflamed ferret dental pulps with recombinant bone morphogenetic protein-7. Eur J Oral Sci. 2000;108:202–6.
27. Rutherford RB. BMP-7 gene transfer to inflamed ferret dental pulps. Eur J Oral Sci. 2001;109:422–4.
28. Iohara K, Nakashima M, Ito M, Ishikawa M, Nakasima A, Akamine A. Dentin regeneration by dental pulp stem cell therapy with recombinant human bone morphogenetic protein 2. J Dent Res. 2004;83: 590–5.
29. Gu K, Smoke RH, Rutherford RB. Expression of genes for bone morphogenetic proteins and receptors in human dental pulp. Arch Oral Biol. 1996;41: 919–23.
30. Nakatomi M, Ida-Yonemochi H, Ohshima H. Lymphoid enhancer-binding factor 1 expression precedes dentin sialophosphoprotein expression during rat odontoblast differentiation and regeneration. J Endod. 2013;39: 612–8.
31. Nakashima M, Mizunuma K, Murakami T, Akmine A. Induction of dental pulp stem cell differentiation into odontoblasts by electroporation-mediated gene delivery of growth/differentiation factor 11(Gdf11). Gene Ther. 2002;9:814–8.
32. Paula-Silva FWG, Ghosh A, Silva LAB, Kapila YL. TNF-α promotes an odontoblastic phenotype in dental pulp cells. J Dent Res. 2009;88:339–44.
33. Muromachi K, Kamio N, Matsumoto T, Matsushima K. Role of CTGF/CCN2 in reparative dentinogenesis in human dental pulp. J Oral Sci. 2012;54:47–54.
34. Lovschall H, Fejerskov O, Flyvbjerg A. Pulp-capping with recombinant human insulin-like growth factor I (rhIGF-I) in rat molars. Adv Dent Res. 2001;15: 108–12.
35. Murakami Masuda Y, Wang X, Yokose S, Yamada Y, Kimura Y, Okano T, Matsumoto K. Effect of glypican-1 gene on the pulp cells during the reparative dentine process. Cell Biol Int. 2010;34:1069–74.
36. Chen L, Liu H, Sun W, Bai X, Karaplis AC, Goltzman D, Miao D. Fibroblast growth factor 23 overexpression impacts negatively on dentin mineralization and dentinogenesis in mice. Clin Exp Pharmacol Physiol. 2011;38:395–402.
37. Nagata K, Huang YH, Ohsaki Y, Kukita T, Nakata M, Kurisu K. Demonstration of type III collagen in the dentin of mice. Matrix. 1992;12:448–55.
38. Yasuda Y, Izumikawa M, Okamoto K, Tsukuba T, Saito T. Dentin phosphophoryn promotes cellular migration of human dental pulp cells. J Endod. 2008;34:575–8.
39. Decup F, Six N, Palmier B, Buch D, Lasfargues JJ, Salih E, Goldberg M. Bone sialoprotein-induced reparative dentinogenesis in the pulp of rat's molar. Clin Oral Investig. 2000;4:110–9.
40. Goldberg M, Six N, Decup F, Buch D, Soheili Majd E, Lasfargues JJ, Salih E, Stanislawski L. Application of bioactive molecules in pulp-capping situations. Adv Dent Res. 2001;15:91–5.
41. Almushayt A, Narayanan K, Zaki AE, George A. Dentin matrix protein 1 induces cyto differentiation of dental pulp stem cells into odontoblasts. Gene Ther. 2006;13:611–20.
42. Six N, Septier D, Chaussain-Miller C, Blacher R, DenBesten P, Goldberg M. Dentonin, a MEPE fragment, initiates pulp-healing response to injury. J Dent Res. 2007;86:780–5.

43. Yoshiba K, Yoshiba N, Nakamura H, Iwaku M, Ozawa H. Immunolocalization of fibronectin during reparative dentinogenesis in human teeth after pulp capping with calcium hydroxide. J Dent Res. 1996; 75:1590–7.
44. Nakamura Y, Slaby I, Spahr A, Pezeshki G, Matsumoto K, Lyngstadaas SP. Ameloblastin fusion protein enhances pulpal healing and dentin formation in porcine teeth. Calcif Tissue Int. 2006;78:278–84.
45. Lacerda-Pinheiro S, Septier D, Tompkins K, Veis A, Goldberg M, Chardin H. Amelogenin gene splice products A+4 and A-4 implanted in soft tissue determine the reorientation of CD45-positive cells to an osteo- chondrogenic lineage. J Biomed Mater Res A. 2006;79:1015–22.
46. Huang Y, Goldberg M, Le T, Qiang R, Warner D, Witkowska HE, Liu H, Zhu L, DenBesten P, Li W. Amelogenin exons 8 and 9 encoded peptide enhances Leucine Rich Amelogenin Peptide (LRAP) mediated dental pulp repair. Cells Tissues Organs. 2012;196:151–60.
47. Zheng L, Amano K, Iohara K, Ito M, Imabayashi K, Into T, Matsushita K, Nakamura H, Nakashima M. Matrix metalloproteinase-3 accelerates wound healing following dental pulp injury. Am J Pathol. 2009;175:1905–14.

11 Genetic Alterations: Heritable Dentin Defects

Agnès Bloch-Zupan

11.1 Introduction

Dentin defects that accompany rare genetic diseases [1, 2] (diseases that by definition affect less than 1/2,000 individuals) can be described phenotypically by various types of anomalies that not only impair the formation and structure of the dentin and per se the color of the teeth but also cause anomalies of the crown, root, and/or pulp space shape and aberrant dentin formation such as pulpolithes or intrapulpal calcifications (Fig. 11.1a–f).

They may also induce or be associated with subsequent enamel, dentin/enamel junction, or cementum/periodontium anomalies as dentinogenesis proceeds and occurs in coordination and interaction with amelogenesis and periodontium formation through epithelio-mesenchymal interactions.

These almost always genetically driven defects are seen in isolation or in association with other defects in syndromes. They mainly occur in conjunction with bone dysplasia, reflecting the similarities between dentin and bone formation, extracellular matrix chemical composition, and mineralization.

In this chapter, we will first describe the archetype of the genetic alterations of dentin formation in dentinogenesis imperfecta and dentin dysplasia type I and type II associated mainly with *DSPP* mutations [3]. However, other rare diseases also display some kind of dentin dysplasia in their clinical synopsis.

We will then review dentin defects encountered in rare diseases related to genes encoding dentin and bone extracellular matrix proteins, such as bone dysplasia caused by mutations on collagen genes, rickets caused by mutations in small integrin-binding ligand N-linked glycoproteins (SIBLINGS), and hypophosphatasia caused by mutations in calcium-binding proteins such as alkaline phosphatase.

We will also review animal models presenting with dentinal defects.

11.2 Genetic Dentin Defects

11.2.1 Dentinogenesis Imperfecta

Dentinogenesis imperfecta is a group of inherited conditions that show autosomal dominant transmission and that encompass dentinogenesis imperfecta type II (DGI-II) or hereditary opalescent dentin (OMIM #125490) also named dentinogenesis imperfecta 1 (DGI-1) or Capdepont teeth, dentinogenesis imperfecta associated with

A. Bloch-Zupan, HDR, PhD, DDS
Department of Sciences Biologiques, Faculty of Dentistry, University of Strasbourg, 8 rue St. Elisabeth, Strasbourg 67000, France

Reference Center for Orodental Manifestations of Rare Diseases, Pôle de Médecine et Chirurgie Bucco-Dentaires, Hôpitaux Universitaires de Strasbourg, Strasbourg, France

Institute of Genetics and Molecular and Cellular Biology, INSERM U964, CNRS-UdS UMR7104, Illkirch, France

Eastman Dental Institute, UCL, London, UK
e-mail: agnes.bloch-zupan@unistra.fr

M. Goldberg (ed.), *The Dental Pulp*,
DOI 10.1007/978-3-642-55160-4_11,

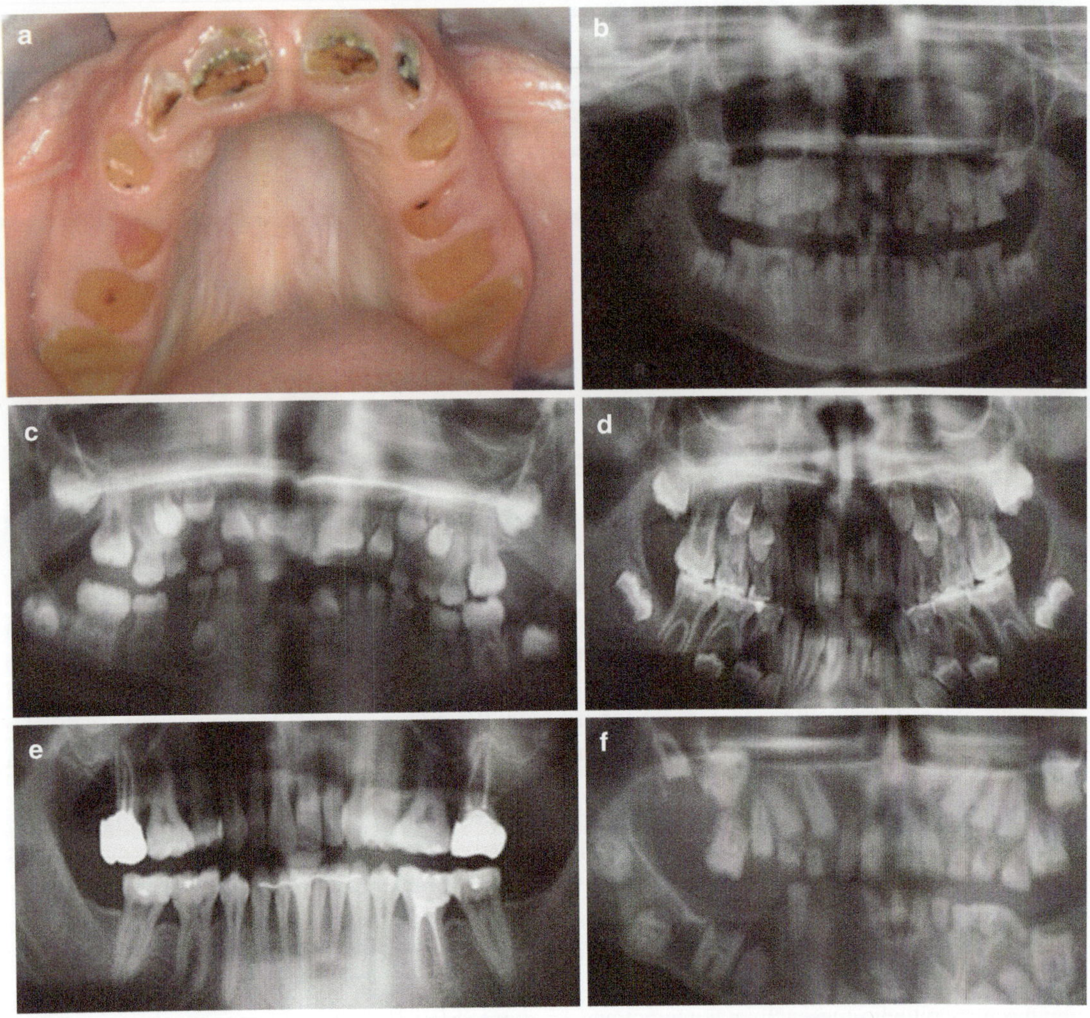

Fig. 11.1 Heritable dentin defects such as dentinogenesis imperfecta (**a** intraoral, the amber dentin color and the severe wear are visible, and **b** radiographic features with obliterated pulp spaces, cervical constrictions, short roots), dentin dysplasia (**c**), intrapulpal calcifications (**f**) encountered in isolation or associated with other symptoms in *SMOC2* mutation-associated rare disease (**c** dentin dysplasia, short roots with extreme microdontia in the primary dentition and oligodontia in the permanent dentition), hypophosphatemic X-linked rickets (**d** enlarged pulps), hypophosphatasia (**e** alveolar bone resorption; dentin and pulp chamber anomalies are present), and enamel renal syndrome (**f** intrapulpal calcifications are associated with absent enamel, retained teeth, and hyperplastic follicles)

progressive sensorineural hearing loss (OMIM # 605594) [4], and dentinogenesis imperfecta type III (DGI-III; OMIM # 125500) [5–7].

11.2.1.1 Dentinogenesis Imperfecta Type II

Dentinogenesis imperfecta type II results from mutations in genes encoding dentin structural matrix proteins, i.e., phosphoproteins and collagens.

As an isolated trait, it is caused by mutations in the dentin sialophosphoprotein *DSPP* gene (4q21.3), which belongs to the SIBLINGs family and encodes three major non-collagenous dentin matrix proteins-dentin sialoprotein (DSP), dentin glycoprotein (DGP), and dentin phosphoprotein (DPP) [8–14]. When associated with osteogenesis imperfecta, DI is caused by mutations in type I collagen genes and other related genes.

The intraoral clinical features are characteristic and lead to the diagnosis. Both primary and permanent dentitions are affected. The primary dentition is usually more severely affected. Dentin is abnormal in color appearing dull and bluish brown, amber, or opalescent. Enamel shedding occurs rapidly due to a defective enamel/dentin junction exposing the colored dentin to bacterial contamination within the oral cavity. The dentin has reduced wear resistance and the teeth rapidly wear from occlusal or biting stresses. The crowns appear bulbous radiographically with a marked cervical constriction. Pulp chambers and root canals are narrow or totally obliterated, and roots are short. Multiple pulp exposures and multiple abscesses may occur. Osteogenesis imperfecta is not a feature.

Enamel defects in addition to dentinogenesis imperfecta and *DSPP* mutations are described in a few families [15, 16].

At the ultrastructural level, enamel presents with an irregular surface and cracks [17]. Irregular dentin tubules, smooth dentinoenamel junction, abnormal enamel structure, and abnormal amounts of fibril bundles around dentin tubules were described in a family with a nonsense mutation in *DSPP* [17].

11.2.1.2 Dentinogenesis Imperfecta Type III

Dentinogenesis imperfecta type III has been described in a US population called the Brandywine isolate from Maryland and Washington, DC (an inbred triracial population of Caucasians, African Americans, and Native Americans) [18]. This disease is a phenotypic variation of dentinogenesis imperfecta DGI-II. The two diseases are allelic conditions sharing all the already described genetic *DSPP* mutations and clinical features [19]. However, in DGI-III, primary teeth often look like "shell" teeth on radiographs. Pulps of developing teeth are larger than normal during early development but rapidly become obliterated. Associated anterior open bite has been described.

11.2.2 Dentin Dysplasia

There are two main varieties of dentin dysplasia, type I (DD-I, radicular) and type II (DD-II, coronal). Four distinct forms of dentin dysplasia type I and one form of dentin dysplasia type II are recognized. The enamel seems normal but wears off.

11.2.2.1 Type I or Radicular Dentin Dysplasia (OMIM #125400)

The disease affects both dentitions and is associated with the premature loss of teeth. Tooth crown morphology and color are normal but teeth might be hypermobile. Upon radiography, the roots are short with pointed ends, and conical apical constrictions. Aberrant dentine growth leads to total pulpal obliteration in the primary dentition and reduced pulp space in permanent teeth. Teeth are lost generally due to trauma, inducing easily exfoliation. Delayed eruption is reported. Periapical radiolucencies are often seen in non-carious teeth.

11.2.2.2 Type II or Coronal Dentin Dysplasia (OMIM #125420)

This condition is allelic to DGI-II and is caused by mutations in the *DSPP* gene [20–22].

Primary teeth are opalescent and amber, resembling the phenotype of DGI-II.

On radiographs, pulp chambers are obliterated by abnormal dentin. However, the permanent teeth have a normal appearance with normal crown shape and color. They may show mild radiographic abnormalities with "thistle tube"-shaped pulp chambers and multiple intra-pulpal calcifications.

The nature of the dentin defects, i.e., dentin dysplasia versus dentinogenesis imperfecta, might be related to the type of *DSPP* mutations [23].

11.2.3 Dentin Dysplasia and Rare Diseases

Radicular dentin dysplasia with short roots or aberrant dentin formation is seen in various syndromes.

11.2.3.1 SMOC2

In this rare autosomal recessive disease caused by mutations in the *SMOC2* gene (6q72) that encodes the SPARC-related modular calcium-binding protein-2 [24, 25], the radicular dentin dysplasia is associated with extreme microdontia in the primary dentition and oligodontia in the permanent dentition.

11.2.4 Pulp Defects

It is noteworthy that dentin developmental defects will affect coronal and radicular pulp shape, size, and structure. Pulp may appear smaller or even absent in dentinogenesis imperfecta as dentin may be regarded as overproduced, for example, or pulp may appear larger in hypophosphatemic rickets phenomenon explained either by a delayed or reduced dentin formation. The shape of the pulp chamber or root pulpal spaces may be affected; a "thistle tube"-shaped appearance of the pulp, for example, is described in coronal dentin dysplasia. Intrapulpal calcifications [26] contribute also to reduce the pulp volume and testify the potential of pulp cells to produce mineralized dentin-like tissue.

Defects affecting the epithelial root sheath of Hertwig will also influence shape, size, structure of root, and root dentin formation.

11.3 Rare Diseases with Dentin Defects

Bone, dentin, and even cartilage extracellular matrix share common proteins. It is noteworthy that rare diseases affecting the composition of structural proteins like collagens or even those affecting the mineralization processes show combined skeletal and teeth defects [27]. The extracellular matrix (ECM) is a complex entity composed of structural proteins (such as fibrillins, collagens, elastin), ground substance (proteoglycans), modifying enzymes (ADAMTS, PLOD, lysyloxidases (LOX), matrix metalloproteinases (MMPs)), and cytokines that regulate morphogenesis, growth, homeostasis, remodeling, and repair (transforming growth factor-beta (TGF-beta), bone morphogenetic protein (BMP)).

11.3.1 Bone Dysplasia and Collagens

Collagens are important components of bone extracellular matrix. Their alterations lead to a variety of genetic diseases.

11.3.1.1 Osteogenesis Imperfecta

Osteogenesis imperfecta (OI), also called brittle bone disease, is a group of generalized heritable autosomal dominant or recessive disorders characterized by bone fragility and deformity, accompanied by osteoporosis, susceptibility to fracture, short stature, laxity of skin and ligaments, blue sclera and hearing loss, and eventually dentinogenesis imperfecta. Over 90 % of OI type I–IV disorders (OMIM #166200, #166240, #166210, #610854, #259420, #166220) are primarily caused by mutations in *COL1A1* (17q21.31-q22.05) and *COL1A2* (7q22.1), genes that encode the two alpha chains of type I collagen, which is the major component of the bone matrix.

COL1A mutations have, however, been described in individuals with DGI-II without skeletal abnormalities [28].

Mutations in genes coding for collagen modifying enzymes and chaperones have been discovered (*SERPINF1* (type VI), *CRTAP*, *LEPRE1*, *PPIB*, *SERPINH1*, *FKBP10*, *SP7*, *BMP1*, *TMEM38B*, *WNT1* (type XV)) in rare autosomal recessive forms of type VI–XV osteogenesis imperfecta (OMIM #613982, #610682, #610915, #259440, #613848, #610968, #613849, #614856, #615220, respectively) [29–36].

In types IB, IC, II, III, IVB, X, and XI, teeth demonstrate features of dentinogenesis imperfecta or eventually a phenotype similar to coronal dentin dysplasia type II (type IC).

Severe ultrastructural changes in dentin from patients affected with OI with clinically obvious dentinogenesis imperfecta show occluded tubules, multiple parallel channels, and occluded pulp chambers [37].

Even in the absence of a clear dental phenotype during clinical and radiographic examinations, histopathologic examination, at the ultrastructural level, disclosed characteristic dentin defects such as unevenly calcified matrixes, irregular tubular patterns, obliterated dentinal tubules, and cellular inclusions in the circumpulpal dentin of primary teeth leading to a diagnosis of OI type IV in a patient [38].

Osteogenesis imperfecta type V is a specific OI (#610967) phenotype with interosseous membrane calcification of the forearm and hyperplastic callus formation as typical features. The causative mutations for OI type V have been recently discovered in the gene encoding interferon-induced transmembrane protein 5 (*IFITM5*). Blue sclera and dentinogenesis imperfecta were not evident in any patient. However, hypodontia in the permanent teeth, ectopic eruption, and short roots in molars were observed [39].

Osteogenesis imperfecta type XI (#610968) is caused by mutations in the *FKBP10* gene (17q21.2) and is associated with dentinogenesis imperfecta.

Class III type craniofacial morphology with open bite and increased incidence of impacted permanent molars is often encountered in osteogenesis imperfecta [40].

Vertical underdevelopment of the dentoalveolar structures and the condylar process were identified as the main reasons for the relative mandibular prognathism in OI [41].

Osteogenesis imperfecta is therefore associated with dysplastic dentin that sometimes presents as dentinogenesis imperfecta [42].

The presence of dentinogenesis imperfecta is determined by the type of collagen mutation. The majority of patients with glycine mutations in a1(I) or a2(I) have clinically recognizable dentinogenesis imperfecta. Dentinogenesis imperfecta is absent in patients who have mutations in the amino-terminal end of the a1(I) or a2(I) triple helical domain [43].

11.3.1.2 Ehlers-Danlos Syndromes

Ehlers-Danlos syndromes (EDS) are a heterogeneous group of autosomal dominant and autosomal recessive disorders sharing joint hypermobility, skin extensibility, abnormal scarring, and tissue friability as hallmark diagnostic features [44]. More than XI forms are described. Type I (classical #130000) and type II (#130010) EDS are associated with mutations in *COL1A1* and *COL5A1* or *COL5A2*, respectively.

EDS type IV (#130050) is caused by *COL3A1* mutations and is associated with periodontal disease and early loss of teeth.

Type VII EDS is linked to *COL1A1*, *COL1A2* (#130060), and *ADAMTS2* mutations (#225410).

EDS type VIII (%130080) is distinguished from other EDS subtypes by severe gingival recession and periodontitis leading to premature loss of permanent teeth and resorption of alveolar bone.

An analysis of the ultrastructure of teeth from patients affected by type VIIC Ehlers-Danlos syndrome (substitution of a codon for tryptophan by a stop codon in *ADAMTS2*) associated with multiple tooth agenesis and focal dysplastic dentin defects and of teeth from a patient affected by type I Ehlers-Danlos syndrome (*COL1A1*) demonstrated abnormal dentin formation [45].

Many case reports describe the presence of dentin dysplasia and abnormal dentin formation in various forms of Ehlers-Danlos syndrome [46, 47].

Mice deficient in the Zn transporter Slc39a13/Zip13 show changes in bone, teeth, and connective tissue reminiscent of the clinical spectrum of human Ehlers-Danlos syndrome (EDS). The *Slc39a13* knockout (*Slc39a13*-KO) mice show defects in the maturation of osteoblasts, chondrocytes, odontoblasts, and fibroblasts [48].

11.3.2 Rickets-Related Diseases

Rickets is responsible for abnormalities in the formation and mineralization of skeletal bone, resulting in bone growth defects and malformations. The phenotype of the rare diseases belonging to this group includes dental anomalies in their clinical synopsis [49].

11.3.2.1 Hypophosphatemic Rickets

Hypophosphatemic vitamin D-resistant rickets is characterized by rickets associated with short stature, bone defects such as bowing of the extremities, frontal bossing in the craniofacial area, and dental anomalies. At the biological level it implies hypophosphatemia (low amount of phosphate), normal calcemia, normal or low levels of vitamin D, normal levels of PTH, and increased activity of alkaline phosphatase in the serum and hypocalciuria in the urine.

Hypophosphatemic rickets can be inherited in an X-linked dominant (XLH) manner (OMIM #300550) due to *PHEX* (phosphate regulating endopeptidase homologue, Xp22.2-1) mutations [50]. The *PHEX* gene codes for a membrane-bound endoprotease (Zn-metalloendopeptidase proteolytic enzyme) which is predominantly expressed in osteoblasts and which regulates phosphates.

This form of rickets can also be transmitted as an autosomal dominant disease (ADHR #193100) due to *FGF23* (12p13.3) mutations or as an autosomal recessive disease (ARHR #241520) with *DMP1* mutations [51] or even as an X-linked recessive disease (#300554) due to mutations in the *CLCN5* gene, which encodes a voltage-gated chloride ion channel [52]. PHEX regulates the function of fibroblast growth factor 23 (FGF23). The absence of functional PHEX leads to abnormal accumulation of ASARM (acidic serine- and aspartate-rich motif) peptide – a substrate for PHEX and a strong inhibitor of mineralization – derived from MEPE (matrix extracellular phosphoglycoprotein) and other matrix proteins. MEPE-derived ASARM peptide accumulates in tooth dentin of XLH patients where it may impair dentinogenesis [53].

1,25 vitamin D regulates DMP-1 expression through a VDR-dependent mechanism, possibly contributing to local changes in bone/tooth mineral homeostasis [52].

DMP1 has a protective antiapoptotic role for odontoblasts and ameloblasts, and the dentin defects might be enhanced by increased Pi level [54].

Beside enamel hypomineralization and delayed tooth eruption, the dental defects affect dentin, which is seen as globular and hypocalcified and presents many clefts and tubular defects. Abnormal large pulp chambers with pulp horns extending into the dentin and reaching the dentin-enamel junction are exposed directly to the oral cavity via the enamel defects and lead to pulpitis and pulp necrosis with recurrent abscesses and periapical pathology [55–58]. The cementum is also abnormal.

Treatment of hypophosphatemic rickets allows normal dentin development and mineralization [59, 60].

The discovery of mutations in *FAM20C* provides a putative new mechanism in human subjects for dysregulated FGF23 levels, hypophosphatemia, hyperphosphaturia, dental anomalies, intracerebral calcifications, and osteosclerosis of the long bones in the absence of rickets [61]. The teeth in an individual with a *FAM20C* mutation initially appeared to be normal but had been removed due to periapical abscesses. Clinical and histopathological examination of the teeth revealed enlarged pulp chambers, elongated pulp horns up to the enamel-dentin junction, and globular defects in the hypomineralized dentin. The enamel was hypoplastic. The roots were partly resorbed [61].

FAM20C knockout mice manifest hypophosphatemic rickets. The KO mice exhibit small malformed teeth, severe enamel defects, and very thin dentin, less cementum than normal, and overall hypomineralization in the dental mineralized tissues. DMP1 and DSPP are downregulated in the odontoblasts of these mice [62].

MEPE (matrix extracellular phosphoglycoprotein) (4q22.1) another SIBLING protein side to osteopontin (OPN) named secreted phosphoprotein-1 (SPP1), dentin matrix protein-1 (DMP1), bone sialoprotein (IBSP; 147563), and dentin sialophosphoprotein (DSPP) are also phosphorylated by FAM20C [63]. ASARM peptide, derived from MEPE, accumulates in the absence of functional PHEX and inhibits the odontogenic differentiation of dental pulp stem cells and impairs mineralization in tooth models of X-linked hypophosphatemia, thus explaining some of the dentin defects observed in X-linked hypophosphatemia [53].

11.3.2.2 Hypophosphatasia

Hypophosphatasia is an inherited disorder caused by a large panel of mutations in the *ALPL* gene (1p36.12) leading to a deficiency in the tissue-nonspecific alkaline phosphatase protein (TNAP) [64–66]. The modes of inheritance are autosomal dominant or autosomal recessive with often compound heterozygous mutations. Dominant negative effects of mutations complicate genetic counseling [67–69].

In the serum, reduced levels of alkaline phosphatase are detected, whereas in the urine, increased levels of phosphoethanolamine, calcium, and inorganic pyrophosphate (PPi) are present. Clinical features observed in the four clinical forms (infantile including the perinatal lethal form (#241500), childhood (#241510), adult (#146300), odontohypophosphatasia (#146300)), defined by the age of onset of the disease, range from complete absence of mineralization of the skeleton to premature loss of primary and eventually permanent teeth as an isolated feature. The main clinical features of this progressive disease are failure to thrive, severe rickets, bone deformities, short stature, craniosynostosis, seizures, gastrointestinal and renal problems, and spontaneous or fatigue fractures leading to mobility impairment [70].

The premature loss of teeth before 3 years of age for primary teeth shed with an intact root in the absence of periodontal disease is a common pathognomonic diagnostic sign whatever the clinical form [71]. The number of teeth lost and their type is also indicative of the severity of the disease, with incisors and canines being lost more frequently compared to primary molars. When confronted by this clinical presentation, it is important to refer the patient to the pediatrician or geneticist for further investigations.

TNAP is present during enamel, dentin, and cement formation [72]. Its deficiency will lead to absent acellular cement and periodontal anomalies in the ligament and alveolar bone leading to the premature loss of teeth and gradual defective dentinogenesis manifesting through small dentinal walls, large pulp chambers, and enamel defects.

Alkaline phosphatase knockout mice recapitulate the metabolic and skeletal defects of infantile hypophosphatasia [73] and demonstrate dentin mineralization defects as well as disturbed root dentinogenesis. The administration of a bioengineered recombinant alkaline phosphatase enzyme targeted to bone and mineralized tissues [74] rescued the skeletal and dentin defects and restored normal root formation mainly via a reduction of PP(i), a potent inhibitor of mineralization [75]. The replacement of alkaline phosphatase also corrected the periodontal/cementum anomalies [76]. This enzyme replacement therapy is currently under clinical trial [77].

Another molecule, PHOSPHO1, a soluble phosphatase with phosphoethanolamine and phosphocholine phosphatase activities, phosphoethanolamine and phosphocholine being present in matrix vesicles, is responsible for initiating hydroxyapatite crystal formation inside matrix vesicles and has functional roles complementary to TNAP during ossification [78]. In wild-type mice, Phospho1 and TNAP co-localize to odontoblasts at early stages of dentinogenesis, coincident with the early mineralization of mantle dentin. Nonredundant roles for both Phospho1 and TNAP in dentin mineralization are demonstrated by the spectrum of severity of dentin mineralization abnormalities in knockout mouse models [79].

11.3.3 Other Syndromes

Other syndromes combine in their clinical synopsis dentin, bone or cartilage defects, and/or other symptoms.

11.3.3.1 Goldblatt Syndrome or Spondylometaphyseal Dysplasia with Dentinogenesis Imperfecta (OMIM 184260)

This chondrodysplasia is characterized by severe shortness of stature and osteoporosis without fractures [80–82] combined with dentinogenesis imperfecta.

11.3.3.2 Elsahy-Waters Branchio-Skeleto-Genital Syndrome (OMIM 211380)

This syndrome is characterized by moderate mental retardation, hypospadias, and characteristic craniofacial morphology, which includes brachycephaly, facial asymmetry, exotropia, hypertelorism/telecanthus, broad nose, concave nasal ridge, underdeveloped midface, prognathism, and radicular dentin dysplasia [83].

11.3.3.3 Microcephalic Osteodysplastic Primordial Dwarfism, Type II, MOPD2 (OMIM #210720)

This condition is caused by mutations in the pericentrin (*PCNT*) gene (21q22.3) [84, 85]. Severe dental anomalies (microdontia, tooth shape anomalies, dentin dysplasia with opalescent and rootless teeth) and hypoplastic alveolar bone are observed in this syndrome [86, 87].

11.3.3.4 Immunoosseous Dysplasia, Schimke Type

This bone dysplasia (OMIM #242900) is an autosomal recessive disorder linked to *SMARCAL1* (2q35) gene mutations and presents with spondyloepiphyseal dysplasia, T-cell deficiency, and focal segmental glomerulosclerosis. *SMARCAL1* encodes the matrix-associated, actin-dependent regulator of chromatin, subfamily a-like 1 protein [88]. The dental anomalies are hypodontia, microdontia, bulbous crown, and short roots [89].

11.3.3.5 Kenny-Caffey Syndrome

Kenny-Caffey syndrome is an osteosclerotic bone dysplasia combining hypocalcemia, short stature, eye defects, and dental anomalies.

Two types of Kenny-Caffey syndromes have been described: (a) type 1, the autosomal recessive form (OMIM #24460), is due to mutations in the gene encoding tubulin-specific chaperone E (*TBCE*) [90]. Short roots are a frequent dental feature associated with teeth agenesis and microdontia [91]. (b) Autosomal dominant type 2 (#127000) is caused by mutation in the family with sequence similarity 111, member A gene (*FAM111A*), and defective dentition has been reported in some patients [92].

11.3.3.6 Congenital Insensitivity to Pain and Anhidrosis

This rare inherited disorder (OMIM #256800) is caused by mutations in the neurotrophic tyrosine kinase receptor, type 1, *NTRK1*, and presents with unexplained fever, the inability to sweat, repeated traumatic injuries, mental retardation, and self-mutilating behavior. NTRK1 is the receptor for nerve growth factor NGF. Severe defects in peripheral nerve fibers formation are observed. Dental anomalies range from missing teeth to hypomineralization with dentin, cement, and periodontal ligament anomalies [93].

11.3.4 Ion Channels and Transporters

Root dysplasia (CLCN7 (chloride channel 7) – osteopetrosis OMIM #166600, #611490; KCNJ2 (potassium channel family J member 2) – Andersen syndrome OMIM #170390) and dentin dysplasia (CLCN5 (chloride channel 5) – X-linked recessive hypophosphatemic rickets; Dent disease 1 OMIM #300009) are seen in various channelopathies [94].

11.3.5 Lipids in Dentinogenesis and Rare Diseases

Phospholipids play an important role during dentinogenesis and amelogenesis and in the formation and mineralization of dental tissues [95, 96]. Sphingomyelin degradation is a key factor involved in dentin and bone mineralization [97].

Krabbe disease (OMIM #245200) is a galactosylceramide lipidosis and leukodystrophy caused by homozygous or compound heterozygous mutations in the galactosylceramidase gene *GALC*. Anomalies of dental mineralized tissues are present at the clinical and ultrastructural level and include enamel hypoplasia, dendritic inclusions of amorphous material inside the mantle dentin, and lysosomal storage inclusions in all the cells of the dental pulp. The myelin sheaths of dental peripheral nerves display severe degenerative changes [98–101].

Osteogenesis imperfecta associated with dentinogenesis imperfecta is observed in a mouse model, named fro/fro (fragilitas ossium), deficient for *Smpd3*, a gene that encodes neutral sphingomyelin phosphodiesterase 3 [102].

11.3.6 Enamel Proteins, Dentin Defects, and Rare Diseases

Dentinogenesis and amelogenesis are interrelated. Amelogenin is also expressed/secreted by odontoblasts, and DSP and DPP are expressed by ameloblasts during enamel-dentin junction formation [103, 104]. Enamel and dentin defects are encountered in various rare diseases including amelogenesis imperfecta [105].

11.3.6.1 Enamel Renal Syndrome

Enamel renal syndrome (OMIM #204690) combines gingival hyperplasia or fibromatosis with hypoplastic amelogenesis imperfecta with almost absent or very thin enamel, intrapulpal calcifications, retained unerupted permanent teeth, and hyperplastic follicles to nephrocalcinosis and nephrolithiasis [106, 107]. This recessive disorder is caused by mutations in the *FAM20A* gene [108, 109].

11.3.6.2 Jalili Syndrome

This autosomal recessive hypoplastic/hypomineralized amelogenesis imperfecta is associated with cone-rod dystrophy and is caused by mutations in *CNMM4*, which encodes a magnesium transporter [110–112]. A similar reduced mineral density is observed in both enamel and dentin [113].

11.3.6.3 Trichodentoosseus Syndrome

This autosomal dominant syndrome (OMIM #190320), caused by mutations in the *DLX3* divergent homeobox gene, presents with curly hair, bone sclerosis, thin brittle nails, and enamel hypoplasia or amelogenesis imperfecta with taurodontism [114–116]. Dentin defects with large pulp chambers and intrapulpal calcifications are also associated with this syndrome [117].

11.4 Animal Models with Dentin Defects

Many transgenic mouse models demonstrate dentin/root defects [118–122]. Although mutations in these genes have not as yet been discovered in human diseases, high-throughput genome and exome sequencing studies may lead to the discovery of mutations in these genes in human disorders.

Conclusion

Dentin constitutes the scaffold of the tooth and represents the link between enamel and the periodontium. Heritable dentin defects ranging from the classical dentinogenesis imperfecta presentation to dentin dysplasia, root anomalies, pulp chamber and canal anomalies, and intrapulpal calcifications are associated with various rare diseases and may not always manifest clinically but appear clearly at a lower ultrastructural scale. They result from impaired dentinogenesis and reflect in their diversity the complexity of dentin extracellular matrix composition, formation, and mineralization as well as the coordination between amelogenesis and dentinogenesis essential to the epithelio-mesenchymal interactions driving tooth development.

The observation of such defects during clinical and radiographic examinations may have important consequences in orienting the diagnosis of rare disease per se and in patient dental management. Patients should be referred to rare disease reference centers to enhance knowledge of dentin heritable disorders at both the phenotype and genotype levels, to provide specialist care if needed and to facilitate the establishment of treatment guidelines allowing the implementation of science-based dentistry [123, 124]. International and European research programs such as the EU-funded project (ERDF) A27 "Oro-dental manifestations of rare diseases," supported by the RMT-TMO Offensive Sciences initiative, INTERREG IV Upper Rhine program (www.genosmile.eu), will contribute to improving the quality of care for special needs patients with these rare diseases.

References

1. Witkop Jr CJ. Hereditary defects of dentin. Dent Clin North Am. 1975;19(1):25–45.
2. Bloch-Zupan A, Sedano H, Scully C. Dento/Oro/Craniofacial anomalies and genetics. 1st ed. London: Elsevier Inc; 2012.
3. Hart PS, Hart TC. Disorders of human dentin. Cells Tissues Organs. 2007;186(1):70–7.
4. Xiao S, Yu C, Chou X, Yuan W, Wang Y, Bu L, et al. Dentinogenesis imperfecta 1 with or without progressive hearing loss is associated with distinct mutations in DSPP. Nat Genet. 2001;27(2):201–4.
5. Barron MJ, McDonnell ST, Mackie I, Dixon MJ. Hereditary dentine disorders: dentinogenesis imperfecta and dentine dysplasia. Orphanet J Rare Dis. 2008;3:31.
6. Kim JW, Simmer JP. Hereditary dentin defects. J Dent Res. 2007;86(5):392–9.
7. Shields ED, Bixler D, el-Kafrawy AM. A proposed classification for heritable human dentine defects with a description of a new entity. Arch Oral Biol. 1973; 18(4):543–53.
8. Kim JW, Hu JC, Lee JI, Moon SK, Kim YJ, Jang KT, et al. Mutational hot spot in the DSPP gene causing dentinogenesis imperfecta type II. Hum Genet. 2005;116(3):186–91.
9. Lee KE, Kang HY, Lee SK, Yoo SH, Lee JC, Hwang YH, et al. Novel dentin phosphoprotein frameshift mutations in dentinogenesis imperfecta type II. Clin Genet. 2011;79(4):378–84.
10. Lee SK, Hu JC, Lee KE, Simmer JP, Kim JW. A dentin sialophosphoprotein mutation that partially disrupts a splice acceptor site causes type II dentin dysplasia. J Endod. 2008;34(12):1470–3.
11. Lee SK, Lee KE, Jeon D, Lee G, Lee H, Shin CU, et al. A novel mutation in the DSPP gene associated with dentinogenesis imperfecta type II. J Dent Res. 2009;88(1):51–5.
12. Lee SK, Lee KE, Song SJ, Hyun HK, Lee SH, Kim JW. A DSPP mutation causing dentinogenesis imperfecta and characterization of the mutational effect. Bio Med Res Int. 2013;2013:948181. [Research Support, Non-U.S. Gov't].
13. MacDougall M. Dental structural diseases mapping to human chromosome 4q21. Connect Tissue Res. 2003;44 Suppl 1:285–91.
14. Zhang X, Zhao J, Li C, Gao S, Qiu C, Liu P, et al. DSPP mutation in dentinogenesis imperfecta shields type II. Nat Genet. 2001;27(2):151–2.
15. Lee SK, Lee KE, Hwang YH, Kida M, Tsutsumi T, Ariga T, et al. Identification of the DSPP mutation in a new kindred and phenotype-genotype correlation. Oral Dis. 2011;17(3):314–9. [Research Support, Non-U.S. Gov't].
16. Wang SK, Chan HC, Rajderkar S, Milkovich RN, Uston KA, Kim JW, et al. Enamel malformations associated with a defined dentin sialophosphoprotein mutation in two families. Eur J Oral Sci. 2011;119 Suppl 1:158–67. [Research Support, N.I.H., Extramural Research Support, Non-U.S. Gov't].
17. Wieczorek A, Loster J. Dentinogenesis imperfecta type II: ultrastructure of teeth in sagittal sections. Folia Histochem Cytobiol. 2013;51(3):244–7.
18. Levin LS, Leaf SH, Jelmini RJ, Rose JJ, Rosenbaum KN. Dentinogenesis imperfecta in the Brandywine isolate (DI type III): clinical, radiologic, and scanning electron microscopic studies of the dentition. Oral Surg Oral Med Oral Pathol. 1983;56(3):267–74.
19. MacDougall M, Jeffords LG, Gu TT, Knight CB, Frei G, Reus BE, et al. Genetic linkage of the dentinogenesis imperfecta type III locus to chromosome 4q. J Dent Res. 1999;78(6):1277–82.
20. McKnight DA, Simmer JP, Hart PS, Hart TC, Fisher LW. Overlapping DSPP mutations cause dentin dysplasia and dentinogenesis imperfecta. J Dent Res. 2008;87(12):1108–11.
21. McKnight DA, Suzanne Hart P, Hart TC, Hartsfield JK, Wilson A, Wright JT, et al. A comprehensive analysis of normal variation and disease-causing mutations in the human DSPP gene. Hum Mutat. 2008;29(12):1392–404.
22. Rajpar MH, Koch MJ, Davies RM, Mellody KT, Kielty CM, Dixon MJ. Mutation of the signal peptide region of the bicistronic gene DSPP affects translocation to the endoplasmic reticulum and results in defective dentine biomineralization. Hum Mol Genet. 2002;11(21):2559–65.
23. Nieminen P, Papagiannoulis-Lascarides L, Waltimo-Siren J, Ollila P, Karjalainen S, Arte S, et al. Frameshift mutations in dentin phosphoprotein and dependence of dentin disease phenotype on mutation location. J Bone Miner Res. 2011;26(4):873–80.
24. Bloch-Zupan A, Jamet X, Etard C, Laugel V, Muller J, Geoffroy V, et al. Homozygosity mapping and candidate prioritization identify mutations, missed by whole-exome sequencing, in SMOC2, causing major dental developmental defects. Am J Hum Genet. 2011;89(6):773–81.
25. Alfawaz S, Fong F, Plagnol V, Wong FS, Fearne J, Kelsell DP. Recessive oligodontia linked to a homozygous loss-of-function mutation in the SMOC2 gene. Arch Oral Biol. 2013;58(5):462–6. [Research Support, Non-U.S. Gov't].
26. Parekh S, Kyriazidou A, Bloch-Zupan A, Roberts G. Multiple pulp stones and shortened roots of unknown etiology. Oral Surg Oral Med Oral Pathol Oral Radiol Endod. 2006;101(6):e139–42.
27. Opsahl Vital S, Gaucher C, Bardet C, Rowe PS, George A, Linglart A, et al. Tooth dentin defects reflect genetic disorders affecting bone mineralization. Bone. 2012;50(4):989–97.
28. Wang SK, Chan HC, Makovey I, Simmer JP, Hu JC. Novel PAX9 and COL1A2 missense mutations causing tooth agenesis and OI/DGI without skeletal abnormalities. PLoS One. 2012;7(12):e51533.
29. Rohrbach M, Giunta C. Recessive osteogenesis imperfecta: clinical, radiological, and molecular find-

ings. Am J Med Genet C: Semin Med Genet. 2012;160C(3):175–89.
30. Basel D, Steiner RD. Osteogenesis imperfecta: recent findings shed new light on this once well-understood condition. Genet Med. 2009;11(6):375–85.
31. Christiansen HE, Schwarze U, Pyott SM, Alswaid A, Al Balwi M, Alrasheed S, et al. Homozygosity for a missense mutation in SERPINH1, which encodes the collagen chaperone protein HSP47, results in severe recessive osteogenesis imperfecta. Am J Hum Genet. 2010;86(3):389–98.
32. Homan EP, Rauch F, Grafe I, Lietman C, Doll JA, Dawson B, et al. Mutations in SERPINF1 cause osteogenesis imperfecta type VI. J Bone Miner Res. 2011;26(12):2798–803.
33. Lapunzina P, Aglan M, Temtamy S, Caparros-Martin JA, Valencia M, Leton R, et al. Identification of a frameshift mutation in Osterix in a patient with recessive osteogenesis imperfecta. Am J Hum Genet. 2010;87(1):110–4.
34. Martinez-Glez V, Valencia M, Caparros-Martin JA, Aglan M, Temtamy S, Tenorio J, et al. Identification of a mutation causing deficient BMP1/mTLD proteolytic activity in autosomal recessive osteogenesis imperfecta. Hum Mutat. 2012;33(2):343–50. [Research Support, Non-U.S. Gov't].
35. Van Dijk FS, Nesbitt IM, Nikkels PG, Dalton A, Bongers EM, van de Kamp JM, et al. CRTAP mutations in lethal and severe osteogenesis imperfecta: the importance of combining biochemical and molecular genetic analysis. Eur J Hum Genet. 2009;17(12): 1560–9.
36. van Dijk FS, Nesbitt IM, Zwikstra EH, Nikkels PG, Piersma SR, Fratantoni SA, et al. PPIB mutations cause severe osteogenesis imperfecta. Am J Hum Genet. 2009;85(4):521–7.
37. Hall RK, Maniere MC, Palamara J, Hemmerle J. Odontoblast dysfunction in osteogenesis imperfecta: an LM, SEM, and ultrastructural study. Connect Tissue Res. 2002;43(2–3):401–5.
38. Koreeda-Miura M, Onishi T, Ooshima T. Significance of histopathologic examination in the diagnosis of dentin defects associated with type IV osteogenesis imperfecta: two case reports. Oral Surg Oral Med Oral Pathol Oral Radiol Endod. 2003;95(1):85–9.
39. Kim OH, Jin DK, Kosaki K, Kim JW, Cho SY, Yoo WJ, et al. Osteogenesis imperfecta type V: clinical and radiographic manifestations in mutation confirmed patients. Am J Med Genet A. 2013;161A(8):1972–9.
40. Kamoun-Goldrat AS, Le Merrer MF. Osteogenesis imperfecta and dentinogenesis imperfecta: diagnostic frontiers and importance in dentofacial orthopedics. Orthod Fr. 2007;78(2):89–99.
41. Waltimo-Siren J, Kolkka M, Pynnonen S, Kuurila K, Kaitila I, Kovero O. Craniofacial features in osteogenesis imperfecta: a cephalometric study. Am J Med Genet A. 2005;133A(2):142–50.
42. Malmgren B, Lindskog S. Assessment of dysplastic dentin in osteogenesis imperfecta and dentinogenesis imperfecta. Acta Odontol Scand. 2003;61(2):72–80.
43. Rauch F, Lalic L, Roughley P, Glorieux FH. Genotype-phenotype correlations in nonlethal osteogenesis imperfecta caused by mutations in the helical domain of collagen type I. Eur J Hum Genet. 2010;18(6): 642–7.
44. Byers PH, Murray ML. Heritable collagen disorders: the paradigm of the Ehlers-Danlos syndrome. J Invest Dermatol. 2012;132(E1):E6–11.
45. De Coster PJ, Cornelissen M, De Paepe A, Martens LC, Vral A. Abnormal dentin structure in two novel gene mutations [COL1A1, Arg134Cys] and [ADAMTS2, Trp795-to-ter] causing rare type I collagen disorders. Arch Oral Biol. 2007;52(2):101–9.
46. Pope FM, Komorowska A, Lee KW, Speight P, Zorawska H, Ranta H, et al. Ehlers Danlos syndrome type I with novel dental features. J Oral Pathol Med. 1992;21(9):418–21.
47. Ferre FC, Frank M, Gogly B, Golmard L, Naveau A, Cherifi H, et al. Oral phenotype and scoring of vascular Ehlers-Danlos syndrome: a case-control study. BMJ Open. 2012;2(2):e000705.
48. Fukada T, Civic N, Furuichi T, Shimoda S, Mishima K, Higashiyama H, et al. The zinc transporter SLC39A13/ZIP13 is required for connective tissue development; its involvement in BMP/TGF-beta signaling pathways. PLoS One. 2008;3(11):e3642.
49. Foster BL, Nociti Jr FH, Somerman MJ. The rachitic tooth. Endocr Rev. 2014;35(1):1–34.
50. Holm IA, Nelson AE, Robinson BG, Mason RS, Marsh DJ, Cowell CT, et al. Mutational analysis and genotype-phenotype correlation of the PHEX gene in X-linked hypophosphatemic rickets. J Clin Endocrinol Metab. 2001;86(8):3889–99.
51. Lorenz-Depiereux B, Bastepe M, Benet-Pages A, Amyere M, Wagenstaller J, Muller-Barth U, et al. DMP1 mutations in autosomal recessive hypophosphatemia implicate a bone matrix protein in the regulation of phosphate homeostasis. Nat Genet. 2006;38(11):1248–50.
52. Nociti Jr FH, Foster BL, Tran AB, Dunn D, Presland RB, Wang L, et al. Vitamin D represses dentin matrix protein 1 in cementoblasts and osteocytes. J Dent Res. 2014;93(2):148–54.
53. Salmon B, Bardet C, Khaddam M, Naji J, Coyac BR, Baroukh B, et al. MEPE-derived ASARM peptide inhibits odontogenic differentiation of dental pulp stem cells and impairs mineralization in tooth models of X-linked hypophosphatemia. PLoS One. 2013;8(2):e56749.
54. Rangiani A, Cao ZG, Liu Y, Voisey Rodgers A, Jiang Y, Qin CL, et al. Dentin matrix protein 1 and phosphate homeostasis are critical for postnatal pulp, dentin and enamel formation. Int J Oral Sci. 2012;4(4):189–95.
55. Pereira CM, de Andrade CR, Vargas PA, Coletta RD, de Almeida OP, Lopes MA. Dental alterations associated with X-linked hypophosphatemic rickets. J Endod. 2004;30(4):241–5.
56. Yamamoto T. Diagnosis of X-linked hypophosphatemic vitamin D resistant rickets. Acta Paediatr Jpn. 1997;39(4):499–502.

57. Baroncelli GI, Angiolini M, Ninni E, Galli V, Saggese R, Giuca MR. Prevalence and pathogenesis of dental and periodontal lesions in children with X-linked hypophosphatemic rickets. Eur J Paediatr Dent. 2006;7(2):61–6.
58. Batra P, Tejani Z, Mars M. X-linked hypophosphatemia: dental and histologic findings. J Can Dent Assoc. 2006;72(1):69–72.
59. Chaussain-Miller C, Sinding C, Septier D, Wolikow M, Goldberg M, Garabedian M. Dentin structure in familial hypophosphatemic rickets: benefits of vitamin D and phosphate treatment. Oral Dis. 2007; 13(5):482–9.
60. Chaussain-Miller C, Sinding C, Wolikow M, Lasfargues JJ, Godeau G, Garabedian M. Dental abnormalities in patients with familial hypophosphatemic vitamin D-resistant rickets: prevention by early treatment with 1-hydroxyvitamin D. J Pediatr. 2003; 142(3):324–31.
61. Rafaelsen SH, Raeder H, Fagerheim AK, Knappskog P, Carpenter TO, Johansson S, et al. Exome sequencing reveals FAM20c mutations associated with fibroblast growth factor 23-related hypophosphatemia, dental anomalies, and ectopic calcification. J Bone Miner Res. 2013;28(6):1378–85.
62. Wang X, Wang S, Lu Y, Gibson MP, Liu Y, Yuan B, et al. FAM20C plays an essential role in the formation of murine teeth. J Biol Chem. 2012;287(43): 35934–42.
63. Tagliabracci VS, Engel JL, Wen J, Wiley SE, Worby CA, Kinch LN, et al. Secreted kinase phosphorylates extracellular proteins that regulate biomineralization. Science. 2012;336(6085):1150–3.
64. Mornet E, Hofmann C, Bloch-Zupan A, Girschick H, Le Merrer M. Clinical utility gene card for: hypophosphatasia – update 2013. Eur J Hum Gen: EJHG. 2014;22(4):e1–6.
65. Mornet E. Hypophosphatasia. Orphanet J Rare Dis. 2007;2:40.
66. Spentchian M, Merrien Y, Herasse M, Dobbie Z, Glaser D, Holder SE, et al. Severe hypophosphatasia: characterization of fifteen novel mutations in the ALPL gene. Hum Mutat. 2003;22(1):105–6.
67. Mornet E, Simon-Bouy B. Genetics of hypophosphatasia. Arch Pediatr. 2004;11(5):444–8.
68. Mornet E. Hypophosphatasia: the mutations in the tissue-nonspecific alkaline phosphatase gene. Hum Mutat. 2000;15(4):309–15.
69. Fauvert D, Brun-Heath I, Lia-Baldini AS, Bellazi L, Taillandier A, Serre JL, et al. Mild forms of hypophosphatasia mostly result from dominant negative effect of severe alleles or from compound heterozygosity for severe and moderate alleles. BMC Med Genet. 2009;10:51.
70. Berkseth KE, Tebben PJ, Drake MT, Hefferan TE, Jewison DE, Wermers RA. Clinical spectrum of hypophosphatasia diagnosed in adults. Bone. 2013;54(1): 21–7.
71. Reibel A, Maniere MC, Clauss F, Droz D, Alembik Y, Mornet E, et al. Orodental phenotype and genotype findings in all subtypes of hypophosphatasia. Orphanet J Rare Dis. 2009;4:6.
72. Hotton D, Mauro N, Lezot F, Forest N, Berdal A. Differential expression and activity of tissue-nonspecific alkaline phosphatase (TNAP) in rat odontogenic cells in vivo. J Histochem Cytochem. 1999;47(12):1541–52.
73. Fedde KN, Blair L, Silverstein J, Coburn SP, Ryan LM, Weinstein RS, et al. Alkaline phosphatase knockout mice recapitulate the metabolic and skeletal defects of infantile hypophosphatasia. J Bone Miner Res. 1999;14(12):2015–26.
74. Millan JL, Narisawa S, Lemire I, Loisel TP, Boileau G, Leonard P, et al. Enzyme replacement therapy for murine hypophosphatasia. J Bone Miner Res. 2008; 23(6):777–87.
75. Foster BL, Nagatomo KJ, Tso HW, Tran AB, Nociti Jr FH, Narisawa S, et al. Tooth root dentin mineralization defects in a mouse model of hypophosphatasia. J Bone Miner Res. 2013;28(2):271–82.
76. McKee MD, Nakano Y, Masica DL, Gray JJ, Lemire I, Heft R, et al. Enzyme replacement therapy prevents dental defects in a model of hypophosphatasia. J Dent Res. 2011;90(4):470–6.
77. Whyte MP, Greenberg CR, Salman NJ, Bober MB, McAlister WH, Wenkert D, et al. Enzyme-replacement therapy in life-threatening hypophosphatasia. N Engl J Med. 2012;366(10):904–13.
78. Millan JL. The role of phosphatases in the initiation of skeletal mineralization. Calcif Tissue Int. 2013;93(4): 299–306.
79. McKee MD, Yadav MC, Foster BL, Somerman MJ, Farquharson C, Millan JL. Compounded PHOSPHO1/ALPL deficiencies reduce dentin mineralization. J Dent Res. 2013;92(8):721–7.
80. Bonaventure J, Stanescu R, Stanescu V, Allain JC, Muriel MP, Ginisty D, et al. Type II collagen defect in two sibs with the Goldblatt syndrome, a chondrodysplasia with dentinogenesis imperfecta, and joint laxity. Am J Med Genet. 1992;44(6): 738–53.
81. Goldblatt J, Carman P, Sprague P. Unique dwarfing, spondylometaphyseal skeletal dysplasia, with joint laxity and dentinogenesis imperfecta. Am J Med Genet. 1991;39(2):170–2.
82. Unger S, Antoniazzi F, Brugnara M, Alanay Y, Caglayan A, Lachlan K, et al. Clinical and radiographic delineation of odontochondrodysplasia. Am J Med Genet A. 2008;146A(6):770–8.
83. Castori M, Cascone P, Valiante M, Laino L, Iannetti G, Hennekam RC, et al. Elsahy-Waters syndrome: evidence for autosomal recessive inheritance. Am J Med Genet A. 2010;152A(11):2810–5.
84. Rauch A. The shortest of the short: pericentrin mutations and beyond. Best Pract Res Clin Endocrinol Metab. 2011;25(1):125–30.
85. Rauch A, Thiel CT, Schindler D, Wick U, Crow YJ, Ekici AB, et al. Mutations in the pericentrin (PCNT) gene cause primordial dwarfism. Science. 2008;319 (5864):816–9.

86. Kantaputra P, Tanpaiboon P, Porntaveetus T, Ohazama A, Sharpe P, Rauch A, et al. The smallest teeth in the world are caused by mutations in the PCNT gene. Am J Med Genet A. 2011;155A(6):1398–403.
87. Kantaputra PN, Tanpaiboon P, Unachak K, Praphanphoj V. Microcephalic osteodysplastic primordial dwarfism with severe microdontia and skin anomalies: confirmation of a new syndrome. Am J Med Genet A. 2004;130(2):181–90.
88. Hunter KB, Lucke T, Spranger J, Smithson SF, Alpay H, Andre JL, et al. Schimke immunoosseous dysplasia: defining skeletal features. Eur J Pediatr. 2010;169(7):801–11.
89. Morimoto M, Kerouredan O, Gendronneau M, Shuen C, Baradaran-Heravi A, Asakura Y, et al. Dental abnormalities in Schimke immuno-osseous dysplasia. J Dent Res. 2012;91(7 Suppl):29S–37.
90. Parvari R, Hershkovitz E, Grossman N, Gorodischer R, Loeys B, Zecic A, et al. Mutation of TBCE causes hypoparathyroidism-retardation-dysmorphism and autosomal recessive Kenny-Caffey syndrome. Nat Genet. 2002;32(3):448–52.
91. Moussaid Y, Griffiths D, Richard B, Dieux A, Lemerrer M, Leger J, et al. Oral manifestations of patients with Kenny-Caffey syndrome. Eur J Med Genet. 2012;55(8–9):441–5. [Review].
92. Unger S, Gorna MW, Le Bechec A, Do Vale-Pereira S, Bedeschi MF, Geiberger S, et al. FAM111A mutations result in hypoparathyroidism and impaired skeletal development. Am J Hum Genet. 2013;92(6):990–5.
93. Gao L, Guo H, Ye N, Bai Y, Liu X, Yu P, et al. Oral and craniofacial manifestations and two novel missense mutations of the NTRK1 gene identified in the patient with congenital insensitivity to pain with anhidrosis. PLoS One. 2013;8(6):e66863.
94. Duan X. Ion channels, channelopathies, and tooth formation. J Dent Res. 2014;93(2):117–25.
95. Goldberg M, Septier D. Phospholipids in amelogenesis and dentinogenesis. Crit Rev Oral Biol Med. 2002;13(3):276–90.
96. Goldberg M, Septier D, Lecolle S, Vermelin L, Bissila-Mapahou P, Carreau JP, et al. Lipids in predentine and dentine. Connect Tissue Res. 1995;33(1–3):105–14.
97. Goldberg M, Opsahl S, Aubin I, Septier D, Chaussain-Miller C, Boskey A, et al. Sphingomyelin degradation is a key factor in dentin and bone mineralization: lessons from the fro/fro mouse. The chemistry and histochemistry of dentin lipids. J Dent Res. 2008;87(1):9–13.
98. Bloch-Zupan A, Lecolle S, Goldberg M. Galactosylceramide lipidosis (Krabbe's disease) and deciduous dental tissues. A case report. J Submicrosc Cytol Pathol. 1994;26(3):425–35.
99. Goldberg M, Gritli A, Bloch-Zupan A, Septier D, Lecolle S, Legrand JM. Effets des maladies de surcharge lysosomale sur l'odontogenèse. Entretiens de Bichat – Oodntologie et Stomatologie, Expansion Scientifique Française 1992:7–11.
100. Goldberg M, Gritli A, Bloch-Zupan A, Septier D, Lecolle S, Legrand JM, et al. Lysosomal storage diseases, genetic or drug-induced? Effect of glycosaminoglycan and sphingolipid disorders on dental tissues. C R Seances Soc Biol Fil. 1993;187(5):596–607.
101. Goldberg M, Gritli A, Bloch-Zupan A, Septier D, Lecolle S, Legrand JM, et al. Maladies de surcharge lysosomale, génétiques ou induites pharmacologiquement: effets de pathologies de glycosaminoglycanes et de sphingolipides sur les tissus dentaires. C R Soc Biol. 1993;187:596–607.
102. Aubin I, Adams CP, Opsahl S, Septier D, Bishop CE, Auge N, et al. A deletion in the gene encoding sphingomyelin phosphodiesterase 3 (Smpd3) results in osteogenesis and dentinogenesis imperfecta in the mouse. Nat Genet. 2005;37(8):803–5.
103. Ye L, Le TQ, Zhu L, Butcher K, Schneider RA, Li W, et al. Amelogenins in human developing and mature dental pulp. J Dent Res. 2006;85(9):814–8.
104. White SN, Paine ML, Ngan AY, Miklus VG, Luo W, Wang H, et al. Ectopic expression of dentin sialoprotein during amelogenesis hardens bulk enamel. J Biol Chem. 2007;282(8):5340–5.
105. Crawford PJ, Aldred M, Bloch-Zupan A. Amelogenesis imperfecta. Orphanet J Rare Dis. 2007;2:17.
106. Cho SH, Seymen F, Lee KE, Lee SK, Kweon YS, Kim KJ, et al. Novel FAM20A mutations in hypoplastic amelogenesis imperfecta. Hum Mutat. 2012;33(1):91–4.
107. Kantaputra PN, Kaewgahya M, Khemaleelakul U, Dejkhamron P, Sutthimethakorn S, Thongboonkerd V, et al. Enamel-renal-gingival syndrome and FAM20A mutations. Am J Med Genet A. 2014;164(1):1–9.
108. Jaureguiberry G, De la Dure-Molla M, Parry D, Quentric M, Himmerkus N, Koike T, et al. Nephrocalcinosis (enamel renal syndrome) caused by autosomal recessive FAM20A mutations. Nephron Physiol. 2013;122(1–2):1–6.
109. O'Sullivan J, Bitu CC, Daly SB, Urquhart JE, Barron MJ, Bhaskar SS, et al. Whole-Exome sequencing identifies FAM20A mutations as a cause of amelogenesis imperfecta and gingival hyperplasia syndrome. Am J Hum Genet. 2011;88(5):616–20.
110. Parry DA, Mighell AJ, El-Sayed W, Shore RC, Jalili IK, Dollfus H, et al. Mutations in CNNM4 cause Jalili syndrome, consisting of autosomal-recessive cone-rod dystrophy and amelogenesis imperfecta. Am J Hum Genet. 2009;84(2):266–73.
111. Jalili IK. Cone-rod dystrophy and amelogenesis imperfecta (Jalili syndrome): phenotypes and environs. Eye (Lond). 2010;24(11):1659–68.
112. Gomez Garcia I, Oyenarte I, Martinez-Cruz LA. Purification, crystallization and preliminary crystallographic analysis of the CBS pair of the human metal transporter CNNM4. Acta Crystallogr Sect F: Struct Biol Cryst Commun. 2011;67(Pt 3): 349–53.
113. Luder HU, Gerth-Kahlert C, Ostertag-Benzinger S, Schorderet DF. Dental phenotype in Jalili syndrome due to a c.1312 dupC homozygous mutation in the CNNM4 gene. PLoS One. 2013;8(10):e78529.

114. Dong J, Amor D, Aldred MJ, Gu T, Escamilla M, MacDougall M. DLX3 mutation associated with autosomal dominant amelogenesis imperfecta with taurodontism. Am J Med Genet A. 2005;133(2):138–41.
115. Haldeman RJ, Cooper LF, Hart TC, Phillips C, Boyd C, Lester GE, et al. Increased bone density associated with DLX3 mutation in the tricho-dento-osseous syndrome. Bone. 2004;35(4):988–97.
116. Lee SK, Lee ZH, Lee SJ, Ahn BD, Kim YJ, Lee SH, et al. DLX3 mutation in a new family and its phenotypic variations. J Dent Res. 2008;87(4):354–7.
117. Nieminen P, Lukinmaa PL, Alapulli H, Methuen M, Suojarvi T, Kivirikko S, et al. DLX3 homeodomain mutations cause tricho-dento-osseous syndrome with novel phenotypes. Cells Tissues Organs. 2011;194(1):49–59.
118. Lim WH, Liu B, Cheng D, Hunter DJ, Zhong Z, Ramos DM, et al. Wnt signaling regulates pulp volume and dentin thickness. J Bone Miner Res. 2014;29(4):892–901.
119. Wang Y, Cox MK, Coricor G, MacDougall M, Serra R. Inactivation of Tgfbr2 in Osterix-Cre expressing dental mesenchyme disrupts molar root formation. Dev Biol. 2013;382(1):27–37. [Research Support, N.I.H., Extramural Research Support, Non-U.S. Gov't].
120. Zhang R, Yang G, Wu X, Xie J, Yang X, Li T. Disruption of Wnt/beta-catenin signaling in odontoblasts and cementoblasts arrests tooth root development in postnatal mouse teeth. Int J Biol Sci. 2013;9(3):228–36.
121. Hayano S, Kurosaka H, Yanagita T, Kalus I, Milz F, Ishihara Y, et al. Roles of heparan sulfate sulfation in dentinogenesis. J Biol Chem. 2012; 287(15):12217–29. [Research Support, Non-U.S. Gov't].
122. Zhang Y, Kim SO, Opsahl-Vital S, Ho SP, Souron JB, Kim C, et al. Multiple effects of the cellular prion protein on tooth development. Int J Dev Biol. 2011;55(10–12):953–60. [Research Support, N.I.H., Extramural].
123. Feierabend S, Bloch-Zupan A, Hellwig E, Frei O, Wolff A, Moog U, et al. Seltene Erkrankungen – relevant für den Zahnarzt? Evidence-based dentistry – current advice for the practitioner. Case 7: rare diseases – relevant to dentists? Deutscher Ärzte-Verlag | DZZ | Deutsche Zahnärztliche Zeitschrift. 2012;67(12):14–18.
124. Bloch-Zupan A, Maniere MC. Manifestations Odontologiques des Maladies Rares. Centre de Référence de Strasbourg. Alpha Omega News. 2008;118:4–6.

12 Pulp Reactions to Dental Materials

Gottfried Schmalz

12.1 Introduction

The dental pulp is – beside the periodontal tissues and the oral mucous membranes – the prime local target organ for direct or indirect biologic interaction with dental materials. Dental materials are not only classical restorative materials and those needed in the course of fabricating the restorations (e.g., impression materials) but also substances like tooth-whitening products or preventive varnishes. Due to its anatomical characteristics containing tubule with odontoblastic processes and lateral processes, the dentin is not only permeable, but it is also itself as a vital tissue biologically responsive, and thus any contact of a material or substance with dentin may possibly interfere with the dental pulp. Even the dental enamel is permeable for certain small molecules like hydrogen peroxide, which is released from tooth-whitening products, and again the pulp is the final target organ.

Clinically, pain, pulp inflammation, or even pulp necrosis may result after contact with such materials, but often pulp damage occurs without overt clinical symptoms. However, through proper precautions, pulp damage can often be prevented. Furthermore, dental materials may not only damage the dental pulp, but they may also inhibit repair and/or regeneration [1]. Thus, dental materials may interfere with the final aim of pulp therapy, namely, to keep the dental pulp vital.

The topic of this chapter covers a large area. Due to the limited space of this chapter, mainly general mechanisms of material-related pulp reactions as well as methods for preventing pulp damage and for stimulating pulp repair/regeneration are outlined. More detailed information especially on material groups is available in the literature [2].

12.2 Cavity/Crown Preparation

In many cases, dental materials are brought into contact with dentin/pulp after cavity or crown preparation with the consequence of dentin liquor outflow [3]. Heat produced by such procedures, especially with high-speed rotary instruments, may cause pulp damage [4], and histologically odontoblastic nuclei displacement into the dentinal tubule can be seen, especially in cases of insufficient water cooling. The reason for this displacement is unclear, and ideas put forward were, e.g., fluid evaporation due to the heat produced by the grinding process or increase of intrapulpal pressure due to traumatic preparation [3].

These effects can be prevented or at least minimized by an *atraumatic preparation* using appropriate water cooling [4, 5]. According to the ISO Standard 14457:2012, a water coolant supply of 50 ml/min from the dental treatment unit must be

G. Schmalz, DDS, DMD, PhD
Department of Operative Dentistry and Periodontology, University Hospital Regensburg, University Clinics, Universitaetsstr. 11, Regensburg 93052, Germany

School of Dental Medicine - ZMK Bern, University of Bern, Switzerland
e-mail: gottfried.schmalz@ukr.de

M. Goldberg (ed.), *The Dental Pulp*,
DOI 10.1007/978-3-642-55160-4_12, © Springer-Verlag Berlin Heidelberg 2014

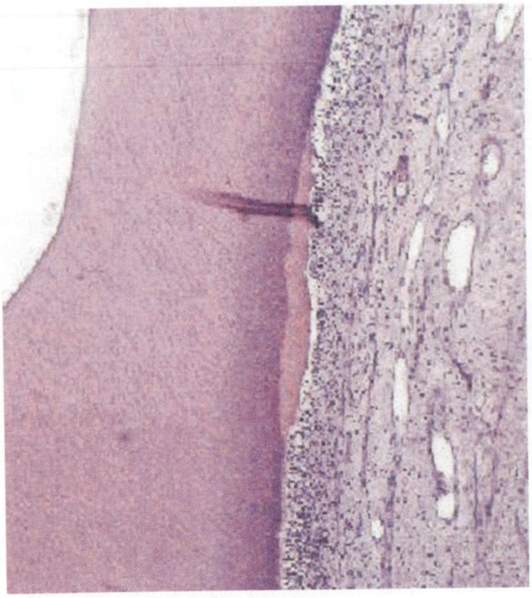

Fig. 12.1 Slight reactionary dentin formation with no persistent inflammation after atraumatic cavity preparation and application of a nontoxic material in a medium deep cavity (magnification ×80) (Used with permission of Springer Science+Business Media from Schmalz and Arenholt-Bindslev [2])

delivered [6], which is then (in the handpiece) mixed with air to generate the spray. However, this does not mean that the amount of water supplied (50 ml/min) reaches the site, where the bur touches the dentin and creates heat [7]. The tubing in the handpieces may be (partially) blocked, e.g., by calcareous deposits. Therefore, regular control of the coolant quantity at the handpiece spray outlet is necessary. Furthermore, the mode of cavity/tooth preparation is important: the tooth itself may block the water spray, especially if handpieces with only one spray aperture are used. High vacuum suction, which is held too close to the coolant exit, may tear the water spray away from the preparation site. Finally, low-pressure intermittent cutting is recommended [3] and extensive drying of cut dentin should be avoided [3].

But even after atraumatic tooth preparation, odontoblast processes are cut during cavity/tooth preparation. However, apparently odontoblasts survive this cell trauma, and "only" reactionary dentin formation is the consequence (Fig. 12.1) – if the odontoblast process is not cut close to the cell body [8].

Non-rotary methods for cavity preparation may comprise alumina air abrasion, lasers, or oscillating instruments. For air abrasion no heat generation was reported [9]. Er:YAG laser-based preparation tools seem to produce less temperature increase than ruby, CO_2, and Nd:YAG lasers, the later above pulpal tolerance when used on mineralized tissues [5, 10]. For oscillating instruments again water coolant is necessary, but 7.3 ml water/min was reported to be sufficient [11] and the relevant ISO Standard 18397:2013 requires a minimum of 20 ml/min cooling liquid.

12.3 Symptoms

Generally, clinical symptoms like pain in its different manifestations are a rather unreliable indicator for pulp disease (see Chap. 9) and, therefore, also for material-related pulp damage. The absence of pain is definitively not a sign that no – histologically detectable – pulp damage has occurred. Histological studies in human teeth after pulp capping with dental adhesives clearly showed that the teeth were clinically asymptomatic although histological evaluation proved severe inflammation [12].

If pain occurs after material application, this indeed may be due to a substance diffusing to the pulp. This is quite often – up to 65 % – observed after the application of (highly concentrated) hydrogen peroxide-releasing tooth-whitening products [13], and it may be related to hydrogen peroxide diffusion through the enamel and dentin to the pulp [14]. On the other hand, similar symptoms can be observed after the application of resin-based composite restorations ("postoperative sensitivity"), which are mainly related to the intratubular fluid movement after masticatory load (see the discussion to come).

Besides pain, other symptoms may be the consequence of material tissue interaction like discoloration of the tooth after pulp necrosis. Furthermore, tertiary dentin formation (biomineralization) can be the consequence of material application. Interestingly, dentin formation can also be inhibited, e.g., by pulp capping with resin materials [8, 12, 15] with the clinical consequence of deficient or lacking dentin bridge formation.

Taken together the clinical diagnosis of material-related pulp damage is very complicated, and absence of clinical symptoms is no indicator for pulp biocompatibility. Therefore, all possible and reasonable measures must be undertaken to prevent pulp damage as best as present knowledge allows.

12.4 Reasons

A prime reason for pulp damage is the direct interaction between a (toxic) substance released from a material and a relevant cellular or another molecule within the pulp tissue [16]. Virtually all dental materials release substances especially during and shortly after setting [17, 18], which then contact vital (often freshly cut) dentin and which can diffuse toward the pulp [19, 20] (Fig. 12.2). In vital teeth, dentinal liquor is diffusing from the pulp outward to the material contact area possibly causing (partial) dissolution of the material and an impairment of the setting reaction. This is especially pronounced, if the material is in direct contact with the exposed pulp: due to the high degree of wetness, comparatively large amounts of substances are released from the material. The amount/concentration of the released substance at the target site (pulp cells) depends on a number of variables, an important one being the thickness of the residual dentin (see the discussion to come).

Further to this direct material tissue interaction, pulp damage after restorative procedures was also related to bacteria at the cavity floor (indirect interaction). Especially under resin-based composite restorations, bacteria at the cavity floor [21] were consistently found in cases of pulp inflammation both in experimental animals and in humans [22], and thus bacteria were assumed to be the cause of the pulp damage [22] (Fig. 12.3a, b). Bacterial penetration can occur through microgaps at the material/dentin interface. The use of adequate dentinal adhesives markedly reduces the bacterial penetration, but still penetration studies reveal quite some amount of microleakage, more in methacrylate-based materials than with siloran-based products [23, 24]. However, in such studies mainly dyes are used for quantifying microleakage, and it is still unclear if this can be related to bacterial penetration in vivo [25]. On the other hand, a material-related reduced bacterial clearance [26, 27] and the ability of monomers to stimulate bacterial growth [28] could further promote bacterial presence at the cavity floor.

Heat is a point of concern not only for cavity preparation (see earlier discussion) but also for light-curing units, which are needed for resin polymerization. Recently, such devices

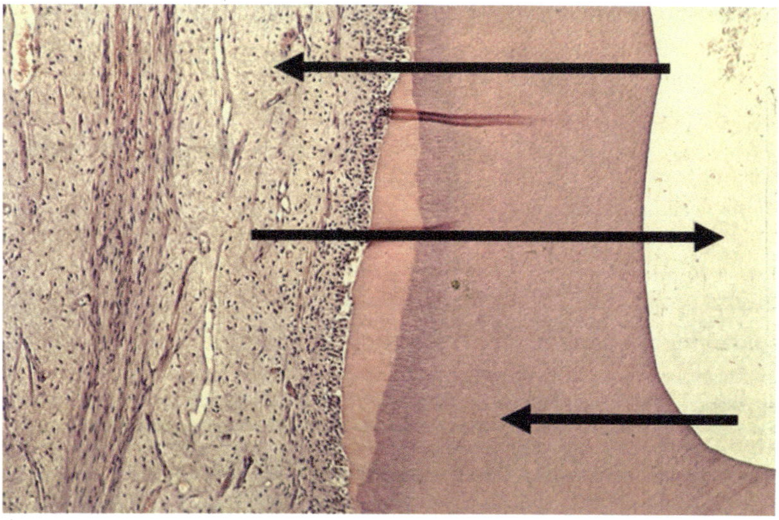

Fig. 12.2 Material-dentin/pulp interaction: release of substances from the applied material which diffuses through the dentin to the pulp (*upper arrow*), outflow of dentin liquor after cavity preparation (*middle arrow*), and adsorption of released substance within the dentin (*lower arrow*) (magnification ×80) (Used with permission of Springer Science+Business Media from Schmalz and Arenholt-Bindslev [2])

Fig. 12.3 Strong pulp inflammation (**a** magnification ×80) associated with bacteria at the cavity floor under the restoration (**b**, magnification ×400) (Used with permission of Springer Science+Business Media from Schmalz and Arenholt-Bindslev [2])

with a high energy output were marketed, and especially close to the pulp, heat-related pulp damage is possible [29] (see also the discussion to come).

Mechanically caused fluid displacement in the dentinal tubule has been described to be the reason for postoperative sensitivity after placement of resin-based composite restorations [2]. This is mainly observed after occlusal loading, and it has been postulated that due to insufficient bonding and the formation of microgaps between the cavity floor and the restoration, a pump effect occurs after masticatory load, which causes a fluid movement in the dentinal tubule with consecutive pain. Using dentinal adhesives reduces markedly sensitivity [2]; some authors favor in this respect self-etch adhesives [30].

12.5 Residual Dentin

Although dentin contains a large number of tubule, in which odontoblastic processes and lateral processes are located [31], it has been reported that dentin (even if cut) has a protective effect on the pulp especially for substances released from materials. To demonstrate this effect, the influence of different dentin thicknesses/distances from the pulp on different parameters has been studied. In vitro, dentin permeability was shown to be dependent upon dentin thickness [31], and permeability exponentially increased with decreasing dentin thickness especially at <500 µm (Fig. 12.4) [31]. This corresponds to the fact that – according to Hagen-Poiseuille's law – permeability is inversely related to the length of the tubule

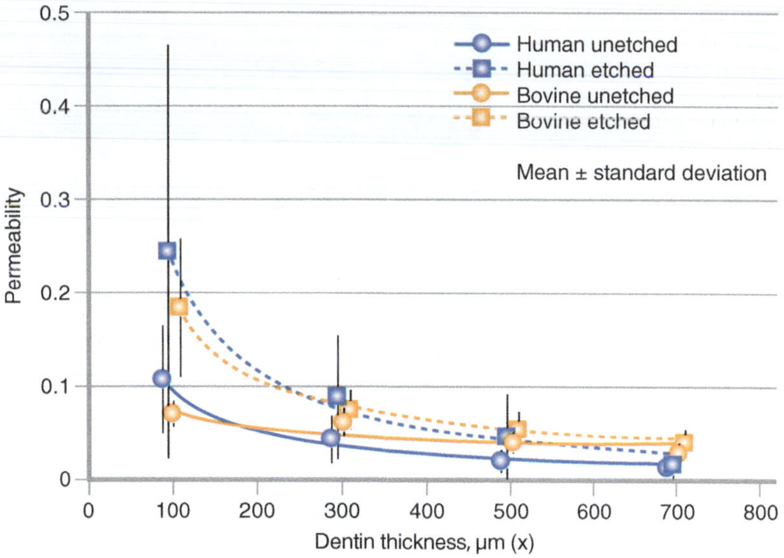

Fig. 12.4 Permeability of dentin at different dentin thicknesses with and without etching (Used with permission of Schmalz et al. [31])

(dentin thickness). Furthermore, the number and the diameter of the tubule close to the pulp are considerably larger than at the dentin-enamel junction [31] (Fig. 12.5a, b), by which permeability is further increased close to the pulp. Therefore, the residual dentin acts as a barrier reducing the amount of substances reaching the target cells (pulp cells) [31] being especially effective for residual dentin thicknesses of >500 µm. In accordance with that cell culture studies showed that dentin thickness between test material and target cells can reduce cell toxicity of test materials [15]. The same is known from in vivo studies on animals and on humans [2].

This effect is modified by several factors: the first being dentinal sclerosis. Mineral deposition within the dentinal tubule occurs as a result of long-term irritation, e.g., by caries. Dentin permeability depends – according to the aforementioned Hagen-Poiseuille's law – to the fourth power on the tubule diameter. Therefore, a reduction of the tubule diameter, sometimes even the total closure of the tubule, decreases considerably dentin permeability and thus the diffusion of possibly toxic molecules toward the pulp.

A second modifying factor is the so-called smear layer, which is produced after grinding dentin, e.g., during caries removal and cavity preparation. Removal of the smear layer increases dentin permeability [32]. However, we found that the removal of the smear layer, e.g., by an acid does effect dentin permeability mainly in thin dentin slices [31], i.e., in deep cavities (<0.5 mm residual dentin) (Fig. 12.4). This may be an explanation of the fact that in median and shallow cavities, acid treatment of dentin does apparently not cause pulp damage. A final modifying factor to be considered is the adsorption of some substances released from materials like Zn (from amalgam) or eugenol (from respective cements) to the inorganic dentin component (hydroxyl apatite) of dentin or to proteins [43, 46].

All these factors influence the concentrations of substances released from the material, which reach the dental pulp with the consequence that material-related pulp reactions are mainly expected in deep cavities (i.e., <0.5 mm residual dentin thickness) and after pulp exposure. This means that dentin indeed has a protective effect in these cases.

Interestingly, pulp reactions toward resin-based materials were observed in medium cavities, if bacteria were present (see earlier discussion). Apparently, the protective effect of dentin, which is observed after material application, is not operating against bacteria. An explanation may be that

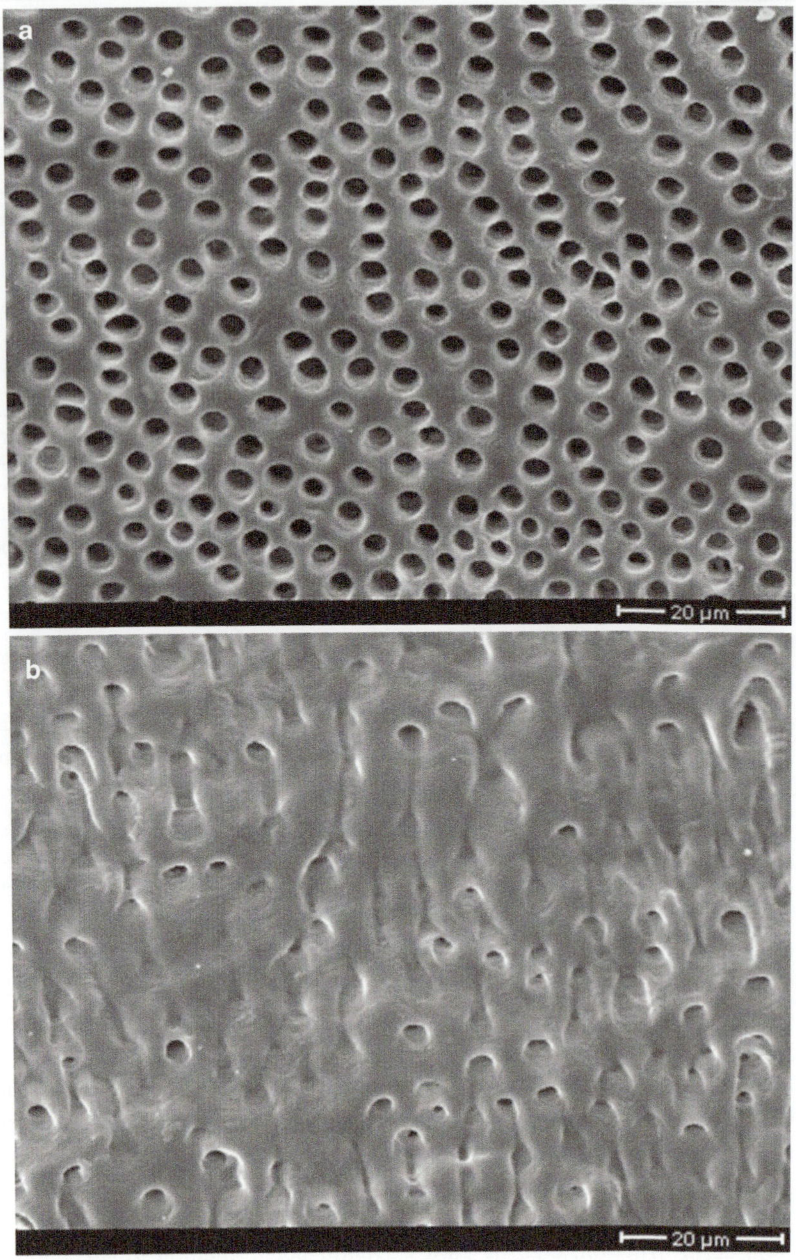

Fig. 12.5 Anatomy of dentin (**a**) close to the pulp more tubules with larger diameter than (**b**) close to the dentin-enamel junction (Used with permission of Springer Science+Business Media from Schmalz and Arenholt-Bindslev [2])

the concentration of substances released from materials decreases with time [17], but the amount of toxicants produced by bacteria on the cavity floor may increase over time by bacterial growth, if, e.g., through gaps, bacteria receive sufficient nutrition from the oral environment. Apparently, the protective capacity of dentin is not high enough to cope with this bacterial challenge.

12.6 Pulp Biocompatibility Testing

Studies on pulp biocompatibility of dental materials are done (1) to elucidate pulp reactions and the mechanisms behind them and (2) in the course of legally required premarket safety testing. Dental restorative materials are – meanwhile worldwide –

regarded as medical devices and legally regulated. Generally, the manufacturer is responsible for the safety of his products, and the manufacturer defines the indication for each product. The dentist is responsible for keeping to the given indication and for the correct processing of the materials.

Medical devices must fulfill certain requirements before they are allowed to be marketed. These requirements also cover safety aspects including pulp biocompatibility. For this, sets of test methods are compiled in ISO standards (ISO 10 993-series, ISO 7405) [2]. In an attempt to fulfill the legal requirements, manufacturers must complete a clinical risk assessment according to ISO 14971, mainly based on these ISO standards [2]. Basically, these tests comprise elution tests (chemical determination of substances released from a material into a liquid) and cell culture tests (for cytotoxicity and for mutagenicity). If necessary, animal studies and studies on humans (e.g., teeth to be extracted for orthodontic reasons) are possible [2]. Today, for assessing the risk of pulp damage, mainly cell culture-based studies are performed, because (large) animal studies are expensive, time consuming, and not without ethical problems. In special cases (completely new chemistry or new materials claiming pulp repair/regeneration), they may become necessary; in cases of only "improved" materials, they can possibly be waived, if cell culture tests give no indication of a possible toxic behavior [2]. The same is true today for long-term animal-based toxicity/carcinogenicity studies.

Classical cell culture methods are fast and comparatively easy to perform, but they do not reflect the clinical situation of a tooth cavity. Therefore, the extrapolation of the obtained data to the patient is problematic, and data must be carefully interpreted. In order to bridge the gap between the classical in vitro cell culture cytotoxicity tests and the patient situation, the so-called dentin-barrier test has been developed, in which a dentin disk of a defined thickness is placed between the material and the target cells [33, 34]. This test method is also accepted as ISO standard (ISO 7405) [35]. Other models mainly being used to elucidate reaction mechanisms are the tooth slice model from human [36] or rat [37] teeth or the whole tooth model [38].

Pulp studies on small laboratory animals, on large animals (e.g., nonhuman primates), or even on human teeth are performed under ideal conditions and thus are not without limitations; e.g., sound teeth of young animals or patients are used for testing, which is in contrast to most patients situations presenting teeth with a caries-affected dentin and pulp. Therefore, preclinical assessment of pulp biocompatibility of dental materials is a first and necessary step, which must be based on a battery of different tests. Even then one cannot assume that all possible side effects can be foreseen by such an evaluation. Therefore, according to legal regulations it is compulsory for the dentist to report observed adverse effects to relevant authorities in the respective country (postmarket surveillance).

12.7 Reaction Patterns

Pulp reactions toward materials/irritants are complex events, and their outcome is determined by the:

- Properties of the material/irritant (e.g., toxicity and change of toxicity over time, diffusibility in an aqueous system)
- Local cavity situation (e.g., the amount and the quality of residual dentin)
- Healing capacity of the pulp (e.g., age, functioning immune system or preexisting caries-induced inflammation)

In the following, pulp reactions in three scenarios (slight irritation, strong irritation, pulp exposure) are described, however, keeping in mind that in reality combinations of these reactions occur.

Slight irritations (gentle cavity preparation in medium cavities filled with materials of only initially medium but then low toxicity or with low diffusibility due to high hydrophobicity and in the absence of bacteria) will be associated with a temporary (few days) inflammatory pulp reaction. If there is no further irritation, the inflammation will resolve, and new dentin ("reactionary dentin") will be formed by survived odontoblasts at the end of the cut dentinal tubule [39] (see also Fig. 12.1). Reactionary dentin is delineated from the preexisting dentin by a calcio-traumatic line [8]. This new dentin (also called tertiary dentin) is more irregular in its structure than primary and secondary dentin, but normally dentin tubule can be clearly seen and tubular continuity with secondary dentin may exist [39]. Apparently, the odontoblasts can repair their cut processes, if injury does not happen close to the cell body [8]. Further to the new dentin formation, apposition of peritubular dentin and precipitation of intratubular crystals occur, which reduce dentin permeability and serve as a further protective shield [39].

Strong and/or persistent irritations lead to a more extended pulpal inflammation, e.g., in cases of bacteria at the cavity floor under the restorative material. The inflammation will not resolve and potentially lead to pulp necrosis (see also Fig. 12.3). Normally, odontoblasts do not survive under these conditions, and the recruitment of new (secondary) odontoblasts (see the discussion to come) will be hampered by the extensive inflammation process.

Pulp exposure is always associated with the loss of the original (primary) odontoblasts, and new odontoblasts cannot regenerate from neighboring odontoblasts by cell division, because odontoblasts are terminally differentiated postmitotic cells [8]. Such tissue damage always leads to a pulp inflammation. If the irritation is timely restricted and after an appropriate treatment (e.g., by application of a calcium hydroxide-releasing preparation), it is apparently possible that new odontoblasts are recruited from pulpal (stem/progenitor) cells. Such cells are located in different areas of the pulp and in the periapical area, often in perivascular niches. These cells now proliferate, migrate to the pulp exposure site, and differentiate into secondary odontoblasts, which then form new dentin (reparative dentin, dentinal bridge formation). This new dentin may be rather irregular and atubular (fibrodentin/osteodentin) and may contain voids and tunnels especially close to the material tissue contact area. However, it is a new hard tissue with the potential to protect the dental pulp from further injury, if treated properly [8] and bacterial invasion is avoided, e.g. by a restoration tighly sealing the cavitiy walls against bacterial penetration.

The mechanisms behind these different pulp responses toward different irritation scenarios are not fully understood (for more details, see, e.g., [8, 39]). Apparently, odontoblasts constitute the first line of defense, and their reaction determines the further consequences reaching from transient inflammation to necrosis. They are assisted by antigen-presenting dendritic cells, of which one population is present in the odontoblastic region and the other dendritic cell population is located in the central part of the pulp. These cells form a continuous reticular network involving the entire pulp tissue, including the odontoblastic layer [40, 41, 60]. They allow a delayed-type hypersensitivity reaction toward relevant materials [40]. The inflammation itself is characterized by an invasion of cells like neutrophils, lymphocytes, or macrophages. In this context it is interesting that dental monomers reduce the antibacterial defense mechanism of macrophages; these cells normally produce inflammatory mediators like interleukin-6 after a bacterial (LPS) challenge with the aim to eliminate the bacteria through an inflammatory process. Dental monomers impair this synthesis and this may interfere with bacterial clearance [26, 27].

For pulp repair a transient inflammation is apparently necessary [39]. Furthermore, for the formation of new dentin, bioactive molecules are involved like members of the TGF-β family, including TGF-β1, TGF-β3, and BMP-7 [4]. These molecules are available, e.g., from residual dentin after injury and after solubilization by substances like calcium hydroxide or EDTA [1] (Fig. 12.6). Material cell interaction may lead to

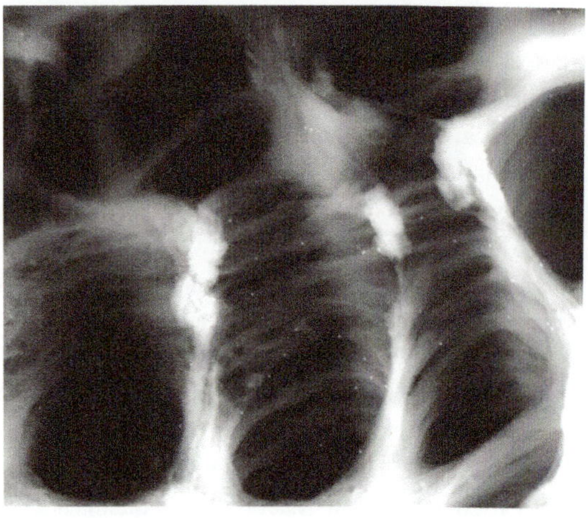

Fig. 12.6 Gold labeling of exposed TGF-β1 (*white dots*) from dentin (after EDTA treatment)

cell necrosis with a consecutive propagation of inflammation, which then may interfere with pulp repair, especially when the pulp is exposed [3]. An existing bacterial-induced inflammation (due to a caries process) impairs pulp repair [41] and thus influences the pulp reaction to dental materials.

Also, substances released from materials may induce apoptosis. For instance, dental monomers increase intracellular production of reactive oxygen species (ROS) with consecutive DNA damage [16]. Cells may be able to repair the DNA damage or they go into apoptosis [42]. Apoptosis is not followed by inflammation, but repair and apoptosis are highly energy consuming. This may interfere with the ability of cells to differentiate and to induce biomineralization [43].

With increasing age of the patient, the size of the pulp decreases, and the number of odontoblasts, the number of the cells in the subodontoblast layer and, the number of pulp fibroblasts decreases [44]. Due to a reduced blood flow, removal of toxicants is reduced [39]. The reduced number of odontoblasts together with a reduced secretory activity suggests a compromised repair capacity for reactionary dentin formation for elderly persons. The same is true for the reduced number of fibroblasts in cases of pulp exposure. Therefore, with increasing age the repair capacity of the pulp in general decreases [44].

In summary, the final pulp response to a material is dependent not only upon the material itself but also on the local situation and the status of the pulp. Concerning the material itself, not only biologic interactions are important but also the sealing ability of a material to prevent bacterial migration under the restoration. And finally, concerning the biologic properties of materials, they should be not only nontoxic but also antibacterial and repair stimulating. A problem is that antibacterial materials are very often toxic to (pulp) cells, because the mechanism of the antibacterial effect of dental materials is mainly unspecific and of a multitarget nature. Solutions of this problem can be (1) that materials are only initially antibacterial/toxic eliminating bacteria, but after setting they are inert, allowing for healing and repair, and that (2) toxic materials with high hydrophobicity (like eugenol) are toxic against bacteria at the cavity floor, but due to the low diffusion through the hydrophilic dentin, the effective concentration in the pulp is much lower, and thus it is nontoxic to pulp cells [45].

12.8 Dental Materials/Substances

12.8.1 Amalgam

Amalgam is cytotoxic when freshly mixed but toxicity is significantly reduced after setting [46]. The same was shown in rat subcutaneous implantation studies [47]. Initial local toxicity of zinc-containing and high copper amalgam was higher than zinc-free or low copper products [2, 48]. Eluted metals are bound to the dentin of the cavity floor and are responsible for dentin discoloration. Small amounts of metals are penetrating to the dental pulp. In median and shallow cavities (>0.5 mm residual dentin), usually only slight or no inflammatory reactions are observed several months after application. An apposition of tertiary dentin may be the only histological indication of a pulpal effect [2].

Postoperative sensitivities after placement disappear within a few days. Dentin discoloration can be prevented by using a cavity varnish or a dentinal adhesive. In deep cavities, a cement layer beyond the amalgam restoration is recommended, because due to the amalgam application technique (condensation pressure) the residual dentin may be destroyed and a material placed into the vital pulp. The topic of systemic toxicity of amalgam is not discussed here, but the reader is referred to the literature (e.g., [2]).

12.8.2 Cements

Cements comprise a large group of materials, which normally consist of a powder (zinc oxide or silicon dioxide) and a liquid (often an acid), which both are mixed and which harden by chemical reactions. Silicate cements and silicophosphate cements are virtually not used any more in dental practice. Both material groups were reported to be associated with pulp damage [2].

Zinc phosphate cements are still used, e.g. for luting indirect restorations although it is today mainly replaced by glass ionomer cements. Zinc phosphate cements are briefly after mixing cytotoxic [49] and cause initial pain when used for luting purposes. Cytotoxicity decreases after total setting. In medium and shallow cavities, no long-term pulp damage has been reported, but in deep cavities a protective layer of a calcium hydroxide preparation is recommended [50].

Glass ionomer cements are less cytotoxic than zinc phosphate cements [49]. In shallow, medium, and even deep cavities, no pulp damage was reported, if bacterial penetration to the cavity floor was prevented [50, 51]. However, if glass ionomer cement directly contacted the exposed pulp, severe inflammatory reactions have been observed [51]. Furthermore, in deep cavities, after several weeks no tertiary dentin formation was observed. This can be interpreted as an inhibition of biomineralization. Therefore, again, in deep cavities a protective calcium hydroxide cement layer has been recommended [2, 51].

Calcium hydroxide products (either as suspension or as setting cement) contain an active ingredient (calcium hydroxide), which since many years has been known to induce tertiary dentin formation when in close contact with the exposed pulp. Calcium hydroxide suspensions have a higher pH (12–13) than calcium hydroxide cements (pH 10–11). Suspensions are associated with a limited layer of necrotic tissue at the contact site on which reparative dentin is formed. With cements this necrotic layer is smaller or nonexistent [3]. It is also known and generally accepted that these products can be used as a protective layer in deep cavities to prevent unwanted pulp reactions due to the restorative material used on top of it. The stimulating effect of calcium hydroxide on dentin formation especially in pulp exposure situations is probably caused by a combination of several factors; among those the release of growth factors like TGF-β1 from dentin after application of the alkaline material is considered to be a key event [52] (Fig. 12.6). A main disadvantage of calcium hydroxide cements is their tendency for disintegration, especially in cases of microleakage at the margins of the final restoration [53]. Therefore, a combination of a calcium hydroxide suspension with a cover (e.g., glass ionomer cement) is recommended.

Tricalcium silicate cements, like mineral trioxide aggregates (MTA), Biodentine, or mixtures from basic materials [54], have been developed in order to have a pulp-compatible material available, which has a sufficient mechanical stability and induces biomineralization. These materials release calcium hydroxide during and after setting. While MTA so far still has some technical problems like a prolonged setting time (over 1 h), Biodentine seems to be more easy to handle. Both materials are reported to be noncytotoxic [55], and they stimulate pulpal (stem/progenitor) cells to express biomineralization markers like alkaline phosphatase [56], dentin sialoprotein (DSP) [54], and TGF-β1 [57]. For both MTA and Biodentine, tertiary dentin formation has been shown in animal experimentation [58]. Studies on patients with teeth scheduled for extraction due to orthodontic reasons showed successful pulp capping with no significant difference between MTA and Biodentine [59]. These (and similar) materials seem to be a good alternative to calcium hydroxide preparations. Recently, a light-curing tricalcium silicate cement was marketed [60]. According to the material safety data sheet of the company, >60 % MTA and <50 % resin monomers are included. Besides elution data (e.g., Ca release), no information is published so far concerning the claimed pulp repair with dentin formation.

Pulp-capping materials with recombinant growth factors have repeatedly been tested [1, 8, 37, 61, 62] and found to stimulate dentin formation, but no such product so far has been placed on the market.

Zinc oxide and eugenol cements are cytotoxic [63] and release for a prolonged time eugenol. They are antibacterial at the cavity floor. However, due to the lipophilic character of eugenol, the concentration drops significantly after diffusion through the hydrophilic dentin [64]. Therefore, no pulp reaction can be seen in light microscopy, if this material is placed on an intact dentin, and it is also used as a nontoxic control material in the ISO 7405 pulp-dentin test [35]. From a clinical point of view, the material has a sedative effect, and eugenol was shown to block nerve transmission in clinically relevant concentrations [65]. On the other hand, direct application of this material on the exposed pulp may – due to the toxicity – lead to pulp necrosis [66]. Therefore, the material is not recommended for direct pulp capping.

12.8.3 Resin-Based Composites/Adhesives

This material group comprises a large variety of different products with inorganic fillers and an organic matrix mainly based on methacrylate chemistry, with substances like bisphenol-a-glycidyl dimethacrylate (bis-GMA), bisphenol-a-dimethacrylate (bis-DMA), 1,6-bis(methacrylyloxy-2-ethoxycarbonylamino)-2,4,4-trimethylhexane (UDMA), triethylene glycol dimethacrylate (TEGDMA), 2-hydroxymethyl-methacrylate(HEMA), and many others. The composition of these materials is – compared to amalgam – very complicated (for more details, see [2]). As a common characteristic resin-based composites are today always used together with a liquid adhesive, which may consist of one or more "bottles." These adhesives improve marginal integrity (with less microleakage) and this may prevent bacterial penetration between the restorative material and the cavity walls (bacterial seal). However, on the other hand, they contain biologically active substances, which come into contact with the dentin and the pulp.

Adhesion to dentin implies the interaction of the adhesive with the smear layer. With one method ("etch-and-rinse adhesives"), the smear layer is removed by an acid (often 38 % phosphoric acid). This increases dentin permeability, but apparently this effect becomes relevant only in cases of little residual dentin (< 0.5 mm from the pulp) [31]. Furthermore, the acid is neutralized by contact with the dentin. In other products, the smear layer is modified by acidic monomers (self-etch adhesives), which only partially dissolve the smear layer but then polymerize without a rinsing step. For both adhesive systems, in shallow and medium cavities, short dentin etching does not damage the dental pulp [2].

Most resin composites are initially cytotoxic, but after setting toxicity significantly decreases [2]. Again, in shallow and medium cavities, no pulp damage is expected. However, even in shallow and medium cavities, pulp damage under resin-based composite fillings occurs, if bacteria are present at the cavity floor [22]. Effective sealing by an adhesive should prevent bacterial invasion. In this sense, the adhesive is pulp protective in shallow and medium cavities. Adhesives containing antibacterial monomers, like methacryloyloxy-dodecylpyridinium bromide (MDPB), are an interesting approach. As the antibacterial (toxic) effect disappears after polymerization, no pulp damage occurs [67–70].

Light-curing units (LCU) commonly used for resin curing are mainly based on quartz-tungsten-halogen (QTH) or light-emitting diodes (LED) technology. While the relative amount of heat generated with LEDs is lower than with halogen units, the heat produced by LEDs is still considerable, especially with the so-called high-power LCUs with a total irradiance of >1,000 mW/cm^2 or even new devices with >3,000 mW/cm^2 [29]. This adds up to the exothermic setting reaction of the resin. A critical rise of pulp temperature is today assumed at 5.5 °C [71].

A number of mainly in vitro studies have been recently published on this topic. Intrapulpal temperature rise was found to be dependent upon the residual dentin thickness [72]: an LCU emitting <1,000 mW/cm^2 irradiance increased the pulp temperature more at 0.5 mm residual dentin than at 1 mm residual dentin. For bulk-filling technique, temperature increase was higher than with incremental technique [73], and curing flowable resin-based composites and of bonding agents increased temperature more than curing conventional ones [74, 75]. Irradiances of <1,000 mW/cm^2 usually did not increase the intrapulpal temperature more than the critical 5.5 °C [71, 72]. However, another in vitro study in a pulp chamber used for cytotoxicity testing with 0.5 mm dentin showed a temperature increase over 6 °C after 30 s irradiation with an LCU of 553 mW/cm^2 with consecutive inhibition of metabolic activity of test cells [76]. Irradiances of more than 1,000 mW/cm^2 lead to a temperature rise in primary teeth to around 4.5 °C with a flowable composite [74] and in other studies to >5.5 °C [75]. For LCUs of >3,000 mW/cm^2, published data are scarce. In one source critical intrapulpal temperature rises are mentioned [77]. Simulated blood flow reduced heat accumulation; otherwise, critical temperature rise occurred [78]. Clinically, a reduced blood flow follows local anesthesia and relevant temperature increase is possible. Furthermore, irradiance distribution across the face of the light tips was found to be inhomogeneous [79], resulting in parts of the light tip with higher irradiances than the reported average value. Interpreting these data, one must take into consideration that LCUs in deep cavities are used in some distance from the cavity floor and that thus some energy may be lost. However, using high-power curing mode (>1,000 mW/cm^2) for curing the first layer in deep cavities, implies the risk of pulp damage and should be avoided.

Pulp capping with dentinal adhesives has been described, and some authors have reported positive results with dentin bridge formation in animal studies [80]. However, other authors have shown opposing data [81]. In many studies on human teeth, both etch-and-rinse and self-etch systems (with control of hemorrhage) evoked moderate to severe inflammatory cell infiltrate involving the coronal pulp with chronic abscesses, but no clinical symptoms were present. Dentin bridging was observed in no or in only few specimens [12, 82, 83]. The lack of bridge formation was also observed when applying an antibacterial adhesive on the pulp [84]. In most studies on human teeth, calcium hydroxide preparations yielded better dentin bridge formation. The reasons for the lack of bridge formation after applying dentinal adhesives on the pulp are unclear. Apparently, no secondary odontoblasts are recruited, despite – as seen in some studies – rather little inflammation [84]. As mentioned earlier, resin monomers increase ROS production in the contacting cells [16, 42, 85]. Both repair of consecutive DNA damage and apoptosis are energy-consuming processes. This may explain the fact that differentiation of pulp cells to express biomineralization markers is inhibited [43].

12.9 Summary and Conclusions

In this chapter it was shown that materials interact with the pulp, especially in deep cavities or after pulp exposure. Bacteria under restorations may also cause pulp damage as well as heat produced by traumatic cavity/tooth preparation and by using high-power light-curing units. Pulp reactions may vary from transient inflammation to chronic and severe inflammation ending in necrosis. The severity of the pulp reaction depends on the material, the residual dentin, and the healing capacity of the pulp.

Preventive measures for pulp protection start with an atraumatic cavity/tooth preparation. Materials should only be used according to the indications given by the manufacturers. In deep cavities a protective layer of a calcium hydroxide-releasing material should be applied. In medium and shallow cavities, bacterial penetration should be avoided by using an effective dentinal adhesive, which is pulp protective due to the inhibition of bacterial invasion. Care should be exerted with high-power LCUs, and for curing resin layers close to the pulp, <1000 mW/cm^2 should be used. No resin adhesives should be applied for direct pulp capping. Following these measures, we can come close to a primary aim of restorative treatment, namely, to keep the pulp vital.

References

1. Smith AJ, Lumley PJ, Tomson PL, Cooper PR. Dental regeneration and materials: a partnership. Clin Oral Investig. 2008;12:103–8.
2. Schmalz G, Arenholt-Bindlev D. Biocompatibility of dental materials. Heidelberg: Springer; 2009.
3. Mjör I. Pulp-dentin biology in restorative dentistry. Chicago/Berlin: Quintessence Publishing Co; 2002.
4. Murray PE, Smith AJ, Garcia-Godoy F, Lumley PJ. Comparison of operative procedure variables on pulpal viability in an ex vivo model. Int Endod J. 2008;41:389–400.
5. Cavalcanti BN, Lage-Marques JL, Rode SM. Pulpal temperature increases with Er:YAG laser and high-speed handpieces. J Prosthet Dent. 2003;90:447–51.
6. International Organization for Standardization: ISO 14457:2012 Dentistry – Handpieces and motors. International Organization for Standardization. Case postale 56 – CH-1211 Geneva 20; 2014.
7. Cavalcanti BN, Serairdarian PI, Rode SM. Water flow in high-speed handpieces. Quintessence Int. 2005;36:361–4.

8. Goldberg M, Smith AJ. Cells and extracellular matrices of dentin and pulp: a biological basis for repair and tissue engineering. Crit Rev Oral Biol Med. 2004; 15:13–27.
9. Banerjee A, Pabari H, Paolinelis G, Thompson JD, Watson TF. An in vitro evaluation of selective demineralised enamel removal using bio-active glass air abrasion. Clin Oral Investig. 2011;15:895–900.
10. Louw NP, Pameijer CH, Ackermann WD, Ertl T, Cappius HJ, Norval G. Pulp histology after Er:YAG laser cavity preparation in subhuman primates–a pilot study. SADJ. 2002;57:313–7.
11. Zesewitz H, Klaiber B, Hotz P, Hugo B. Heat propagation in dentin during cavity preparation in vitro with oscillating instruments. Schweiz Monatsschr Zahnmed. 2005;115:536–41.
12. Silva GA, Lanza LD, Lopes-Junior N, Moreira A, Alves JB. Direct pulp capping with a dentin bonding system in human teeth: a clinical and histological evaluation. Oper Dent. 2006;31:297–307.
13. Ziebolz D, Helms K, Hannig C, Attin T. Efficacy and oral side effects of two highly concentrated tray-based bleaching systems. Clin Oral Investig. 2007;11:267–75.
14. de Oliveira Duque CC, Soares DG, Basso FG, Hebling J, de Souza Costa CA. Bleaching effectiveness, hydrogen peroxide diffusion, and cytotoxicity of a chemically activated bleaching gel. Clin Oral Investig. 2013. doi: 10.1007/s00784-013-1147-4.
15. Galler K, Hiller KA, Ettl T, Schmalz G. Selective influence of dentin thickness upon cytotoxicity of dentin contacting materials. J Endod. 2005;31:396–9.
16. Schweikl H, Spagnuolo G, Schmalz G. Genetic and cellular toxicology of dental resin monomers. J Dent Res. 2006;85:870–7.
17. Ferracane JL. Elution of leachable components from composites. J Oral Rehabil. 1994;21:441–52.
18. Geurtsen W. Substances released from dental resin composites and glass ionomer cements. Eur J Oral Sci. 1998;106:687–95.
19. Gerzina TM, Hume WR. Diffusion of monomers from bonding resin-resin composite combinations through dentine in vitro. J Dent. 1996;24:125–8.
20. Noda M, Wataha JC, Kaga M, Lockwood PE, Volkmann KR, Sano H. Components of dentinal adhesives modulate heat shock protein 72 expression in heat-stressed THP-1 human monocytes at sublethal concentrations. J Dent Res. 2002;81:265–9.
21. Splieth C, Bernhardt O, Heinrich A, Bernhardt H, Meyer G. Anaerobic microflora under Class I and Class II composite and amalgam restorations. Quintessence Int. 2003;34:497–503.
22. Bergenholtz G. Evidence for bacterial causation of adverse pulpal responses in resin-based dental restorations. Crit Rev Oral Biol Med. 2000;11:467–80.
23. Krifka S, Federlin M, Hiller KA, Schmalz G. Microleakage of silorane- and methacrylate-based class V composite restorations. Clin Oral Investig. 2012;16:1117–24.
24. Bagis YH, Baltacioglu IH, Kahyaogullari S. Comparing microleakage and the layering methods of silorane-based resin composite in wide Class II MOD cavities. Oper Dent. 2009;34:578–85.
25. Heintze SD. Systematic reviews: I. The correlation between laboratory tests on marginal quality and bond strength. II. The correlation between marginal quality and clinical outcome. J Adhes Dent. 2007;9 Suppl 1:77–106.
26. Eckhardt A, Harorli T, Limtanyakul J, Hiller KA, Bosl C, Bolay C, Reichl FX, Schmalz G, Schweikl H. Inhibition of cytokine and surface antigen expression in LPS-stimulated murine macrophages by triethylene glycol dimethacrylate. Biomaterials. 2009; 30:1665–74.
27. Schmalz G, Krifka S, Schweikl H. Toll-like receptors, LPS, and dental monomers. Adv Dent Res. 2011;23:302–6.
28. Hansel C, Leyhausen G, Mai UE, Geurtsen W. Effects of various resin composite (co)monomers and extracts on two caries-associated micro-organisms in vitro. J Dent Res. 1998;77:60–7.
29. Jandt KD, Mills RW. A brief history of LED photopolymerization. Dent Mater. 2013;29:605–17.
30. Unemori M, Matsuya Y, Akashi A, Goto Y, Akamine A. Self-etching adhesives and postoperative sensitivity. Am J Dent. 2004;17:191–5.
31. Schmalz G, Hiller KA, Nunez LJ, Stoll J, Weis K. Permeability characteristics of bovine and human dentin under different pretreatment conditions. J Endod. 2001;27:23–30.
32. Reeder Jr OW, Walton RE, Livingston MJ, Pashley DH. Dentin permeability: determinants of hydraulic conductance. J Dent Res. 1978;57:187–93.
33. Schmalz G, Schuster U, Nuetzel K, Schweikl H. An in vitro pulp chamber with three-dimensional cell cultures. J Endod. 1999;25:24–9.
34. Schuster U, Schmalz G, Thonemann B, Mendel N, Metzl C. Cytotoxicity testing with three-dimensional cultures of transfected pulp-derived cells. J Endod. 2001;27:259–65.
35. International Organization for Standardization: ISO 7405: Dentistry. Preclinical evaluation of the biocompatibility of medical devices used in dentistry: test methods for dental materials. Geneva: InternationalOrganization for Standardization; 2012, 2013.
36. Magloire H, Joffre A, Bleicher F. An in vitro model of human dental pulp repair. J Dent Res. 1996;75:1971–8.
37. Sloan AJ, Smith AJ. Stimulation of the dentine-pulp complex of rat incisor teeth by transforming growth factor-beta isoforms 1–3 in vitro. Arch Oral Biol. 1999;44:149–56.
38. About I. Dentin regeneration in vitro: the pivotal role of supportive cells. Adv Dent Res. 2011;23:320–4.
39. Tziafas D, Smith AJ, Lesot H. Designing new treatment strategies in vital pulp therapy. J Dent. 2000; 28:77–92.
40. Jontell M, Okiji T, Dahlgren U, Bergenholtz G. Immune defense mechanisms of the dental pulp. Crit Rev Oral Biol Med. 1998;9:179–200.

41. Bhingare AC, Ohno T, Tomura M, Zhang C, Aramaki O, Otsuki M, Tagami J, Azuma M. Dental pulp dendritic cells migrate to regional lymph nodes. J Dent Res. 2014;93:288–93.
42. Krifka S, Spagnuolo G, Schmalz G, Schweikl H. A review of adaptive mechanisms in cell responses towards oxidative stress caused by dental resin monomers. Biomaterials. 2013;34:4555–63.
43. Galler KM, Schweikl H, Hiller KA, Cavender AC, Bolay C, D'Souza RN, Schmalz G. TEGDMA reduces mineralization in dental pulp cells. J Dent Res. 2011;90:257–62.
44. Murray PE, Stanley HR, Matthews JB, Sloan AJ, Smith AJ. Age-related odontometric changes of human teeth. Oral Surg Oral Med Oral Pathol Oral Radiol Endod. 2002;93:474–82.
45. Schmalz G, Hoffmann M, Weis K, Schweikl H. Influence of albumin and collagen on the cell mortality evoked by zinc oxide-eugenol in vitro. J Endod. 2000;26:284–7.
46. Schedle A, Franz A, Rausch-Fan XH, Samorapoompichit P, Boltz-Nitulescu G, Slavicek R. Cell culture study of dental materials: composite compared to amalgam. Z Stomatol. 1994;91(Suppl. 6):39–42.
47. Schmalz G, Schmalz C. Toxicity tests on dental filling materials. Int Dent J. 1981;31:185–92.
48. Meryon SD. The effect of zinc on the biocompatibility of dental amalgams in vitro. Biomaterials. 1984;5:293–7.
49. Schmalz G, Hiller K-A, Aslan-Dorter F. New developments in the filter test system for cytotoxicity testing. Mater Sci Mater Med. 2013;5:43–51.
50. Pameijer CH, Stanley HR, Ecker G. Biocompatibility of a glass ionomer luting agent. 2. Crown cementation. Am J Dent. 1991;4:134–41.
51. Schmalz G, Thonemann B, Riedel M, Elderton RJ. Biological and clinical investigations of a glass ionomer base material. Dent Mater. 1994;10:304–13.
52. Graham L, Cooper PR, Cassidy N, Nor JE, Sloan AJ, Smith AJ. The effect of calcium hydroxide on solubilisation of bio-active dentine matrix components. Biomaterials. 2006;27:2865–73.
53. Rehfeld RL, Mazer RB, Leinfelder KF, Russell CM. Evaluation of various forms of calcium hydroxide in the monitoring of microleakage. Dent Mater. 1991;7:202–5.
54. Peng W, Liu W, Zhai W, Jiang L, Li L, Chang J, Zhu Y. Effect of tricalcium silicate on the proliferation and odontogenic differentiation of human dental pulp cells. J Endod. 2011;37:1240–6.
55. Laurent P, Camps J, De MM, Dejou J, About I. Induction of specific cell responses to a Ca(3) SiO(5)-based posterior restorative material. Dent Mater. 2008;24:1486–94.
56. Eid AA, Niu LN, Primus CM, Opperman LA, Pashley DH, Watanabe I, Tay FR. In vitro osteogenic/dentinogenic potential of an experimental calcium aluminosilicate cement. J Endod. 2013;39:1161–6.
57. Laurent P, Camps J, About I. Biodentine(TM) induces TGF-beta1 release from human pulp cells and early dental pulp mineralization. Int Endod J. 2012;45:439–48.
58. Tran XV, Gorin C, Willig C, Baroukh B, Pellat B, Decup F, Opsahl VS, Chaussain C, Boukpessi T. Effect of a calcium-silicate-based restorative cement on pulp repair. J Dent Res. 2012;91:1166–71.
59. Nowicka A, Lipski M, Parafiniuk M, Sporniak-Tutak K, Lichota D, Kosierkiewicz A, Kaczmarek W, Buczkowska-Radlinska J. Response of human dental pulp capped with biodentine and mineral trioxide aggregate. J Endod. 2013;39:743–7.
60. Gandolfi MG, Siboni F, Prati C. Chemical-physical properties of TheraCal, a novel light-curable MTA-like material for pulp capping. Int Endod J. 2012;45:571–9.
61. Decup F, Six N, Palmier B, Buch D, Lasfargues JJ, Salih E, Goldberg M. Bone sialoprotein-induced reparative dentinogenesis in the pulp of rat's molar. Clin Oral Investig. 2000;4:110–9.
62. Goldberg M, Farges JC, Lacerda-Pinheiro S, Six N, Jegat N, Decup F, Septier D, Carrouel F, Durand S, Chaussain-Miller C, Denbesten P, Veis A, Poliard A. Inflammatory and immunological aspects of dental pulp repair. Pharmacol Res. 2008;58:137–47.
63. Hensten-Pettersen A, Helgeland K. Evaluation of biologic effects of dental materials using four different cell culture techniques. Scand J Dent Res. 1977;85:291–6.
64. Meryon SD, Jakeman KJ. An in vitro study of the role of dentine in moderating the cytotoxicity of zinc oxide eugenol cement. Biomaterials. 1986;7: 459–62.
65. Trowbridge H, Edwall L, Panopoulos P. Effect of zinc oxide-eugenol and calcium hydroxide on intradental nerve activity. J Endod. 1982;8:403–6.
66. Sela J, Ulmansky M. Reaction of normal and inflamed dental pulp to Calxyl and zinc oxide and eugenol in rats. Oral Surg Oral Med Oral Pathol. 1970;30:425–30.
67. Imazato S, Kinomoto Y, Tarumi H, Torii M, Russell RR, McCabe JF. Incorporation of antibacterial monomer MDPB into dentin primer. J Dent Res. 1997;76:768–72.
68. Imazato S. Bio-active restorative materials with antibacterial effects: new dimension of innovation in restorative dentistry. Dent Mater J. 2009;28:11–9.
69. Muller R, Eidt A, Hiller KA, Katzur V, Subat M, Schweikl H, Imazato S, Ruhl S, Schmalz G. Influences of protein films on antibacterial or bacteria-repellent surface coatings in a model system using silicon wafers. Biomaterials. 2009;30:4921–9.
70. Schmalz G, Ergucu Z, Hiller KA. Effect of dentin on the antibacterial activity of dentin bonding agents. J Endod. 2004;30:352–8.
71. Zach L, Cohen G. Pulp response to externally applied heat. Oral Surg Oral Med Oral Pathol. 1965;19:515–30.
72. Guiraldo RD, Consani S, Consani RL, Berger SB, Correr AB, Sinhoreti MA, Correr-Sobrinho L. Comparison of silorane and methacrylate-based composites on the polymerization heat generated with different light-curing units and dentin thicknesses. Braz Dent J. 2013;24:258–62.
73. Oberholzer TG, Makofane ME, du Preez IC, George R. Modern high powered led curing lights and their effect on pulp chamber temperature of bulk and incrementally cured composite resin. Eur J Prosthodont Restor Dent. 2012;20:50–5.

74. Baroudi K, Silikas N, Watts DC. In vitro pulp chamber temperature rise from irradiation and exotherm of flowable composites. Int J Paediatr Dent. 2009;19:48–54.
75. Santini A, Watterson C, Miletic V. Temperature rise within the pulp chamber during composite resin polymerisation using three different light sources. Open Dent J. 2008;2:137–41.
76. de Souza PP, Hebling J, Scalon MG, Aranha AM, Costa CA. Effects of intrapulpal temperature change induced by visible light units on the metabolism of odontoblast-like cells. Am J Dent. 2009;22:151–6.
77. Rueggeberg FA: heat development and temperature rise around the pulp. Cited in Bluephase Scientific Documentation 2013; www.ivoclarvivadent.us. Accessed Jan 2014.
78. Kodonas K, Gogos C, Tziafa C. Effect of simulated pulpal microcirculation on intrachamber temperature changes following application of various curing units on tooth surface. J Dent. 2009;37:485–90.
79. Michaud PL, Price RB, Labrie D, Rueggeberg FA, Sullivan B. Localised irradiance distribution found in dental light curing units. J Dent. 2014;42(2):129–39.
80. Cox CF, Hafez AA, Akimoto N, Otsuki M, Suzuki S, Tarim B. Biocompatibility of primer, adhesive and resin composite systems on non-exposed and exposed pulps of non-human primate teeth. Am J Dent. 1998;11:S55–63.
81. Pameijer CH, Stanley HR. The disastrous effects of the "total etch" technique in vital pulp capping in primates. Am J Dent. 1998;11:S45–54.
82. Accorinte ML, Loguercio AD, Reis A, Costa CA. Response of human pulps capped with different self-etch adhesive systems. Clin Oral Investig. 2008;12:119–27.
83. Nowicka A, Parafiniuk M, Lipski M, Lichota D, Buczkowska-Radlinska J. Pulpo-dentin complex response after direct capping with self-etch adhesive systems. Folia Histochem Cytobiol. 2012;50:565–73.
84. Tziafas D, Koliniotou-Koumpia E, Tziafa C, Papadimitriou S. Effects of a new antibacterial adhesive on the repair capacity of the pulp-dentine complex in infected teeth. Int Endod J. 2007;40:58–66.
85. Stanislawski L, Lefeuvre M, Bourd K, Soheili-Majd E, Goldberg M, Perianin A. TEGDMA-induced toxicity in human fibroblasts is associated with early and drastic glutathione depletion with subsequent production of oxygen reactive species. J Biomed Mater Res A. 2003;66:476–82.

Effects of Bisphenol A on the Dental Pulp

13

Michel Goldberg

13.1 Introduction

The xenoestrogen bisphenol A (BPA) is widely used in various food and packaging consumer products, in the manufacture of polycarbonate plastics and epoxy resins. It is released from food and beverage containers, baby's bottles, children's toys, and dental restorative materials, including occlusal sealants supposed to prevent the development of carious decays. It produces numerous adverse endocrine and developmental effects in rodents, resulting in general cytotoxic and pathologic outcomes. Some local effects are closely associated with this family of endocrine-disrupting compounds. What is induced is related to the estrogenic properties of BPA and resulting from alterations of synthesis of estradiol and testosterone. These effects are interfering with receptor binding. Irregular cycles, multiple ovarian cysts, reduction in primary follicles, neonatal mortality, sexual dysfunctions, and decreased libido have been reported as undesirable or adverse effects. Epigenetic effects are associated with an increased risk of cancer, namely, breast and prostate malignancies. Low-dose BPA exposure seems to increase adipogenesis in female animals, obesity, non-insulin-dependent diabetes mellitus, allergies, asthma, autism, cognitive decline, memory impairment, depression, and anxiety [1]. In addition to these well-documented general effects, many questions are related to the risks due to release of BPA after the dental restorations after a carious lesion or after the sealing of pits and fissures.

13.2 Cytotoxic Effects and Induced General Pathologies

The lower dose inducing cell damage is determined by the *no observed adverse effect level* (*NOAEL*). It was evaluated by the Food and Drug Administration to be as low as 5 mg BPA/kg body weight (bw)/day. However, according to safety authorities and protection agencies, the *tolerable daily intake* (*TDI*) considered as a reference would be a dose of 0.05 mg/kg bw/day. The issue of the dose is still a matter of debate, but it is clear that doses below the NOAEL have significant effects. According to Moon et al. [2] doses of BPA below the NOAEL induce mitochondrial dysfunctions in the liver and are associated with an increase in oxidative stress and inflammation.

Low concentrations of BPA induce lipid accumulation in hepatic cells, mediated by the production of reactive oxygen species in the mitochondria of HepG2 cells. Mitochondrial dysfunctions,

M. Goldberg, DDS, PhD
Department of Oral Biology,
Institut National de la Santé et de la
Recherche Médicale, Université Paris Descartes,
45 Rue des Saints Pères, Paris 75006, France
e-mail: mgoldod@gmail.com,
michel.goldberg@parisdescartes.fr

M. Goldberg (ed.), *The Dental Pulp*,
DOI 10.1007/978-3-642-55160-4_13, © Springer-Verlag Berlin Heidelberg 2014

including ROS production, lipoperoxidation, and the release of proinflammatory cytokines, are contributing to steatosis. They result from low concentration of BPA [3].

Traditional classical dogma in toxicology was "the dose makes the poison." Evolution of the concept suggests that effects may be detected with low doses below that used for traditional toxicological studies. In addition non-monotonic dose-response should also be taken into account. The effects of low doses cannot predict the effects observed at higher doses [4]. This implies that the effects of low doses have to be taken into consideration in terms of undesirable effects and of possible induced pathologies.

Cabaton et al. [5] have reported the effects of low doses of bisphenol A on the metabolome of perinatally exposed CD-1 mice (a method used for determining the metabolic changes to nutritional, pharmacological, and toxic stimuli). Dose-dependent variations in glucose, pyruvate, some amino acids, and neurotransmitters were identified, supporting that low dose of the endocrine disruptor BPA administered from day 1 up to day 21 interferes and disrupts the global metabolism.

13.2.1 Determination of Blood, Urine, Saliva, and Sweat Parameters of Excretion

- Normal values and concentrations found after BPA treatment were reported in body fluids.
- *Blood*: No BPA was found in blood samples prior or after dental treatment.
- *Saliva*: Olea et al. [6] collected the saliva 1 h before the application of cured sealants. After treatment, all saliva samples contained BPA in amount ranging from 90 to 931 μg. In control patients, BPA was detected in the saliva of all patients prior to the placement of the sealants and ranged between 0.07 and 6.00 ng/ml at baseline. Three hours after treatment, the salivary concentration peaked and returned to the baseline level within 24 h. Low peak levels were 3.98 ng/ml (one sealant application alone), whereas 9.08 ng/ml in the high-dose group (more than four sealants) [7]. Altogether, the different clinical studies available conclude that the highest level of BPA reported in saliva from dental sealants is more than 50,000 lower than the lethal dose 50 (LD50) values reported for BPA. This allows some researchers to conclude that human exposure to BPA from dental resins is minimal and poses no known health risk [8]. This contradicts some findings establishing that some low-dose effects of BPA are at the origin of undesirable effects.
- The daily *urinary* BPA excretion gave a median value of 1.2 μg/day, far below the tolerable daily intake recommended by the European Commission in 2002 [9].
- *Human excretion of sweat*: Monitoring the bioaccumulation of BPA in blood, urine, and sweat, Genuis et al. [1] concluded that blood and urine testing might underestimate the total body burden of the potential toxicant. By contrast, they considered that sweat analysis should be considered as an additional method to follow accurately the accumulation of BPA and its elimination.

Therefore, metrologic evaluations should keep attention on the false body fluids and not on what should be considered as significant.

13.2.2 Toxicogenomics and Adverse Health Effects

BPA exhibits toxicogenomics and undesirable effects on human health owning to the 89 common interacting genes/proteins. These genes/proteins may serve as biomarkers to assay the toxicities of the different chemicals leached out from the widely used plastics.

Bisphenol A acting as an endocrine disruptor is implicated in the feminization in various organs and displays various estrogenic effects. Due to a competitive ligand binding, BPA is bound to estrogen receptors α and β. The density of mammary buds was increased in BPA-exposed monkeys, leading to precancerous forms [10].

13.2.3 Precancerous and Cancerous Effects

Exposure to low doses of BPA resulted in significant alterations in gland morphology, which varied to subtle effects on mammary gland development when the exposure period occurs in adulthood, leading from precancerous to cancerous lesions. Prenatal exposure to relevant doses of BPA increases the number of intraductal hyperplasia and ductal carcinoma. Acevedo et al. [11] reported that the environmental levels of BPA during gestation and lactation induce mammary gland neoplasms even in the absence of any additional carcinogenic treatment.

13.2.4 Other Adverse Effects

13.2.4.1 Brain Development

- Prenatal and lactational exposures to low doses of BPA show effects on brain development in mice.
- In the adult mice brains, abnormal neocortical architecture and abnormal corticothalamic projections persisted in the group exposed to the BPA. Epigenetic alterations might trigger some of the effects on brain development after exposure to BPA [12].
- High-dose BPA impairs hippocampal neurogenesis in female mice across generations. This shed lights on another important feature: the *transgenerational effect*. The evaluation of transgenerational effects of BPA on hippocampal neurogenesis showed that when pregnant female mice were exposed to BPA (F0), the offspring (F2) from F1 generation display a decrease of newly generated cells in the hippocampi of F2 female mice. BPA adversely affects hippocampal neurogenesis of future generation by modulating ERK and BDNF-CREB signaling cascades [13]. The fact that the second or third generation of mice shows epigenetic alterations even without any contact with BPA is important for the potential development of pathologies of BPA-treated patients.

13.2.4.2 Effects on Type 2 Diabetes

- Short-term treatment with BPA leads to metabolic abnormalities in insulin-sensitive peripheral tissues. Mice treated with BPA were insulin resistant and had increased glucose-stimulated insulin release. It was concluded that short-term treatment with low dose of BPA slows down whole body energy and disrupts insulin signaling in peripheral tissues. Therefore, BPA can be considered as a risk factor for the development of type 2 diabetes [14].

13.2.4.3 Obesity

- Exposure of 3T3-L1 preadipocytes for 14 days to BPA reduced the amount of triglyceride accumulation and suppressed the gene transcription of the lipogenic enzyme lipoprotein lipase. BPA can reduce triglyceride accumulation during adipogenesis [15].

13.2.4.4 Transgenerational Actions of Environmental Compounds

- After transient exposure of F0 gestating female rats during the period of embryonic gonadal sex determination, the subsequent F1–F3 generations were obtained in the absence of any environmental exposure. Spermatogenesis cell apoptosis was affected transgenerationally. Ovarian primordial follicle pool size was significantly decreased. Different DNA methylation of the F3 generation supports the altered epigenetic transgenerational inheritance [16].

13.3 Bisphenol A in Dental Materials

Since 40 years the pulp response to bisphenol A-releasing restorative materials was investigated. Comparison was made between a methyl methacrylate monomer and a dimethacrylate thinner material [17]. Two materials (a cement and a composite resin) were evaluated. The pulp reaction shows that they were within the limits of tolerance. After a strong initial response seen for the two materials, the long-term response (45 days) showed a well-defined repair and regeneration of the underlying pulp tissue.

One of the comonomer used to decrease the viscosity of the monomer forming the backbone of the composites is the bisphenol A-diglycidylether methacrylate (Bis-GMA). Dental products release BPA in some very particular conditions. Triethyleneglycol-dimethacrylate (TEGDMA) is ended by two functional methacrylate groups. Between the two methacrylate groups, the molecule is linear. The proportions of Bis-GMA and TEGDMA vary among the different products. BPA does not exist as such in composites, adhesives, or sealants, but is used in the synthesis of the main backbone molecule of the composite resins [18]. The existence of three different chemical forms has been reported: (1) the bisphenol A-diglycidylether methacrylate (Bis-GMA), (2) the monomer-like bisphenol A-diglycidylether (BADGE), and (3) the bisphenol A-dimethacrylate (Bis-DMA), used in some adhesives and sealants. The polymerization of the monomers and comonomers is never complete. The conversion rate (proportion of polymerized molecules compared to the initial amount of unpolymerized molecules) varies between 30 and 80 %, depending on the resins. However, there are always free monomers and comonomers released from a composite resin. Nevertheless, it is highly unlikely that TEGDMA can be produced by degradation of the polymerized matrix. Concerning BPA it is mostly found as an impurity in BADGE and Bis-DMA, which have less estrogenicity than the resins containing bisphenol A [19].

BPA is released from dental resins through salivary enzymatic hydrolysis of BPA derivatives. BPA is detectable in saliva for up to 3 h after resin placement. The majority of dental composites release TEGDMA in vitro and in vivo. This component is toxic. The compound induces allergies and cytotoxicity. Many reports describe allergic dermatitis in dental personnel, but far less in the oral cavity of patients. In an estrogen-sensitive cell line, estrogenic effects were found with BPA, Bis-DMA, and Bis-GMA, but not with TEGDMA.

Unpolymerized monomeric resin components from dental composites act on the function of accessory cells derived from the rat incisor pulp. Accessory cells and T lymphocytes reacted to low concentrations of urethane dimethacrylate, bis-glycidyl methacrylate, triethylene glycol dimethacrylate, and bisphenol A. They increased spleen cell proliferation to concanavalin A [20]. Transdentinal diffusion of BPA released from the resin composite restoration may demonstrate adverse effects on the dental pulp.

In vitro the resin component BPA was acting on the viability and substrate adherence capacity of macrophages. Viability was determined by trypan blue exclusion. The adherence index of macrophage decreased in the presence of 10^{-8} M BPA. It was concluded that the resin component BPA has the capacity to inhibit macrophage function and modulate immune and inflammatory response in dental pulp and periapical tissues [21].

13.4 Genetic and Cellular Toxicology of Dental Resin Monomers

Monomers cause adverse biological effects in mammalian cells. TEGDMA causes gene mutations in vitro. The formation of micronuclei indicates chromosomal damages, and monomers such as TEGDMA and HEMA induce DNA strand breaks. The comet assay quantified the DNA single-strand breaks, alkali labile, and incomplete excision repair sites [22]. The impairment of cellular pro- and antioxidant redox balance is caused by monomers. Monomers reduced the level of the radical scavenger glutathione (GSH) that protects the cells against reactive oxygen species (ROS). Cytotoxic and genotoxic effects of TEGDMA and HEMA are inhibited by the presence of ROS scavengers like N-acetylcysteine (NAC), ascorbate, and Trolox (vitamin E). Pathways regulating cellular homeostasis, dentinogenesis, or tissue repair may be modified by monomers at very low concentrations, in any case below those that induce acute toxicity [23].

13.4.1 Effects on Human Dental Pulp of Adhesive Resins and Monomers

Direct pulp capping increases the blood vessel density near the pulp exposure. VEGF expression was upregulated primarily at post-transcriptional level [24].

There was a concentration-dependent decrease in cell proliferation and an increase in cell number after exposure to Bis-GMA. Cells showed typical characteristics of apoptotic cells after exposure to high concentrations of Bis-GMA. In contrast, cells exposed to low concentrations recovered their viability [25].

13.5 Summary

Bisphenol A is released by many resin components, including restorative materials, pit and fissure sealants, and resins aiming to seal orthodontic appliances. In vitro and in vivo adverse effects have been noted and the severity of these effects has been evaluated. A few toxic, genotoxic, and allergic reactions have been shown, displaying minor to severe responses. Many health concerns are documented, with increasing severity. During rodent development, cardiovascular, brain, and developmental deficiencies; obesity; and adverse effects of BPA have been well documented. In adult animals, as well as in humans, severe pathologies have been identified, such as diabetes, defective male and female genital tracts, ovarian cysts, and/or precancerous and cancerous lesions. However, it is difficult to extrapolate from animal pathologies to human. Therefore, the question of the potential adverse effects of resins releasing BPA remains open. Although some answers deny any adverse effects on public health in view of the small quantities released by BPA from restorative resins or sealants, the level being below the "non-detectable adverse effect level," four issues raise new insights and lead to a reappraisal of the safety of BPA in dentistry:

- Firstly, for a long period of time, the dose level was the most important point. Above a certain level, BPA was considered as a potential inducer of adverse effects. It is now clear that noxious effects are detectable even below a very low dose.
- Secondly, BPA effects appear to be transgenerational and they are observed even at the third generation of animals that have never received directly BPA. This means that our BPA-treated patients may not present immediately adverse effects. Two generations later, epigenetic effects might appear and noxious effects might influence some induced pathologies appearing at the third generation or later, even in BPA-untreated patients.
- Thirdly, the level of BPA in blood and urine levels, which was systematically measured up to now, may not be significant. In contrast, sweat analysis seems to provide more indicative information.
- Fourthly, it was recently shown that the possible high systemic bioavailability of BPA (70–90 %) is controlled by sublingual supervision. Along this line, the transmucosal absorption of BPA within the oral cavity led to much higher BPA internal exposure than the absorption resulting from the gastrointestinal tract. This focuses on the responsibility of BPA released from dental restorative material. The absorption through the oral mucosa may be an efficient systemic entry route, more efficient than orogastric gavage [26].

Altogether, these four points lead to reformulate the crucial question of the safety of BPA used in dentistry.

References

1. Genuis SJ, Beesoon S, Birkholz D, Lobo RA. Human excretion of Bisphenol A: blood, urine, and sweat (BUS) study. J Environ Publ Health 2012 (2012); article ID 185731.
2. Moon MK, Kim MJ, Jung IK, Koo YD, Ann HY, Lee KJ, Kim SH, Yoon YC, Cho B-J, Park KS, Jang HC, Park YJ. Bisphenol A impairs mitochondrial function in the live rat doses below the no observed adverse effect level. J Korean Med Sci. 2012;27:644–52.
3. Huc L, Lemarié A, Guéraud F, Héliès-Toussaint C. Low concentrations of bisphenol A induce lipid accumulation mediated by the production of reactive oxygen species in the mitochondria of HepG2 cells. Toxicol In Vitro. 2012;26:709–17.
4. Vandenberg LN, Colborn T, Hayes TB, Heindel JJ, Jacobs DR, Jr Lee D-H, Shioda T, Soto AM, vom Saal FS, Welshons WV, Zoeller RT, Myers JP. Hormones and endocrine-disrupting chemicals: low-dose effects and nonmonotonic dose responses. Endocr Rev. 2012;33(3):378–455.
5. Cabaton NJ, Canlet C, Wadia PR, Tremblay-Franco M, Gautier R, Molina J, Sonnenschein C, Cravedi J-P, Rubin BS, Soto AM, Zalko D. Effects of low doses of bisphenol A on the metabolome of perinatally exposed

CD-1 mice. Environ Health Perspect. 2013;121: 586–93.

6. Olea N, Pulgar R, Perez P, Olea-Serrano F, Rivas A, Novillo-Fertrell A, et al. Estrogenicity of resin-based composites and sealants used in dentistry. Environ Health Perspect. 1996;104:298–305.
7. Zimmerman JM, Shuman D, Stull SC, Ratzlaff RE. Bisphenol A blood and saliva levels prior to and after dental sealant placement in adults. J Dent Hyg. 2010;84:145–50.
8. Rathee M, Malik P, Singh J. Bisphenol A in dental sealants and its estrogen like effect. Indian J Endocrinol Metab. 2012;16:339–42.
9. Arakawa C, Fujimaki K, Yoshinaga J, Imai H, Serizawa S, Shiraishi H. Daily urinary excretion of bisphenol A. Environ Health Prev Med. 2004;9:22–6.
10. Tharp AP, Maffini MV, Hunt PA, Vandevoort CA, Sonnenschein C, Soto AM. Bisphenol A alters the development of the rhesus monkey mammary gland. Proc Natl Acad Sci U S A. 2012;109:8190–5.
11. Acevedo N, Davis B, Schaeberle CM, Sonnenschein C, Soto AM. Perinatally administered bisphenol A acts as a mammary gland carcinogen in rats. Environ Health Perspect. 2013;121(9):1040–6.
12. Itoh K, Yaoi T, Fushiki S. Bisphenol A, an endocrine-disrupting chemical, and brain development. Neuropathology. 2012;32(4):447–57.
13. Jang YJ, Park HR, Kim TH, Yang WJ, Lee JJ, Choi SY, Oh SB, Lee E, Park JH, Kim HP, Kim HS, Lee J. High dose bisphenol A impairs hippocampal neurogenesis in female mice across generations. Toxicology. 2012;296:73–82.
14. Batista TM, Alonso-Magdalena P, Vieira E, Amaral MEC, Cederroth CR, Nef S, Quesada I, Carneiro EM, Nadal A. Short-term treatment with bisphenol-A leads to metabolic abnormalities in adult male mice. PLoS ONE. 2012:7(3):e33814.
15. Linehan C, Gupta S, Samall A, O'Connor L. Bisphenol A-mediated suppression of LPL gene expression inhibits triglyceride accumulation during adipogenic differentiation of human adult stem cells. PLoS ONE. 2012;7(5):e36109.
16. Manikkam M, Guerrero-Bosagna C, Tracey R, Haque MM, Skinner MK. Transgenerational actions of environmental compounds on reproductive disease and identification of epigenetic biomarkers of ancestral exposures. PLoS ONE. 2012;7(2): e31901.
17. Auvenshine RC, Eames WB. Pulpal response of monkeys to modifications of a bisphenol A composite resin and cement. J Dent Res. 1972;51:1062–6.
18. Meyer J-M. TEGDMA et Bisphenol-A. même niveau de risque en médecine dentaire? Autredent. 2010;56: 81–6.
19. Fleisch AF, Sheffield PE, Chinn C, Edelstein BL, Landrigan PJ. Bisphenol A and related compounds in dental materials. Pediatrics. 2010;126:760–8.
20. Jontell M, Hanks CT, Bratel J, Bergenholtz G. Effects of unpolymerized resin components on the function of accessory cells derived from the rat incisor pulp. J Dent Res. 1995;74:1162–7.
21. Segura JJ, Jiménez-Rubio A, Pulgar R, Olea N, Guerrero JM, Calvo JR. In vitro effect of the resin component bisphenol A on substrate adherence capacity of macrophages. J Endod. 1999;25:341–4.
22. Kleinsasser NH, Wallner BC, Harréus UA, Kleinjung T, Folwaczny M, Hickel R, Kehe K, Reichl FX. Genotoxicity and cytotoxicity of dental materials in human lymphocytes as assessed by the single cell microgel electrophoresis (comet) assay. J Dent. 2004; 32:229–34.
23. Schweikl H, Spagnuolo G, Schmalz G. Genetic and cellular toxicology of dental resin monomers. J Dent Res. 2006;85:870–7.
24. Mantellini MG, Botero T, Yaman P, Dennison JB, Hanks CT, Nör JE. Adhesive resin and the hydrophilic monomer HEMA induce VEGF expression on dental pulp cells and macrophages. Dent Mater. 2006;22: 434–40.
25. Yano J, Kitamura C, Nishihara T, Tokuda M, Washio A, Chen K-K, Terashita M. Apoptosis and survivability of human dental pulp cells under exposure to Bis-GMA. J Appl Oral Sci. 2011;19:218–22.
26. Gayrard V, Lacroix MZ, Collet SH, Viguié C, Bousquet-Melou A, Toutain P-L, Picard-Hagen N. High bioavailability of bisphenol A from sublingual exposure. Environ Health Perspect. 2013;121: 951–6.

Fluoride Effects on the Dentin-Pulp Complex

14

Yukiko Nakano and Pamela Den Besten

14.1 Introduction: Facts of Fluoride

14.1.1 Fluoride and Dental Uses

Fluoride is known to protect against dental caries by being incorporated into demineralized hydroxyapatite crystals to make them less soluble to acid dissolution. Fluoride incorporation into biofilms has also been related to the inhibition of lactate production, possibly related to its incorporation into bacterial biofilms where it inhibits enolases [1, 2]. Fluoride has been administered to inhibit dental caries by adding it to drinking water or in topical forms including in toothpastes, gels, or varnishes. Ingestion of excess amounts of fluoride during tooth formation results in the formation of dental fluorosis.

14.1.2 Dental Fluorosis

Dental fluorosis generally refers to the obvious effects of excess fluoride ingestion on tooth enamel formation [1]. Enamel fluorosis ranges from mildly fluorosed teeth, which have more accentuated perikymata, to more severe fluorosis where teeth have a white opaque appearance and more severely affected teeth which are characterized by pitting and loss of enamel, with increasing porosity relative to the severity of fluorosis [3, 4]. Pitting can lead to secondary staining causing a yellow to brown discoloration [5]. The severity of fluorosis in mineralized tissue is related to the concentration of fluoride in the serum [6].

All degrees of severity of enamel fluorosis are characterized by subsurface enamel hypomineralization. In the maturation stage, this hypomineralization may be in large part due to the effects of fluoride to enhance mineral precipitation in the highly mineralized enamel matrix, with a resulting increased formation of protons leading to acidification and hypomineralization [7–10]. This cycle of fluoride-mediated mineralization may explain the bands of hypo- and hypermineralized mineral that are found in fluorosed enamel.

The effect of fluoride on mineralized tissues may be due to fluoride-related changes in the mineralizing extracellular matrix, which then influences cell function. However, it is also possible that fluoride may have direct cellular effects which cause alterations in the formation of the extracellular mineralizing matrix.

Y. Nakano, DDS, PhD
Department of Orofacial Sciences,
School of Dentistry, University of California,
San Francisco, 513 Parnassus Avenue, RM S704,
San Francisco, CA 94143-0422, USA
e-mail: yukiko.nakano@ucsf.edu

P. Den Besten, DDS, MS (✉)
Division of Pediatric Dentistry,
Department of Orofacial Sciences,
School of Dentistry, University of California,
San Francisco, 513 Parnassus Avenue, RM S612,
San Francisco, CA 94143-0422, USA
e-mail: pamela.denbesten@ucsf.edu

M. Goldberg (ed.), *The Dental Pulp*,
DOI 10.1007/978-3-642-55160-4_14,

Examination of the effects of fluoride on dentin formation may provide better understanding of the effects of fluoride on mineralizing tissues.

14.2 Dentin Fluorosis

14.2.1 Hypomineralization and Dentin Fluorosis

The effects of chronic fluoride exposure on human and rodent tooth formation were first compared by Fejerskov et al. [11]. They noted that while dentin is generally considered to be affected in dental fluorosis, consistent with previous studies, human dentin samples collected from an area with chronic endemic dental fluorosis (with 3.5 ppm F in drinking water) exhibited inhibition of mineralization, similar to the incisor dentin from rats receiving chronic high dose of fluoride (56.5 and 113 ppm F in drinking water for 1–3 months). The dentin-related changes included striations, hypoplastic defects, and hypomineralized interglobular spaces, which increased with increasing levels of fluoride and time of exposure. They further stressed that dental fluorosis is characterized by changes in both enamel and dentin [4, 12] (Fig. 14.1).

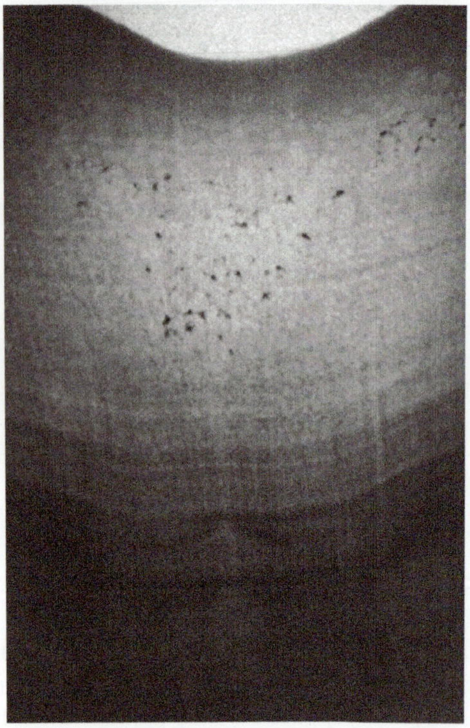

Fig. 14.1 Micrograph of dentin from a severely fluorosed tooth showing distinct layering of hypomineralized lines (Used with permission of John Wiley and Sons from Fejerskov et al. [4])

Moreover, numbers of studies have found that the severity of dentin fluorosis is related to fluoride levels in the drinking water, dentin fluoride concentration, and dentin microhardness [13–15]. Studies by Vieira et al. [13] measured microhardness and dentin mineralization in tooth samples from Montreal and Toronto, Canada, and Fortaleza, Brazil, where water fluoride levels were 0.2, 1.0, and 0.7 ppm, respectively. They found that dentin fluoride concentrations correlated with dental fluorosis severity. Dental fluorosis severity correlated with dentin microhardness, while enamel fluoride concentrations correlated with both dentin microhardness and dentin mineralization. Interestingly, the teeth from Brazil exposed to 0.7 ppm fluoride in the drinking water presented with the highest dentin fluoride concentration values and were also harder and less mineralized than the Montreal (0.2 ppm) and Toronto (1.0 ppm) teeth. The Montreal (0.2 ppm) teeth had lower levels of dental fluorosis when compared with teeth from Toronto (1 ppm) and Fortaleza (0.7 ppm).

Severely fluorosed deer [15] have a disrupted dentin crystal structure. The disrupted crystal formation may relate to the findings in a recent *in vitro* study by Waidyasekera and co-workers [16], who described crystallites in fluorosed human dentin as larger, but not homogenously arranged, and found a lower density of crystallites in fluorosed dentin.

14.2.2 Alteration in Extracellular Matrix Properties

Similar to the enamel matrix, the physicochemical effect of fluoride incorporated into the crystal lattice could directly regulate the growth of hydroxyapatite crystals in dentin. However, the molecular mechanism of hypomineralization in fluorotic dentin is not yet fully elucidated. Several studies have suggested the changes in the inorganic phase of mineralization in dentin are due to structural alteration in the extracellular matrix resulting from changes in synthesis or modification of collagenous and noncollagenous extracellular matrix proteins.

Type I collagen is the major organic component of dentin extracellular matrix constructing structural basis of the dentin matrix. Moseley et al. reported that 6 mM sodium fluoride supplemented in dentin organ culture inhibited collagen synthesis [17]. A further *in vivo* study by Maciejewska et al. showed that type I collagen expression was downregulated in the early stage of tooth germs from pulps (P1 or P5) from dams that ingested 110 ppm fluoride [18].

Proteoglycans are a class of glycosylated proteins consists of sulfated glycosaminoglycans (i.e., chondroitin sulfate, dermatan sulfate, heparan sulfate, heparin, keratan sulfate) covalently linked to core proteins. In mineralized tissues like dentin, proteoglycans are essential for maintenance of the stabilization of collagen fibers and act as a framework for mineralization in association with type I collagen. The proteoglycan extracted from rat teeth received 20 ppm fluoride for 10 weeks exhibited a more anionic profile due to changes in glycosaminoglycan characteristics and the presence of additional dermatan sulfate and heparin sulfate [19]. Changes in production of proteoglycans have also been demonstrated *in vitro*. Tooth organs cultured for 14 days in media supplemented with higher levels of fluoride than would be found in serum *in vivo* (1–6 mM) showed a reduction in biglycan, but no change in decorin or versican. In this same study, the levels of dermatan sulfate and sulfation of glycosaminoglycans in predentin and pulp extract were reduced by supplementation with 6 mM fluoride [20].

It is known that the each proteoglycan has different roles in matrix formation and mineralization. Chondroitin sulfate-substituted small leucine-rich proteoglycan/SLRP (including decorin and biglycan) within dentin may function to regulate the mineralization process, while dermatan sulfate-substituted SLRPs inhibit collagen fibril formation [21]. The functions of these different proteoglycans are partially explained by the different binding abilities of these molecules to hydroxyapatite [22, 23]. Septier et al. have demonstrated that the expression of glycosaminoglycan and dermatan sulfate/chondroitin sulfate in predentin decreases toward the predentin (unmineralized)/dentine (mineralized) interface, indicating regulation of the mineralization status of dentin as one of the roles of proteoglycans [6]. If fluoride at physiological levels (micromolar) can alter proteoglycans in the dentin matrix, it would certainly alter dentin mineralization.

14.2.3 Alteration in Mineralization Front

Dentin phosphoprotein (DPP, a protein hydrolysis product of dentin sialophosphoprotein/DSPP) is the most abundant noncollagenous protein in the dentin matrix [24, 25] and is reported to have diverse effects on mineralization [26–30]. Because it is highly phosphorylated [31, 32], DPP has very high binding capacity for calcium ions and an affinity for crystalline hydroxyapatite surfaces, suggesting its role in nucleating mineral initiation at the mineralization front [33]. The phosphate content of DPP has been reported to be significantly reduced in dentin of rats receiving 20 ppm fluoride in drinking water for 17 weeks [34]. The possible mechanism of this finding was suggested by an *in vitro* study [35] showing that physiological concentration of fluoride (0.04–4 μM) inhibits casein kinase II, a potent kinase that phosphorylates DPP [24, 25].

Appositional mineralization of circumpulpal dentin has been shown to progress rapidly at the mineralization front. A lightly mineralized layer (LL) is observed in predentin immediately adjacent to the mineralization front of dentin [36, 37]. Crystal alignment in LL is within narrow spaces between the individual non-mineralized collagen fibers, thereby making LL distinct from the bulk of fully mineralized dentin where crystals are deposited in and around collagen fibers. This indicates that LL is where dentin mineralization initiates before more rapid mineralization of the

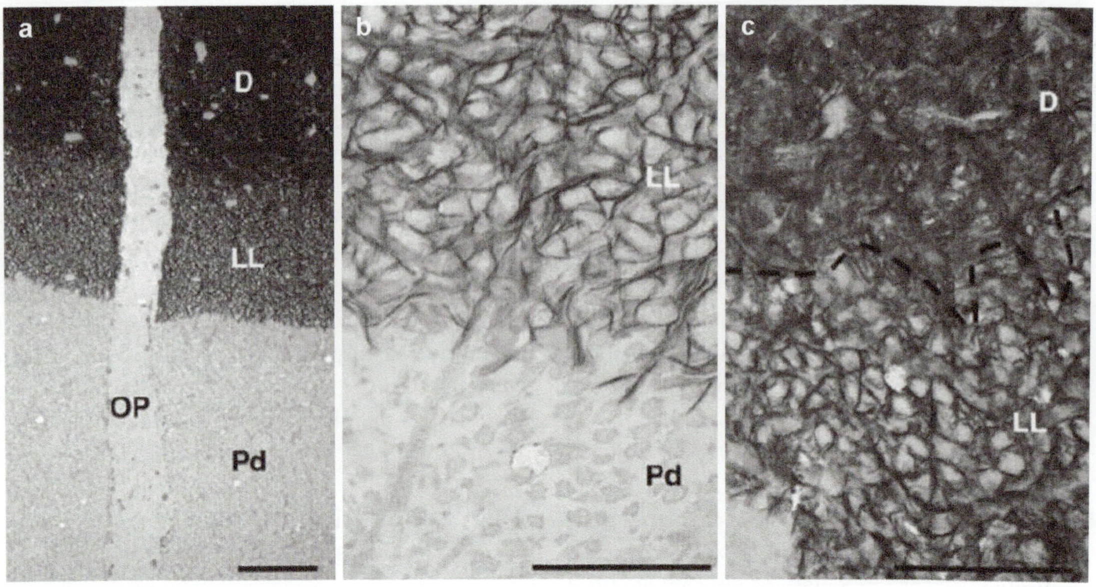

Fig. 14.2 Lightly mineralized layer (*LL*). (**a**) LL along the mineralization front of normal rat dentin. (**b, c**) LL at high magnification (Used with permission of Oxford University Press from Ahmad et al. [36] *OP* odontoblast process, *PD* predentin)

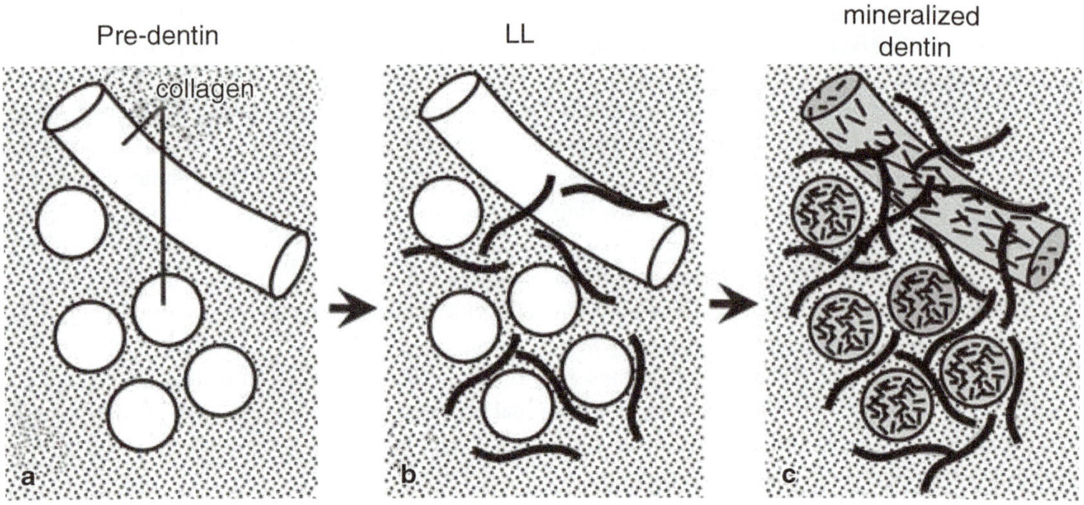

Fig. 14.3 Diagrammatic model of stepwise mineralization at the mineralization front of dentin. (**a**) Predentin without mineral deposits. (**b**) LL where mineral crystals deposited in noncollagenous matrix between collagen fibers. (**c**) Mineralized dentin after massive mineral deposition in and around collagen fibers (Used with permission of Oxford University Press from Ahmad et al. [36])

matrix occurs (Figs. 14.2a–c and 14.3a–c). The LL along the mineralization front is clearly recognized at the crown-analogue dentin of the continuously erupting mouse and rat incisors as compared to root-analogue dentin. This is likely due to the differences in both organic and inorganic components between the crown- and root-analogue dentin matrix [36, 38–42].

DPP is present in the crown-analogue dentin at about four times greater levels than in the root-analogue dentin [39]. DPP has been reported to delay the mineralization of collagen fibers in the rodent incisor crown-analogue dentin, though at lower concentrations DPP promotes the nucleation of hydroxyapatite [36, 42–44]. These findings are consistent with those

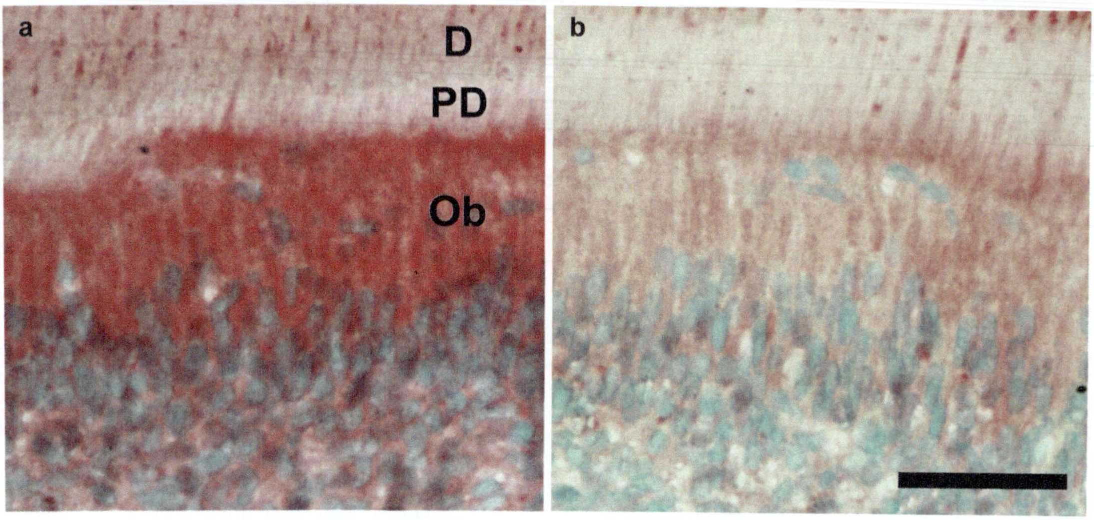

Fig. 14.4 (**a**, **b**) Immunostaining of DSPP/DSP. DSPP/DSP is expressed in cytoplasm of odontoblasts and adjacent pulp cells. DSPP/DSP expression in cytoplasm of odontoblasts is downregulated in fluoride (50 ppm)-exposed A/J mice. *Ob* odontoblast, *D* dentin, *PD* predentin, bar 100 μm

of Boskey et al., who showed that high concentrations of DPP can inhibit the growth of nucleated crystals.

Mice ingesting 50 ppm fluoride in drinking water appear to lose LL in dentin from mice as compared to dentin from control mice (our unpublished data, Takano Y. 2014), with an abrupt change in the mineralization status of collagen fibers at the border between predentin and mineralized dentin under the influence of fluoride. These results, combined with our findings of reduced immunostaining for DSP (another hydrolysis product of DSPP to create DPP) in odontoblasts of the fluoride treated mice, further support the possibility that the loss of LL is caused by a reduction in the synthesis of DPP (Fig. 14.4a, b) (unpublished data, Nakano Y. 2014), allowing more rapid mineralization at the dentin mineral initiation front. These results support the possibility that the fluoride can have a direct effect on odontoblasts to reduce either the synthesis of DSPP or the hydrolysis of DSPP to form DPP.

14.3 Effects of Fluoride on the Dental Pulp

Few studies have specifically looked at the effects of fluoride on dental pulp. However, the changes that have been identified are consistent with those found in odontoblasts. *In vitro* studies of human dental pulp cells exposed to fluoride show that below 5 ppm fluoride, cell proliferation and alkaline phosphatase synthesis increase [45, 46]. However, at higher fluoride concentrations, alkaline phosphatase and type 1 collagen synthesis decrease, with no similar effects on fibronectin synthesis [47] or change in cell morphology [17]. Fluoride incorporation into pulp capping agents has been shown to reduce thermal conductivity and dentin acid solubility, with no irritating effects on the pulp [48]. At high concentrations, fluoride can also inhibit lysosomal enzymes found in the pulp, with a possible application of limiting cellular destruction in damaged pulp tissue [49].

14.4 Clinical Significance of Dentin Fluorosis

These relatively few studies showing that fluoride can alter dentin and pulp formation suggest that more attention should be given to possible changes in fluorosed dentin. Possible effects of fluorosed dentin include alterations in caries susceptibility, alterations in the dentin structure to affect bonding of composite based dental materials, the capability of dentin to remineralize, secondary dentin formation, and dentin bridge formation. One aspect of fluoride-related effects on dentin formation has been explored in analyzing the effects of fluoride in drinking water on the so-called hidden caries.

14.4.1 Fluorosis and Dentin Caries

In the 1980s the term "hidden caries" was introduced to dentistry. This term is used to describe occlusal pit and fissure-type carious lesions seen in the dentin on a bitewing radiograph, where clinically the occlusal enamel appears intact or is minimally demineralized [50] (Fig. 14.5). These hidden caries create the risk that after cariogenic bacteria penetrate into the enamel through a minimal opening in the enamel surface, when they reach the less mineralized dentin, caries progresses at an increased rate. Meanwhile the fluoridated enamel undergoes remineralization, while caries progresses at an increased rate in the dentin. Hidden carious lesions present a challenge to dentists in diagnosis, treatment planning, and research, due to their "hidden" nature, resulting in misdiagnosis and treatment of dental caries [51].

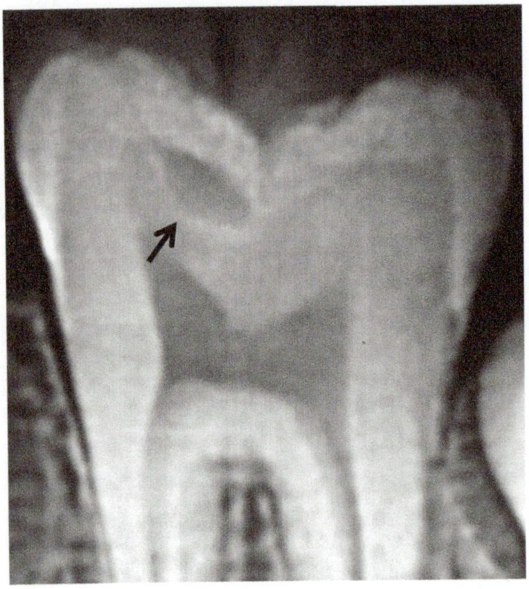

Fig. 14.5 Hidden caries. Radiolucency in the dentin (*arrow*) under an intact enamel surface suggests hidden caries

In a clinical study to investigate the possible role of fluoride in the formation of hidden caries, Weerheijm et al. showed that children ingesting "optimally" fluoridated water at a level of 1.1 ppm from birth until the day of data collection had significantly fewer hidden caries than the group of children drinking water containing 0.1 ppm. Hashizume et al. examined the role of public water fluoridation and fluoride dentifrice on the prevalence of hidden caries in 8–10-year-old children and, similar to the study by Weerheijm et al., found a statistically significant reduction in the prevalence of hidden caries by water fluoridation at currently recommended level (0.7 ppm) [52]. These results suggest that the ingestion of fluoride at optimal levels can reduce the presence of radiographically detectable hidden caries.

Epidemiological data are not available to determine effect of dentin fluorosis on correlating hidden caries. Nevertheless, there is *in vitro* evidence to indicate that fluorosed dentin demineralizes more rapidly than normal dentin. Waidyasekera and colleagues grouped 33 human molars according to the modified Thylstrup-Fejerskov index (TFI) into normal (N, TFI 0), mild fluorosis (ML, TFI 1–3), and moderate fluorosis (MD, TFI 4–6) and exposed the cut enamel and dentin surfaces to acid to create artificial caries. They found that dentin from either mildly fluorosed or moderately fluorosed teeth demineralized to a significantly greater depth than normal teeth, while moderately fluorosed enamel was more acid resistant [53]. Further ultrastructural studies on fluorosed dentin showed large more randomly formed crystals with a more loosely formed collagen network in fluorosed as compared to normal dentin [54]. Differences were also found in optimal dentin bonding systems in fluorosed as compared to control dentin.

Conclusion

Dentin formation is regulated by signaling molecules, which direct matrix formation and mineralization. The effects of fluoride on the formation of collagenous and noncollagenous dentin and pulp proteins show that, at high levels, fluoride can inhibit matrix formation resulting in hypomineralized dentin, possibly related to a disruption in initial dentin mineralization. Studies of *in vitro* organ culture systems with very high (millimolar) levels of fluoride suggest that fluoride mediates changes in matrix protein synthesis. Further studies of fluoride effects on dentin formation at lower, more physiological fluoride concentrations are needed to better understand the mechanisms by which fluoride alters dentin formation. Though fluoride at optimal levels in drinking water reduces the presence of hidden caries, *in vitro* studies show increased demineralization of dentin from even mildly fluorosed enamel as compared with nonfluorosed dentin. Strategic research approaches to combine *in vivo* and *in vitro* studies using human fluorosed teeth and mice/rat incisor models will allow us to examine the effect of fluoride on odontoblasts at different stages of odontoblast differentiation. *In vitro* and organ culture of dentin-pulp complex with physiological level of fluoride supplementation will enable us to analyze detailed pathways to provide further understanding of dentin fluorosis and its relevance to oral health.

References

1. Denbesten P, Li W. Chronic fluoride toxicity: dental fluorosis. Monogr Oral Sci. 2011;22:81–96.
2. Levine RS. Fluoride and caries prevention: 1. Scientific rationale. Dent Updat. 1991;18(3):105–6, 108–10.
3. Fejerskov O, Silverstone LM, Melsen B, Moller IJ. Histological features of fluorosed human dental enamel. Caries Res. 1975;9(3):190–210.
4. Fejerskov O, Thylstrup A, Larsen MJ. Clinical and structural features and possible pathogenic mechanisms of dental fluorosis. Scand J Dent Res. 1977;85(7):510–34.
5. Fejerskov O, Yanagisawa T, Tohda H, Larsen MJ, Josephsen K, Mosha HJ. Posteruptive changes in human dental fluorosis–a histological and ultrastructural study. Proc Finn Dent Soc. 1991;87(4):607–19.
6. Angmar-Mansson B, Ericsson Y, Ekberg O. Plasma fluoride and enamel fluorosis. Calcif Tissue Res. 1976;22(1):77–84.
7. Bronckers AL, Lyaruu DM, DenBesten PK. The impact of fluoride on ameloblasts and the mechanisms of enamel fluorosis. J Dent Res. 2009;88(10):877–93.
8. DenBesten PK. Effects of fluoride on protein secretion and removal during enamel development in the rat. J Dent Res. 1986;65:1272–7.
9. DenBesten PK, Crenshaw MA. The effects of chronic high fluoride levels on forming enamel in the rat. Arch Oral Biol. 1984;29(9):675–9.
10. DenBesten PK, Yan Y, Featherstone JD, Hilton JF, Smith CE, Li W. Effects of fluoride on rat dental enamel matrix proteinases. Arch Oral Biol. 2002; 47(11):763–70.
11. Fejerskov O, Yaeger JA, Thylstrup A. Microradiography of the effect of acute and chronic administration of fluoride on human and rat dentine and enamel. Arch Oral Biol. 1979;24(2): 123–30.
12. Fejerskov O, Larsen MJ, Josephsen K, Thylstrup A. Effect of long-term administration of fluoride on plasma fluoride and calcium in relation to forming enamel and dentin in rats. Scand J Dent Res. 1979;87(2):98–104.
13. Vieira A, Hancock R, Dumitriu M, Schwartz M, Limeback H, Grynpas M. How does fluoride affect dentin microhardness and mineralization? J Dent Res. 2005;84(10):951–7.
14. Vieira AP, Hancock R, Limeback H, Maia R, Grynpas MD. Is fluoride concentration in dentin and enamel a good indicator of dental fluorosis? J Dent Res. 2004;83(1):76–80.
15. Kierdorf U, Kierdorf H, Fejerskov O. Fluoride-induced developmental changes in enamel and dentine of European roe deer (Capreolus capreolus L.) as a result of environmental pollution. Arch Oral Biol. 1993;38(12):1071–81.
16. Waidyasekera K, Nikaido T, Weerasinghe D, Watanabe A, Ichinose S, Tay F, et al. Why does fluorosed dentine show a higher susceptibility for caries: an ultra-morphological explanation. J Med Dent Sci. 2010;57(1):17–23.
17. Moseley R, Sloan AJ, Waddington RJ, Smith AJ, Hall RC, Embery G. The influence of fluoride on the cellular morphology and synthetic activity of the rat dentine-pulp complex in vitro. Arch Oral Biol. 2003;48(1):39–46.
18. Maciejewska I, Spodnik JH, Domaradzka-Pytel B, Sidor-Kaczmarek J, Bereznowski Z. Fluoride alters type I collagen expression in the early stages of odontogenesis. Folia Morphol (Warsz). 2006;65(4): 359–66.
19. Hall RC, Embery G, Waddington RJ. Modification of the proteoglycans of rat incisor dentin-predentin during in vivo fluorosis. Eur J Oral Sci. 1996;104(3): 285–91.
20. Waddington RJ, Moseley R, Smith AJ, Sloan AJ, Embery G. Fluoride-induced changes to proteoglycan structure synthesised within the dentine-pulp complex in vitro. Biochim Biophys Acta. 2004;1689(2): 142–51.

21. Hedbom E, Heinegard D. Interaction of a 59-kDa connective tissue matrix protein with collagen I and collagen II. J Biol Chem. 1989;264(12):6898–905.
22. Embery G, Hall R, Waddington R, Septier D, Goldberg M. Proteoglycans in dentinogenesis. Crit Rev Oral Biol Med. 2001;12(4):331–49.
23. Sugars RV, Milan AM, Brown JO, Waddington RJ, Hall RC, Embery G. Molecular interaction of recombinant decorin and biglycan with type I collagen influences crystal growth. Connect Tissue Res. 2003;44 Suppl 1:189–95.
24. Dimuzio MT, Veis A. Phosphophoryns-major noncollagenous proteins of rat incisor dentin. Calcif Tissue Res. 1978;25(2):169–78.
25. Butler WT. The chemistry and biology of mineralized tissues: proceedings of the Second International Conference on the Chemistry and Biology of Mineralized Tissues, held in Gulf Shores, Alabama, Sept 9–14, 1984. [S.l.: s.n.]; 1985. 436 p.
26. Saito T, Arsenault AL, Yamauchi M, Kuboki Y, Crenshaw MA. Mineral induction by immobilized phosphoproteins. Bone. 1997;21(4):305–11.
27. Linde A, Lussi A, Crenshaw MA. Mineral induction by immobilized polyanionic proteins. Calcif Tissue Int. 1989;44(4):286–95.
28. George A, Bannon L, Sabsay B, Dillon JW, Malone J, Veis A, et al. The carboxyl-terminal domain of phosphophoryn contains unique extended triplet amino acid repeat sequences forming ordered carboxyl-phosphate interaction ridges that may be essential in the biomineralization process. J Biol Chem. 1996; 271(51):32869–73.
29. Butler WT. Dentin matrix proteins. Eur J Oral Sci. 1998;106 Suppl 1:204–10.
30. Boskey AL, Maresca M, Doty S, Sabsay B, Veis A. Concentration-dependent effects of dentin phosphophoryn in the regulation of in vitro hydroxyapatite formation and growth. Bone Miner. 1990;11(1):55–65.
31. Jonsson M, Fredriksson S, Jontell M, Linde A. Isoelectric focusing of the phosphoprotein of rat-incisor dentin in ampholine and acid pH gradients. Evidence for carrier ampholyte-protein complexes. J Chromatogr. 1978;157:235–42.
32. Stetler-Stevenson WG, Veis A. Bovine dentin phosphophoryn: composition and molecular weight. Biochemistry. 1983;22(18):4326–35.
33. Nawrot CF, Campbell DJ, Schroeder JK, Van Valkenburg M. Dental phosphoprotein-induced formation of hydroxylapatite during in vitro synthesis of amorphous calcium phosphate. Biochemistry. 1976; 15(16):3445–9.
34. Milan AM, Waddington RJ, Embery G. Altered phosphorylation of rat dentine phosphoproteins by fluoride in vivo. Calcif Tissue Int. 1999;64(3):234–8.
35. Milan AM, Waddington RJ, Embery G. Fluoride alters casein kinase II and alkaline phosphatase activity in vitro with potential implications for dentine mineralization. Arch Oral Biol. 2001;46(4):343–51.
36. Ahmad M, Iseki H, Abduweli D, Baba O, Tabata MJ, Takano Y. Ultrastructural and histochemical evaluation of appositional mineralization of circumpulpal dentin at the crown- and root-analog portions of rat incisors. J Electron Microsc. 2011;60(1): 79–87.
37. Goldberg M, Septier D. A comparative study of the transition between predentin and dentin, using various preparative procedures in the rat. Eur J Oral Sci. 1996;104(3):269–77.
38. Takagi Y, Nagai H, Sasaki S. Difference in noncollagenous matrix composition between crown and root dentin of bovine incisor. Calcif Tissue Int. 1988; 42(2):97–103.
39. Steinfort J, van den Bos T, Beertsen W. Differences between enamel-related and cementum-related dentin in the rat incisor with special emphasis on the phosphoproteins. J Biol Chem. 1989;264(5):2840–5.
40. Steinfort J, Deblauwe BM, Beertsen W. The inorganic components of cementum- and enamel-related dentin in the rat incisor. J Dent Res. 1990;69(6):1287–92.
41. Beertsen W, Niehof A, Everts V. Effects of 1-hydroxyethylidene-1, 1-bisphosphonate (HEBP) on the formation of dentin and the periodontal attachment apparatus in the mouse. Am J Anat. 1985; 174(1):83–103.
42. Beertsen W, Niehof A. Root-analogue versus crown-analogue dentin: a radioautographic and ultrastructural investigation of the mouse incisor. Anat Rec. 1986;215(2):106–18.
43. Weinstock M, Leblond CP. Radioautographic visualization of the deposition of a phosphoprotein at the mineralization front in the dentin of the rat incisor. J Cell Biol. 1973;56(3):838–45.
44. Inage T, Toda Y. Phosphoprotein synthesis and secretion by odontoblasts in rat incisors as revealed by electron microscopic radioautography. Am J Anat. 1988;182(4):369–80.
45. Nakade O, Koyama H, Arai J, Ariji H, Takada J, Kaku T. Stimulation by low concentrations of fluoride of the proliferation and alkaline phosphatase activity of human dental pulp cells in vitro. Arch Oral Biol. 1999;44(1):89–92.
46. Thaweboon S, Thaweboon B, Chunhabundit P, Suppukpatana P. Effect of fluoride on human dental pulp cells in vitro. Southeast Asian J Trop Med Public Health. 2003;34(4):915–8.
47. Veron MH, Couble ML, Magloire H. Selective inhibition of collagen synthesis by fluoride in human pulp fibroblasts in vitro. Calcif Tissue Int. 1993;53(1): 38–44.
48. Seltzer S. Advances in biology of the human dental pulp. Newer restorative materials and the pulp. Oral Surg Oral Med Oral Pathol. 1971;32(3):454–60.
49. Spott RJ, Rosett T. Lysosomes and the dental pulp. Oral Surg Oral Med Oral Pathol. 1973;36(4):569–79.
50. Weerheijm KL. Occlusal 'hidden caries'. Dent Updat. 1997;24(5):182–4.

51. Zadik Y, Bechor R. Hidden occlusal caries lesion – a diagnostic challenge. Refuat Hapeh Vehashinayim. 2008;25(1):34–9, 83.
52. Hashizume LN, Mathias TC, Cibils DM, Maltz M. Effect of the widespread use of fluorides on the occurrence of hidden caries in children. Int J Paediatr Dent. 2013;23(1):72–6.
53. Waidyasekera PG, Nikaido T, Weerasinghe DD, Wettasinghe KA, Tagami J. Caries susceptibility of human fluorosed enamel and dentine. J Dent. 2007; 35(4):343–9.

Part III

Pulp Repair and Regeneration

15 Experimental In Vivo Approaches of Pulp Regeneration

Misako Nakashima and Koichiro Iohara

15.1 Introduction

Deep caries and pulp exposure have been treated by pulp capping or partial pulp amputation to preserve the pulp tissue, with limited success. The pulp tissue has to be entirely removed in case of irreversible pulpitis, and the root canal space is filled with various materials after disinfection in the traditional root canal treatment. Despite the success in early years, it gradually decreases as incomplete disinfection, and coronal leakage leads to periapical disease. The treatment of endodontically failed teeth with periapical lesions and/or other clinical symptoms generally decreases the success rate to 50–70 % [1]. In addition, the loss of pulp vitality, excessive enlargement, and debridement of infected root canals and post preparation increase the possibility of root fracture and tooth loss [2, 3]. Therefore, the ideal approaches for endodontic treatment are the following: (1) minimum removal of caries to conserve healthy tooth structure with prevention of infection; (2) prevention of microleakage from oral cavity; and (3) maintenance of the properties and mechanical strength of the tooth structure. There has been no synthetic filling material superior to natural pulp and dentin, which is highly organized and complex. Thus, regenerative therapy for dentin/pulp is the ideal goal to restore tooth functions and morphology compromised by pulp injury and/or inflammation [4–6].

Attempts for pulp regeneration, which started in the 1990s based on concepts of stem cell biology and regenerative medicine [7], were accelerated by the discovery of dental pulp stem cells (DPSCs) and their potential to ectopic pulp/dentin formation [8]. Revitalization procedures hardly regenerate pulp/dentin-like tissue. Experimental in vivo approaches including ectopic tooth transplantation and orthotopic transplantation, however, have demonstrated pulp/dentin regeneration successfully. Stem cell therapy represents a potential strategy to regenerate the dentin-pulp complex, enabling conservation and restoration of teeth. On the other hand, acellular morphogen therapy with homing/migration factors or morphogenetic and related signaling molecules may be a potential alternative for stem cell therapy. Thus, the objective of this section is to introduce the experimental ectopic and orthotopic models for pulp regeneration, including the concept of revitalization, stem cell therapy, and acellular morphogen therapy with or without stem/progenitor cells and morphogenetic signaling molecules in addition to extracellular matrix scaffold for partial and complete regeneration.

M. Nakashima, PhD, DDS (✉)
K. Iohara, DDS, PhD
Department of Dental Regenerative Medicine, Center of Advanced Medicine for Dental and Oral Diseases, National Center for Geriatrics and Gerontology, Research Institute, 35 Gengo, Morioka, Obu 474-8522, Japan
e-mail: misako@ncgg.go.jp; iohara@ncgg.go.jp

M. Goldberg (ed.), *The Dental Pulp*,
DOI 10.1007/978-3-642-55160-4_15, © Springer-Verlag Berlin Heidelberg 2014

15.2 Revitalization/ Revascularization of Immature Teeth

Based on a series of clinical case reports, revitalization/revascularization of infected necrotic pulp tissue and either apical periodontitis or abscess in immature teeth has been introduced to stimulate ingrowth of vital tissues into root canal spaces followed by hard-tissue deposition resulting in a narrowing of the root canal [9–15] (Fig. 15.1a, b). This treatment offers immense potential to avoid the need for traditional apexification with calcium hydroxide or the need for achievement of an artificial apical barrier with mineral trioxide aggregate (MTA), by optimizing physiological strength of the tooth root. The revitalization/revascularization procedures are as follows: (1) minimal mechanical debridement, (2) gentle irrigation with 1.25–5.25 % NaOCl, (3) disinfection of the canal with topical antibiotic paste, and (4) filling of the canal space with blood clot which serves as a rich source of growth factors. In experimental work in dogs, it was shown that the new ingrowth tissue had little similarity to normal pulp tissue in most cases and was a mosaic of cementum, periodontal ligament, and bone [9, 16]. The exact nature of this tissue growing into the canal and its fate in the long term is not known [11]. Further investigations are necessary to determine a more evidence-based protocol for revitalization/revascularization procedures and the clinical outcome.

15.3 Ectopic Approach for Pulp Regeneration

The concept of pulp regeneration is distinct from revitalization/revascularization. The tissue engineering triad of stem/progenitor cells, morphogenetic signaling molecules, and extracellular scaffold is utilized for pulp regenerative therapy [4, 17] (Fig. 15.1a, b). The first evidence for de novo pulp regeneration has been demonstrated in ectopic tooth transplantation approach in which human tooth slices or root fragments were injected with stem cells/scaffolds into the root canal and were subsequently transplanted into the subcutaneous space of immunocompromised mice [18, 19] (Fig. 15.2a). The ectopic transplantation approach has advantages to neglect a direct blood supply essential for regeneration. It is a useful model to examine regeneration of pulp tissue and dentin by dental pulp/progenitor stem cells and its mechanisms (Fig. 15.3a, b). It may also allow for the study of molecular and cellular events involved

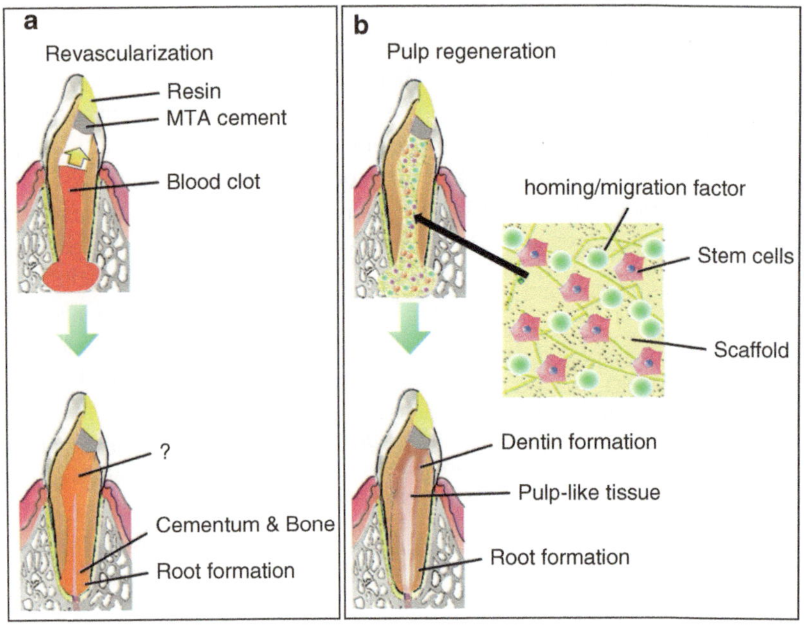

Fig. 15.1 Difference between revascularization and pulp regeneration in immature teeth. (**a**) The revitalization/revascularization procedures of filling of the canal space with blood clot. (**b**) The pulp regeneration procedures of filling of the canal space with stem cells with homing/migration factor

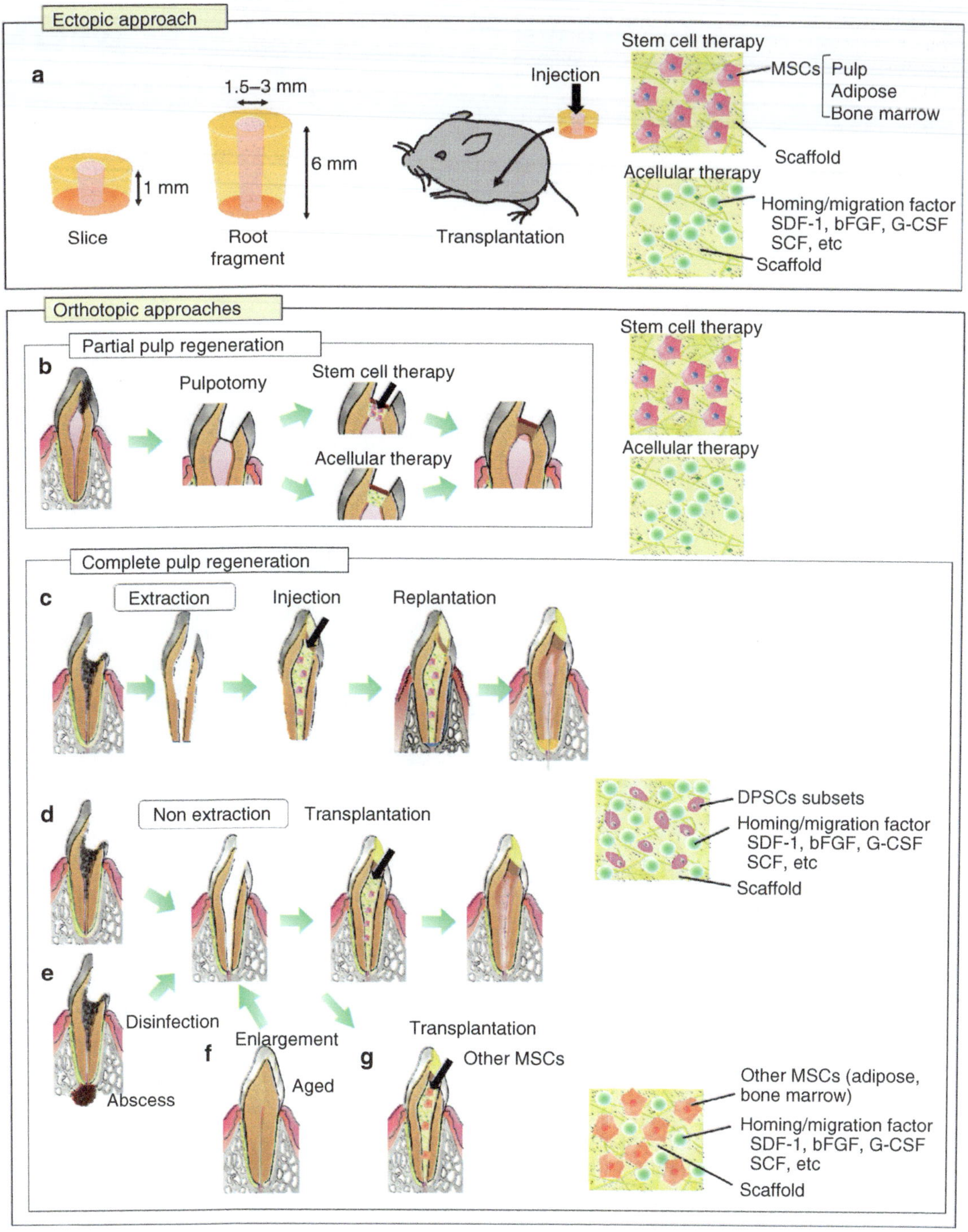

Fig. 15.2 Experimental in vivo approaches for pulp regeneration. (**a**) Ectopic approach. (**b**–**g**) Orthotopic approaches. (**b**) Partial pulp regeneration model. (**c**–**g**) Complete pulp regeneration model. **c** Tooth extraction model in case of pulpectomy. (**d**–**g**) Tooth non-extraction method in case of pulpectomy. (**e**) Complete pulp regeneration with periapical disease. (**f**) Complete pulp regeneration in the aged. (**g**) Complete pulp regeneration harnessing MSCs from other tissue sources

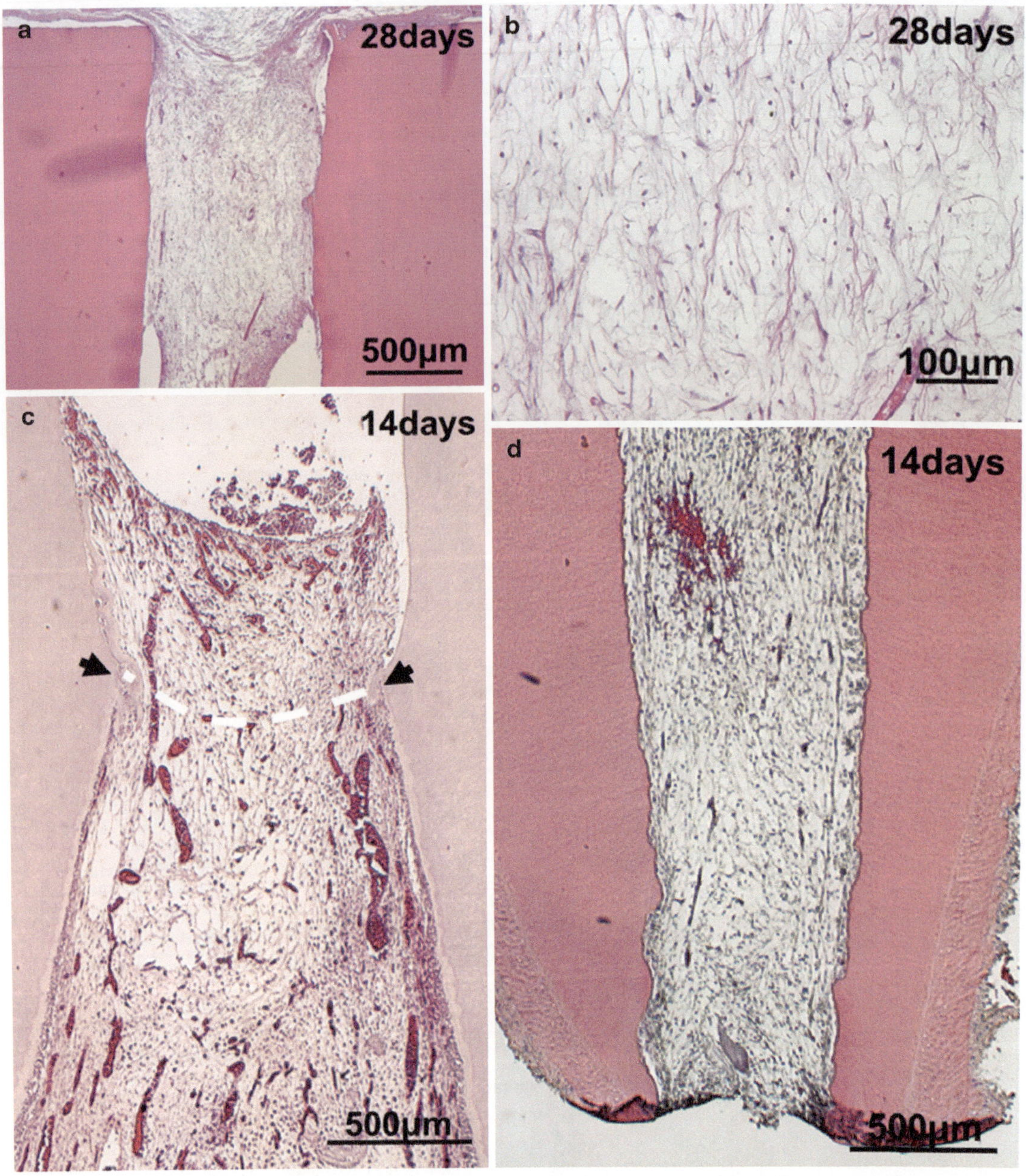

Fig. 15.3 Experimental in vivo approaches of pulp regeneration. (**a**, **b**) Ectopic pulp regeneration after transplantation of $CD31^{-}/CD146^{-}$ SP cells with tooth root into subcutaneous sites of mouse. (**c**) Partial pulp regeneration in the cavity of the amputated pulp of dogs 14 days after autologous transplantation of $CD31^{-}/CD146^{-}$ SP cells of a three-dimensional pellet with type I and type III collagen. (**d–g**) Complete regeneration of pulp-like tissue after autologous cell transplantation in the extraction model of dogs. (**d**) Fourteen days after $CD31^{-}CD146^{-}$ SP cell transplantation with 10 ng/μl SDF-1. (**e**) Thirty-five days after $CD105^{+}$ cell transplantation. Odontoblastic cells (*arrows*) lining to newly formed dentin. (**f**) Immunostaining with BS-1 lectin on day 14. v: Newly formed capillaries. (**g**) Immunostaining with PGP 9.5 on day 14. Neuronal process (*arrows*). (**h–l**) Complete regeneration of pulp-like tissue after autologous $CD105^{+}$ cell transplantation with 10 ng/μl SDF-1 in the non-extraction model of dogs. Fourteen days after transplantation. (**j**) Ninety days after transplantation. Odontoblastic cells lining to newly formed dentin. (**k**) Immunostaining with BS-1 lectin on day 14. v: Newly formed capillaries. (**l**) Immunostaining with PGP 9.5 on day 14

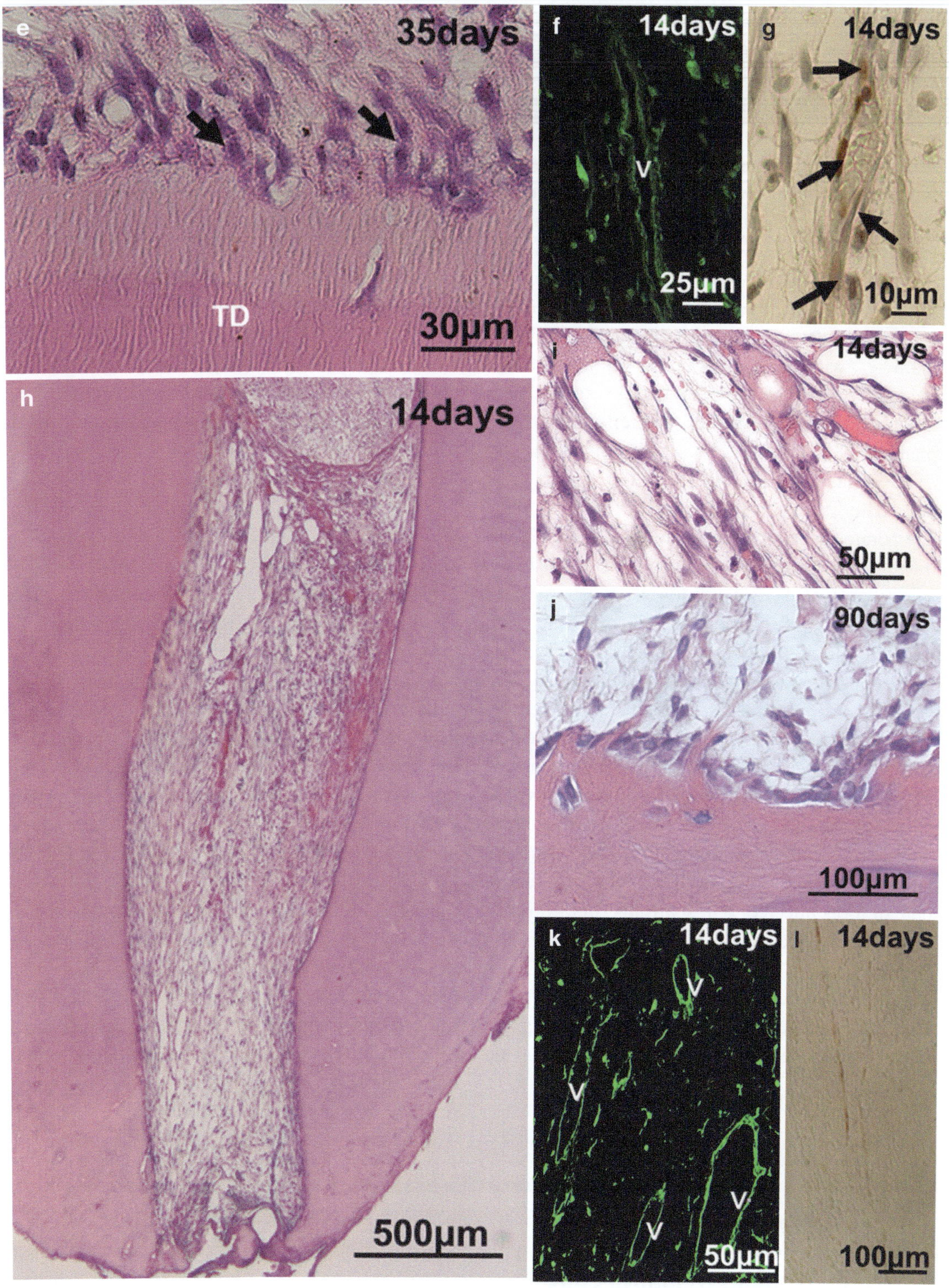

Fig. 15.3 (continued)

in the fate of stem cell therapy [19]. Pulp-like tissue is induced in a tooth slice model (1-mm-thick tooth slice (cross sections)) by filling of pulp stem cells and dentin matrix protein 1 (DMP1) together [20] or stem cells from exfoliated deciduous teeth (SHED) only [21]. SHED can differentiate into functional odontoblast-like cells that generate tubular dentin along the dentinal wall and angiogenic endothelial cells in the tooth slice model [22]. Similarly pulp-like tissue is regenerated in subcutaneously transplanted tooth root fragment (6–7-mm long) with an enlarged root canal (2–3 mm in diameter) with one end sealed with mineral trioxide aggregate (MTA) cement. In this root fragment model, it is unlikely to occur shortly after transplantation that the blood vessel growth into the end of the canal space provides nutrients for the stem cells. Thus, the cells not only survive well but also regenerate tissue by the nutrients which are able to diffuse into the canal space [23]. PLG poly(DL-lactide-co-glycolide) scaffolds seeded with stem cells from apical papilla (SCAP) or DPSCs are injected into the root canal. Three to four months after transplantation, the canal space is filled with vascularized pulp-like tissue, and a newly generated dentin-like layer is deposited onto the existing dentinal walls and the MTA cement surface [23]. The regenerated dentin-like tissue has less continuity and thickness in DPSCs transplantation with PLG compared with that in SCAP transplantation with PLG [23]. More recently, full-length human roots injected with scaffolds (PuraMatrix™ (3-D Matrix Medical Technology, Waltham, MA) or recombinant collagen) containing SHED are transplanted subcutaneously and resulted in pulp-like tissue regeneration with odontoblasts capable of generating new tubular dentin throughout the root canals [24].

15.4 Orthotopic Model for Pulp Regeneration in Mature Teeth

15.4.1 Pulp Stem/Progenitor Cells

Dental pulp tissue is rich in vasculature and innervation, and their intimate association is involved in pulp homeostasis including regulation of inflammation and enhancing pulp defense mechanisms [25]. Angiogenesis and reinnervation are also prerequisites for pulp regeneration to stimulate migration/homing, proliferation of stem cells, and stromal matrix formation in the root canal [4]. Thus, for optimal stem cell therapy, pulp stem/progenitor cells need to not only have high proliferative and multi-differentiation potential in vitro as typical stem cell properties but also to enhance angiogenesis/vasculogenesis and neurogenesis in vivo [5]. Subfractions of dental pulp stem cells (DPSCs) with high angiogenic/vasculogenic and neurogenic potential, pulp $CD31^-$ side population (SP) cells, and $CD105/endoglin^+$ cells can be isolated from human, porcine, and canine adult teeth [5, 26–31]. Pulp SP cells are enriched in the "true" or "mother" adult stem cells, which exhibit lower level of the DNA-binding fluorescent dye, Hoechst 33342, than the rest of pulp cells [32]. CD105/endoglin is a component of the transforming growth factor-beta receptor complex which is widely expressed on mesenchymal stem cells (MSCs) [33] and has been used for isolation of MSCs [34, 35]. Pulp $CD31^-$ SP cells and $CD105^+$ cells are positive for CD29, CD44, CD73, and Thy-1/CD90 and negative for CD31 and CD45 as unfractionated colony-derived DPSCs and other MSCs [36]. Their positive rates of CD105, CXCR4, and G-CSF receptor (G-CSFR), however, are higher compared with colony-derived DPSCs [29]. Pulp $CD31^-$ SP cells and $CD105^+$ cells stimulate the blood flow recovery and angiogenesis/vasculogenesis after transplantation into mouse hindlimb ischemic models [5, 28, 30]. Those cells also enhance angiogenesis and neurogenesis in rat cerebral ischemic models in peri-infarct area by promoting migration and differentiation of the endogenous neuronal stem/progenitor cells, which resulted in the functional recovery [5, 30, 37]. The conditioned medium of DPSC subfractions contain high concentration of angiogenic/neurotrophic factors such as brain-derived neurotrophic factor (BDNF), nerve growth factor (NGF), vascular endothelial growth factor (VEGF-A), glial cell-line-derived neurotrophic factor (GDNF), granulocyte-macrophage colony-stimulating factor (GM-CSF), granulocyte colony-stimulating factor (G-CSF), and matrix metalloproteinase 3 (MMP3). The stimulatory role of these trophic factors on proliferation, migration, and anti-apoptosis has also been reported in vitro in endothelial cells, neuroprogenitor cells, and fibroblasts [28, 30, 37]. The transplanted DPSC subfractions are successfully engrafted in proximity to the newly formed vasculature and neuron, releasing angiogenic/neurotrophic factors without direct incorporation into the vessels or nerves, suggesting their trophic effect [28, 30, 37, 38] as reported in other MSCs [39–41]. These findings demonstrate cell sources of the DPSC subfractions, pulp $CD31^-$ SP cells, and $CD105^+$ cells for stem cell therapy in pulp regeneration by promoting cell survival, angiogenesis, neurogenesis, and neuroprotection and by activating endogenous stem cells in a paracrine fashion.

15.4.2 Partial Pulp Regeneration

With the recent advances in tissue engineering and regenerative medicine, partial pulp regeneration therapy in case of pulp exposure or partial pulpitis has been developed (Fig. 15.2b). The two potential methods in partial pulp regeneration are (1) transplantation of engineered pulp/dentin tissue or DPSC subfractions with scaffold and (2) application of homing/migration factors with scaffolds.

It has been reported to engineer pulp-like tissue in vitro by harnessing pulp fibroblasts adhered to polyglycolic acid (PGA) scaffolds [7, 42, 43]. The pulp cells seeded in PGA for 24 h survive and produce extracellular matrix after subcutaneous transplantation in immunocompromised mice [42], although their engraftment and resultant partial pulp regeneration have not been demonstrated in any orthotopic models until recently. After that, the successful induction of the partial pulp regeneration as indicated in the first method has been demonstrated after autologous transplantation of three-dimensional culture of pulp $CD31^-$ SP cells with collagen scaffold in the cavity on the amputated pulp (Fig. 15.3c). The regenerated pulp tissue contains well-developed vasculature and innervation, and the tubular dentin formed along the dentinal wall [27]. As seen in the cerebral and hindlimb ischemic models, the transplanted $CD31^-$ SP cells localize closely to the newly formed capillaries with expression of angiogenic/neurotrophic factors, implicating their trophic effects on neovascularization also in partial pulp regeneration. In contrast with $CD31^-$ SP transplantation, $CD31^+$ SP cell transplantation induces regenerated tissue in less volume with fewer capillaries [27], suggesting distinct regenerative potential for pulp tissue among subfractions of DPSCs.

The second method is to recruit endogenous progenitor cells by some homing/migration factor, subsequently participating in partial pulp regeneration. It is advantageous that the second method lacks excessive costs including cell isolation, handling, storage and shipping, ex vivo manipulation, and immune rejection (for allogeneic cells) and is also free from liabilities of potential contamination, pathogen transmission, and tumorigenesis that may be associated with cell transplantation [44]. Prominent migration/homing factors that have recently emerged as aids in regenerative medicine are fibroblast growth factor-2 (FGF-2) [45] and stromal cell-derived factor-1α (SDF-1, CXCL12/pre-B-cell growth-stimulating factor) [46]. SDF-1 is known as a chemokine for CXCR4-positive stem cells. When DPSCs seeded on the surface of three-dimensional collagen gel cylinders are incubated in chemically defined media supplemented with FGF-2 or SDF-1, more cells migrate into collagen gel than in the absence of these cytokines [47]. Re-cellularization and revascularization have been demonstrated in endodontically treated human teeth with FGF-2 followed by ectopic transplantation [47]. A noncontrolled release of free FGF-2 only accelerates reparative dentin formation in the residual dental pulp, whereas a controlled release of FGF-2 from gelatin hydrogels induces formation of osteodentin in the pulp proliferating in the dentin defects. The controlled release of an appropriate dosage of FGF-2 from gelatin hydrogels induced osteodentin formation on the surface of the regenerated pulp [48–50]. Another migration/homing factor related to the second method is matrix metalloproteinase-3 (MMP-3). MMP-3 has high proliferation, migration, and anti-apoptotic effects on endothelial cells in vitro and promotes angiogenesis and pulp wound healing when applied on the amputated pulp of rat incisors [51]. The pulp tissue is regenerated by anti-inflammatory effects of MMP-3 even in partial irreversible pulpitis in a canine experimental model [52]. Furthermore, the migration of DPSCs is also promoted by the extracellular matrix proteins (EMPs), particularly laminin, and chemoattractants, such as sphingosine-1-phosphate (S1P) and TGF-β1, suggesting the highly regulated migration by the interplay between the EMPs and chemoattractants [53]. More recently, stem cell factor (SCF) which binds to the c-Kit receptor CD117 is applied in pulp regeneration, resulting in induced cell homing and promoting angiogenesis, remodeling of the implanted collagen scaffold, and new collagen matrix synthesis [54]. Thus, these findings demonstrate that selective homing/migration factors, such as bFGF, SDF-1, MMP3, S1P, TGF-β, and SCF, are potentially useful to recruit endogenous dental pulp cells including stem/progenitor cells from the remaining pulp tissue in partial pulp regeneration.

15.4.3 Complete Pulp Regeneration with DPSC Subfractions and Homing/Migration Factors

The discovery of a new population of mesenchymal stem cells residing in the apical papilla (SCAPs) of the developing teeth with incomplete apical closure [55–57] has raised the possibility of whole pulp regeneration after pulpectomy in immature teeth (Fig. 15.1b). Furthermore, the existence of mesenchymal stem cells in the surrounding tissue of the teeth, bone marrow, and periodontal ligament (BMSCs and PDLSCs) [18] makes considerable optimism for whole pulp regeneration even in case of mature teeth with complete apical closure (Fig. 15.2c, d). The transplanted cells need nutrition and oxygen in the empty root canals and some DPSC subfractions with high migration potential including $CD105^+$ cells and $CD31^-$ SP cells have potential utility. They induce angiogenesis and reinnervation and are not incorporated into the vasculature or nerve [5, 28]. Thus, homing/migration factors are potentially useful for whole pulp regeneration to recruit endogenous stem/progenitor cells into the pulpectomized root canal [5]. SDF-1 is one of the prominent homing/migration factors known to be involved in hematopoietic stem cells (HSC) homing to the bone marrow niche [58, 59] and in subventricular zone (SVZ) cells homing to areas of ischemic injury in central nervous system (CNS) [60–62]. MSCs highly express CXCR-4 on their surface, and CXCR-4-SDF-1 axis is used by MSCs in homing and engraftment at injured or inflammation sites [63]. The DPSC subfractions, pulp $CD31^-$ SP cells, and $CD105^+$ cells are also CXCR4-positive with high migratory and proliferative activity with SDF-1 [5, 28]. A recent report on SDF-1 has demonstrated that GFP-labeled bone marrow-derived cells engraft more in the dental pulp than other tissues/organs after transplantation into irradiated wild-type mice under normal conditions, since dental pulp expresses SDF-1 significantly higher than other tissues/organs [64]. These results shed new light on the possibility of complete pulp regeneration by application of SDF-1 together with DPSC subfractions in the pulpectomized root canal. Combinatorial delivery of bFGF, VEGF, or PDGF with NGF and BMP7 has shown to promote cell migration into pulpectomized root canals of human teeth to induce revascularization and pulp regeneration after ectopic transplantation [44]. Furthermore, most recently, stem cell factor (SCF) which binds to the c-Kit receptor CD117 has shown to increase in the cell number and capillaries in subcutaneously implanted collagen sponge, suggesting the suitability of SCF as a potent aid in pulp regeneration [54]. Thus, other possible homing/migration factors might be bFGF, VEGF, PDGF, and SCF.

An optimal scaffold with the following properties has been developed for efficient and safe complete pulp regeneration: (1) optimal flow for injection to be devoid of bubble formation and little contraction after hardening in the root canal; (2) biocompatibility with high bioactivity to be impregnated with trophic factors secreted by transplanted stem cells for homing, engraftment, anti-apoptosis and proliferation of endogenous stem/progenitor cells; (3) biodegradability to release trophic factors and replace the damaged tissue by the regenerated tissue [65–67]; (4) close resemblance of the cell's physiological environment, natural extracellular matrix (ECM), providing the chemical signals to modulate cellular behavior and reinforce a particular phenotype [65]; and (5) no stimulation of differentiation of odontoblasts/osteodentinoblasts and mineral deposition inside the root canal except along the dentinal wall and in the dentin defect [5]. The two categories of materials are natural polymers such as collagen, gelatin, dextran and fibronectin, and synthetic polymers such as poly(lactic) acid (PLA) and poly(glycolic) acid (PGA). These materials exhibit satisfactory results in terms of biocompatibility and degradation. Collagen is biocompatible and degradable by enzymes, but its processing and customizing for specific application are often difficult. It is also affiliated with the risk of transmitting animal-associated pathogens or provoking an immunoresponse [65]. In contrast, PLA and PGA and their copolymer are nontoxic and biocompatible and degrade by hydrolysis. They lack the morphogenetic cues that are physiologically present in the extracellular matrix [65]. Ectopic transplantation of SHED seeded onto PLA into tooth slice forms vascularized soft connective, pulp-like tissue [21, 22]. SCAP and DPSCs seeded onto PLGA into root canals also form pulp-like tissue [23]. A recent promising class of biomaterials, self-assembling peptide hydrogels, generate extracellular matrix-like materials with cell adhesion motifs, controlled release of bioactive molecules, enhanced cell migration, and enzyme-cleavable sites for cell-mediated degradation [68]. The combination of inductive scaffold with stem cells might optimize the approaches for pulp regeneration. Thus, based on the concepts described previously, complete pulp regeneration has been developed using the tissue engineering triad, stem/progenitor cells, homing/migration factors, and optimal scaffolds. To mimic the endodontic treatment, transplantation has been performed in mature teeth with complete apical closure into the alveolar bone in dogs. The first experimental model for complete pulp regeneration is a tooth extraction model in case of pulpectomy (Fig. 15.2c). The tooth is extracted, and its apical portion of the root (1 mm in length) is cut out. The apical foramen of the root canal is enlarged to 0.8 mm in width after whole pulp removal. Ex vivo injection of autologous DPSC subfractions, $CD31^-$ SP cells, or $CD105^+$ cells, 1×10^6 cells, is performed with collagen types I and III (1:1) scaffold (collagen TE, Nitta Gelatin) in the lower part of the root canals. The upper part is further filled with SDF-1 at the final concentration of 10 ng/mL with collagen TE. And finally, the tooth is retransplanted into the alveolar bone. Pulp-like tissue with well vasculature and nerves is formed 14 days after transplantation (Fig. 15.3d, f, g). The odontoblast-like cells extending their process into the dentin tubules and producing tubular dentin along the dentinal wall are seen on day 35 (Fig. 15.3e). The regenerated pulp-like tissue does not mineralize inside the root canal except along the dentinal wall and in dentin defect (Fig. 15.3d), unlike transplantation of unfractionated DPSCs which yields mineralized tissue in the entire regenerated tissue [69].

The second experimental model is tooth non-extraction method in case of pulpectomy. Pulpectomy with enlarged apical portion, 0.6~0.7 mm in width, is also performed without tooth extraction in mature teeth with complete apical closure in an experimental model in dogs (Fig. 15.2d). Autologous transplantation of DPSC subfractions, 1×10^6 cells with collagen TE in the lower part of the root canal and SDF-1 (15 ng/mL) with collagen TE in the upper part, is performed [26]. Furthermore, the modified non-extraction method in which DPSC subfractions and SDF-1 are injected together without separation in both parts of the root canal has been developed [31]. Although non-extraction methods yield identical pulp regeneration on day 14 as in the extraction method (Fig. 15.3h, i, k, l), it is noteworthy that in the non-extraction method neither internal nor external root resorption is observed in the tooth. Induced vascularization in the regenerated tissue is similar in density and orientation to those in the normal pulp analyzed by the three-dimensional image of two-photon microscopy. Numerous transplanted cells are in the vicinity of newly formed capillaries and express angiogenic/neurotrophic factors, implicating a trophic role in neovascularization [26, 31]. DiI labelling on the regenerated pulp in the lower incisor has demonstrated the neuronal process from the regenerated pulp connecting to the inferior alveolar nerve [26]. Additional dentin formation along the dentinal wall and in the enlarged apical portion is detected on day 90 (Fig. 15.3j). It is noteworthy that transplantation of DPSC subfractions alone, or SDF-1 alone, yields significantly less pulp tissue. In addition, after transplantation of unfractionated DPSCs, the regenerated tissue is significantly less and undergoes mineralization [26]. The regenerated tissue by transplantation of the DPSC subfractions and SDF-1 has been demonstrated to be identical to normal functional pulp tissue by similar expression of the pulp tissue markers, *Syndecan* and *TRH-DE* mRNA. Furthermore, two-dimensional electrophoretic analyses and microarray analyses have shown that the qualitative and quantitative protein and mRNA expression patterns of the regenerated tissue are virtually identical to normal pulp [26, 38].

The possible mechanisms of pulp regeneration based on DPSC subfractions and SDF-1 are as follows: the CXCR4/SDF-1 axis may induce CXCR4-positive endogenous stem/progenitor cells to migrate into the root canal and proliferate and differentiate into the endothelial cells, pulp cells, and odontoblasts. Angiogenesis and reinnervation are stimulated by angiogenic/neurotrophic factors secreted by the transplanted stem cells. However, the precise mechanisms of recruitment and crucial molecules for cell migration are still unclear and need additional experimental scrutiny.

15.4.4 Preclinical Trial of Complete Pulp Regeneration

With the successful results on complete pulp regeneration described previously, a preclinical trial by harnessing pulp stem/progenitor cells with high angiogenic/neurogenic potential and homing/migration factors with proper scaffold has assessed its efficacy and safety as a preparation for the impending clinical trials. First, for clinical translation, it is essential to manufacture clinical-grade DPSC subsets according to good manufacturing practice (GMP) conditions without using conventional flow cytometry. To isolate DPSC subsets, a new cost-effective method has been devised by employing an optimized granulocyte colony-stimulating factor (G-CSF)-induced mobilization [29]. The DPSCs mobilized by G-CSF (MDPSCs) are enriched for CD105, CXCR4, and G-CSF receptor (G-CSFR)-positive cells, demonstrating stem cell properties including high proliferation rate, high migratory activity, high expression of multiple trophic factors, and stability. The absence of contamination, abnormalities/aberrations in karyotype, and tumor formation after transplantation in an immunodeficient mouse assures excellent quality control of MDPSCs [29]. For preclinical efficacy test, the second experimental model of complete pulp regeneration was used except the apical foramen enlarged to open 0.6 mm in width. As homing/migration factors, clinical-grade G-CSF which has been approved by the US Food and Drug Administration was used. Augmentation of tissue regeneration in combined strategies, that is, transplantation of MSCs together with G-CSF, has already been known in peripheral nerve injury [70], spinal cord injury [71, 72], and cerebral ischemia [73]. Thus, autologous transplantation of MDPSCs and G-CSF (Neutrogin®, Chugai Pharmaceutical, Tokyo, Japan) with atelocollagen (Koken, Tokyo, Japan), which are all in clinical grade, was evaluated and resulted in complete regeneration of pulp tissue with vasculature and innervation in the root canal. Dentin was also formed in the coronal part to prevent microleakage up to day 180 [74]. Pulp regeneration was confirmed by gene expression analysis of biomarkers of pulp tissue, such as *tenascin C*, *syndecan 3*, and *thyrotropin-releasing hormone (TRH)-degrading enzyme* (*TRH-DE*), and by hierarchical clustering based on affymetrix data. A positive reaction in electric pulp testing and DiI-labeled nerves extending to the trigeminal ganglion indicated that regenerated tissue can transmit sensory signals perceived as pain. For safety evaluation, there was no evidence of toxicity and adverse events [74]. The regenerated dentin-pulp complex was significantly larger in volume in transplantation of MDPSCs with G-CSF compared with that in transplantation of each alone. Inflammatory cells and apoptotic cells were reduced in number, and neurite outgrowth was significantly increased compared to those without G-CSF. The transplanted stem cells were in the vicinity of newly formed capillaries and outgrowth neuritis, expressing angiogenic/neurotrophic factors. G-CSF together with conditioned medium of MDPSCs stimulated cell migration and neurite outgrowth, prevented cell death, and promoted immunosuppression in vitro. These findings suggest that G-CSF has combinatorial trophic effects with MDPSCs including increased migration and proliferation of endogenous stem/progenitor cells, suppression of apoptotic death of transplanted MDPSCs, attenuation of inflammatory response, and induction of extrinsic and endogenous neurogenesis. Thus, based on these results of preclinical trials, scientific evidence of the safety and efficacy critical for clinical applications has been presented.

15.5 Complete Pulp Regeneration with Periapical Disease

The third experimental model is complete pulp regeneration with periapical disease (Fig. 15.2e). In this model, the root canal of mature tooth with complete apical closure is kept open for more than 2 weeks, and infection has been confirmed by microbiological and X-ray analyses. Pulp tissue was regenerated after transplantation of DPSC subfractions with SDF-1 as described previously in case of pulpectomy. The only difference is that the regenerated tissue is less in volume and contains a few of inflammatory cells. Further progress of complete disinfection method is needed for promising stem cell therapy in periapical disease.

15.6 Complete Pulp Regeneration for the Aged

The fourth experimental model is complete pulp regeneration in the aged (Fig. 15.2f). It is the challenge for endodontic therapy to preserve the teeth and facilitate optimal oral function in aged patients [75]. Tissue regeneration and maintenance is dependent on MSCs [76], and to understand the influence of the age of MSCs on their cell properties and regenerative potential is important for stem cell therapy for the aged. The supply of autologous pulp tissue is very limited with age [77]. Teeth from aged patients have pulpal tissue of smaller size and supplied fewer colonies of human DPSCs than those from young patients [78]. Human DPSCs from aged donors lose their proliferative activities and differentiation capabilities after repeated expansion [79, 80]. However, the possibility of obtaining human pulp stem cells from a small amount of tissue in aged donors has been reported in a hypoxic culture condition under 3 % O_2 [78]. Recently, MDPSCs from aged donors were compared with young donors to determine the changes in the biological properties and stability depending on age. The aged MDPSCs were enriched in stem cells, expressing similar high levels of trophic factors as the young MDPSCs. In contrast, colony-derived DPSCs showed significant differences in those properties between aged and young donors. A little age-dependent increase in expression of senescence markers, such as p16 and p21 in long-term culture, has been demonstrated in MDPSCs. The regenerative potential of aged MDPSCs was similar to that of young MDPSCs in an ischemic hindlimb model and an ectopic tooth transplantation model in immunodeficient mice. In the fourth experimental pulpectomized tooth model in dogs, however, pulp regeneration was retarded after autologous transplantation of aged MDPSCs. The isolated periodontal ligament stem cells (PDLSCs) from aged dogs, representative of migrating endogenous stem/progenitor cells from outside of the tooth, had lower proliferation, migration, and anti-apoptotic abilities. The mechanisms involved in the age-dependent decline in pulp regeneration might be attributed to a decrease in the regenerative potential of resident stem cells. Future investigations should focus on the cellular and molecular mechanisms underlying the age-dependent changes in the stem cell niche influencing stem cell quiescence and function and in the extrinsic factors emanating from extracellular matrix.

15.7 Other Sources of Tissue Stem Cells for Pulp Regeneration

The fifth experimental model is complete pulp regeneration harnessing MSCs from other tissue sources (Fig. 15.2g). Supply of autologous pulp tissue declines with age, and alternative sources of MSCs might contribute to stem cell therapy for median- and elderly aged patients in pulp regeneration. Transcriptional and epigenetic analyses demonstrate very similar profiles among MSCs from many different tissues including the bone marrow, adipose tissue, placenta, umbilical cord, and amnion [81, 82]. Expression profile of trophic factors and growth factors, however, is distinct among MSCs [81, 83, 84]. Thus, the requirements and optimal preconditions of MSCs for efficient stem cell therapy in pulp regeneration need to be elucidated [38]. Autologous MSCs derived from the bone marrow and adipose tissues have neither ethical nor immunoreactive issues. There are many identical characteristics between the two types MSCs [85]. There are differences, however, in protein expression and function [83, 84]. A similar level of gene expression between pulp and bone marrow MSCs is detected for more than 4,000 known human genes, except a few differentially expressed genes [86]. The differential proteomic expression profiles of ovine pulp and bone marrow MSCs from an individual donor identified 23 proteins upregulated in pulp MSCs [87]. Expression patterns of gene markers and the broad differentiation potentials induced by specific methods in vitro are similar between rabbit pulp and adipose MSCs analyzed by side-by-side comparisons. Differences in the biological properties of MSC populations, however, have been previously ascribed to donor-associated variability [88] and to cell isolation and cultivation methods [89]. Thus, to evaluate whether bone marrow- and adipose tissue-derived MSCs might be alternative to pulp tissue-derived MSCs for pulp regeneration, biological characteristics were compared in pulp, bone marrow, and adipose MSCs which associated with a defined protocol for isolation (SP subfraction, $CD31^-$ SP cells) and cultivation and originated from a single donor [38]. The results demonstrated higher migration activity and higher expression of angiogenic/neurotrophic factors in pulp $CD31^-$ SP cells compared with bone marrow and adipose $CD31^-$ SP cells. Two-dimensional electrophoretic analyses demonstrated that the qualitative and quantitative protein pattern of pulp $CD31^-$ SP cells was similar to others [38]. Transplantation of adipose $CD31^-$ SP cells induced a similar amount of regenerated tissue compared with transplantation of pulp $CD31^-$ SP cells, although bone marrow $CD31^-$ SP cell transplantation induced significantly less amount. The regenerated tissues, however, all revealed identical to pulp tissue analyzed by morphological and functional similarity, mRNA expression patterns of microarray, and two-dimensional electrophoresis [38]. Thus, MSCs from different tissues are comparable in pulp regeneration in vivo, and their potential for pulp regeneration may depend not on the niche of their origin but on the local microenvironment. The local microenvironment may provide signals driving the fate of homing endogenous stem/progenitor cells originated from surrounding tissues [30, 38]. Transplantation of adipose $CD31^-$ SP cells is enhanced in matrix formation leading to root canal obliteration compared with transplantation of pulp $CD31^-$ SP cells, suggesting that some trophic factors specifically released from adipose MSCs may influence on active pathway for differentiation. An additional method for differentiation and mineralization control might contribute to the elimination of obliteration in the root canal after pulp regeneration.

15.8 Future Perspectives

The goal of pulp regeneration is the complete restoration of tooth function. Regeneration of pulp tissue by stem cell therapy resulted in (1) well-vascularized and innervated pulp tissue, (2) the cell density and architecture of the extracellular matrix that is identical to normal pulp, and (3) the lining of functional and aligned odontoblasts along the dentinal wall with nascent dentin, demonstrating functional regeneration. Although in the past the small apical canal opening was deemed difficult for vascularization, in the canine models pulp tissue can be completely regenerated even when the apical opening is about 0.6 mm [74]. Thus, the ingrowth of blood vessels into the engineered pulp tissue is not dependent on the apical opening. The angiogenesis and neovascularization may stimulate optimal cell density and assembly of extracellular matrix and tissue organization of the dental pulp.

For the clinical success of the pulp/dentin regeneration, the stem/progenitor cells should be isolated and expanded in a safe, stable, and efficient manner. Good manufacturing practice (GMP) facilities are needed infrastructure to make cell-based therapy practical [90]. The clinical-grade DPSC subsets, MDPSCs capable of confirming safety and efficiency from a small amount of pulp tissue, are useful in clinical trials. In addition to stem cells, homing/migration factors and the scaffold should be a GMP product. Therefore, the manufacturing methods need to be evaluated before regenerative endodontics can replace the conventional endodontic treatment. The availability of an off-the-shelf tissue banked allogeneic stem/progenitor cells will have a profound effect on clinical practice of endodontics and conservative dentistry.

Abbreviations

BDNF	Brain-derived neurotrophic factor
DMP1	Dentin matrix protein 1
DPSCs	Dental pulp stem cells
FGF-2	Fibroblast growth factor-2
G-CSF	Granulocyte colony-stimulating factor
GDNF	Glial cell-line-derived neurotrophic factor
GM-CSF	Granulocyte-macrophage colony-stimulating factor
MDPSCs	Mobilized dental pulp stem cells
MMP3	Matrix metalloproteinase 3
MSCs	Mesenchymal stem cells
NGF	Nerve growth factor
PDLSCs	Periodontal ligament stem cells
PGA	Poly(glycolic) acid
PLA	Poly(lactic) acid
SCAP	Stem cells from apical papilla
SDF-1α	Stromal cell-derived factor-1α
SHED	Stem cells from exfoliated deciduous teeth
SP	Side population
TRH-DE	Thyrotropin-releasing hormone-degrading enzyme
VEGF-A	Vascular endothelial growth factor

References

1. Matchou P, Reit C. Non-surgical retreatment. In: Bergenholtz G, Horsted-Bindslev P, Reit C, editors. Textbook of endodontology. 1st ed. Oxford: Wiley-Blackwell; 2003. p. 300–10.
2. Wilcox LR, Van Surksum R. Endodontic retreatment in large and small straight canals. J Endod. 1991; 17(3):119–21.
3. Fuss Z, Lustig J, Katz A, Tamse A. An evaluation of endodontically treated vertical root fractured teeth: impact of operative procedures. J Endod. 2001;27(1): 46–8.
4. Nakashima M, Akamine A. The application of tissue engineering to regeneration of pulp and dentin in endodontics. J Endod. 2005;31(10):711–8.
5. Nakashima M, Iohara K, Sugiyama M. Human dental pulp stem cells with highly angiogenic and neurogenic potential for possible use in pulp regeneration. Cytokine Growth Factor Rev. 2009;20(5–6): 435–40.
6. Huang GT. Apexification: the beginning of its end. Int Endod J. 2009;42(10):855–66.
7. Mooney DJ, Powell C, Piana J, Rutherford B. Engineering dental pulp-like tissue in vitro. Biotechnol Prog. 1996;12(6):865–8.
8. Gronthos S, Mankani M, Brahim J, Robey PG, Shi S. Postnatal human dental pulp stem cells (DPSCs) in vitro and in vivo. Proc Natl Acad Sci U S A. 2000;97(25):13625–30.
9. Thibodeau B, Teixeira F, Yamauchi M, Caplan DJ, Trope M. Pulp revascularization of immature dog teeth with apical periodontitis. J Endod. 2007;33(6): 680–9.
10. Thomson A, Kahler B. Regenerative endodontics–biologically-based treatment for immature permanent teeth: a case report and review of the literature. Aust Dent J. 2010;55(4):446–52.
11. Andreasen JO, Bakland LK. Pulp regeneration after non-infected and infected necrosis, what type of tissue do we want? A review. Dent Traumatol. 2011;28(1):13–8.
12. Nosrat A, Seifi A, Asgary S. Regenerative endodontic treatment (revascularization) for necrotic immature permanent molars: a review and report of two cases with a new biomaterial. J Endod. 2011;37(4):562–7.
13. Chen MY, Chen KL, Chen CA, Tayebaty F, Rosenberg PA, Lin LM. Responses of immature permanent teeth

with infected necrotic pulp tissue and apical periodontitis/abscess to revascularization procedures. Int Endod J. 2012;45(3):294–305.
14. Law AS. Considerations for regeneration procedures. J Endod. 2013;39(3 Suppl):S44–56.
15. Trope M. Treatment of the immature tooth with a non-vital pulp and apical periodontitis. Dent Clin North Am. 2010;54(2):313–24.
16. Wang X, Thibodeau B, Trope M, Lin LM, Huang GT. Histologic characterization of regenerated tissues in canal space after the revitalization/revascularization procedure of immature dog teeth with apical periodontitis. J Endod. 2010;36(1):56–63.
17. Murray PE, Garcia-Godoy F, Hargreaves KM. Regenerative endodontics: a review of current status and a call for action. J Endod. 2007;33(4):377–90.
18. Huang GT, Gronthos S, Shi S. Mesenchymal stem cells derived from dental tissues vs. those from other sources: their biology and role in regenerative medicine. J Dent Res. 2009;88(9):792–806.
19. Sakai VT, Cordeiro MM, Dong Z, Zhang Z, Zeitlin BD, Nor JE. Tooth slice/scaffold model of dental pulp tissue engineering. Adv Dent Res. 2011;23(3):325–32.
20. Prescott RS, Alsanea R, Fayad MI, Johnson BR, Wenckus CS, Hao J, et al. In vivo generation of dental pulp-like tissue by using dental pulp stem cells, a collagen scaffold, and dentin matrix protein 1 after subcutaneous transplantation in mice. J Endod. 2008;34(4): 421–6.
21. Cordeiro MM, Dong Z, Kaneko T, Zhang Z, Miyazawa M, Shi S, et al. Dental pulp tissue engineering with stem cells from exfoliated deciduous teeth. J Endod. 2008;34(8):962–9.
22. Sakai VT, Zhang Z, Dong Z, Neiva KG, Machado MA, Shi S, et al. SHED differentiate into functional odontoblasts and endothelium. J Dent Res. 2010; 89(8):791–6.
23. Huang GT, Yamaza T, Shea LD, Djouad F, Kuhn NZ, Tuan RS, et al. Stem/progenitor cell-mediated de novo regeneration of dental pulp with newly deposited continuous layer of dentin in an in vivo model. Tissue Eng Part A. 2010;16(2):605–15.
24. Rosa V, Zhang Z, Grande RH, Nor JE. Dental pulp tissue engineering in full-length human root canals. J Dent Res. 2013;92(11):970–5.
25. Olgart L, Kerezoudis NP. Nerve-pulp interactions. Arch Oral Biol. 1994;39(Suppl):47S–54.
26. Iohara K, Imabayashi K, Ishizaka R, Watanabe A, Nabekura J, Ito M, et al. Complete pulp regeneration after pulpectomy by transplantation of CD105+ stem cells with stromal cell-derived factor-1. Tissue Eng Part A. 2011;17(15–16):1911–20.
27. Iohara K, Zheng L, Ito M, Ishizaka R, Nakamura H, Into T, et al. Regeneration of dental pulp after pulpotomy by transplantation of CD31(-)/CD146(-) side population cells from a canine tooth. Regen Med. 2009;4(3):377–85.
28. Iohara K, Zheng L, Wake H, Ito M, Nabekura J, Wakita H, et al. A novel stem cell source for vasculogenesis in ischemia: subfraction of side population cells from dental pulp. Stem Cells. 2008;26(9): 2408–18.
29. Murakami M, Horibe H, Iohara K, Hayashi Y, Osako Y, Takei Y, et al. The use of granulocyte-colony stimulating factor induced mobilization for isolation of dental pulp stem cells with high regenerative potential. Biomaterials. 2013;34(36):9036–47.
30. Ishizaka R, Hayashi Y, Iohara K, Sugiyama M, Murakami M, Yamamoto T, et al. Stimulation of angiogenesis, neurogenesis and regeneration by side population cells from dental pulp. Biomaterials. 2013;34(8):1888–97.
31. Ishizaka R, Iohara K, Murakami M, Fukuta O, Nakashima M. Regeneration of dental pulp following pulpectomy by fractionated stem/progenitor cells from bone marrow and adipose tissue. Biomaterials. 2012;33(7):2109–18.
32. Iohara K, Zheng L, Ito M, Tomokiyo A, Matsushita K, Nakashima M. Side population cells isolated from porcine dental pulp tissue with self-renewal and multipotency for dentinogenesis, chondrogenesis, adipogenesis, and neurogenesis. Stem Cells. 2006;24(11):2493–503.
33. Barry FP, Boynton RE, Haynesworth S, Murphy JM, Zaia J. The monoclonal antibody SH-2, raised against human mesenchymal stem cells, recognizes an epitope on endoglin (CD105). Biochem Biophys Res Commun. 1999;265(1):134–9.
34. Kastrinaki MC, Andreakou I, Charbord P, Papadaki HA. Isolation of human bone marrow mesenchymal stem cells using different membrane markers: comparison of colony/cloning efficiency, differentiation potential, and molecular profile. Tissue Eng Part C Methods. 2008;14(4):333–9.
35. Jarocha D, Lukasiewicz E, Majka M. Advantage of mesenchymal stem cells (MSC) expansion directly from purified bone marrow CD105+ and CD271+ cells. Folia Histochem Cytobiol. 2008;46(3):307–14.
36. Mafi R, Hindocha S, Mafi P, Griffin M, Khan WS. Sources of adult mesenchymal stem cells applicable for musculoskeletal applications – a systematic review of the literature. Open Orthop J. 2011;5 Suppl 2:242–8.
37. Sugiyama M, Iohara K, Wakita H, Hattori H, Ueda M, Matsushita K, et al. Dental pulp-derived CD31(-)/CD146(-) side population stem/progenitor cells enhance recovery of focal cerebral ischemia in rats. Tissue Eng Part A. 2011;17(9–10):1303–11.
38. Ishizaka R, Hayashi Y, Iohara K, Sugiyama M, Murakami M, Yamamoto T, et al. Stimulation of angiogenesis, neurogenesis and regeneration by side population cells from dental pulp. Biomaterials. 2012; 34(8):1888–97.
39. Kassis I, Vaknin-Dembinsky A, Karussis D. Bone marrow mesenchymal stem cells: agents of immunomodulation and neuroprotection. Curr Stem Cell Res Ther. 2011;6(1):63–8.
40. Parr AM, Tator CH, Keating A. Bone marrow-derived mesenchymal stromal cells for the repair of central nervous system injury. Bone Marrow Transplant. 2007;40(7):609–19.

41. Doorn J, Moll G, Le Blanc K, van Blitterswijk C, de Boer J. Therapeutic applications of mesenchymal stromal cells: paracrine effects and potential improvements. Tissue Eng Part B Rev. 2012;18(2):101–15.
42. Buurma B, Gu K, Rutherford RB. Transplantation of human pulpal and gingival fibroblasts attached to synthetic scaffolds. Eur J Oral Sci. 1999;107(4):282–9.
43. Bohl KS, Shon J, Rutherford B, Mooney DJ. Role of synthetic extracellular matrix in development of engineered dental pulp. J Biomater Sci Polym Ed. 1998;9(7):749–64.
44. Kim JY, Xin X, Moioli EK, Chung J, Lee CH, Chen M, et al. Regeneration of dental-pulp-like tissue by chemotaxis-induced cell homing. Tissue Eng Part A. 2010;16(10):3023–31.
45. Schmidt A, Ladage D, Schinkothe T, Klausmann U, Ulrichs C, Klinz FJ, et al. Basic fibroblast growth factor controls migration in human mesenchymal stem cells. Stem Cells. 2006;24(7):1750–8.
46. Wynn RF, Hart CA, Corradi-Perini C, O'Neill L, Evans CA, Wraith JE, et al. A small proportion of mesenchymal stem cells strongly expresses functionally active CXCR4 receptor capable of promoting migration to bone marrow. Blood. 2004;104(9):2643–5.
47. Suzuki T, Lee CH, Chen M, Zhao W, Fu SY, Qi JJ, et al. Induced migration of dental pulp stem cells for in vivo pulp regeneration. J Dent Res. 2011;90(8):1013–8.
48. Kitamura C, Nishihara T, Terashita M, Tabata Y, Washio A. Local regeneration of dentin-pulp complex using controlled release of fgf-2 and naturally derived sponge-like scaffolds. Int J Dent. 2012;2012:190561.
49. Ishimatsu H, Kitamura C, Morotomi T, Tabata Y, Nishihara T, Chen KK, et al. Formation of dentinal bridge on surface of regenerated dental pulp in dentin defects by controlled release of fibroblast growth factor-2 from gelatin hydrogels. J Endod. 2009;35(6):858–65.
50. Kikuchi N, Kitamura C, Morotomi T, Inuyama Y, Ishimatsu H, Tabata Y, et al. Formation of dentin-like particles in dentin defects above exposed pulp by controlled release of fibroblast growth factor 2 from gelatin hydrogels. J Endod. 2007;33(10):1198–202.
51. Zheng L, Amano K, Iohara K, Ito M, Imabayashi K, Into T, et al. Matrix metalloproteinase-3 accelerates wound healing following dental pulp injury. Am J Pathol. 2009;175(5):1905–14.
52. Eba H, Murasawa Y, Iohara K, Isogai Z, Nakamura H, Nakashima M. The anti-inflammatory effects of matrix metalloproteinase-3 on irreversible pulpitis of mature erupted teeth. PLoS One. 2012;7(12):e52523.
53. Howard C, Murray PE, Namerow KN. Dental pulp stem cell migration. J Endod. 2010;36(12):1963–6.
54. Pan S, Dangaria S, Gopinathan G, Yan X, Lu X, Kolokythas A, et al. SCF promotes dental pulp progenitor migration, neovascularization, and collagen remodeling – potential applications as a homing factor in dental pulp regeneration. Stem Cell Rev. 2013;9(5):655–67.
55. Sonoyama W, Liu Y, Yamaza T, Tuan RS, Wang S, Shi S, et al. Characterization of the apical papilla and its residing stem cells from human immature permanent teeth: a pilot study. J Endod. 2008;34(2):166–71.
56. Sonoyama W, Liu Y, Fang D, Yamaza T, Seo BM, Zhang C, et al. Mesenchymal stem cell-mediated functional tooth regeneration in swine. PLoS One. 2006;1:e79.
57. Huang GT, Sonoyama W, Liu Y, Liu H, Wang S, Shi S. The hidden treasure in apical papilla: the potential role in pulp/dentin regeneration and bioroot engineering. J Endod. 2008;34(6):645–51.
58. Kaplan RN, Psaila B, Lyden D. Niche-to-niche migration of bone-marrow-derived cells. Trends Mol Med. 2007;13(2):72–81.
59. Chute JP. Stem cell homing. Curr Opin Hematol. 2006;13(6):399–406.
60. Thored P, Arvidsson A, Cacci E, Ahlenius H, Kallur T, Darsalia V, et al. Persistent production of neurons from adult brain stem cells during recovery after stroke. Stem Cells. 2006;24(3):739–47.
61. Robin AM, Zhang ZG, Wang L, Zhang RL, Katakowski M, Zhang L, et al. Stromal cell-derived factor 1alpha mediates neural progenitor cell motility after focal cerebral ischemia. J Cereb Blood Flow Metab. 2006;26(1):125–34.
62. Ohab JJ, Fleming S, Blesch A, Carmichael ST. A neurovascular niche for neurogenesis after stroke. J Neurosci. 2006;26(50):13007–16.
63. Kucia M, Reca R, Miekus K, Wanzeck J, Wojakowski W, Janowska-Wieczorek A, et al. Trafficking of normal stem cells and metastasis of cancer stem cells involve similar mechanisms: pivotal role of the SDF-1-CXCR4 axis. Stem Cells. 2005;23(7):879–94.
64. Zhou J, Shi S, Shi Y, Xie H, Chen L, He Y, et al. Role of bone marrow-derived progenitor cells in the maintenance and regeneration of dental mesenchymal tissues. J Cell Physiol. 2011;226(8):2081–90.
65. Galler KM, Cavender AC, Koeklue U, Suggs LJ, Schmalz G, D'Souza RN. Bioengineering of dental stem cells in a PEGylated fibrin gel. Regen Med. 2011;6(2):191–200.
66. Yuan Z, Nie H, Wang S, Lee CH, Li A, Fu SY, et al. Biomaterial selection for tooth regeneration. Tissue Eng Part B Rev. 2011;17(5):373–88.
67. Hayashi Y, Yamada S, Yanagi Guchi K, Koyama Z, Ikeda T. Chitosan and fish collagen as biomaterials for regenerative medicine. Adv Food Nutr Res. 2012;65:107–20.
68. Galler KM, Aulisa L, Regan KR, D'Souza RN, Hartgerink JD. Self-assembling multidomain peptide hydrogels: designed susceptibility to enzymatic cleavage allows enhanced cell migration and spreading. J Am Chem Soc. 2010;132(9):3217–23.
69. Nakashima M, Iohara K. Regeneration of dental pulp by stem cells. Adv Dent Res. 2011;23(3):313–9.
70. Pan HC, Wu HT, Cheng FC, Chen CH, Sheu ML, Chen CJ. Potentiation of angiogenesis and regeneration by G-CSF after sciatic nerve crush injury. Biochem Biophys Res Commun. 2009;382(1):177–82.

71. Pan HC, Cheng FC, Lai SZ, Yang DY, Wang YC, Lee MS. Enhanced regeneration in spinal cord injury by concomitant treatment with granulocyte colony-stimulating factor and neuronal stem cells. J Clin Neurosci. 2008;15(6):656–64.
72. Luo J, Zhang HT, Jiang XD, Xue S, Ke YQ. Combination of bone marrow stromal cell transplantation with mobilization by granulocyte-colony stimulating factor promotes functional recovery after spinal cord transection. Acta Neurochir (Wien). 2009;151(11):1483–92.
73. Zhang XM, Du F, Yang D, Wang R, Yu CJ, Huang XN, et al. Granulocyte colony-stimulating factor increases the therapeutic efficacy of bone marrow mononuclear cell transplantation in cerebral ischemia in mice. BMC Neurosci. 2011;12:61.
74. Iohara K, Murakami M, Takeuchi N, Osako Y, Ito M, Ishizaka R, et al. A novel combinatorial therapy with pulp stem cells and granulocyte colony-stimulating factor for total pulp regeneration. Stem Cells Transl Med. 2013;2(7):521–33.
75. Allen PF, Whitworth JM. Endodontic considerations in the elderly. Gerodontology. 2004;21(4):185–94.
76. Kumar S, Chanda D, Ponnazhagan S. Therapeutic potential of genetically modified mesenchymal stem cells. Gene Ther. 2008;15(10):711–5.
77. Murray PE, Stanley HR, Matthews JB, Sloan AJ, Smith AJ. Age-related odontometric changes of human teeth. Oral Surg Oral Med Oral Pathol Oral Radiol Endod. 2002;93(4):474–82.
78. Iida K, Takeda-Kawaguchi T, Tezuka Y, Kunisada T, Shibata T, Tezuka K. Hypoxia enhances colony formation and proliferation but inhibits differentiation of human dental pulp cells. Arch Oral Biol. 2010;55(9):648–54.
79. Bressan E, Ferroni L, Gardin C, Pinton P, Stellini E, Botticelli D, et al. Donor age-related biological properties of human dental pulp stem cells change in nanostructured scaffolds. PLoS One. 2012;7(11):e49146.
80. Takeda T, Tezuka Y, Horiuchi M, Hosono K, Iida K, Hatakeyama D, et al. Characterization of dental pulp stem cells of human tooth germs. J Dent Res. 2008;87(7):676–81.
81. Boeuf S, Richter W. Chondrogenesis of mesenchymal stem cells: role of tissue source and inducing factors. Stem Cell Res Ther. 2010;1(4):31.
82. Aranda P, Agirre X, Ballestar E, Andreu EJ, Roman-Gomez J, Prieto I, et al. Epigenetic signatures associated with different levels of differentiation potential in human stem cells. PLoS One. 2009;4(11):e7809.
83. Philippe B, Luc S, Valerie PB, Jerome R, Alessandra BR, Louis C. Culture and use of mesenchymal stromal cells in Phase I and II clinical trials. Stem Cells Int. 2010;2010:503593.
84. Noel D, Caton D, Roche S, Bony C, Lehmann S, Casteilla L, et al. Cell specific differences between human adipose-derived and mesenchymal-stromal cells despite similar differentiation potentials. Exp Cell Res. 2008;314(7):1575–84.
85. Zuk PA, Zhu M, Mizuno H, Huang J, Futrell JW, Katz AJ, et al. Multilineage cells from human adipose tissue: implications for cell-based therapies. Tissue Eng. 2001;7(2):211–28.
86. Shi S, Robey PG, Gronthos S. Comparison of human dental pulp and bone marrow stromal stem cells by cDNA microarray analysis. Bone. 2001;29(6):532–9.
87. Mrozik KM, Zilm PS, Bagley CJ, Hack S, Hoffmann P, Gronthos S, et al. Proteomic characterization of mesenchymal stem cell-like populations derived from ovine periodontal ligament, dental pulp, and bone marrow: analysis of differentially expressed proteins. Stem Cells Dev. 2010;19(10):1485–99.
88. Phinney DG, Kopen G, Righter W, Webster S, Tremain N, Prockop DJ. Donor variation in the growth properties and osteogenic potential of human marrow stromal cells. J Cell Biochem. 1999;75(3):424–36.
89. Vemuri MC, Chase LG, Rao MS. Mesenchymal stem cell assays and applications. Methods Mol Biol. 2011;698:3–8.
90. Sensebe L, Krampera M, Schrezenmeier H, Bourin P, Giordano R. Mesenchymal stem cells for clinical application. Vox Sang. 2010;98(2):93–107.

Pulp Stem Cells: Niches of Stem Cells

16

Michel Goldberg

16.1 Stem Cells in General

Stem cells are found in all multicellular organisms. They can divide and differentiate into diverse specialized cell types. They can self-renew to produce more cells and they have been used extensively in the field of repair and regeneration of defective tissues and organs.

Stem cells may be obtained either from the embryo (ESCs) or from the adult (ASCs). At early embryonic stages, stem cells are multipotent. At postnatal stages and in adult, they display more restricted phenotypes. There is no specific marker, but the association of a series of transcription factors constitutes a major characteristic of stem cells.

In addition, induced pluripotent stem cells (iPSCs) may be obtained from already differentiated cells. Under the influence of a limited number of transcription factors, they may dedifferentiate and redifferentiate shifting their original phenotype to another. These successive steps may provide tools to adapt adult cells to therapeutic needs. Although this strategy is promising, the phenotype obtained seems to be not stable after a series of passages. For these reasons, up to now ESCs may be involved in the correction of genetic alterations, but ASCs provide the best tools to regenerate wounded adult tissues, and along this line, iPSCs have still to be explored.

To summarize a long list of reports, *embryonic stem cells* (*ECSs*) are initially derived from the inner cell mass of the blastocyst from which many tissues of the embryo arise. They are initially totipotent cells, clonogenic, and capable of self-renewal. At later stages, they can differentiate into each of the more than 200 cell types of the adult body. As an example, during tooth development at the cap and bell stages, Oct4, Nanog, Stat-3, and Sox-2 display nuclear or cytoplasmic expression, suggesting an interaction between the transcription factors that are involved in self-renewal process and cell differentiation [1].

At later stages, *postnatal or adult stem cells* (*ASCs*) are no longer totipotent [2]. Stem cells display multilineage differentiation potential when sporadically activated. They maintain the capacity to self-renewal and "stemness." They become glia, neurons, osteogenic, odontogenic, myogenic, tendons, melanocytes, chondrocytes, endocrine cells, adipose cells, and also a few other phenotypes. They demonstrated transdifferentiation into corneal cells and islet of pancreas.

They have a more restricted potential although they are still multipotent. They are derived from the neural ectomesenchyme, as most of the craniofacial tissues. They are classified according to their origin and differentiation potential. Some stem cells keep pluripotency, while monopotent cells display restricted differentiated progenies.

M. Goldberg, DDS, PhD
Department of Oral Biology, Institut National de la Santé et de la Recherche Médicale, Université Paris Descartes, 45 Rue des Saints Pères, Paris 75006, France
e-mail: mgoldod@gmail.com, michel.goldberg@parisdescartes.fr

M. Goldberg (ed.), *The Dental Pulp*,
DOI 10.1007/978-3-642-55160-4_16,

The concomitant presence of pluripotent and monopotent cell lines issued from the pulp of SV40 transgenic mice was established and the dual attendance may play a role in cell differentiation [3]. The combined advantages of multipotency/pluripotency and the ease of access of pulp tissue render adult stem cells attractive in regenerative dentistry and medicine [4].

16.1.1 Stem Cell Markers

Adult stem cells have been shown to undergo asymmetric cell division, one daughter remaining in the stem cell compartment and the other daughter cell undergoing further cell divisions giving rise to differentiated cells [5].

Stem cells have been isolated as clonogenic, rapidly proliferative population of cells, including cells from the dental pulp. A new c-kit+/CD34+/CD45- cell population of stromal bone producing cells was identified, selected by using a FACSorting, expanded, and cultured [6].

Hematopoietic cells generate derivatives of the blood. The bone marrow stromal cells (BMSCs) give rise to muscle and neuron-like cells in the brain. Among the bone marrow and blood cells, a small number of pluripotent cells were found [7, 8]. Mesenchymal cells arise from the neural crest or from the mesoderm. Both neural crest-derived cells and mesoderm-derived cells contribute to the development of dental, thymic, and bone marrow mesenchyme from the fetal age to the adult stage. These cells are partly responsible for the transformation of their phenotype into perivascular cells and endothelial cells. Komada et al. [9] concluded that mesenchymal stem cells found in the different tissues had different origin, but similar properties in each organ. Subsets of stem cells are called *mesenchymal stem cells* (*MSCs*) or *skeletal stem cells*. Many of these cells are a rich source of osteogenic progenitors, promoting repair or regeneration of skeletal defects.

Dental pulp stem cells (DPSCs) may be obtained by two widely adapted isolation procedures. They may be obtained either after enzyme digestion of the pulp tissue or by the explant method. The two groups of cells that are obtained showed no differences with regard to proliferation rate and colony formation. They are both positive for CD29, CD44, CD90, CD105, CD117, and CD146 expression. They may both differentiate into adipogenic, chondrogenic, and osteogenic cell types [10].

Positive markers used to identify MSC-like cells are STRO-1, CD13, CD44, CD24, CD29, CD73, CD90, CD105, CD106, CD146, Oct4, Nanog, and beta-2 integrin, whereas negative markers are CD14, CD34, CD45, and HLA-DR. Subpopulations of MSCs are expressing c-kit$^+$/CD34$^+$/STRO-1$^+$ DPSCs or exfoliating deciduous teeth (SHED) and were reported to be multipotent stem, whereas c-kit and CD34 are markers for hematopoietic lineage of cells [11].

Stem cells derived from human adult dental pulp of healthy 30–45-year-old patients were selected using a FACSorter. c-kit$^+$/CD34$^+$/CD45$^-$ cell population was produced, expanded, and cultured. Stromal bone producing cells (SBP/DPSCs) are highly clonogenic and differentiate into osteoblast precursors (CD44$^+$/RUNX-2$^+$), which are self-renewing and then transform into osteoblasts [12].

Two populations of stem cells were discovered in the bone marrow. A first population of hematopoietic stem cells contributes to the formation of all types of blood cells in the body. A second population was shown to be bone marrow stromal cells. Cells from human exfoliated deciduous teeth are capable of extensive proliferation and multipotential differentiation [13].

16.2 How Many Stem Cells Are Found in the Dental Pulp?

Stem cells are found in a number of tissues; however, huge variations in number occur depending on the species and age of the animals, where and how the sampling was done, which culture medium was used, and the tissue environment. For these reasons, it is difficult to get precisely a quantitative evaluation of the number of stem cells present within the dental pulp.

Multipotency, self-renewal, and clonogenicity are the three characteristic features retained for SC criteria and subsequent selection. The lack of specific marker makes these evaluations more confusing. Pluripotent and unipotent cells are not explicitly labeled before their phenotypic differentiation [14]. This is easier at later stages, when genes, transcription factors, and extracellular molecules are expressed.

Concerning a peripheral blood sample of mice containing 9,000 nucleated cells, only ten were labeled with 3H-thymidine and consequently may be considered as stem cells [# 0.01 %] [15].

Human marrow multipotent mesodermal progenitor cells were found to be rare cells living in various mesenchymal tissues. For example, in the bone marrow stroma, 2–5 stem cells were found per million of nucleated cells [16].

Mesenchymal stem cells (MSCs) were also named colony-forming unit fibroblast (CFU-F) or marrow stromal fibroblasts (MSFs). Despite difference in naming, they were also recognized as mesenchymal progenitor cells, which give rise to several mesenchymal tissues (bone, cartilage, muscle, ligament, tendon, adipose, and stroma). They display restricted potential and have a fibroblastic appearance. However, no evidence was found for self-renewal capability or multilineage differentiation capacity [17]. Mesenchymal stem cells express mRNA for osteocalcin but not for dentin sialophosphoprotein. They do not express CD11b, glycophorin-A, or CD45. In contrast, they express positive SC markers such as Stro-1, CD73, and CD106, which combine altogether and constitute a good human MSC marker.

Multipotent MSC or a mixture of committed progenitor cells, each with a restricted potential, was evaluated by Pittenger et al. [18] as being about 0.001–0.01 % of the grand total.

Diverging views were expressed in the different reports and a high variability was noted. According to D'Aquino et al., the DPSCs represent roughly 10 % of the dental pulp cells [19]. Nine percent of human exfoliated deciduous teeth stem cells (SHED) expressed a STRO-1 protein, which potentially define a multipotent progenitor considered to be a marker of odontoblast precursor. Some important differences were also recorded between young immature dental pulp stem cells (IDPSCs) and late pulp populations [20]. The former express embryonic stem cell markers (Oct ¾, Nanog, and Sox-2) and differentiate in vitro. IDPSCs display a very low percentage of Oct3/4 positive cells: ~ 0.75 % in early population. Conversely this percentage was increased to 10.03 % in late population. This report provides some clarification to the conflicting results expressed between publications.

According to Sloan and Waddington [21], the subset of undifferentiated cells can represent in the dental pulp as little as 1 % of the total cell population. After staining the cells with Hoechst 33342, side population (SP) cells in human deciduous dental pulp were evaluated as 2 % of the total cells [22]. The SP marker ABCG2 protein was localized to DPSs in the cell membrane by immunofluorescence staining. Flow cytometry analysis demonstrates that 3.6 % of DPCs were ABCG2 positive. The authors concluded that a higher proportion of SP was present in the deciduous teeth, compared with permanent teeth. In another report approximately 0.2 % of the population was identified as SCs. [23]. Along this line, according to Kenmotsu et al. [24] approximately 0.40 % of the pulp cells may be stem cells or side population when they are found in young rats, whereas only 0.11 % is found in adult rats.

Altogether, it may be concluded that only a small number of stem cells are present initially in the dental pulp. The clonogenic stem cells present at an immature stage should rapidly proliferate and produce a larger population, which acquire a terminal phenotype and contribute to tissue regeneration.

16.3 Adult Postnatal Stem Cells

In the field of repair and regeneration of defective tissues and organs, the potential of adult postnatal multipotent and monopotent stem cells has been investigated. The postnatal dental pulp constitutes the main reservoir of cells investigated either in vitro or in vivo in the frame of preclinical animal investigations.

Table 16.1 Origin of adult postnatal stem cells

Stem cells permanently present in the adult tooth
Dental pulp stem cells (DPSCs) isolated from the pulp of permanent teeth
Exfoliated deciduous teeth stem cells (SHEDs) and immature dental pulp stem cells (IDPSC) from deciduous teeth
Apical papilla stem cells (SCAPs)
Stem cells present in periodontal tissues
Periodontal ligament stem cells (PDLSCs)
Dental follicle stem cells (DFSCs), differentiating in odontoblast-like cells or endothelial cells
SM taking origin from other tissues
Induced pluripotent stem cells (iPS)
Hematopoietic stem cells (HSCs)
Neuronal stem cells (NSCs)

Table 16.1 shows the origin of adult postnatal stem cells.

16.3.1 Stem Cells Permanently Present in the Adult Tooth

Adult human dental pulp stem cells (DPSCs) are derived from the pulp of human permanent third molar. They take origin from the neural crest cells (NCCs), which are multipotential cells, migrate, and contribute to different tissues ranging from the peripheral nervous system to the craniofacial skeleton. Pre-migratory cells display plasticity. The cranial NCCs give rise to the majority of the bone and cartilage of the head and face [25].

The relative ease of access of the pulp to autologous use renders attractive the *dental pulp stem cells and/or human exfoliated deciduous teeth stem cells (DPSC/SHED)* as a valid option in regenerative dentistry [26].

In the dental pulp, DPSCs have been found both in adults and in children, in the superficial "cell-rich zone," located beneath the Hoehl's cell layer. They take origin from the neural crest and may differentiate into multiple cell lineages. Under specific stimuli, they differentiate into adipocytes, neurons, chondrocytes, and mesenchymal stem cells.

Adult DPSCs and SHEDs may regenerate a dentin-pulp-like complex. DPSCs form sporadic densely calcified nodules in vitro. When the stem cells are transplanted with hydroxyapatite/tricalcium phosphate into immunocompromised mice, a dentin-like structure is formed within 6 weeks, lined by odontoblast-like cells, with cellular processes into a mineralized matrix [27, 28]. Furthermore, SHEDs are able to induce differentiation into bone-forming cells. SHEDs have a higher proliferation rate compared with DPSCs of permanent teeth. They are easy to expand in vitro. They have high plasticity and may differentiate into neurons, adipocytes, osteoblasts, and odontoblasts [29].

Compared with human *bone marrow stromal stem cells (BMSSCs)*, known as osteoblast precursors, DPSCs share similar phenotype in vitro, but they produced only sporadic densely calcified nodules. They did not form adipocytes, whereas BMSCs form adherent cell layer with clusters of lipid-loaded adipocytes. DPSCs transplanted into immunocompromised mice generated a dentin-like structure lined with human odontoblast-like cells surrounding a pulp-like structure. BMSCs formed lamellar bone-containing osteocytes and surface-lining osteoblasts. Therefore, DPSCs have the ability to form a dentin-pulp-like complex [30].

Other groups of cells issued from dental tissues may contribute as a source of odontoblasts; this is the case for the *stem cells from the apical papilla (SCAP)*. According to Huang et al. [31], these cells seem to be responsible for the continuous formation of the root dentin. They are involved in "bioroot engineering."

16.3.2 Stem Cells Taking Origin from Periodontal Tissues and from Other Tissues

In addition, mesenchymal cells originating from the dental surrounding tissues may differentiate and be used for tooth regeneration. *Stem cells of the apical part of the papilla (SCAPs), stem cells from the dental follicle (DFSCs), periodontal ligament stem cells (PDLSCs), and bone marrow-derived mesenchymal stem cells (BMSCs)* may be also a source of stem cells that may be implicated in tooth regeneration [32]. Perivascular multilineage progenitor cells and migrating mesenchymal stem cells may contribute to pulp regeneration. *Adipose-derived stromal cells (ADSCs)* contain a group of pluripotent mesenchymal stem cells. These cells manifest multilineage differentiation capacity. They exhibit stable growth and proliferation kinetics in vitro. They express bone marker proteins including alkaline phosphatase, type I collagen, osteopontin, and osteocalcin and produce mineralized matrix. Therefore, dental and non-dental stem cells share the following properties: high proliferation rate, multi-differentiation ability, and easy to be induced. They may be instrumental in tissue-engineered organ replacement [29, 30, 33–36].

16.4 Niches of Stem Cells

Stem cells are located in particular microenvironments known as niches. Schofield introduced the concept of a stem cell "niche" in 1978 [37]. The niche encompasses all of the elements immediately surrounding the stem cells, which are in their naïve state. This imaginary space includes the non-stem cells, which are in direct contact with the stem cells, as well as ECM and soluble molecules found in that location [38].

Stem cell behavior is regulated by a local microenvironment referred as "the stem cell niche" characterized by three essential properties: (1) A niche provides an anatomic space where the number of stem cell is regulated; (2) it is the place where stem cells are instructed to control the maintenance, quiescence, self-renewal, and recruitment toward differentiation, fate determination, and long-term regenerative capacity; (3) the niche will influence the cell motility [39]. Hallmarks of a niche include the stem cell itself, stromal supporting cells that interact directly with the stem cells via secreted factors, and cell surface molecules. The extracellular matrix (ECM) provides a structural support to the niche and allows the diffusion of mechanical and chemical signals. Systemic signals are implicated in the recruitment of inflammatory and circulating cells into the niche.

In the teeth, as in the adult blood system, multiple niches may exist; however, specific markers allowing the definitive identification of stem cells within the pulp are still lacking. Specific niches have been identified in the pulp. Niches provide unique microenvironments that regulate stem cells. Most investigations are related to the basal forming part of the rodent incisor, a situation that does not apply to the teeth of limited growth as they are found in humans.

In the rodent incisor, the production of progenitor cells is under the control of molecular signals such as Notch1, Lunatic fringe, and fibroblast growth factor-10 expressed in the apical bud, both in the surrounding mesenchyme and in the epithelium [40]. FGFR2 signaling axis maintains the stem cell niche required for incisor development and lifelong growth [41].

In the incisor, the term "apical bud" indicates a specialized epithelial structure, whereas in the guinea pig molars, which are continuously growing teeth, specific proliferative regions were identified, implicated in the formation of "apical bud" at the apical end [42].

The neuronal marker expressions of undifferentiated DPSCs are α-III-tubulin, S100 protein, and synaptophysin. A subset of the population showed a positive immunolabeling for galactocerebroside, neurofilament, and nerve growth factor receptor p75. Immunolabeling of young dental pulp tissue demonstrated the presence of a perivascular cell niche and a second stem cell niche at the cervical area. In adult dental pulp only the perivascular niche could be observed [43].

Lizier et al. [20] compared stem cells at early passages (2–5 passages) to a late population (6 months after several transfers). No morphological difference was noted between the two periods. It was the same for the expression of stem cell markers (nestin, vimentin, fibronectin, SH2, SH3, and Oct ¾ and chondrogenic, myogenic, and differentiation potential). Bromo-2′-deoxyuridine (BrdU)-positive cells were observed in the central part of the dissected pulp after 6 h, whereas after 72 h, BrdU-positive cells increased in number and were located at the pulp periphery. This suggested a migration of labeled cells from the central pulp toward the outer pulp border. Multiple niches were identified expressing nestin, vimentin, and Oct3/4, while STRO-1 protein localization was restricted to perivascular niche.

Perivascular niches were identified as expressing STRO-1-positive markers, identical to human bone marrow stromal stem cells (BMSSCs) and dental pulp stem cells (DPSCs). They were negative for von Willebrand factor and positive for α-smooth muscle actin and CD146. DPSCs expressed the pericyte marker 3G5 [44]. However, this localization is still controversial, and the lack of perivascular immunolabeling with an α-smooth actin antibody does not corroborate such assumptions [45].

In the primate molars, which are the teeth of limited growth, cervical class V cavities were prepared. After pulp exposure, pulp capping was carried out. The monkey received a single injection of 3H-thymidine. Six, 8, and 12 days after the treatment, the teeth were extracted and the number of cells labeled after radioautography was scored. Labeled cells were seen first in the deeper central pulp. Two cell replications were dependent on the timing of the injection of tritiated thymidine. Injected 84 and 96 h after capping, for the shorter period cells were radiolabeled in the central pulp, whereas the second injection occurring after a longer period of time was mostly located beneath the Hoehl's cells and odontoblast layers. It was shown that odontoblasts are clearly postmitotic cells, never labeled after such procedure. Replacement cells differentiate into functioning odontoblast-like, and a continuous influx of differentiating cells was evidenced. They have initially replicated their DNA in the deep zone of the pulp. Then they migrate toward the terminal migration point, beneath the wounded odontoblasts, where the terminal duplication occurred. Hence, fibroblast-like cells, perivascular cell population, and progenitor cell are candidate to participate to the continuous influx of new candidates as stem cells [46].

Although this pioneer publication shed some light on the renewal and migration of cells within the dental pulp, it is noteworthy that (1) only two time points were taken into account, and therefore a more detailed kinetic study is still needed; (2) the cell migration phenomenon was partially elucidated in the crown, (3) but this was not the case for the root.

16.5 The A4 Clone Induces "In Vivo" the Formation of Reparative Dentin

Our aim was to explore for the first time the potential of pulp-derived "stem" cells to improve dental repair upon injury. In order to characterize the dental pulp progenitors and elucidate the molecular and cellular mechanisms governing their differentiation, we used clonal cell lines obtained in our laboratory from tooth germs of day 18 mouse embryos transgenic for an adenovirus-SV40 recombinant plasmid (pK4) [47]. Among the 50 clones obtained from the molar dental pulp, the clone named A4 represents a homogenous population of precursor cells which display in vitro properties of multipotent mesoblastic cells [3], which after differentiation were shown to mineralize.

Moreover, our in situ approach was essential since the definition of a "progenitor" implies an attempt to outline its in vivo potential into a repair/regeneration context.

16.5.1 In Vivo Experimental Cell Implantation

16.5.1.1 Implantation into the Mouse Incisor

In an in vivo pilot study, the A4 cells were implanted in the mouse mandibular incisor, and this led to the formation of a mineralized osteodentin within the dental pulp. Since a mouse incisor is very small and its diameter is about the same as the smaller dental burr, lateral perforation occurred establishing uncontrolled and undesirable contact between the dental pulp and the periodontal environment. The incisor is a continuously growing tooth. As a result from the continuous tooth eruption, the damaged tissue was moved forward. Serial sections allowed identifying a newly formed dental tissue in the basal part of the incisor, the lesion area itself and what was formed in the distal part of the tooth before the pulp exposure. As the results were inconclusive, it was necessary to choose another model closer to the human tooth with a limited growth. This is why another set of experiments was carried out on rat's molar.

16.5.1.2 Implantation into the Rat Molar

The small size of the mouse molars makes such experimental procedures difficult, whereas the rat molar provides an easier experimental approach involving the preparation of a cavity on the mesial aspect of the teeth. The initial protocol was carried out on the first maxillary molar and used to study

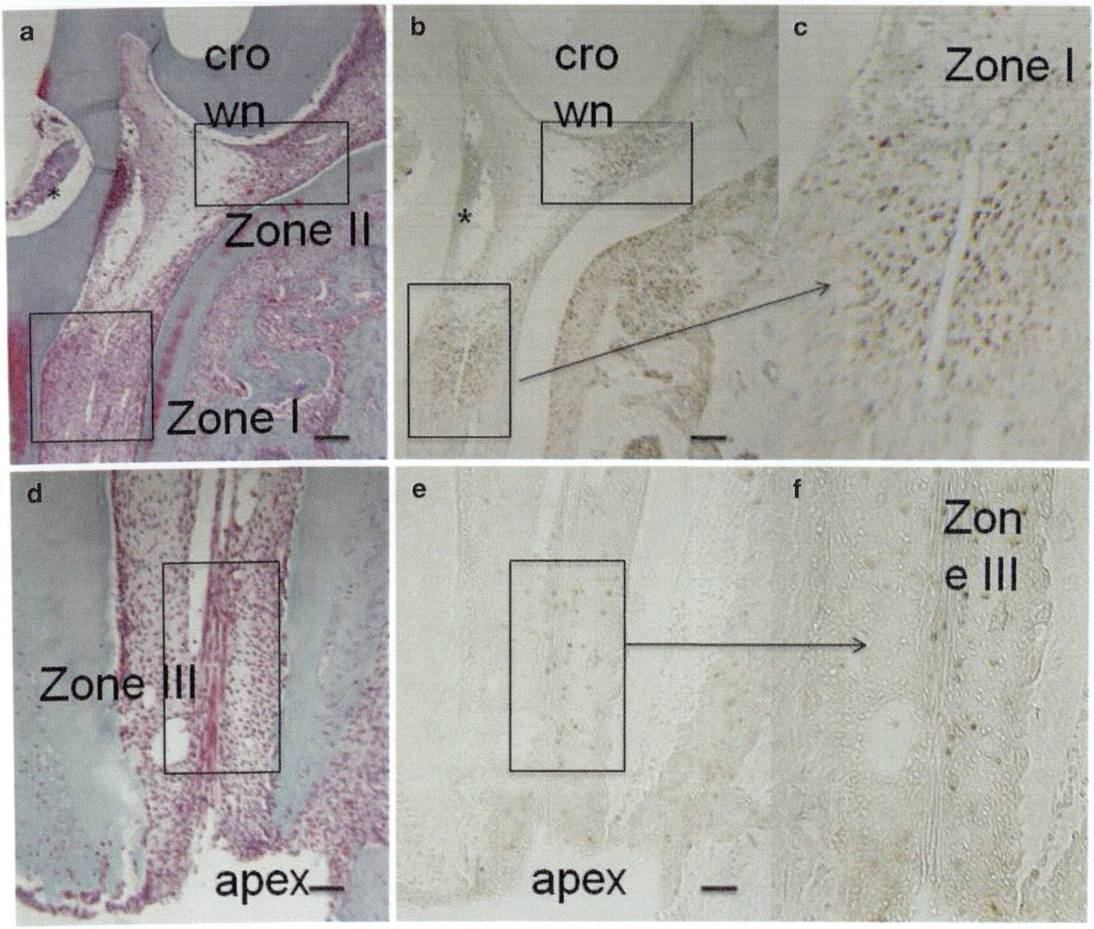

Fig. 16.1 (**a**, **d**) Hematoxylin/eosin. After the preparation of a cavity in the mesial aspect of the first mandibular molar, after proliferating cell nuclear antigen (*PCNA*), three zones were identified. Two zones in the coronal part, one near the cavity (zone I) and another in the isthmus between the mesial and central parts of the pulp (zone II). The zone III was located apically near the root ending. (**b**, **c**, **e**, **f**): PCNA staining. * is the place where the cavity was drilled

the successive events leading to the formation of reactionary dentin [48]. We modified slightly the protocol, and we too used the first maxillary molar of rat to explore the feasibility of a pulp capping therapy after a surgical exposure. Bioactive molecules or molecular fragments or pellets of stem cells were implanted within the dental pulp [49]. Agarose beads 75–150 μm in diameter were used as carrier.

16.5.1.3 Materials and Methods

Following institutionally approved protocols for animal experimentation research, the rats were anesthetized by intraperitoneal injection of 20 % ketamine (Imalgene; Alcyon) and 5 % xylazine (Rompun; Alcyon) solution. A cavity was drilled on the mesial aspect of the tooth using a tungsten dental bur (ISO 006; Dentsply) with a low-speed handpiece. Pulp perforation was accomplished by pressure with the tip of a steel probe. This method avoids rolling the pulp around the dental bur. The surgical approach was followed or not by agarose bead implantation. The beads were loaded or not by extracellular matrix molecules or solely acting as ECM or cell carriers. They were implanted into the pulp exposure of the first maxillary molar of Sprague-Dawley rats, 9 weeks old. About 10^5 A4 multipotent stem cells were collected in a tube and centrifuged to form a pellet, which was subsequently implanted into the pulp of rat's molars (Figs. 16.1a–f and 16.2a–f).

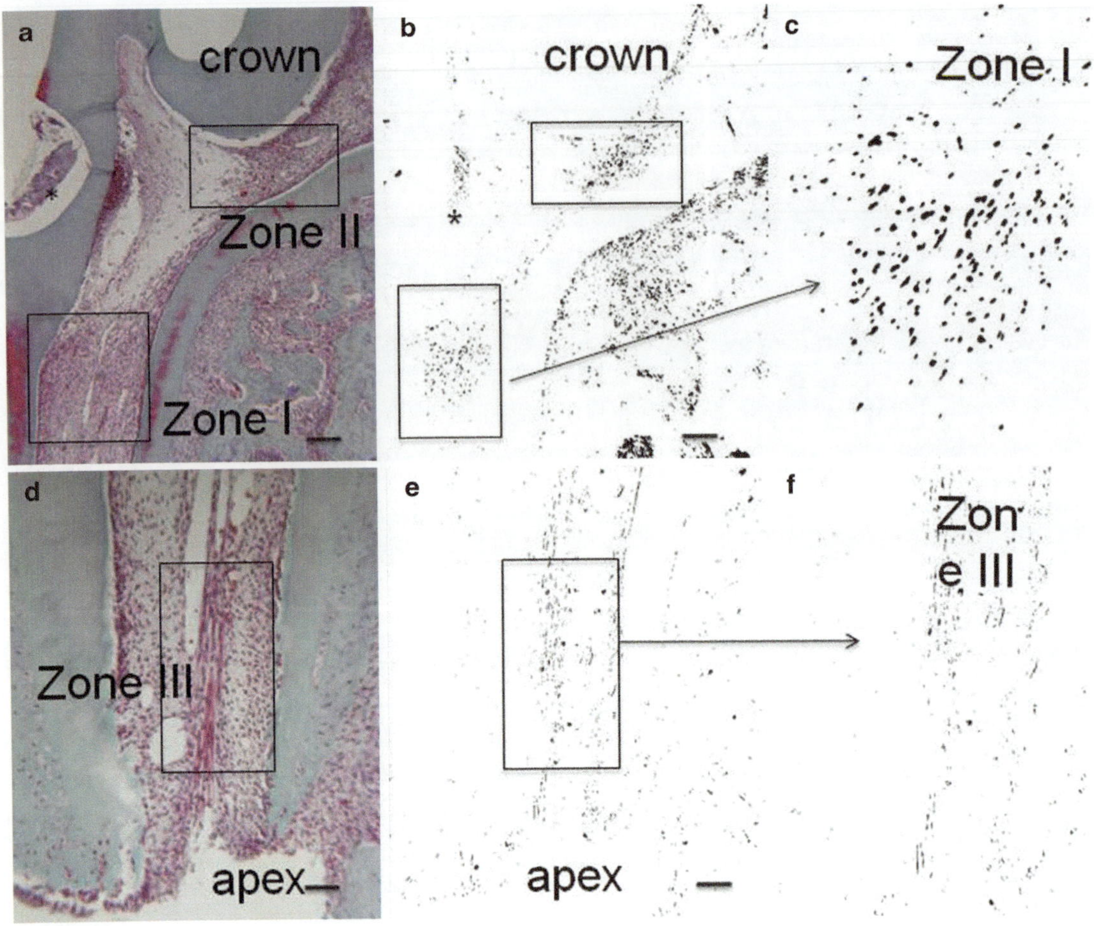

Fig. 16.2 (**a**, **d**) Hematoxylin/eosin. (**b**, **c**, **e**, **f**): negative staining of the same labeled sections. * is the place where the cavity was drilled

In the *control group* of rats, cavities were not prepared. At the end of tooth morphogenesis, odontoblasts and the subodontoblastic Hoehl's cells are postmitotic cells. Accordingly, in sound teeth, after immunostaining with the proliferating cell nuclear antigen (PCNA) or with the silver-binding nucleolar organizer region (AgNOR), odontoblasts remain unlabeled. In physiological conditions, no proliferating cells are found in the human coronal pulp and only a low proliferative activity is confined to the radicular pulp [50]. Upon advanced dental caries, PCNA-stained cells become detectable in the coronal pulp [51]. After a dental lesion, the staining seen after 5-bromo-2′-deoxyuridine (BrdU) labeling disappears from the pulp-dentin border, and a wide range of proliferative cells is seen within the dental pulp. The pulp of a few animals was exposed but not implanted. The three other groups of rats are indicated in Table 16.2.

The pulps were either exposed but not implanted (*control groups*) or implanted with unloaded agarose beads (*sham groups*) or with agarose beads loaded with bioactive ECM molecules (A+4, A-4, or LRAP amelogenin peptides) or with pellets of A4 cell lines. Agarose beads loaded by one amelogenin peptide A-4, A+4 (generous gifts from Prof. Art Veis, Northwestern University, Chicago) [52], or LRAP fragment (generous gift from Prof. Pamela Denbesten, University of California at San Francisco) were implanted in the exposed dental pulp (*experimental group with bioactive molecule implantation*).

To protect the pulp from bacterial contamination, after the surgery, the cavities were filled with glass ionomer cement (Fuji IX; GC). One, 3, 8, 15, or 30 days after the preparation of cavities alone, or after pulp implantation of unloaded or bioactive ECM molecule-loaded agarose beads, rats were anesthetized and killed by intracardiac perfusion of the fixative solution (4 % paraformaldehyde solution overnight at +4 °C). For each subgroups, three rats were killed for each period of time. Block sections including the three hemi-maxillary molars were dissected, fixed by immersion, demineralized in a 4.13 % EDTA, and embedded in paraffin (Paraplast Plus; Kendall).

Table 16.2 The three groups of rats: sham group and two experimental groups

The sham group: A surgical pulp exposure was performed in rat molars. They were implanted with agarose beads, loaded or not with ECM molecules or A4 cells. One to 3 days after bead implantation, the pulp tissue appeared normal, without inflammatory process. In the absence of A4 cells, inflammatory cells were recruited near the implantation site. One month after the surgery, the mesial pulp chamber was fibrotic, but was never mineralized
Two *experimental groups* were prepared 1. *The experimental A4 implantation group*: After 1 month, the mesial part of the pulp chamber was massively filled by mineralized osteodentin, separated from the previously formed dentin by a calciotraumatic line
2. *Implantation of agarose beads loaded with or without A4 cells*: In order to improve the implantation protocol, i.e., precisely visualize the implantation site and control the number of cells implanted within the pulp, we used agarose beads as a carrier

Seven thick sections were collected on glass slides and observed after Masson's trichrome staining (Fig. 16.3a, b). Other slides were immunostained with the proliferative cell nuclear antigen (PCNA), and all the mitotic cells were labeled. We used also antibodies raised against CD68 used to label monocytes/macrophages, MHCIIB identifying mainly dendritic cells, and α-smooth muscle actin (α-SMA) [53]. As control, we omit the primary antibody and employ only the secondary antibody. Western blots were used to ascertain the specificity of the antibodies. Using antibodies against actin to label microfilaments

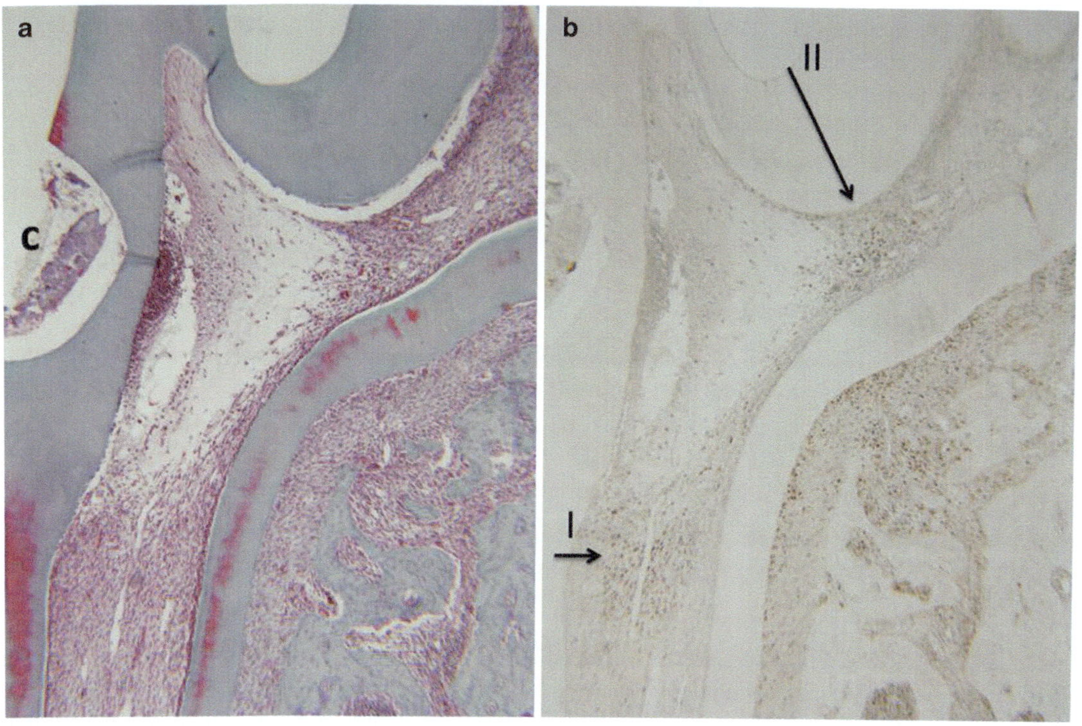

Fig. 16.3 (**a**) Hematoxylin/eosin staining. (**b**) PCNA labeling of zones I and II. *C* cavity

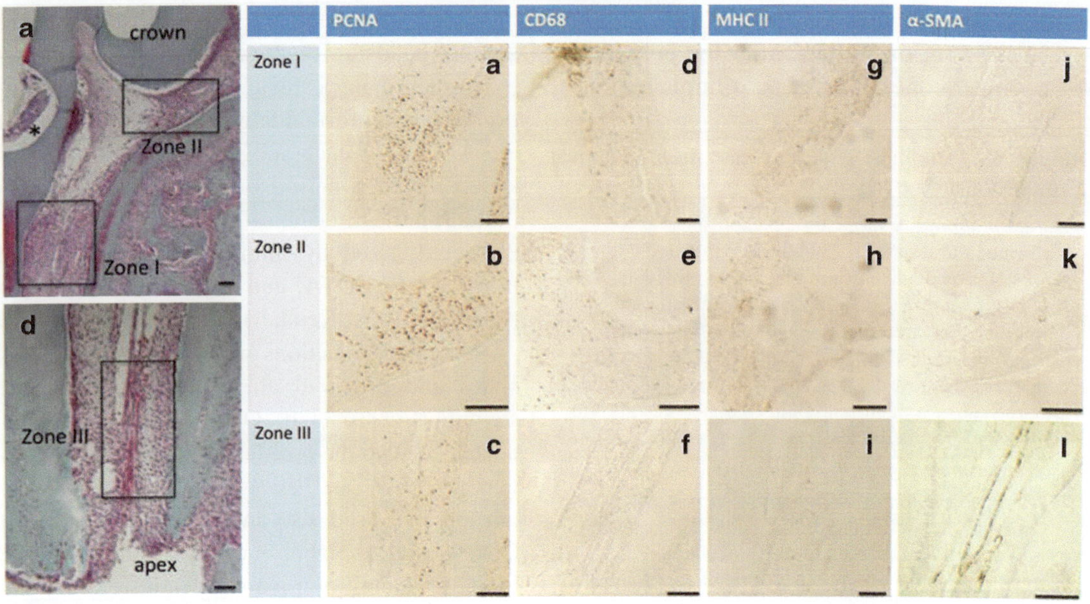

Fig. 16.4 (**a–l**) First column PCNA (the three zones are labeled), second column CD68, and third column MHC II (a few cells are labeled in zones I and II; in zone III MHC II alone label this zone): inflammatory cells; fourth column: alpha-smooth muscle actin (α-*SMA*) endothelial vascular cell (only zone III is labeled)

and alpha-acetyl tubulin revealed microtubules and cilia (Figs. 16.2a–f and 16.4a–l).

Ultrathin sections of rat's molars were obtained after rapid freezing followed by freeze-substitution, and replicas were prepared after freeze fracture. Junctional complexes were identified by electron microscopy in the pulp of rat molars. In the extracellular matrix, thin collagen fibrils, glycosaminoglycans, and proteoglycans were visualized.

16.5.1.4 Results

When no cavity was prepared or prior to a pulp exposure, no PCNA positively labeled pulp cell was detectable, both in the crown and in the root. Consequently, cell divisions appeared to be associated with a repair process during the formation of reactionary dentin. The mitotic reaction was enhanced and produced reparative dentin mostly after a pulp exposure.

Forty-eight hours after a pulp exposure of the first maxillary molar, three different PCNA-labeled areas were observed in the dental pulp [45].

Zones I and II were located in the coronal part of the molar. Zone I was facing the pulp exposure in the mesial part of the crown, whereas zone II was located in the isthmus between the mesial and the central pulp chamber (Fig. 16.3a, b). Both zones presented large clusters of PCNA-labeled cells. Inflammatory processes may contribute to the cell recruitment in these two zones, as inflammatory cells labeled with CD68 or MHCIIB antibodies are located also in the same territories as PCNA-positive cells (Fig. 16.4a–l). This pointed out also at the potential diffusion of mediators implicated in stem cell commitments and proliferation. Of note, in the crown, endothelial and perivascular cells revealed by alpha-smooth actin immunostaining were barely detectable (Fig 16.5).

Zone III was found in the apical part of the mesial root indicating that a recruitment of mitotic cells takes place some distance away from the lesion site (Figs. 16.6a, b, 16.7, 16.8, 16.9, and 16.10a, b). Anatomically, in the apical part implicated in root formation, three distinct zones have been identified: the apical cell-rich zone, the apical papilla mesenchyme, and the radicular dental pulp [54]. It is noteworthy that no PCNA-labeled cell has been detected in the

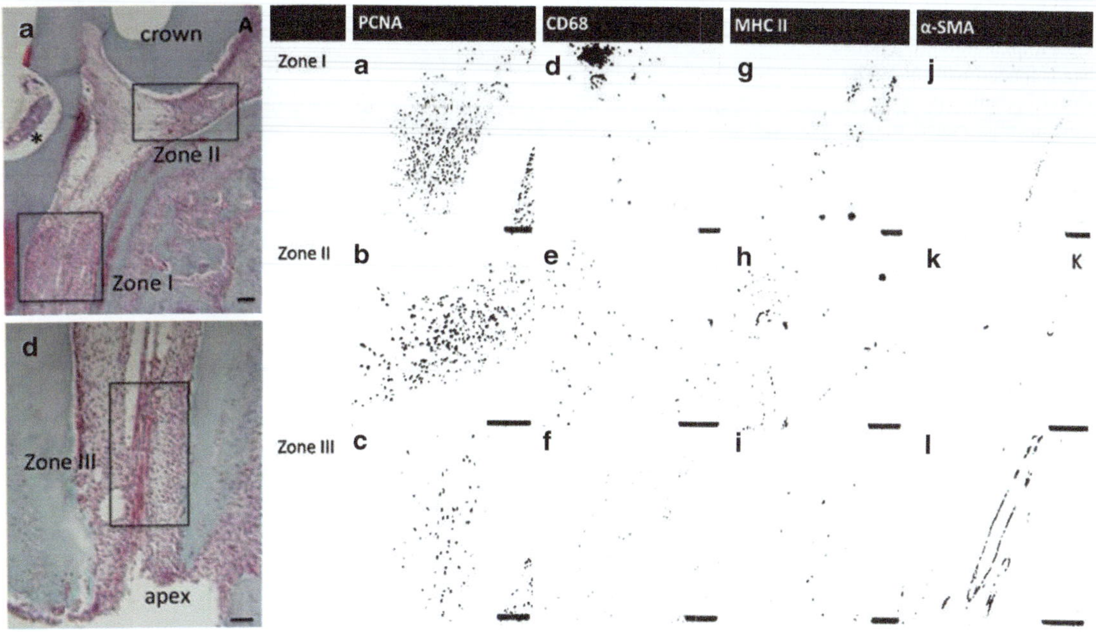

Fig. 16.5 (**a–l**) First column PCNA (the three zones are labeled), second column CD68, and third column MHC II (a few cells are labeled in zones I and II; in zone III MHC II alone label this zone): inflammatory cells; fourth column: alpha-smooth muscle actin (α-*SMA*) endothelial vascular cell (only zone III is labeled). * is the place where the cavity was drilled

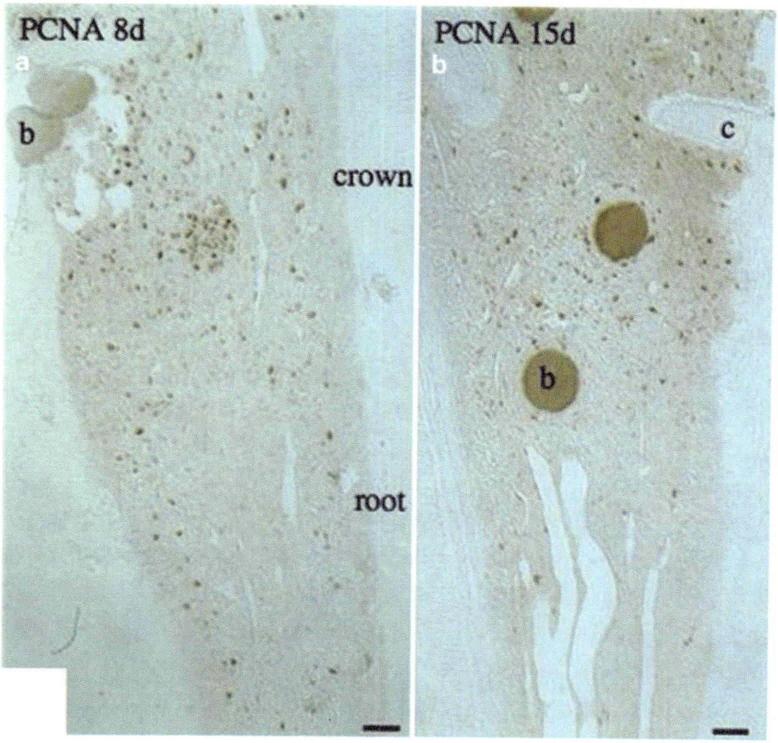

Fig. 16.6 PCNA labeling after the implantation of agarose beads loaded with A-4. (**a**) At day 8 after implantation, labeled cells are present in the root at the periphery of the pulp, beneath the odontoblast/Hoehl's cell complex. Many cells load the crown part. (**b**) At day 15 after implantation, fewer cells a re labeled. No labeling is detectable in the root and most of the labeled cells are located in the coronal part of the pulp. Beads (*b*), cavity (*c*)

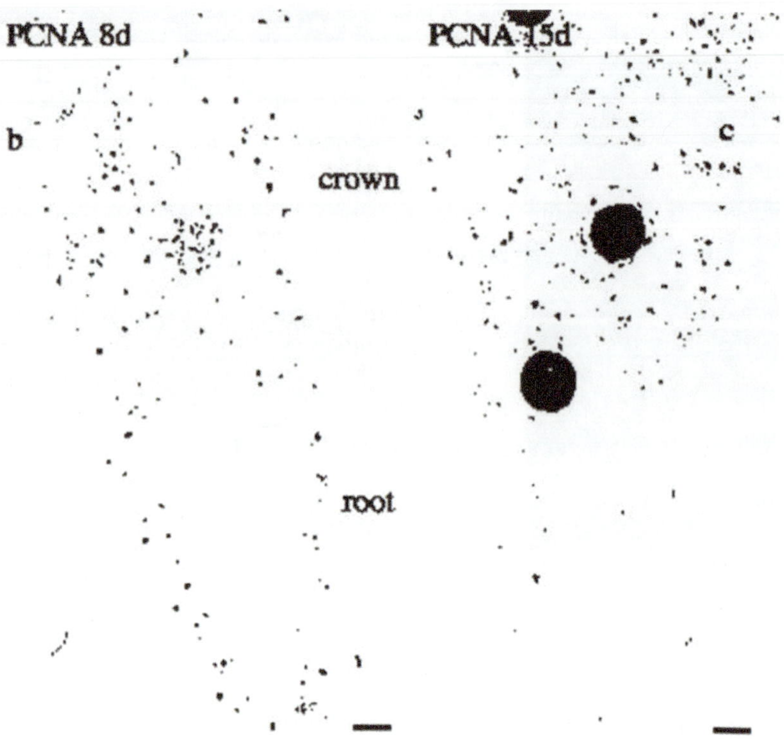

Fig. 16.7 PCNA labeling after implantation of agarose beads loaded with A-4. The shift of labeling is more obvious after negative staining compared with the positive staining. This suggests that cells divide in the root and migrate from the root to the coronal part. Beads (*b*), cavity (*c*)

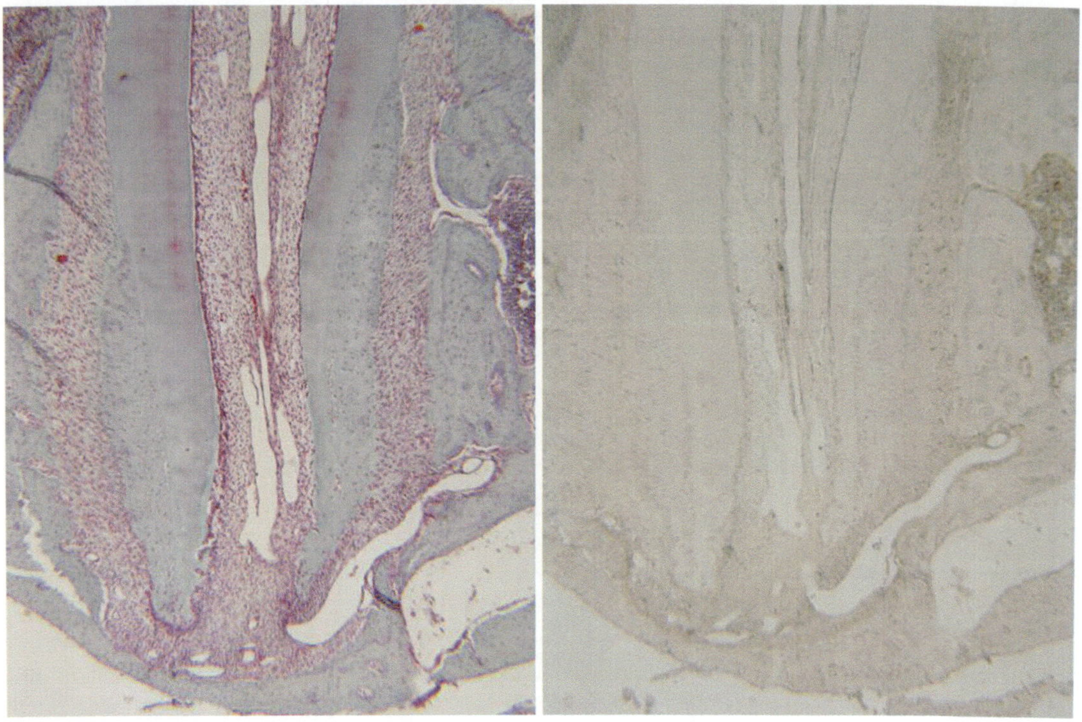

Fig. 16.8 (**a**, **b**). Implantation of bead alone establishes that no PCNA labeling is seen in the dental pulp, in contrast with the periodontal ligament where PCNA-positive cells are always present (cell turnover)

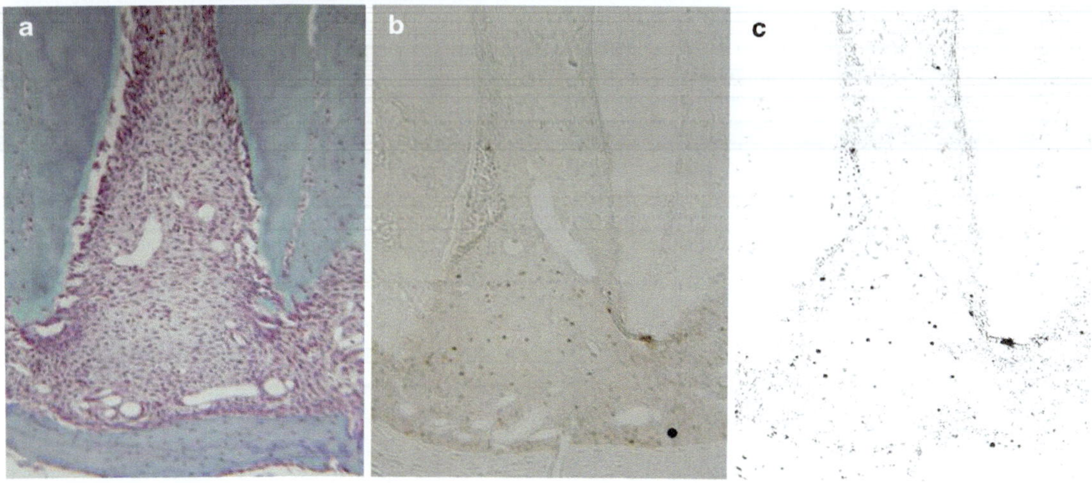

Fig. 16.9 (**a–c**) After the preparation of a cavity followed by PCNA labeling, hematoxylin/eosin labeling allows visualization of the apical part of the pulp. Positive and negative staining establishes the presence of labeled cells in the apical papilla mesenchyme and in the cell-rich zone

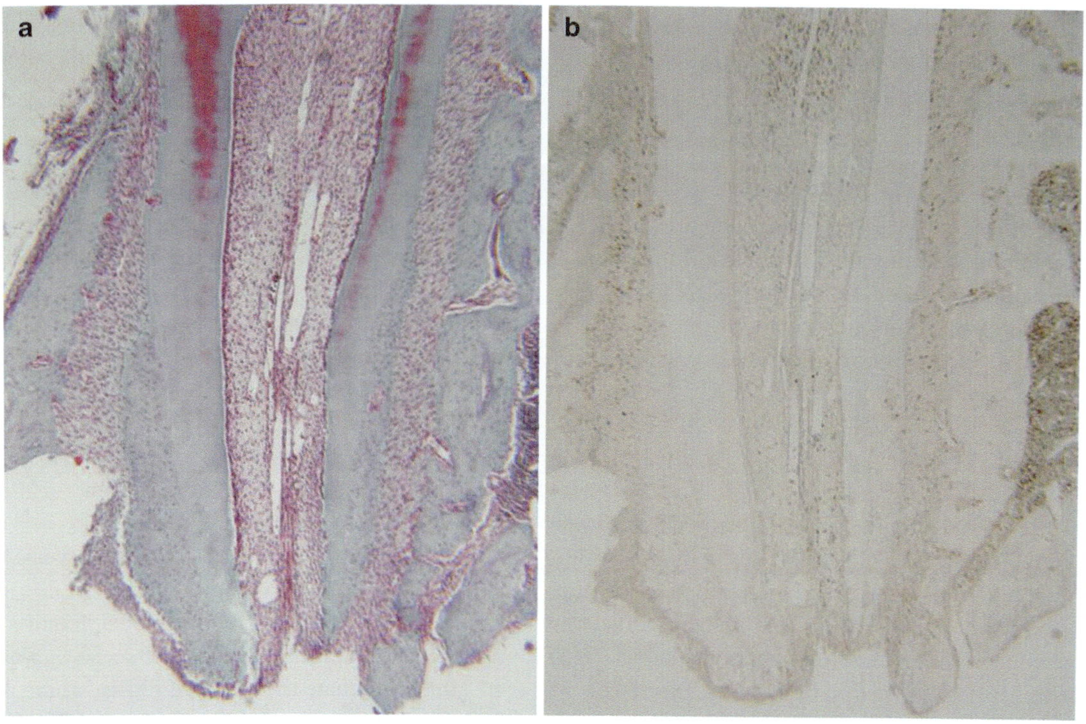

Fig. 16.10 Implantation of LRAP-loaded beads (**a**) and PCNA (**b**) immunostaining shows labeling in the radicular dental pulp after 2 days

apical zone of molars in the absence of pulp lesion, in contrast with the periodontal ligament where mitotic cells were clearly labeled. The high capacity for proliferation detected in these areas after pulp exposure indicates that the apical cells may be a source of stem cells implicated in pulp regeneration [31]. The preliminary data presented here on the apical niche suggest that pulp

Table 16.3 Density of PCNA-positive cells/cm^2 scored at day 3 in the apical cell-rich zone and in the root pulp, in the control, sham, and experimental groups using bioactive amelogenin peptides (three rats per subgroup)

Zone III	Apical cell-rich zone	Root pulp
Control (pulp exposure without bead implantation	3.4	0.43
Sham (pulp exposure and implantation of unloaded bead)	2	1.33
Experimental group: implantation of bioactive molecule in the pulp [52]	1	4.23

cells are mobile and perform a double dynamic sliding. Further experiments will be required for a statistically significant analysis. Variations in the number of labeled cells scored in the three different sub-compartments of the apical zone supported the sliding hypothesis. Some pulp cells are committed in the apical cell-rich zone where they divide and, during the first 1–3 days following implantation, migrate toward the root pulp. The reasons for this ascending sliding are not yet understood. Chemotaxis may play a crucial role in this event. The diffusion of inflammatory molecules released in zones I and II may influence the adhesion gradient (also named haptotaxis) and/or the cell migration in relation with the stiffness gradient of the ECM (durotaxis) [55].

Here, we focus our attention on zone III to shed some light on the potential contribution of this area during tooth repair. One to 3 days after pulp exposure, PCNA-positive cells were found in the central part of the radicular pulp (Table 16.3). In the control group, most stem cells were committed in the apical cell-rich zone as the density of PCNA-positive cells was sevenfold higher in this area compared to the root pulp. The distribution of PCNA-labeled cells was changed after bead implantation (Table 16.3) (Fig. 16.11a–h). In the sham group, the density of PCNA-positive cells in the apical cell-rich zone was only 1.5-fold higher than in the root pulp. After implantation of ECM bioactive molecules, we observed an increase of PCNA-labeled cells in the radicular dental pulp compared to the apical papilla mesenchyme (Table 16.3 and Fig. 16.1a–f). After agarose bead implantation, the labeling was decreasing in the apical cell-rich zone, whereas an increased density was found in the root pulp. These differences of PCNA-positive cell repartition between control and loaded and unloaded bead-implanted groups suggest that the cell recruitment is controlled by the pulp exposure, a phenomenon that may be related to the cytokines discharged during the burst of inflammatory molecules released by zones I and II.

In the zone III, endothelial and perivascular cells labeled with alpha-smooth actin did not overlap with PCNA-positive cells indicating that mitotic cells were located in the pulp mesenchyme. In contrast to zones I and II, no positively immunostained cells were detected in zone III using antibodies raised against inflammatory molecules such as CD68 and MHCII. This diverging immunolabeling between the coronal zones I and II and the apical zone III suggests that in the root pulp, the cell recruitment is independent from the inflammatory process. However, we cannot rule out the potential chemotactic effects of molecules released in zones I and II on cell recruitment in the apical part.

Eight days after implantation of agarose beads loaded with bioactive amelogenin peptides, we did not observed PCNA labeling in the central part of the root pulp. Most of the PCNA-positive cells were located along the subodontoblastic radicular border. These results suggest that 8 days after a pulp lesion, positively labeled PCNA cells had migrated from the center toward the periphery of the radicular pulp. Two weeks following pulp exposure, PCNA-positive cells were not detectable in the root and were present in the coronal pulp, near the surgery site.

Electron microscopy analysis also revealed that a loose network of thin collagen fibrils formed the ECM. After conventional fixation, proteoglycan appeared as electron-dense aggregates that punctuate the collagen fibrils, whereas after physical fixation avoiding artifactual shrinkage, the occurrence of an expanded gel-like glycosaminoglycan-rich structure was recognized, favoring cell sliding [56].

Doing so, we get the first evidence that implantation of pulp precursor cells within the pulp promotes the formation of a dentin barrier separating the residual pulp from the reparative area.

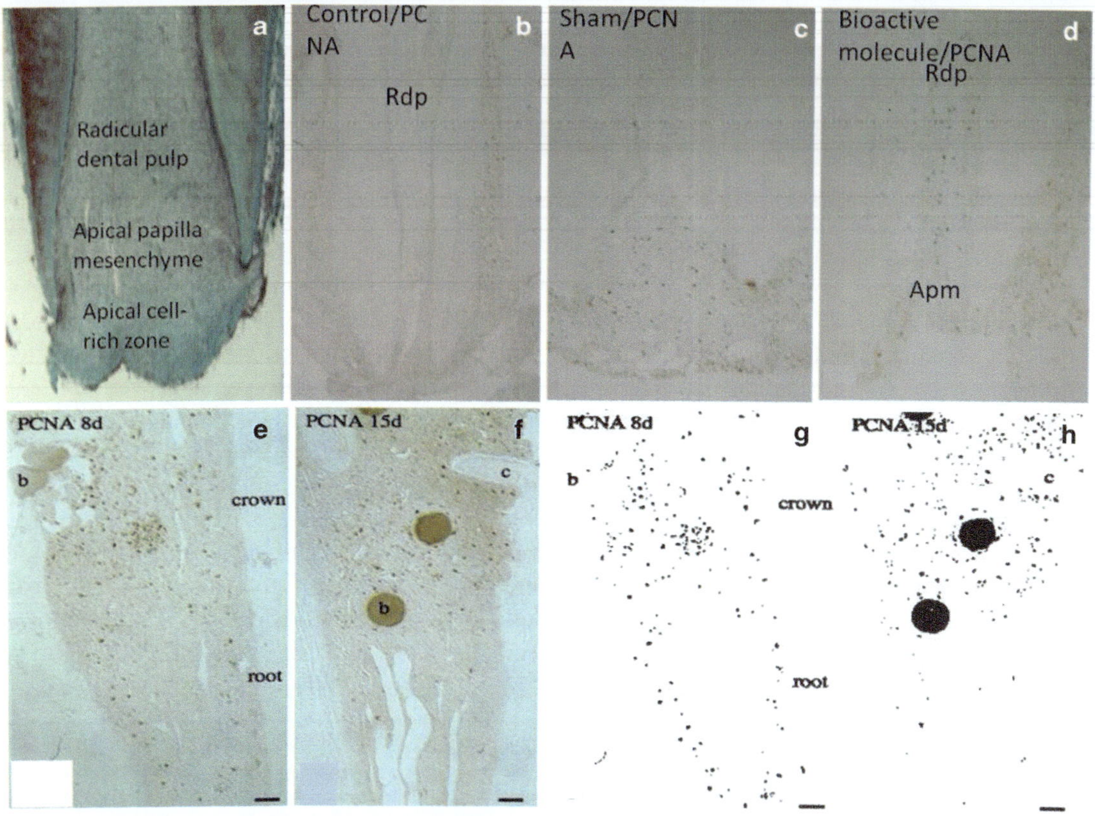

Fig. 16.11 The three apical zones display labeling according to the bead implantation. Controls do not reveal any pulp labeling, whereas in the sham group cells located in the cell-rich zone are labeled (**a**, **b**). Labeling of the apical papilla mesenchyme occurs when the sham animals are analyzed (**c**). Implantation of bioactive molecules shows labeling in the radicular dental pulp (**d**). The migration from the root to the crown continues between 8 and 15 days (**e**–**h**). Beads (*b*), cavity (*c*)

16.5.1.5 Discussion

The ability of the pulp to respond to a variety of pathological conditions and injuries by the deposition of reparative dentin by pulp stem cells is now well admitted, even though the localization and the identity of these pulp cells are largely unknown. After pulp exposure of rat molar in addition to proliferating cell nuclear antigen (PCNA) immunostaining, we identified an area of mitotic cells in the root. The PCNA-positive pulp radicular cells may migrate from the central part of the radicular pulp to the subodontoblastic cell layer and then from the apical root to the crown.

Along these lines of evidence, we have previously shown that after prostaglandin and leukotriene inhibition consequential to an essential fatty acid deficient diet (EFAD) [57], pulp cells migrate from the central part of the crown toward the lateral subodontoblastic area, where they underwent apoptosis [58]. This pointed out at equilibrium between the emergences of new pulp cells located centrally and their destruction afterward in the outer pulp border. Although the mechanisms reported here are almost related to a heterogeneous population of pulp cells, the observations carried out on PCNA-labeled cells, which are presumably pulp stem cells, lead to similar conclusions on a possible cell sliding.

Ultrastructural observations establish that a dense network of cytoskeletal connects pulp cells. Gap junctions are implicated in intercellular communications and in the electrotonic cell-to-cell transfer of low molecular weight coupling molecules. Desmosome-like junctions play a mechanical role in maintaining firmly the morphological integrity between cells and preventing eventual changes in shape. Actin microfilaments are distributed along the inner surface of the plasma membrane of pulp cells contributing to the plasmalemmal undercoat and are grouped near adherens junctions. This organization implies that pulp cells form a continuum, reported as a syncytium-like structure [59]. Consequently, the cells cannot slide individually, but rows of cells firmly bound migrate altogether. The experiments reported here suggest that some pulp cells committed in the apical cell-rich zone proliferate and migrate toward the root pulp.

The junctional complex binding together pulp cells observed by electron microscopy may contribute to what has been named plithotaxis, implying a collective cell migration [55]. This new concept refers to a sliding effect that needs to be taken into account to promote pulp regeneration. The clinical implications are alternatively the following: (1) either implantation of calcium hydroxide or molecules in the coronal part (conventional pulp capping) leads to the formation of a dentinal bridge followed by pulp healing (2) or cells or molecules taking origin in the apical part of the pulp penetrate the ECM network and regenerate the damaged dental pulp. This alternative strategy invalidates pulp capping or pulpotomy. Pulp cells may recolonize gradually a natural or artificial scaffold or eventually a residual pulp. This process initiated in the apical part may regenerate the dental pulp until completion in the crown. Recolonization of a scaffold by apical stem cells may be an efficient approach, less traumatic than pulp capping. It leads to the healing and regeneration of the dental pulp, at least if the apical part of the root is widely open as it was here in the case. These results need to be reinvestigated after apical closure.

The new emergent and dynamic concept highlights the sliding of pulp cell in the pulp. This is occurring first inside the root canal, cells moving from the center of the pulp to the periphery, and secondly from the root toward the crown. These approaches shed light on the apical niches of stem cells present in molars, and they seem to provide new tools for future regenerative pulp therapies.

The stem cells appear as tools to get a better understanding of the cellular mechanisms of pulp repair. They display innovating potentials in dental therapies. The present results indicate also for the first time that the direct implantation of mouse progenitor cells in the dental pulp of a rat molar leads to the formation of reparative osteodentin.

It is important to determine (1) whether precursor cells reintroduced in a pulp "natural" environment differentiate into osteo-odontogenic cells or (2) whether the implanted cells recruit resident pulp stem cells toward osteo-odontogenic differentiation and indirectly promote the formation of the dentinal bridge. The direct or indirect implication of implanted stem cells in reparative dentin formation remains yet unanswered.

In conclusion, our preclinical experimental approach has identified three different locations of niches of stem cells (two coronal niches and one radicular niche). It paves the way for the development of cellular therapies after pulp injury. The long-term goal should provide new clinical strategies to restore the functionality of an injured tooth by using pulp stem cells.

References

1. Da Cunha JM, Da Costa-Neves A, Kerkis I, Da Silva MC. Pluripotent stem cell transcription factors during human odontogenesis. Cell Tissue Res. 2013;353(3):435–41.
2. Jo YY, Lee HJ, Kook SY. Isolation and characterization of postnatal stem cells from human dental tissues. Tissue Eng. 2007;13:67–73.
3. Lacerda-Pinheiro S, Dimitrova-Nakov S, Harichane Y, Souyri M, Petit-Cocault L, Legrès L, Marchadier A, Baudry A, Ribes S, Goldberg M, Kellermann O, Poliard A. Concomitant multipotent and unipotent dental pulp progenitors and their respective contribution to mineralised tissue formation. Eur Cells Mater. 2012;23:371–86.
4. Ranganathan K, Lakshminarayanan V. Stem cells of the dental pulp. Indian J Dent Res. 2012;23:558.
5. Ouellet J, Barral Y. Organelle segregation during mitosis : lessons from asymmetrical dividing cells. J Cell Biol. 2012;196:305–13.
6. Laino G, d'Aquino R, Graziano A, Lanza V, Carinci F, Naro F, Pirozzi G, Papaccio G. A new population of human adult dental pulp stem cells: a useful source of living autologous fibrous bone tissue (LAB). J Bone Miner Res. 2005;20:1394–402.
7. Friedenstein AJ, Chailakhjan RK, Lalykina KS. The development of fibroblast colonies in monolayer cultures of guinea-pig bone marrow and spleen cells. Cell Tissue Kinet. 1970;3:393–403.
8. Masthan KMK, Sankari SL, Aravindha Babu N, Gopalakrishnan T. Mystery inside the tooth : the dental pulp stem cells. J Clin Diagn Res. 2013;7:945–7.

9. Komada Y, Yamane T, Kadota D, Isono K, Takakura N, Hayashi S, Yamazaki H. Origins and properties of dental, thymic, and bone marrow mesenchymal cells and their stem cells. PLoS ONE. 2012;7:e46436.
10. Hilkens P, Gervois P, Fanton Y, Vanormelingen J, Martens W, Struys T, Politis C, Lambrichts I, Bronckaers A. Effect of isolation methodology on stem cell properties and multilineage differentiation potential of human dental pulp stem cells. Cell Tissue Res. 2013;353:65–78.
11. Huang GT-J. Dental pulp and dentin tissue engineering and regeneration- advancement and challenge. Front Biosci (Elite Ed). 2012;3:788–800.
12. Laino G, Graziano A, d'Aquino R, Pirozzi G, Lanza V, Valiante S, De Rosa A, Naro F, Vivarelli E, Papaccio G. An approachable human adult stem cell source for hard-tissue engineering. J Cell Physiol. 2006;206:693–701.
13. Telles PD, Machado MAAM, Sakai VT, Nör JE. Pulp tissue from primary teeth : new source of stem cells. J Appl Oral Sci. 2011;19:189–94.
14. Bianco P, Gehron Robey P, Simmons PJ. Mesenchymal stem cells: revisiting history, concepts, and assays. Cell Stem Cell. 2008;2:313–9.
15. Goodman JW, Hodgson GS. Evidence for stem cells in the peripheral blood of mice. Blood. 1962;19: 702–14.
16. Reyes M, Lund T, Lenvik T, Aguiar D, Koodie L, Verfaillie CM. Purification and ex vivo expansion of postnatal human marrow mesodermal progenitor cells. Blood. 2001;98(9):2615–25.
17. Otaki S, Ueshima S, Shiraishi K, Sugiyama K, Hamada S, Yorimoto M, Matsuo O. Mesenchymal progenitor cells in adult human dental pulp and their ability to form bone when transplanted into immunocompromised mice. Cell Biol Int. 2007;31:1191–7.
18. Pittenger MF, Mackay AM, Beck SC, Jaiswal RK, Douglas R, Mosca JD, Moorman MA, Simonetti DW, Craig S, Marshak DR. Multilineage potential of adult human mesenchymal stem cells. Science. 1999;284: 143–7.
19. D'Aquino R, Papaccio G, Laino G, Graziano G. Dental pulp stem cells: a promising tool for bone regeneration. Stem Cell Rev. 2008;4:21–6.
20. Lizier NF, Kerkis A, Gomes CM, Hebling J, Oliveira CF, Caplan AI, Kerkis I. Scaling-up of dental pulp stem cells isolated from multiple niches. Plos ONE. 2012;7(6):e39885.
21. Sloan AJ, Waddington RJ. Dental pulp stem cells: what, where, how ? Int J Paediat Dent. 2009;19:61–70.
22. Wang J, Wei X, Ling J, Huang Y, Huo Y, Zhou Y. The presence of a side population and its marker ABCG2 in human deciduous dental pulp cells. Biochem Biophys Res Commun. 2010;400:334–9.
23. Iohara K, Zheng L, Ito M, Tomokiyo A, Matsushita K, Nakashima M. Side population cells isolated from porcine dental pulp tissue with self-renewal and multipotency for dentinogenesis, chondrogenesis, adipogenesis, and neurogenesis. Stem Cells. 2006;24: 2493–503.
24. Kenmotsu M, Matsuzaka K, Kokubu E, Azuma T, Inoue T. Analysis of side population cells derived from dental pulp tissue. Int Endo J. 2010;43: 1132–42.
25. Achilleos A, Trainor PA. Neural crest stem cells: discovery, properties and potential for therapy. Cell Res. 2012;22:288–304.
26. Demarco FF, Conde MCM, Cavalcanti BN, Casagrande L, Sakai VT, Nor JE. Dental pulp tissue engineering. Braz Dent J. 2011;22:3–14.
27. Gronthos S, Brahim J, Li W, Fisher LW, Cherman N, Boyde A, et al. Stem cell properties of human dental pulp stem cells. J Dent Res. 2002;81(8):531–5.
28. Kresbach PH, Robey PG. Dental and skeletal stem cells: potential cellular therapeutics for craniofacial regeneration. J Dent Educ. 2002;66:766–73.
29. Miura M, Gronthos S, Zhao M, Lu B, Fisher LW, Robey PG, et al. SHED: Stem cells from human exfoliated deciduous teeth. Proc Natl Acad Sci U S A. 2003;100(10):5807–12.
30. Gronthos S, Mankani M, Brahim J, Gehron Robey P, Shi S. Postnatal human dental pulp stem cells (DPSCs) in vitro and in vivo. Proc Natl Acad Sci U S A. 2000;97:13625–30.
31. Huang GT-J, Sonoyama W, Liu Y, Liu H, Wang S, Shi S. The hidden treasure in apical papilla: the potential role in pulp/dentin regeneration and bioroot engineering. J Endod. 2008;34:645–51.
32. Peng L, Ye L, Zhou X-D. Mesenchymal stem cells and tooth engineering. Int J Oral Sci. 2009;1:6–12.
33. Seo BM, Miura M, Gronthos S, et al. Investigation of multi-potent postnatal stem cells from human periodontal ligament. Lancet. 2004;364:149–55.
34. Sonoyama W, Liu Y, Fang D, et al. Mesenchymal stem cell- mediated functional tooth regeneration in swine. PLoS ONE. 2006;1:e79.
35. Morsczeck C, Gotz W, Schierholz J, et al. Isolation of precursor cells (PCs) from human dental follicle of wisdom teeth. Matrix Biol. 2005;24:155–65.
36. Wada N, Menicanin D, Shi S, Barthold PM, Gronthos S. Immunomodulatory properties of human periodontal ligament stem cells. J Cel Physiol. 2009;219:667–76.
37. Schofield R. The relationship between the spleen colony-forming cell and the haemopoietic stem cell. Blood Cells. 1978;4:7–25.
38. Kolf CM, Cho E, Tuan RS. Mesenchymal stromal cells: biology of adult mesenchymal stem cells: regulation of niche, self-renewal and differentiation. Arthritis Res Ther. 2007;9:204–14.
39. Ema H, Suda T. Two anatomically distinct niches regulate stem cell activity. Blood. 2012;120:2174–81.
40. Tummers M, Thesleff I. Root or crown: a developmental choice orchestrated by the different regulation of the epithelial stem cell niche in the tooth of two rodent species. Development. 2003;130:1049–57.
41. Lin Y, Cheng Y-S L, Qin C, Lin C, D'Souza R, Wang F. FGFR2 in the dental epithelium is essential for development and maintenance of the maxillary cervical loop, a stem cell niche in mouse incisors. Dev Dyn. 2009;238:324–30.

42. Harada H, Ohshima H. New perspectives in tooth development and the dental stem niche. Arch Histol Cytol. 2004;67:1–11.
43. Martens W, Wolfs E, Struys T, Politis C, Bronckaers A, Lambrichts I. Expression pattern of basal markers in human dental pulp stem cells and tissue. Cells Tissues Organs. 2012;196:490–500.
44. Shi S, Gronthos S. Perivascular niche of postnatal mesenchymal stem cells in human bone marrow and dental pulp. J Bone Miner Res. 2003;18:696–704. Magloire H, Couble ML, Thivichon-Prince B, Maurin JC, Bleicher F (2009). Odontoblast: a mechanosensory cell. J Exp Zool Part B: Molecular Dev Evolut. 312B:416–424.
45. Harichane Y, Hirata A, Dimitrova-Nakov S, Granja I, Goldberg M, Kellermann O, Poliard A. Pulpal progenitors and dentin repair. Adv Dent Res J Dent Res. 2011;23:307–12.
46. Fitzgerald M, Chiego DJ, Heys DR. Autoradiographic analysis of osteoblast replacement following pulp exposure in primate teeth. Arch Oral Biol. 1990;35: 707–15.
47. Priam F, Ronco V, Locker M, Bourd K, Bonnefoix M, Duchêne T, Bitard J, Wurtz T, Kellermann O, Goldberg M, Poliard A. New cellular models for tracking the odontoblast phenotype. Arch Oral Biol. 2005;50:271–7.
48. Oshima H. Ultrastructural changes in odontoblasts and pulp capillaries following cavity preparation in rat molars. Arch Histol Cytol. 1990;53:423–38.
49. Decup F, Six N, Palmier B, Buch D, Lasfargues J-J, Salih E, Goldberg M. Bone sialoprotein-induced reparative dentinogenesis in the pulp of rat's molar. Clin Oral Investig. 2000;4:110–9.
50. Casasco A, Casasco M, Calligaro A, Ferrieri G, Brambilla E, Strohmenger L, Alberici R, Mazzini G. Cell proliferation in developing human dental pulp. A combined flow cytometric and immunohistochemical study. Eur J Oral Sci. 1997;105:609–13.
51. Kobayashi I, Izumi T, Okamura K, Matsuo K, Ishibashi Y, Sakai H. Biological behavior of human dental pulp cells in response to carious stimuli analyzed by PCNA immunostaining and AgNOR staining. Caries Res. 1996;30:225–30.
52. Six N, Tompkins K, Septier D, Veis A, Goldberg M. Recruitment and characterization of the cells involved in reparative dentin formation in the exposed rat molar pulp after implantation of amelogenin gene splice products A+4 and A-4. Oral Biosci Med. 2004; 1:35–44.
53. Magloire H, Couble ML, Thivichon-Prince B, Maurin JC, Bleicher F. Odontoblast: a mechano-sensory cell. J Exp Zool Part B: Molecular Dev Evolut. 2009;312B: 416–24.
54. Sonoyama W, Liu YI, Yamaza T, Tuan RS, Wang S, Shi S, Huang GT-J. Characterization of apical papilla and its residing stem cells from human immature permanent teeth – a pilot study. J Endod. 2008;34:166–71.
55. Trepat X, Fredberg JJ. Plithotaxis and emergent dynamics in collective cellular migration. Trends Cell Biol. 2011;21:638–46.
56. Friedl P, Bröcker EB. The biology of cell locomotion within three-dimensional extracellular matrix. Cell Mol Life Sci. 2000;57:41–64.
57. Vermelin L, Ayanoglou C, Septier D, Carreau JP, Bissila-Mapahou P, Goldberg M. Effects of essential fatty acid deficiency on rat molar pulp cells. Eur J Oral Sci. 1995;103:219–24.
58. Vermelin L, Lécolle S, Septier D, Lasfargues J-J, Goldberg M. Apoptosis in human and rat dental pulp. Eur J Oral Sci. 1996;104:547–53.
59. Ikeda H, Suda H. Odontoblastic syncytium through electrical coupling in the human dental pulp. J Dent Res. 2013;92:371–5.

17 Regeneration of the Living Pulp

Tracy L. de Peralta and Jacques Eduardo Nör

17.1 Regenerative Potential of the Dental Pulp

The dental pulp is a complex tissue endowed with inherent regenerative potential that can be engaged to protect the pulp against challenges such as caries or dental trauma. Increasing understanding of the inherent protective mechanisms employed by the dental pulp has raised possibility of exploiting similar processes to induce pulp regeneration therapeutically. In this chapter, we will review inherent mechanisms of dental pulp tissue regeneration and discuss approaches that exploit such processes for the therapeutic regeneration of living dental pulps.

By definition, the goal of regenerative medicine is to replace or restore the normal function of cells, tissues, and organs that are damaged by intrinsic or extrinsic factors [1]. Most human tissues and organs contain undifferentiated stem cells that can be induced to differentiate to replace injured cells and regenerate the tissue. In addition, there are circulating progenitor cells (e.g., circulating endothelial cell progenitors) that can home to sites of injury and participate in regenerative processes. Notably, the discovery of stem cells in the dental pulp of permanent and primary teeth provided a mechanistic explanation for the superb regenerative potential of this tissue [2, 3]. In addition, this discovery unveiled the possibility of using dental pulp stem cells (DPSC) in the engineering of a new dental pulp to treat necrotic teeth, particularly young permanent teeth with open apex.

Dental pulp tissue is derived from migrating neural crest cells during development and has been shown to contain multipotent stem/progenitor cells [2–5]. Indeed, DPSC exhibit key stem-like qualities, such as self-renewal and multi-lineage differentiation potential [2, 6]. These cells are designed by nature to be quiescent in pulp homeostasis and activated to differentiate into cells (e.g., odontoblasts) that orchestrate tissue response to injury caused by deep caries [7]. These undifferentiated cells can migrate from the perivascular region toward the injured site [8], where they have the potential to replace dead odontoblasts and synthesize new dentin. Self-renewal ensures that at each DPSC cell division, at least one new stem cell is generated. Notably, the ability to self-renew ensures the prolonged presence of undifferentiated stem cells in the pulp tissue.

DPSC express high levels of mesenchymal markers (e.g., STRO1, CD29, CD44, CD73, CD90, CD166), but not hematopoietic markers such as CD34 or CD45 [9]. Furthermore, neural stem cell markers (e.g., nestin, nucleostemin) are expressed by DPSCs, possibly reflecting their neural crest origin [9, 10]. The finding that DPSCs

T.L. de Peralta, DMD, PhD, MClinEd
J.E. Nör, DDS, MS, PhD (✉)
Department of Cariology, Restorative Sciences, and Endodontics, University of Michigan, 1011 N University Ave, Ann Arbor, MI 48109-1078, USA
e-mail: tdeperal@umich.edu; jenor@umich.edu

M. Goldberg (ed.), *The Dental Pulp*,
DOI 10.1007/978-3-642-55160-4_17, © Springer-Verlag Berlin Heidelberg 2014

also express embryonic stem cell markers (e.g., Sox-2) supports the possibility that DPSC have a primitive origin [1, 10].

DPSC exhibit multipotency consistent with mesenchymal stem cells, with capacity to give rise to at least three distinct cell lineages: (1) osteo/odontogenic, (2) adipogenic and (3) neurogenic [2, 11, 12]. Indeed, when dental pulp stem cells are exposed to dentin-derived factors, they readily differentiate into functional odontoblasts expressing dentin sialophosphoprotein (DSPP) and dentin matrix protein (DMP)-1 that are capable of synthesizing new tubular dentin [13–15]. These cells can also differentiate into vascular endothelial cells when exposed to dentin matrix proteins supplemented by vascular endothelial growth factor (VEGF), as shown [15–17]. Notably, these DPSC-derived blood vessels are capable of anastomosing with the host vasculature and carry blood, demonstrating that they participate in the perfusion of the newly engineered dental pulp tissue.

Dental pulp stem cells represent an accessible source of undifferentiated cells that can be derived from healthy human teeth from patients of different ages [18]. Interestingly, recent studies have shown that dental pulp stems cells can also be isolated from clinically compromised dental pulps with irreversible pulpitis [19, 20]. The stem cells derived from inflamed dental pulps demonstrated favorable proliferative potential, ex vivo odonto/osteogenic differentiation capacity, and retained capacity for tissue regeneration in vivo. These findings indicate that teeth with pulpitis have not exhausted their supply of stem cells despite the high demands for DPSC to replace compromised odontoblasts and suggest that these teeth can still serve as a potential source of undifferentiated cells for regenerative procedures.

We believe that the search for markers that identify dental pulp stem cells and that can be used for their isolation is of paramount importance for the translation of stem cell-based therapies into the clinic. Furthermore, the development of culture conditions that maintain their stemness while enabling proliferation in absence of animal-derived reagents, such as fetal bovine serum, will be critical for the clinical use of these DPSC in cell-based tissue engineering and in tissue/organ regeneration. Such technical issues are areas of much research today and will be critically important to enable the clinician to harness the inherent regenerative potential of dental pulp stem cells in therapeutic applications.

17.2 Dental Pulp Responses to Noxious Stimuli

Pulp response to noxious stimuli, such as bacterial infection related to deep carious lesions, constitutes a defense mechanism that aims at the maintenance of tissue vitality and function. Understanding innate mechanisms of pulp tissue response to injury provides information that can be applicable to the development of approaches for dental pulp tissue engineering and regeneration. Furthermore, engineered or regenerated dental pulps should ideally be capable of responding to noxious stimuli.

Understanding key mechanisms involved in the pathobiology of pulpitis has unveiled new targets that can be exploited in dental pulp tissue regeneration. Angiogenesis, a critical process in tissue development and wound healing [21], is tightly regulated in the dental pulp tissue. It is well known that pulp microvessel density can be quickly augmented at sites of injury (e.g., deep caries) to provide effective nutrient and oxygen delivery for the high metabolic demands of cells engaged in regeneration processes.

Odontoblasts are the first cells encountered by the pathogens and their by-products, followed by the more centrally located fibroblasts and immature dendritic cells [22]. Cariogenic bacteria and their by-products diffuse through the dentin tubules and cause pulp inflammation. Pulpitis is a disease primarily caused by infection of both gram-positive and gram-negative bacteria and is a dynamic process by which both bacterial virulence and host immunity determine the outcome of the dental pulp [23, 24]. As the carious infection progresses from the tooth surface toward the pulp, both the innate and adaptive immunities are activated [25]. Of note, the low-compliance nature of the dental pulp that is encased within non-expanding mineralized walls can inhibit natural tissue repair responses [26]. Indeed, it is known that bacterial by-products such as lipoteichoic acid (LTA) from gram-positive cariogenic bacteria [27] and lipopolysaccharide (LPS) from gram-negative bacteria [28–30] induce expression of vascular endothelial growth factor (VEGF) by pulp cells and that VEGF is a potent inducer of vascular permeability and edema [31, 32]. It has been postulated that VEGF contributes to the vasodilation and the enhanced vascular permeability observed in pulpitis [27], which can potentially be associated with an increase in intrapulpal pressure [33, 34]. Therefore, while some VEGF is important to induce angiogenesis that is necessary for tissue regeneration, too much VEGF can be problematic because it may lead to high interstitial pressure within the root canal. These observations suggest that one will have to use caution when delivering angiogenic factors (e.g., VEGF) in attempts to regenerate dental pulp tissues.

During dentin-pulp regeneration, the interplay between pulp stem cells and signaling molecules derived from the dentin matrix drives the process of reparative dentinogenesis [35, 36]. In addition to dental pulp cells, the dentin matrix also represents an important source of growth factors and cytokines that may be released during bacteria-induced acidic demineralization. These molecules have been shown to play a major role in dental pulp regenerative responses [26, 37, 38]. Indeed, sequestration of a number of important morphogenic and angiogenic growth factors in the dentin matrix provides a "fossilization" of these molecules which, in turn, protects their bioactivity and allows for sustained release during matrix breakdown in carious episodes or with the application of restorative materials [39–42]. These dentin-derived morphogenic and angiogenic signals can be engaged during dental pulp tissue regeneration to guide differentiation of stem cells. For example, it has been reported that short-term treatment of dentin with mild acids (e.g., EDTA) potentiates the odontoblastic differentiation of dental pulp stem cells [13]. Indeed, a well-orchestrated symphony of cellular and matrix interactions plays a critical role in the regulation of dental pulp tissue repair and regeneration.

17.3 Factors that Regulate Pulp Regeneration

Understanding the process of wound healing in the dental pulp is essential to successfully regenerate pulp tissue [26]. In reconstructing the cellular and extracellular environment in pulp regeneration processes, one must consider the natural pulp physiology and how individual pulp components interact to support each other into a highly functional tissue. For example, a regulated interplay between tissue vasculature and nerves is critical to maintaining homeostasis. Indeed, strategies for pulp regeneration must recreate a microenvironment that supports the cellular cross talk needed to maintain healthy and functional cells endowed with the ability to respond to challenges.

Vasculogenesis is defined as the de novo formation of blood vessels. It has been noted that stem cells from exfoliated deciduous teeth (SHED) have the potential to differentiate into functional vascular endothelial cells via a process that closely resembles embryonic vasculogenesis [15, 16]. This finding suggests that dental pulp stem cells can serve as a source of functional blood vessels, which might enable influx of oxygen and nutrients to support the newly formed pulp-like tissue. Notably, it is widely accepted that the success of tissue engineering relies on rapid establishment of microvascular networking to provide blood and nutrients for cells that are engaged in the tissue regeneration process. In addition, blood vessels allow for the influx of immune cells that provide a mechanism of defense as well as further regulation of the angiogenic response in tissues undergoing processes of repair and/or regeneration [43].

It has been shown that injured pulp fibroblasts and endothelial cells are capable of releasing chemotactic factors and signaling molecules to initiate healing processes [44–46]. Chemokines, including interleukin-8 (IL-8) and monocyte chemoattractant protein-1 (MCP-1), are capable of directing the migration of polymorphonuclear leukocytes, monocytes, and macrophages to the site of pulp infection [47]. It is also known that IL-8 and MCP-1 production can be stimulated by endodontic pathogens [48]. Elevated levels of IL-8 have been reported in the pulp tissues and gingival crevicular fluid from patients with pulpitis [49, 50]. Furthermore, neurogenic inflammation induces the production of IL-8 and MCP-1 in human dental pulps [51]. Therefore, endothelial cells and fibroblasts can initiate signaling events that regulate the influx of immune cells required for tissue defense. Ideally, regenerated dental pulps will contain endothelial cells and fibroblasts that retain such features.

Dental pulp is a tissue densely innervated by sensory nerve fibers originating from the trigeminal ganglion, parasympathetic fibers, and sympathetic fibers originating from the cervical sympathetic ganglia [52, 53]. This complexity in pulp innervation has motivated numerous investigations regarding how these three major neuronal systems regulate pulp physiology and pathology [54]. For example, the glial cell line-derived neurotrophic factor (GDNF) was originally characterized as a potent trophic factor, which promotes the survival and differentiation of neurons [55]. It has been demonstrated that GDNF counteracts tumor necrosis factor (TNF)-alpha-induced dental pulp cell cytotoxicity, suggesting that GDNF may be cytoprotective under disease conditions. These studies have concluded that GDNF promotes cell survival and proliferation of dental pulp cells and suggest that it plays a multifunctional role in the regulation of dental pulp homeostasis [56]. It is also known that nestin and glial fibrillary acidic protein (GFAP) are expressed by pulp cells [57, 58] and that pulp cells are capable of producing a variety of neurotrophins [5].

Neuropeptides play major regulatory roles in the function of dental pulp cells. Neuropeptides known to be involved in neurogenic pulpal inflammation are calcitonin gene-related peptide (CGRP), substance P (SP), neurokinin A (NKA), neuropeptide Y (NPY), vasoactive intestinal polypeptide (VIP), among many others [53, 59, 60]. The release of the neuropeptide CGRP in dental pulp plays a very important role in the clinical inflammatory condition of acute irreversible pulpitis [54, 61]. It has been demonstrated that neuropeptides such as CGRP and SP interact with mastocytes, inducing the release of histamine and thereby causing elevated vascular permeability and increased interstitial pressure in the dental pulp [62]. Neuropeptides such as SP and NKA activate macrophages and lymphocytes to release inflammatory mediators such as prostaglandins and thromboxanes [63]. The neuropeptide CGRP exerts stimulatory effects on the proliferation of pulpal cells, such as fibroblasts and odontoblast-like cells [64]. Interestingly, studies have shown that the addition of CGRP to cell culture can increase expression of bone morphogenetic protein (BMP)-2 transcripts in human dental pulp cells, leading to odontoblast-like cell differentiation [65]. Emerging evidence suggests that neuropeptides play important roles in the regulation of dental pulp stem cell differentiation. However, there have not been many attempts to incorporate neuropeptides into strategies for dental pulp tissue engineering.

It is known that DPSC express neuronal markers and may be induced to acquire a neuron-like morphology following transplantation into the mesencephalon of embryonic day-2 chicken embryos [66]. Further, it has been shown that transplantation of dental pulp stem cells might promote functional recovery after spinal cord injury in murine models [67, 68]. These findings suggest that adult human DPSC provide a readily accessible source of stem/precursor cells that have the potential for use in cell-therapeutic paradigms to treat neurological disease. While this is a very advantageous use of DPSC in regenerative medicine, the mechanisms underlying the neuronal differentiation of dental pulp stem cells are still unclear, and therefore we currently do not have much control of this process when dental pulp stem cells are transplanted in regenerative endodontics-based approaches.

In the physiological architecture of a healthy dental pulp, there is a close working relationship between its vascular and neuronal components. Reconstructing an architecture that allows for this relationship is challenging, as the signaling events mediating neural-vascular cross talk are not fully understood. Interestingly, endothelial cells express the tyrosine kinase receptor neuropilin (NRP)-1, which is critical for VEGF signaling and regulation of VEGF's effect on endothelial cell migration [69]. NRP1 also plays a major role in semaphorin signaling by regulating its effect on axonal guidance [70]. Indeed, NRP1 constitutes an example of how the molecular signaling events regulating angiogenesis and tissue enervation might be interconnected and mutually relevant. Further studies are necessary to understand how the fate of dental pulp stem cells, i.e., cellular decisions in regard to differentiation into odontoblast, vascular endothelial cell, or neural cell, can be optimally regulated in the context of cell-based dental pulp tissue engineering.

17.4 Strategies for Pulp Regeneration

Traumatized immature permanent teeth that developed pulp necrosis pose a significant challenge to dental practitioners. Limited effectiveness of instrumentation and difficulties associated with canal obturation due to lack of apical stop are known challenges. In addition, the traditional approach involving apexification with calcium hydroxide is typically associated with poor long-term outcomes [71]. The regeneration of a functional dental pulp tissue has been proposed as an alternative strategy to eliminate the infection while enabling completion of root formation [72]. Broadly speaking, there are two strategies for dental pulp tissue regeneration: (a) cell-based approaches, which involve the transplantation of cells into the root canal, and (b) cell-free approaches, which is centered around the recruitment of host cells into the pulp chamber. The following is a discussion about advantages and challenges of both approaches.

17.4.1 Cell-Based Approaches

It has been demonstrated that the transplantation of human dental pulp stem cells into human root canals can regenerate a dental pulp-like tissue containing odontoblasts, blood vessels, and innervation in an arrangement similar to pulps found in normal human teeth [2, 16, 73–75]. Indeed, dental pulp stem cell-derived odontoblasts are capable of generating new tubular dentin, as demonstrated by tetracycline staining-positive dentin surrounding engineered dental pulps [15]. Such studies suggest that dental pulp stem cells constitute an adequate cell source for regeneration of dental pulp tissues.

Most protocols suggested to date for cell-based pulp engineering include the use of scaffolds to provide adequate environment for responsive stem cells that can proliferate and differentiate into a functional dental pulp upon appropriate morphogenic signaling. These strategies follow the basic principles of tissue engineering proposed by Langer and Vacanti [76] and adapted to the context of dental tissue engineering by Nakashima and Reddi [77]. The search for the ideal combination of scaffold, stem cells, and morphogenic signaling has been a continuous evolving process over the last 10 years [77–82].

In general, cell-based approaches for dental pulp tissue regeneration include one of the following stem cell sources: (a) DPSC from permanent teeth [2]; (b) SHED harvested from exfoliated deciduous teeth [11]; (c) SCAP cells, i.e., stem cells from the apical papilla [83]; or (d) CD31(−) side population (SP) cells [84]. Cell-based approaches require the retrieval and ex vivo expansion of the stem cells, followed by their transplantation into the recipient tooth using a process that maintains pathogen-free conditions. Ideally, the cellular expansion in the laboratory should ideally happen in serum-free conditions to avoid use of animal products in this process and minimize the risk for transmissibility of disease [85, 86]. After ex vivo expansion, these cells are transplanted into the pulp chamber, ideally using an injectable scaffold material that contains the necessary morphogenic proteins to promote spatial and temporal differentiation of stem cells for optimal tissue replacement.

The development of scaffolds ideally suited for dental pulp tissue engineering is quickly becoming an important new area in the field of

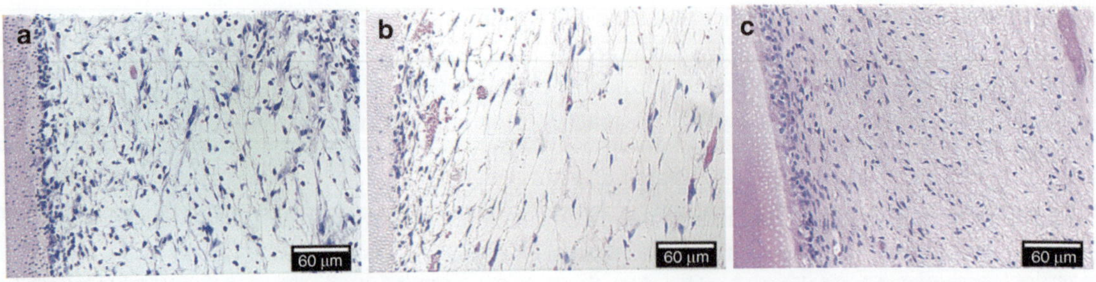

Fig. 17.1 Photomicrographs of dental pulp-like tissues formed when SHED cells mixed with scaffolds (Puramatrix™, 3-D Matrix Technology, Waltham, MA (**a**) or rhCollagen type I (**b**)) were injected into the root canals of human premolars and transplanted into immunodeficient mice. A vascularized connective tissue with histological features similar to those observed in normal human dental pulps occupied the pulp chamber. Higher cell density was observed along the dentin walls. Freshly extracted human premolars were used as positive controls (**c**) (Courtesy of Dr. Vinicius Rosa, National University of Singapore)

dental materials. These scaffolds should ideally be biocompatible, biodegradable, injectable, and contain and/or enable recruitment of morphogenic signals. Poly-L-(lactic) acid (PLLA) and/or polyglycolic acid (PGA)-based scaffolds have been very useful for proof-of-principle studies on dental pulp tissue regeneration [87]. However, the fact that these scaffolds need to be casted within the root canal (i.e., are not injectable) limits their usefulness in regenerative endodontics. In contrast, self-assembling hydrogels are injectable and biocompatible and have been successfully used in dental pulp tissue engineering approaches. A leader in the development of such scaffolds, the Galler laboratory, has reported on the development of customized self-assembling peptide hydrogels for dental pulp tissue engineering [88–90]. A major advantage of the Galler scaffolds for dental pulp tissue engineering is the flexibility of adding cell adhesion molecules and morphogenic factors to improve the binding of cells to the scaffold and regulate their differentiation potential. There are also commercially available self-assembling peptide hydrogel (e.g., Puramatrix™, 3-D Matrix Technology, Waltham, MA) that has been used in preclinical trials for dental pulp tissue engineering [74, 91]. A pulp-like tissue was observed throughout the entire extension of the root canal when dental pulp stem cells were seeded in these scaffolds and injected into human root canals that were transplanted into immunodeficient mice (Fig. 17.1a–c).

Knowing that VEGF is a key proangiogenic factor [92, 93], and that VEGF induces the differentiation of dental pulp stem cells into endothelial cells [15, 17], much interest has been placed in the functionalization of scaffolds with VEGF [90, 94]. Alternatively, it has been shown that a scaffold of collagen type I gel containing dental pulp cells and angiogenic growth factors (i.e., fibroblast growth factor-2, VEGF, and PDGF conjugated to gelatin microspheres) injected into pulpless tooth chambers mediated revascularization and allowed for dental pulp tissue regeneration in vivo [95]. Ongoing studies are attempting to define the optimal concentration of VEGF that promotes the differentiation of dental pulp stem cells into vascular endothelial cells and the recruitment of blood vessels from the periapical region, while at the same time does not elicit excessive vascular edema and increased interstitial pressure with deleterious consequences to the overall viability of the pulp tissue.

A recent study has challenged the need for scaffolds in dental pulp tissue engineering applications. The Sfeir group has recently developed self-assembled, scaffoldless, three-dimensional tissues that were engineered from dental pulp cells [96]. Tissue sheets formed by dental pulp cells are induced to "roll" in culture to form cylinders that are then transplanted into the root canals. In this case, the cells themselves secrete their extracellular matrix forming their own three-dimensional microenvironment and precluding the need for scaffold materials.

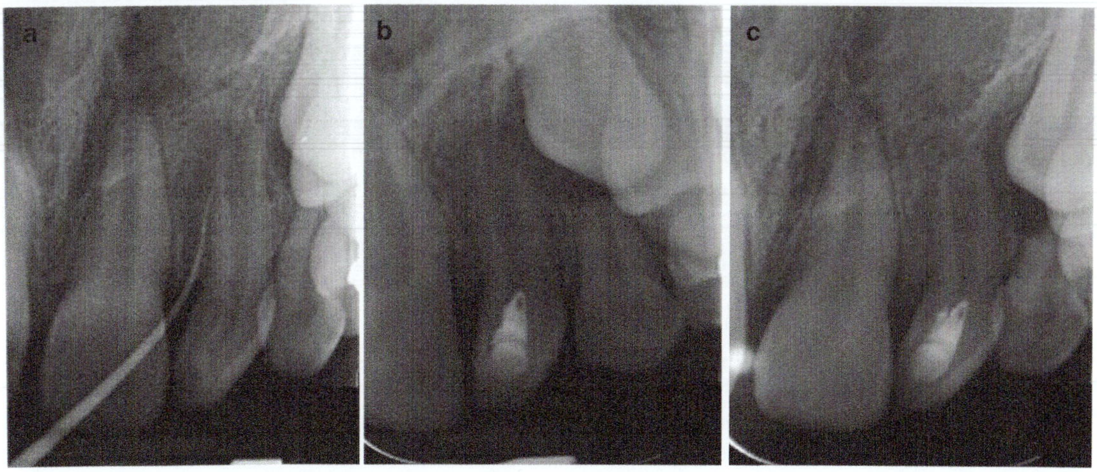

Fig. 17.2 Clinical case demonstrating positive outcome in cell-free regenerative endodontic therapy. (**a**) Illustrates pretreatment phase of upper left lateral incisor with dens invaginatus. (**b**) Illustrates at-treatment (immediate post-operative) phase with stimulation of blood clot at apical foramen and coronal seal with MTA and resin restoration. And (**c**) illustrates apical closure at 6 months posttreatment phase (All courtesy of Dr. Matthew G. Healy and Dr. Tatiana Botero, Endodontic Department, University of Michigan, School of Dentistry)

In general, cell-based approaches have the advantage of enabling the engineering of tissues with morphology and function that closely resemble those observed in normal human dental pulps. However, these approaches have the intrinsic challenge of requiring the transplantation of cells, which is always associated with increased risks of transmission of pathogens or ex vivo transformation of cells leading to uncontrolled tissue growth [97]. Both are serious, potentially life-threatening issues that must be taken into account before such procedures are routinely used in humans. In an attempt to minimize such risks, several cell-free approaches have been proposed, as follows.

17.4.2 Cell-Free Approaches

Cell-free approaches rely on the recruitment of cells from the periapical region into the pulp chamber as a means to regenerate the pulp of necrotic teeth. Pioneering work by Trope and colleagues designed a protocol in which they create a blood clot by intentional instrumentation of the root canal beyond the apex [98, 99]. This procedure, also called revascularization or regenerative endodontics, has the advantages of not requiring cell harvesting, ex vivo expansion, and transplantation (Fig. 17.2a–c).

In general, the regenerative endodontics approach involves the opening of the pulp chamber, followed by instrumentation with copious irrigation. After disinfection, a blood clot is generated by instrumentation beyond the apex with the intent of forming a "scaffold" for the ingrowth of new tissue in the pulp chamber via cell homing induced by molecules naturally released from the blood clot. This is then followed by coronally sealing the canal with mineral trioxide aggregate (MTA), which acts as a hard barrier from pathogenic agents in the oral cavity after the canal has been disinfected. Case reports and clinical studies in humans are currently exploring the safety and efficacy of the revascularization procedure in young permanent teeth [98, 100–102].

Recently, the Mao laboratory suggested a method to exploit chemotaxis as a strategy to guide cells from the periapical region into the pulp chamber as a strategy for regeneration of dental pulps [103]. They showed that injection of a collagen gel solution containing basic fibroblast growth factor (bFGF), VEGF, and platelet-derived growth factor (PDGF) together with nerve growth factor (NGF) and bone morphogenetic protein-7

(BMP7) as a chemotactic was capable of attracting cells into the pulp chamber and regenerating a dental pulp tissue. Using this type of approach, it is possible that dental pulp regeneration is achieved with an injectable scaffold containing a predetermined "cocktail" of chemotactic factors that recruits host cells capable of organizing themselves into a newly formed pulp-like tissue into the root canal.

In general, strategies for dental pulp tissue regeneration that do not require cell transplantation have less regulatory constraints and can be more quickly translated into clinical practice. However, so far the attempts to use cell-free based approaches have not consistently resulted in the formation of a functional dental pulp tissue throughout the full extension of the root canal, when teeth are examined histologically in humans. Under these circumstances, the long-term outcome of teeth treated with cell-free approaches for dental pulp tissue regeneration of necrotic teeth is still rather unclear.

17.5 Challenges Ahead

The seminal discovery that dental pulps contain a subpopulation of proliferative stem cells endowed with multipotency and self-renewal capacities opens the possibility that such cells can be harnessed to regenerate a living pulp in necrotic teeth. However, before this "dream" can be fulfilled, there are several challenges that will have to be overcome (Fig. 17.3), which include:

A. Stem cell transplantation has to be done in a highly controlled process that ensures *safety* in each step. This multistep process includes the retrieval of stem cells from the donor tooth, ex vivo expansion of these cells using pathogen-free processes and reagents, and transplantation to the recipient patient. At each one of these steps, there are potential risks involved, such as contamination of the cells with pathogenic bacteria/fungi and transformation of the cells during ex vivo expansion. Nevertheless, stem cell transplantation has been increasingly used in medicine as standard of care for select procedures, and more recently it began to be used in the context of oral and craniofacial disease [104, 105].
B. The *efficacy* of dental pulp stem cell transplantation in the regeneration of new functional dental pulps in humans will have to be verified in randomized, controlled feasibility trials. Preclinical studies have suggested that it is possible to regenerate a dental pulp throughout the length of human roots transplanted in immunodeficient mice [74]. However, before such procedure can become clinically acceptable, well-designed clinical trials will have to be performed showing that DPSC transplantation consistently and effectively generates new dental pulps that remain viable and functional for extended periods of time.
C. Better *mechanistic understanding* of the processes regulating dental pulp stem cell differentiation upon transplantation is necessary to optimize the procedure. For example, it is widely accepted that rapid vascularization is a critical step for successful tissue engineering. However, the anatomical constraints of root canals pose significant challenges to the establishment of a functional capillary network in an engineered dental pulp, since vascularization is exclusively achieved through the apical foramina system. It is known that dental pulp stem cells can differentiate into functional blood vessels, in addition to their known differentiation potential into odontoblasts [15, 17]. While this multipotency can be exploited to use dental pulp stem cells as a single cellular source for pulp tissue engineering, one will have to find optimal ways to control the timing of this differentiation process in such a way that enables vascularization while not exhausting the source of odontoblastic precursor cells. Such temporal and spatial control of the differentiation process will require exquisite understanding of the mechanisms involved.
D. Considering the fact that pulp tissue engineering would typically be considered in the treatment of necrotic/infected teeth, one will have to develop ways to clean up the root canal and

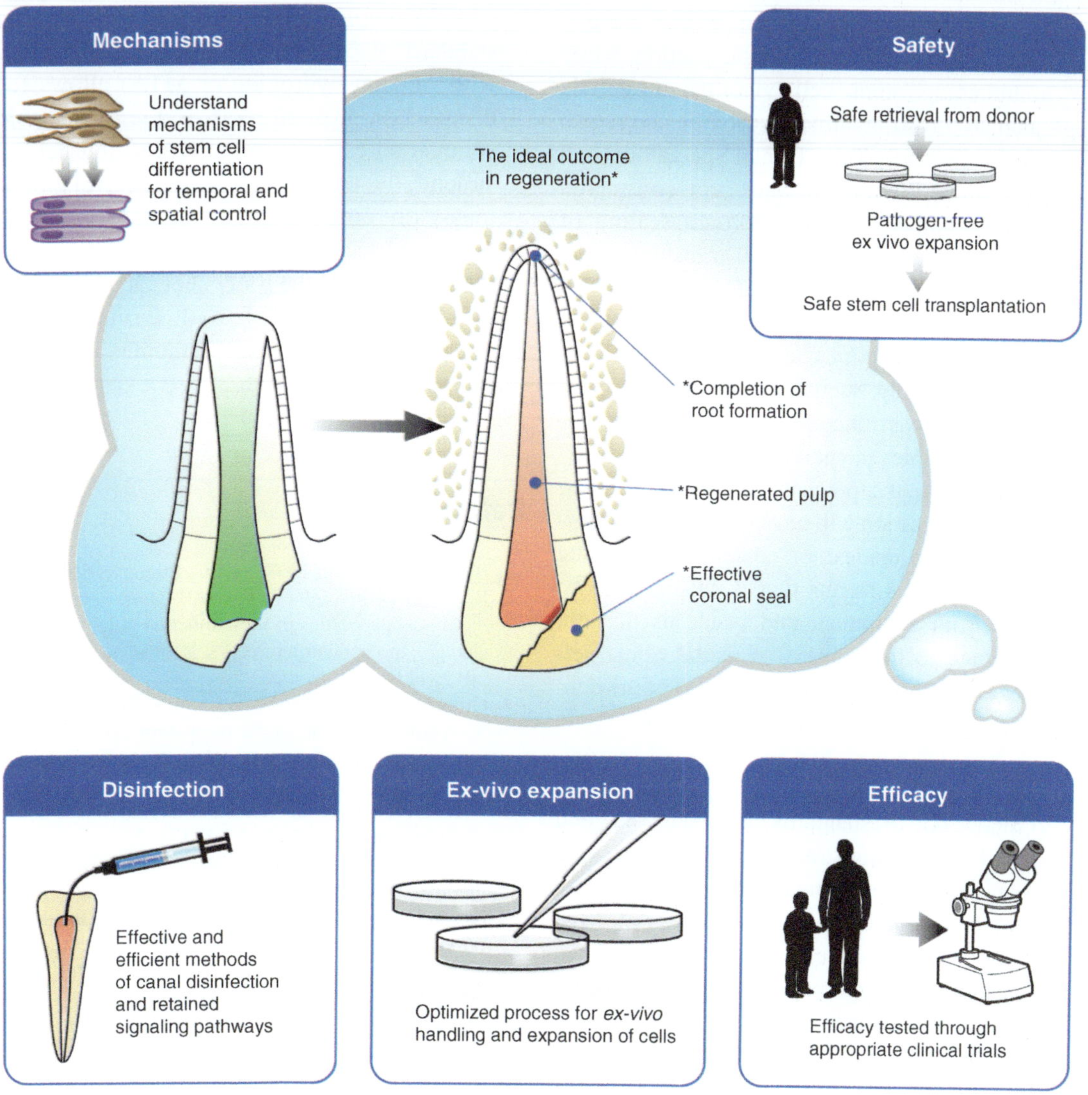

Fig. 17.3 Diagram depicting challenges to the clinical translation of regenerative endodontics procedures: ensure safety of cell-based approaches for dental pulp tissue regeneration; confirm efficacy and long-term outcomes of these procedures through randomized prospective clinical trials; understand mechanisms of stem cell differentiation to enable optimization of dental pulp tissue regeneration in the clinic; effective methods for canal disinfection without compromising dentin-derived morphogenic signals will have to be developed; development of appropriate processes (i.e., ex vivo expansion) that are clinically relevant in the context of contemporary dental practices

eliminate bacterial infection without disrupting the morphogenic signaling pathways that are initiated by dentin-derived proteins (e.g., BMP2) and that are required for odontoblastic differentiation of dental pulp stem cells. For example, it has been shown that excessive use of sodium hypochlorite degrades dentin proteins and prevents dentin-induced odontoblastic differentiation of SHED cells [14]. Recent studies have explored the effects of alternate disinfectant solutions for irrigation of root canals on the survival and differentiation of stem cells [106–108]. Importantly, once a disinfected environment has been prepared within the root canal to receive stem cells, one will also need to maintain it sterile through the use of a coronal sealing strategy that impedes bacterial contamination while maintaining stem cell viability. Materials such as ProRoot mineral trioxide aggregate (MTA, Dentsply Tulsa Dental Specialties) and Biodentine (Septodont) have been considered good candidates for the coronal sealing of teeth that received stem cell transplantation. However, long-term clinical studies are yet to be performed to determine the outcomes of disinfectant solutions and sealing materials. Unquestionably, such studies constitute critical steps toward the clinical use of stem cell-based therapies in endodontics.

E. And finally, the process for stem cell-based pulp regeneration will have to be *clinically relevant* in the context of contemporary dental practices. This means that it will have to be done with equipment and materials that are adequate to the context of dental clinic settings. One envisions the possibility of the dental clinician sending the donor tooth to an external laboratory that is capable of expanding the stem cells under Good Manufacturing Practice (GMP) standards, defined by the Food and Drug Administration (FDA) as ex vivo manipulation of clinical-grade cells that are safe and effective for human use. After the ex vivo expansion, cells would be resuspended in an injectable matrix and sent back to the clinician for transplantation into the root canal. Notably, this procedure would have to be done in a cost-efficient way to become an attractive treatment option to the patient.

Conclusions

While recognizing the many challenges facing the clinical translation of approaches aiming at the regeneration of functional pulp tissues, the authors express guarded enthusiasm that within a reasonable time frame this procedure may become a clinical reality. The need for safe and effective regenerative endodontics procedures comes from the realization that today's dentistry does not have ideal and long-lasting solutions for some clinical problems, such as necrosis of young permanent teeth with incomplete root formation. While the standard of care in 2014 (i.e., apexification with calcium hydroxide or MTA) allows for disinfection of these roots, it does not enable completion of root formation leaving behind structurally weak teeth with relatively poor long-term prognosis. The engineering of a functional and living dental pulp throughout the full extent of the root canal can potentially enable completion of root formation and improve the long-term outcomes of these immature necrotic teeth.

References

1. Bressan E, Ferroni L, Gardin C, Pinton P, Stellini E, Botticelli D, Sivolella S, Zavan B. Donor age-related biological properties of human dental pulp stem cells change in nanostructured scaffolds. PLoS One. 2012;7(11):e49146.
2. Gronthos S, Mankani M, Brahim J, Robey PG, Shi S. Postnatal human dental pulp stem cells (DPSCs) in vitro and in vivo. Proc Natl Acad Sci U S A. 2000;97(25):13625–30.
3. Thesleff I, Aberg T. Molecular regulation of tooth development. Bone. 1999;25(1):123–5.
4. Peters H, Balling R. Teeth. Where and how to make them. Trends Genet. 1999;15(2):59–65.
5. Nosrat IV, Widenfalk J, Olson L, Nosrat CA. Dental pulp cells produce neurotrophic factors, interact with trigeminal neurons in vitro, and rescue motoneurons after spinal cord injury. Dev Biol. 2001;238(1):120–32.
6. Gronthos S, Brahim J, Li W, Fisher LW, Cherman N, Boyde A, DenBesten P, Gehron Robey P, Shi S. Stem cell properties of human dental pulp Stem cells. J Dent Res. 2002;81(8):531–5.
7. Fitzgerald M, Chiego Jr DJ, Heys DR. Autoradiographic analysis of odontoblast replacement following pulp exposure in primate teeth. Arch Oral Biol. 1990;35:707–15.
8. Yamamura T. Differentiation of pulpal cells and inductive influences of various matrices with reference to pulpal wound healing. J Dent Res. 1985;64:530–40.
9. Kerkis I, Kerkis A, Dozortsev D, Stukart-Parsons GC, Gomes Massironi SM, Pereira LV, Caplan AI, Cerruti HF. Isolation and characterization of a population of immature dental pulp stem cells expressing OCT4 and other embryonic stem cell markers. Cells Tissues Organs. 2006;184(3–4):105–16.
10. Estrela C, Alencar AH, Kitten GT, Vencio EF, Gava E. Mesenchymal stem cells in the dental tissue: perspectives for tissue regeneration. Braz Dent J. 2011;22(2):91–8.

11. Miura M, Gronthos S, Zhao M, Lu B, Fisher LW, Robey PG, Shi S. SHED: stem cells from human exfoliated deciduous teeth. Proc Natl Acad Sci U S A. 2003;100(10):5807–12.
12. Perry BC, Zhou D, Wu X, Yang FC, Byers MA, Chu TM, Hockema JJ, Woods EJ, Goebel WS. Collection, cryopreservation and characterization of human dental pulp-derived mesenchymal stem cells for banking and clinical use. Tissue Eng Part C Methods. 2008;14(2):149–55.
13. Casagrande L, Cordeiro MM, Nör SA, Nör JE. Dental pulp stem cells in regenerative dentistry. Odontology. 2011;99(1):1–7.
14. Casagrande L, Demarco FF, Zhang Z, Araujo FB, Shi S, Nör JE. Dentin-derived BMP-2 and odontoblast differentiation. J Dent Res. 2010;89(6):603–8.
15. Sakai VT, Zhang Z, Dong Z, Neiva KG, Machado MA, Shi S, Santos CF, Nör JE. SHED differentiate into functional odontoblasts and endothelium. J Dent Res. 2010;89:791–6.
16. Cordeiro MM, Dong Z, Kaneko T, Zhang Z, Miyazawa M, Shi S, Smith AJ, Nör JE. Dental pulp tissue engineering with stem cells from exfoliated deciduous teeth. J Endod. 2008;34(8):962–9.
17. Bento LW, Zhang Z, Imai A, Nör F, Dong Z, Shi S, Araujo FB, Nör JE. Endothelial differentiation of SHED requires MEK1/ERK signaling. J Dent Res. 2013;92(1):51–7.
18. Atari M, Gil-Racio C, Fabregat M, Garcia-Fernandez D, Barajas M, Carrasco MA, Jung HS, Alfaro FH, Prosper F, Ferres-Padro E, Giner L. Dental pulp of the third molar: a new source of pluripotent-like stem cells. J Cell Sci. 2012;125:3343–56.
19. Alongi DJ, Yamaza T, Song Y, Fouad AF, Romberg EE, Shi S, Tuan RS, Huang GT. Stem/progenitor cells from inflamed human dental pulp retain tissue regeneration potential. Regen Med. 2010;5(4):617–31.
20. Wang Z, Pan J, Wright JT, Bencharit S, Zhang S, Everett ET, Teixeira FB, Preisser JS. Putative stem cells in human dental pulp with irreversible pulpitis: an exploratory study. J Endod. 2010;36(5):820–5.
21. Folkman J, Shing Y. Angiogenesis. J Biol Chem. 1992;267(16):10931–4.
22. Love RM, Jenkinson HF. Invasion of dentinal tubules by oral bacterial. Crit Rev Oral Biol Med. 2002;13(2):171–83.
23. Nair PN. Pathogenesis of apical periodontitis and the causes of endodontic failures. Crit Rev Oral Biol Med. 2004;15(6):348–81.
24. Hahn CL, Liewehr FR. Relationships between caries bacteria, host responses and clinical signs and symptoms of pulpitis. J Endod. 2007;33:213–9.
25. Zhong S, Zhang S, Bair E, Nares S, Khan AA. Differential expression of microRNAs in normal and inflamed human pulps. J Endod. 2012;38(6):746–52.
26. Smith AJ, Smith JG, Shelton RM, Cooper PR. Harnessing the natural regenerative potential of the dental pulp. Dent Clin North Am. 2012;56(3):589–601.
27. Soden RI, Botero TM, Hanks CT, Nör JE. Angiogenic signaling triggered by cariogenic bacteria in pulp cells. J Dent Res. 2009;88(9):835–40.
28. Botero TM, Mantellini MG, Song W, Hanks CT, Nör JE. Effect of lipopolysaccharides on vascular endothelial growth factor expression in mouse pulp cells and macrophages. Eur J Oral Sci. 2003;111:28–234.
29. Botero TM, Shelburne CE, Holland GR, Hanks CT, Nör JE. TLR4 mediates LPS-induced VEGF expression in odontoblasts. J Endod. 2006;32:951–5.
30. Botero TM, Son JS, Vodopyanov D, Hasegawa M, Shelburne CE, Nör JE. MAPK signaling is required for LPS-induced VEGF in pulp stem cells. J Dent Res. 2010;89(3):264–9.
31. Senger DR, Galli SJ, Dvorak AM, Perruzzi CA, Harvey VS, Dvorak HF. Tumor cells secrete a vascular permeability factor that promotes accumulation of ascites fluid. Science. 1983;219:983–5.
32. Nagy JA, Benjamin L, Zeng H, Dvorak AM, Dvorak HF. Vascular permeability, vascular hyperpermeability and angiogenesis. Angiogenesis. 2008;11(2):109–19.
33. Heyeraas KJ, Berggreen E. Interstitial fluid pressure in normal and inflamed pulp. Crit Rev Oral Biol Med. 1999;10:328–36.
34. Heyeraas KJ, Kvinnsland I. Tissue pressure and blood flow in pulpal inflammation. Proc Finn Dent Soc. 1992;88 Suppl 1:393–401.
35. Smith AJ, Lesot H. Induction and regulation of crown dentinogenesis—embryonic events as a template for dental tissue repair. Crit Rev Oral Biol Med. 2001;12(5):425–37.
36. Goldberg M, Smith AJ. Cells and extracellular matrices of dentin and pulp: biological strategies for repair and tissue engineering. Crit Rev Oral Biol Med. 2004;15(1):13–27.
37. Smith AJ, Tobias RS, Plant CG, Browne RM, Lesot H, Ruch JV. In vivo morphogenetic activity of dentine matrix proteins. J Biol Buccale. 1990;18(2):123–9.
38. Smith AJ, Tobias RS, Plant CG, Browne RM, Lesot H, Ruch JV. Morphogenetic proteins from dentine extracellular matrix and cell-matrix interactions. Biochem Soc Trans. 1991;19(2):187S.
39. Roberts-Clark D, Smith AJ. Angiogenic growth factors in human dentine matrix. Arch Oral Biol. 2000;45(11):1013–6.
40. Graham L, Cooper PR, Cassidy N, Nör JE, Sloan AJ, Smith AJ. The effect of calcium hydroxide on solubilisation of bio-active dentine matrix components. Biomaterials. 2006;27(14):2865–73.
41. Tomson PL, Grover LM, Lumley PJ, Sloan AJ, Smith AJ, Cooper PR. Dissolution of bio-active dentine matrix components by mineral trioxide aggregate. J Dent. 2007;35(8):636–42.
42. Zhang R, Cooper P, Smith G, Nör JE, Smith AJ. Angiogenic activity of dentin matrix components. J Endod. 2011;37(1):26–30.
43. Guo S, DiPietro LA. Factors affecting wound healing. J Dent Res. 2010;89(3):219–29.

44. Edwards PC, Mason JM. Gene-enhanced tissue engineering for dental hard tissue regeneration: (2) dentin-pulp and periodontal regeneration. Head Face Med. 2006;2:16.
45. Friedlander LT, Cullinan MP, Love RM. Dental stem cells and their potential role in apexogenesis and apexification. Int Endod J. 2009;42:955–62.
46. About I. Dentin regeneration in vitro: the pivotal role of supportive cells. Adv Dent Res. 2011;23(3): 320–4.
47. Hahn CL, Liewehr FR. Update on the adaptive immune response of the dental pulp. J Endod. 2007;33(7):773–81.
48. Jiang Y, Russell TR, Schilder H, Graves DT. Endodontic pathogens stimulate monocyte chemoattractant protein-1 and interleukin-8 in mononuclear cells. J Endod. 1998;24(2):86–90.
49. Karapanou V, Kempuraj D, Theoharides TC. Interleukin-8 is increased in gingival crevicular fluid from patients with acute pulpitis. J Endod. 2008;34(2):148–51.
50. Huang GT, Potente AP, Kim JW, Chugal N, Zhang X. Increased interleukin-8 expression in inflamed human dental pulps. Oral Med Oral Pathol Oral Radiol Endod. 1999;88(2):214–20.
51. Park SH, Hsiao GY, Huang GT. Role of substance P and calcitonin gene related peptide in the regulation of interleukin-8 and monocyte chemotactic protein-1 expression in human dental pulp. Int Endod J. 2004;37(3):185–92.
52. Avery JK, Cox CF, Chiego Jr DJ. Presence and location of adrenergic nerve endings in the dental pulps of mouse molars. Anat Rec. 1980;198(1):59–71.
53. Wakisaka S. Neuropeptides in the dental pulp: distribution, origins and correlations. J Endod. 1990;16(2): 67–9.
54. Caviedes-Bucheli J, Camargo-Beltrán C, Gómez-la-Rotta AM, Moreno SC, Abello GC, González-Escobar JM. Expression of calcitonin gene related peptide (CGRP) in irreversible acute pulpitis. J Endod. 2004;30(4):201–4.
55. Airaksinen MS, Saarma M. The GDNF family: signaling, biological functions and therapeutic value. Nat Rev Neurosci. 2002;3:383–94.
56. Gale Z, Cooper PR, Scheven BA. Effects of glial cell line-derived neurotrophic factor on dental pulp cells. J Dent Res. 2011;90(10):1240–5.
57. Davidson RM. Neural form of voltage-dependent sodium current in human cultured dental pulp cells. Arch Oral Biol. 1994;39(7):613–20.
58. About I, Bottero MJ, de Denato P, Camps J, Franquin JC, Mitsiades TA. Human dentin production in vitro. Exp Cell Res. 2000;258:33–41.
59. Goodis H, Saeki K. Identification of bradykinin, substance P and Neurokinin A in human dental pulp. J Endod. 1997;23(4):201–4.
60. Casasco A, Calligaro A, Casasco M, Springall DR, Polak JM, Poggi P, Marchetti C. Peptidergic nerves in human dental pulp. Histochemistry. 1990;95: 115–21.
61. Caviedes-Bucheli J, Arenas N, Guiza O, Moncada NA, Moreno GC, Diaz E, Munoz HR. Calcitonin gene related peptide receptor expression in healthy and inflamed human pulp tissue. Int Endod J. 2005; 38(10):712–7.
62. Hargreaves KM, Swift JQ, Roszkowski MT, Bowles W, Garry MG, Jackson DL. Pharmacology of peripheral neuropeptide and inflammatory mediator release. Oral Surg Oral Med Oral Pathol. 1994;78(4):503–10.
63. Patel T, Park SH, Lin LM, Chiapelli F, Huang GT. Substance P induces interleukin-8 secretion from human dental pulp cells. Oral Surg Oral Med Oral Pathol Oral Radiol Endod. 2003;96(4):478–85.
64. Trantor IR, Messer HH, Birner R. The effects of neuropeptides (calcitonin gene related peptide and substance P) on cultured human pulp cells. J Dent Res. 1995;74(4):1066–71.
65. Calland JW, Harris SE, Carnes DL. Human pulp cells respond to calcitonin gene-related peptide in vitro. J Endod. 1997;23(8):485–9.
66. Arthur A, Rychkov G, Shi S, Koblar SA, Gronthos S. Adult human dental pulp stem cells differentiate toward functionally active neurons under appropriate environmental cues. Stem Cells. 2008;26(7): 1787–95.
67. Sakai K, Yamamoto A, Matsubara K, Nakamura S, Naruse M, Yamagata M, Sakamoto K, Tauchi R, Wakao N, Imagama S, Hibi H, Kadomatsu K, Ishiguro N, Ueda M. Human dental pulp-derived stem cells promote locomotor recovery after complete transection of the rat spinal cord by multiple neuro-regenerative mechanisms. J Clin Invest. 2012;122(1):80–90.
68. Taghipour Z, Karbalaie K, Kiani A, Niapour A, Bahramian H, Nasr-Esfahani MH, Baharvand H. Transplantation of undifferentiated and induced human exfoliated deciduous teeth-derived stem cells promote functional recovery of rat spinal cord contusion injury model. Stem Cells Dev. 2012;21(10): 1794–802.
69. Herzog B, Pellet-Many C, Britton G, Hartzoulakis B, Zachary IC. VEGF binding to NRP1 is essential for VEGF stimulation of endothelial cell migration, complex formation between NRP1 and VEGFR2, and signaling via FAK Tyr407 phosphorylation. Mol Biol Cell. 2011;22:2766–76.
70. Rohm B, Ottemeyer A, Lohrum M, Püschel AW. Plexin/neuropilin complexes mediate repulsion by the axonal guidance signal semaphorin 3A. Mech Dev. 2000;93(1–2):95–104.
71. Cvek M. Prognosis of luxated non-vital maxillary incisors treated with calcium hydroxide and filled with gutta-percha. A retrospective clinical study. Endod Dent Traumatol. 1992;8(2):45–55.
72. Nör JE. Tooth regeneration in operative dentistry. Oper Dent. 2006;31(6):633–42.
73. Huang GT, Yamaza T, Shea LD, Djouad F, Kuhn NZ, Tuan RS, Shi S. Stem/progenitor cell-mediated de novo regeneration of dental pulp with newly deposited continuous layer of dentin in an in vivo model. Tissue Eng A. 2010;16(2):605–15.

74. Rosa V, Zhang Z, Grande RH, Nör JE. Dental pulp tissue engineering in full-length human root canals. J Dent Res. 2013;92(11):970–5.
75. Ravindran S, Zhang Y, Huang CC, George A. Odontogenic induction of dental stem cells by extracellular matrix-inspired three-dimensional scaffold. Tissue Eng A. 2014;20(1–2):92–102.
76. Langer R, Vacanti JP. Tissue engineering. Science. 1993;260(5110):920–6.
77. Nakashima M, Reddi AH. The application of bone morphogenetic proteins to dental tissue engineering. Nat Biotechnol. 2003;21(9):1025–32.
78. Murray PE, Garcia-Godoy F, Hargreaves KM. Regenerative endodontics: a review of current status and a call for action. J Endod. 2007;33(4):377–90.
79. Sloan AJ, Smith AH. Stem cells and the dental pulp: potential roles in dentine regeneration and repair. Oral Dis. 2007;13(2):151–7.
80. Hargreaves KM, Giesler T, Henry M, Wang Y. Regeneration potential of the young permanent tooth: what does the future hold? J Endod. 2008;34(7):51–6.
81. Huang GT. Pulp and dentin tissue engineering and regeneration: current progress. Regen Med. 2009;4(5):697–707.
82. Goldberg M. Pulp healing and regeneration: more questions than answer. Adv Dent Res. 2011;23(3):270–4.
83. Sonoyama W, Liu Y, Yamaza T, Tuan RS, Wang S, Shi S, Huang GT. Characterization of the apical papilla and its residing stem cells from human immature permanent teeth: a pilot study. J Endod. 2008;34(2):166–71.
84. Ishizaka R, Hayashi Y, Iohara K, Sugiyama M, Murakami M, Yamamoto T, Fukuta O, Nakashima M. Stimulation of angiogenesis, neurogenesis and regeneration by side population cells from dental pulp. Biomaterials. 2013;34:1888–97.
85. Bernardo ME, Cometa AM, Pagliara D, Vinti L, Rossi F, Cristantielli R, Palumbo G, Locatelli F. Ex vivo expansion of mesenchymal stromal cells. Best Pract Res Clin Haematol. 2011;24(1):73–81.
86. Jung S, Panchalingam KM, Rosenberg L, Behie LA. Ex vivo expansion of human mesenchymal stem cells in defined serum-free media. Stem Cells Int. 2012;2012:123030.
87. Sakai VT, Cordeiro MM, Dong Z, Zhang Z, Zeitlin BD, Nör JE. Tooth slice/scaffold model of dental pulp tissue engineering. Adv Dent Res. 2011;23(3):325–32.
88. Galler KM, Aulisa L, Regan KR, D'Souza RN, Hartgerink JD. Self-assembling multidomain peptide hydrogels: designed susceptibility to enzymatic cleavage allows enhanced cell migration and spreading. J Am Chem Soc. 2010;132(9):3217–23.
89. Galler KM, Cavender AC, Koeklue U, Suggs LJ, Schmalz G, D'Souza RN. Bioengineering of dental pulp cells in a PEGylated fibrin gel. Regen Med. 2011;6(2):191–200.
90. Galler KM, Hartgerink JD, Cavender AC, Schmalz G, D'Souza RN. A customized self-assembling peptide hydrogel for dental pulp tissue engineering. Tissue Eng Part A. 2012;18(1–2):176–84.
91. Cavalcanti BN, Zeitlin BD, Nör JE. A hydrogel scaffold that maintains viability and supports differentiation of dental pulp stem cells. Dent Mater. 2013;29(1):97–102.
92. Carmeliet P, Collen D. Molecular analysis of blood vessel formation and disease. Am J Physiol. 1997;273:H2091–104.
93. Ferrara N. Vascular endothelial growth factor. Arterioscler Thromb Vasc Biol. 2009;29(6):789–91.
94. Zieris A, Prokoph S, Levental KR, Welzel PB, Grimmer M, Freudenberg U, Werner C. FGF-2 and VEGF functionalization of starPEG-heparin hydrogels to modulate biomolecular and physical cues of angiogenesis. Biomaterials. 2010;31:7985–94.
95. Srisuwan T, Tilkorn DJ, Al-Benna S, Abberton K, Messer HH, Thompson EW. Revascularization and tissue regeneration of an empty root canal space is enhanced by a direct blood supply and stem cells. Dent Traumatol. 2013;29(2):84–91.
96. Syed-Picard FN, Ray Jr HL, Kumta PN, Sfeir C. Scaffoldless tissue-engineered dental pulp cell constructs for endodontic therapy. J Dent Res. 2014;93:250–5.
97. Cavalcanti BN, Campos NS, Nör JE. Stem cells in health and disease. Rev Assoc Paul Cir Dent. 2011;65(2):92–7.
98. Banchs F, Trope M. Revascularization of immature permanent teeth with apical periodontitis: new treatment protocol? J Endod. 2004;30(4):196–200.
99. Trope M. Treatment of the immature tooth with a non-vital pulp and apical periodontitis. Dent Clin North Am. 2010;54(2):313–24.
100. Ding RY, Cheung GS, Chen J, Yin XZ, Wang QQ, Zhang CF. Pulp revascularization of immature teeth with apical periodontitis: a clinical study. J Endod. 2009;35(5):745–9.
101. Jung IY, Lee SJ, Hargreaves KM. Biologically based treatment of immature permanent teeth with pulpal necrosis: a case series. Tex Dent J. 2012;129(6):601–16.
102. Jeeruphan T, Jantarat J, Yanpiset K, Suwannapan L, Khewsawai P, Hargreaves KM. Mahidol study 1: comparison of radiographic and survival outcomes of immature teeth treated with either regenerative endodontic or apexification methods: a retrospective study. J Endod. 2012;38(10):1330–6.
103. Kim JY, Xin E, Moioli EK, Chung J, Lee CH, Chen M, Fu SY, Koch PD, Mao JJ. Regeneration of dental-pulp-like tissue by chemotaxis-induced cell homing. Tissue Eng A. 2010;16(10):3023–31.
104. Kaigler D, Pagni G, Park CH, Braun TM, Holman LA, Yi E, Tarle SA, Bartel RL, Giannobile WV. Stem cell therapy for craniofacial bone regeneration: a randomized, controlled feasibility trial. Cell Transplant. 2013;22(5):767–77.

105. Mason S, Tarle SA, Osibin W, Kinfu Y, Kaigler D. Standardization and safety of alveolar bone-derived stem cell isolation. J Dent Res. 2014;93(1):55–61.
106. Trevino EG, Patwardhan AN, Henry MA, Perry G, Dybdal-Hargreaves N, Hargreaves KM, Diogenes A. Effect of irrigants on the survival of human stem cells of the apical papilla in a platelet-rich plasma scaffold in human root tips. J Endod. 2011;37(8):1109–15.
107. Ruparel NB, Teixeira FB, Ferraz CC, Diogenes A. Direct effect of intracanal medicaments on survival of stem cells of the apical papilla. J Endod. 2012;38(10):1372–5.
108. Martin DE, De Almeida JF, Henry MA, Khaing ZZ, Schmidt CE, Teixeira FB, Diogenes A. Concentration-dependent effect of sodium hypochlorite on stem cells of apical papilla survival and differentiation. J Endod. 2014;40(1):51–5.

18 Scaffolds for Pulp Repair and Regeneration

Kerstin M. Galler

18.1 Introduction

Scaffold-based tissue engineering concepts involve the combination of competent (stem) cells and bioactive molecules with a structural matrix to support cell adhesion, proliferation, differentiation, and tissue formation. The goal is to improve or, ideally, fully restore the functions of diseased tissues or organs. Advances in tissue engineering are already translating into medical practice and changing therapeutic strategies today. The area of material science has contributed considerably to this process. For tissue engineering approaches, the choice of an appropriate scaffold is the first and certainly a crucial step. Traditionally, bioinert materials have been used mainly as carriers and delivery vehicles, relying on the intrinsic capability of stem cells to form tissues. As this may not suffice to induce regeneration, there is a need for novel biomimetic scaffolds that can provide mechanical and biochemical cues to promote a variety of specific interactions between cells and matrix. With the development of more versatile and sophisticated biomaterials, scaffolds have transitioned from inert, passive cell carriers and mere delivery vehicles to inductive and instructive matrices, which can be controlled in different aspects of material behavior and can thus elicit a desired cellular behavior by means of stiffness, degradation rate and pattern, bioactive motifs, or controlled release of growth and differentiation signals. That way, tailor-made materials for specific applications can be created.

A vast variety of biomaterials is available to the tissue engineer; each material offers a unique chemistry, composition and structure, degradation profile, and possibility for modification. Recent approaches to generate dental pulp rely mainly on established materials, such as hydroxyapatite [1, 2], collagen [3–5], or polyester [6, 7]. Results after transplantation show soft connective tissue formation and newly generated tubular dentin. Generation of a customized scaffold for dentin–pulp-complex engineering, however, should enable and support vascularization, cell–matrix interactions, biodegradation, growth factor incorporation, biomineralization, and contamination control. These aspects can be accounted for by programming specific features into a tailor-made material. Matrices which enable control of material stiffness, induction of mineral nucleation, or introduction of antibacterial activity and evoke cellular responses by incorporation of cell adhesion motifs, enzyme-cleavable sites, and suitable growth factors can be fabricated. With a wide range of biomaterial choices, the question is in how far strategies for dentin–pulp complex engineering can be optimized with the help of novel and inductive scaffolds which are tunable and tailored toward this specific approach.

K.M. Galler, DDS, PhD
Department of Restorative Dentistry
and Periodontology, University Hospital Regensburg,
Franz-Josef-Strauss-Allee 11, Regensburg
93053, Germany
e-mail: kerstin.galler@ukr.de

M. Goldberg (ed.), *The Dental Pulp*,
DOI 10.1007/978-3-642-55160-4_18,

18.2 Biomaterials for Tissue Engineering

Many materials have been designed and constructed for tissue engineering, mainly natural and synthetic polymers or inorganic materials and composites, which have been fabricated into porous scaffolds, nanofibrous materials, microparticles, and hydrogels. Natural materials include collagen, elastin, fibrin, alginate, silk, glycosaminoglycans such as hyaluronan, and chitosan. They offer a high degree of structural strength, are compatible with cells and tissues and biodegradable, but are often difficult to process and afflicted with the risk of transmitting animal-associated pathogens or inducing an immunoresponse. Collagen has been of special interest and used for manifold approaches in bone [8–13] and tooth tissue engineering [3–5]; it has been fabricated as gels, nanofibers, porous scaffolds, and films [9, 10]. However, it is mechanically weak and undergoes rapid degradation [10]. Synthetic polymers, on the other hand, provide excellent chemical and mechanical properties and allow high control over the physicochemical characteristics, such as molecular weight, configuration of polymer chains, or the presence of functional groups [14–16]. Disadvantages of synthetic polymers can be chronic or acute inflammatory response in the host and localized pH decrease due to relative acidity of hydrolytically degraded by-products. Commonly used synthetic scaffolds are fabricated from polylactic acid (PLA), polyglycolic acid (PGA), and their copolymer, polylactic-co-glycolic acid (PLGA). PLA is a very strong polymer and has found many applications where structural strength is an important criterion. PGA has been used as an artificial scaffold for cell transplantation; it degrades as the cells excrete extracellular matrix. Both PLA and PGA are nontoxic and biocompatible; they degrade by simple hydrolysis and have gained FDA approval for a number of applications (for further review, see [17] and [18]). Recently, hydrogels have been explored for tissue engineering applications in more detail. These are made from natural materials such as collagen [10], fibrin [19], proteoglycans [20], and hyaluronic acid [21]; they can be derived from other biological sources such as chitosan [22] or alginate [23] or synthesized from polyethylene glycol [24] or self-assembling peptide molecules [25]. Hydrogels offer numerous interesting properties including high biocompatibility, a tissue-like water content, viscoelastic properties similar to soft tissues, efficient transport of nutrients and metabolic products, uniform cell encapsulation, and the possibility of injection and gelation in situ [26]. Based on their chemistry, they can be chemically or physically cross-linked, and modifications such as incorporation of biofunctional molecules or growth and differentiation factors are possible [26]. An overview of materials for tissue engineering is given in Table 18.1.

18.3 Requirements for Scaffold Materials

For most regenerative strategies, an organic scaffold is used which should facilitate the attachment, migration, proliferation, and three-dimensional spatial organization of the cell population required for structural and functional replacement of the target tissue. As the science of tissue engineering is progressing, the definition of an ideal scaffold material is still unclear due to complex considerations which involve scaffold architecture and geometry, structural mechanics, surface properties, degradation characteristics, and composition of biological components as well as the change of these factors in in vitro or in vivo systems over time. It is unlikely that a single scaffold could serve as a universal foundation for the regeneration of different tissues. However, general requirements regarding physical, chemical, and biochemical properties can be defined. In either case, a scaffold material has to be nontoxic, biocompatible, and non-immunogenic to avert damage to neighboring cells and prevent adverse tissue reactions. Biodegradability by enzymes or hydrolysis is desired if the scaffold serves as a temporary template to provide support for the transplanted cells, and the degradation rate should match

Table 18.1 Overview of scaffold materials for tissue engineering

Biomaterial	Source	Advantages	Limitations
Natural polymers			
Polysaccharides Alginate Dextran Chitosan Cellulose Starch Hyaluronan	Plant, animal Sea algae Bacteria Crustaceans Plant cell walls Crops ECM	Derived from renewable sources Large diversity Unique but complex structures Functional groups, tailorable chemistry Specific recognition domains Intrinsic biodegradability Composition of hybrid materials	Number of biologically derived polymers is limited Difficult to process Properties may differ Undesirable immunoresponse Pathogen transmission
Extracellular matrix	Tissue specific	Temporary-controlled growth factor release Favorable environment for constructive tissue remodeling Processing into many forms possible	Difficult to process and sterilize Batch variations
Synthetic polymers			
Polyester PLA: poly(lactic) acid PGA: poly(glycolic) acid PLGA: copolymer of PLA and PGA PCL: poly(ε-caprolactone) Poly(urethanes) Poly(ether ester) PEG: poly(ethylene glycol) PBT: poly(butylene terephthalate)		Minimal foreign body reaction Tailorable mechanical properties and degradation rate, wettability, and protein adsorption Degradation products present in the metabolic pathways Functional groups to attract cells or bind growth factors Cheap and reproducible production Easily processed into any shape	Environment unlike ECM Accumulation of acidic degradation products
Hydrogels			
Natural materials Collagen Fibrin Proteoglycans, HA Chitosan Alginate Synthetic materials PEG: poly(ethylene glycol) SAP: self-assembling peptides		High biocompatibility Tissue-like water content Viscoelastic properties similar to soft connective tissue Efficient transport of nutrients and waste Uniform cell encapsulation Injection and gelation in situ Chemical or physical cross-linking Modifications with biofunctional molecules or growth factors possible	Disadvantages of the individual material source Due to low mechanical stiffness not suitable for certain applications, e.g., bone grafts
Bioceramics			
Calcium phosphates Hydroxyapatite Tricalcium phosphate Biphasic calcium phosphate (HA/TCP) Bioactive glasses Silica-based glasses		Biocompatible Excellent bone-bonding properties Biodegradable Osteoconductive Variable rates of degradation	Brittle Subject to fatigue Decrease of mechanical strength under humid conditions

the cells' rate of extracellular matrix production to enable constructive remodeling. This is characterized by scaffold degradation, cellular infiltration, vascularization, differentiation and spatial organization of the cells, and eventually replacement of the scaffold by the appropriate tissues. As carriers for drugs and differentiation factors, the materials should be versatile enough to enable the incorporation and controlled release of bioactive molecules.

Furthermore, any scaffold material must provide sufficient mechanical strength to substitute for the mechanical function of the diseased or damaged tissue that should be regenerated. The three-dimensional scaffold needs to maintain sufficient mechanical support to the transplanted cells during the growth and remodeling process, where the degree of remodeling depends on the cell number and cell type as well as on the tissue itself. The scaffold architecture and chemistry have to enable initial cell attachment, subsequent migration into or through the scaffold, transfer of nutrients and metabolites, as well as provision of sufficient space for the development and later reconstruction of the organized tissue. From a clinical point of view, it must be possible to fabricate the scaffold material under GMP (good manufacturing practice) conditions in a quality-controlled and reproducible fashion.

In one scenario, cells can be seeded onto the scaffold and cultured in vitro to generate the desired tissue before transplantation. A different approach is the design of materials for transplantation of a primarily cell-free system, which will, due to a combination of signaling molecules incorporated in the scaffold, induce the homing of stem cells residing in the respective tissues and promote their differentiation to support regeneration. Cell-free scaffolds are especially attractive because of an easier handling process that eliminates the issues associated with the use of stem cells and their expansion in vitro, with storage and shelf life, cost aspects, immunoresponse of the host, and transmission of diseases [27].

18.4 Dental Pulp Tissue Engineering

Although dentistry is one of the disciplines which have capitalized on the use of biomaterials for the longest time, these have served mainly to replace lost tissues and restore esthetics and function. However, tissue engineering approaches in dentistry are evolving. In the fields of periodontology and oral surgery, regenerative strategies have already been implemented in daily practice. Commercially available products for bone and periodontal tissue regeneration are available to clinicians and have improved treatment outcomes and success rates [28]. More recently, engineering of dental pulp and dentin using pulp-derived stem cells has made considerable progress.

The fact that dental pulp possesses regenerative capabilities has been known for several decades. Stimuli such as bacterial toxins or tissue damage will lead to an upregulation of odontoblast activity, and these cells will produce a reparative dentin as an active defense mechanism to separate the soft tissue from the site of injury. Even after exposure of pulp tissue in deep cavities and disruption of the odontoblast layer, regeneration is possible after application of medicaments such as calcium hydroxide or mineral trioxide aggregate (MTA), which disinfect due to a high pH, cause necrosis in the adjacent cell layer, and stimulate defense mechanisms and reparative dentin formation [29, 30]. However, pulp's capacity to regenerate is limited, unless we find ways to more effectively exploit its intrinsic healing potential. If regeneration fails and inflammation persists, non-regenerative endodontic treatment can keep the tooth functional but terminates dentin formation and root maturation and will leave the tooth deprived of its soft tissue (devital) and weakened.

Several groups have begun to develop strategies to engineer dental pulp. The two categories of materials that are most commonly used in tissue engineering are synthetic polymers such as PLA and PGA [31] and matrices derived from biological sources such as reconstituted collagen [10]. Table 18.2 provides an overview of biomaterials that have been utilized particularly for dental pulp tissue engineering.

In 1996, a pulp-like tissue was first engineered in vitro after seeding pulp fibroblasts on PGA fibers, where cells formed new tissue after 60 days in culture [32]. Using a similar approach, the ability of different scaffold materials to support pulp tissue formation from pulp fibroblasts was evaluated 2 years later. PGA, collagen hydrogels, and alginate were tested, where culturing of cells on PGA resulted in tissue formation and collagen synthesis, whereas only moderate cell proliferation was observed on the collagen scaffold and no proliferation on alginate [33].

Table 18.2 Biomaterials that have been utilized particularly for dental pulp tissue engineering

Material	Engineering approach	Result	Reference
PGA, collagen I, alginate	Pulp fibroblasts seeded onto different materials, cell culture *in vitro*	Pulp-like tissue after 45–60 days on PGA	Mooney et al. [32] Bohl et al. [33]
HA/TCP	Stem cells from dental pulp (SHED, DPSC) mixed with HA/TCP powder transplanted into nude mice	Generation of dentin or bone (SHED) and dentin–pulp-like complexes (DPSC)	Gronthos et al. [1] Miura et al. [2]
Collagen I and III, chitosan, gelatin	Human dental pulp cells seeded into different materials for comparison in vitro	Adhesion and proliferation: Col I > Col III > gelatin> > chitosan ALP activity: Col I > Col III > gelatin> > chitosan Mineralization: Col I > Col III > gelatin Oc, Dspp, and Dmp-1 expression on collagen	Kim et al. [100]
Collagen I with Dmp-1	Collagen scaffolds laden with Dmp-1 and dental pulp stem cells were placed in dentin disks with a simulated furcal perforation and transplanted subcutaneously into nude mice	Formation and organization of new pulp tissue	Prescott et al. [3]
PLA	SHED seeded onto PLA scaffolds into tooth slices, subcutaneous transplantation into nude mice	Formation of vascularized soft connective, pulp-like tissue and new tubular dentin	Cordeiro et al. [6] Sakai et al. [7]
PLGA	SCAP and DPSC seeded onto PLGA into root canals sealed with MTA on one side, subcutaneous implantation into nude mice for 3–4 months	Formation of a pulp-like tissue, deposition of dentin along the root canal wall	Huang et al. [34]

Recently, formation of pulpal tissue could be demonstrated in vivo. After isolation and in vitro characterization of dental pulp stem cells from deciduous teeth and third molars, these cells were mixed with hydroxyapatite/tricalcium phosphate (HA/TCP), and the formation of dentin, bone, and dentin–pulp-like complexes was observed [1, 2]. Highly promising outcomes have been reported by Nör's group, where dental stem cells were seeded on PLA scaffolds and inserted into tooth slices after removal of the original pulp tissue. These constructs were implanted subcutaneously into immunodeficient mice. Formation of a vascularized pulp-like tissue with odontoblast-like cells and newly generated dentin was shown [6, 7]. Mobilization and release of growth factors and proteins from the dentin due to a locally decreased pH by degradation of PLA scaffolds might promote the differentiation process in this case [7]. Similarly, Huang et al. observed soft tissue and deposition of new dentin after transplantation of stem cells from apical papilla (SCAP) on PGLA into an empty root canal space [34]. In previous own work, dental pulp stem cells were seeded into self-assembling peptide hydrogels together with growth factors TGFβ1, FGF-2, and VEGF. The cell–gel mixture was injected into dentin cylinders prepared from extracted human teeth and transplanted subcutaneously into immunocompromised mice. After 5 weeks in situ, a vascularized pulp-like tissue had formed within the dentin cylinders, the cells lining the dentin expressed dentin sialoprotein as an indicator of differentiated odontoblast-like cells, and these cells extended processes into the dentinal tubules similar to the physiological situation [35].

Nakashima et al. developed a pulpotomy model in canines, where the pulp chamber was accessed, the coronal pulp was removed, and the pulp chamber was filled with an angiogenic subpopulation of dental pulp stem cells on a collagen type I carrier laden with the growth factor SDF-1 (stromal-derived factor 1). After 6 weeks, the pulp chamber was filled with pulp-like tissue which could not be distinguished from the original pulp, it was vascularized, and tubular dentin formation had taken place [4]. Taking it one step further, the group extracted teeth in dogs and replanted them after drilling an access cavity to the pulp chamber, pulpectomy and apexectomy, and filling of the root canal with a collagen scaffold, stem cells, and SDF-1. A pulp-like tissue could be found along the root canal, showing that after transplantation of stem cells, dental pulp tissue engineering was possible [5].

In summary, collagen I and the synthetic polymers showed the most favorable results among the materials studied for this particular application. In terms of biocompatibility and degradation, all the previously described materials exhibit satisfactory results. Synthetic polyester such as PLA, PGA, and their copolymers are nontoxic and biocompatible; they degrade by hydrolysis and have gained FDA approval for various applications [17]. Collagen is biocompatible and degradable by enzymes, but natural polymers are often difficult to process and to modify and generally afflicted with the risk of transmitting animal-associated pathogens or provoking an immunoresponse. Alginate, a polysaccharide derived from red algae, offers a mild cell encapsulation process as it can be cross-linked via Ca^{2+}. However, it degrades in a rather uncontrolled manner via dissolution, as the material is sensitive to calcium chelating compounds [36]. Chitosan is derived from chitin, a polysaccharide found in crustaceans. Due to its biocompatibility and degradability via naturally occurring enzymes, it has been used for numerous tissue engineering applications [37].

18.5 Growth and Differentiation Factors

Many signaling pathways and molecules are similar during tooth development and tooth reparative responses. Increased knowledge of the biological cues mediating these processes enables investigators to mimic or supplement regeneration or repair of dental tissues.

Growth factors, especially those of the transforming growth factor-beta (TGFb) family, are important in cellular signaling for proliferation, differentiation, and induction of odontoblast secretory activity. These growth factors are secreted by dentin-forming cells during tooth development and correlate with specific events regarding morphogenesis, histogenesis, and cytodifferentiation [38, 39]. Another important family of growth factors in tooth development and regeneration is the bone morphogenetic proteins (BMPs). Recombinant human BMP-2 stimulates differentiation of postnatal pulp stem cells into odontoblast-like cells in culture [40, 41]. In addition, recombinant BMP-2, BMP-4, and BMP-7 have been shown to induce formation of reparative dentin in vivo [42, 43]. Apart from growth factors, other molecules have been demonstrated to induce pulp cell differentiation. Dentin matrix protein-1, a non-collagenous protein involved in the mineralization process, stimulated cytodifferentiation, collagen production, and deposition of calcified tissue in dental pulp in a rat model [44]. Dexamethasone, a synthetic glucocorticoid, reduced cell proliferation and stimulated the expression of mineralization-associated markers such as alkaline phosphatase and dentin sialophosphoprotein in primary human pulp cells [45]. Addition of β-glycerophosphate to explants from human teeth induced a change in cell morphology, collagen synthesis, and mineral formation [46]. Combinations of inorganic phosphate and dexamethasone have been used as standard osteogenic supplements to drive the differentiation of mesenchymal stem cells into osteoblasts as well odontoblast-like cells followed by mineral deposition [1, 2]. This may be explained by the fact that osteogenesis and dentinogenesis are similar processes, and bone-forming osteoblasts and dentin-forming odontoblasts are closely related cell lineages. However, they remain distinct cell types, as observed by their slightly different gene expression profile and the obvious structural differences of their respective products. Optimal conditions permissive for dental stem cell differentiation into odontoblasts rather than osteoblasts remain to be elucidated. The increasing knowledge about the underlying biological processes enables us to develop materials which allow for the incorporation of biological cues in order to stimulate a desired cellular response.

Whereas the addition of exogenous growth and differentiation factors is one way to drive cell proliferation, differentiation, and tissue formation, questions and problems regarding the optimum concentrations and the possibility of undesirable side effects including tumorogenesis remain. An alternative is the recruitment of endogenous growth factors, which are present in the dentin matrix itself. During tooth development, the odontoblasts secrete a variety of growth factors, neurotrophic factors, and cytokines which are deposited within the organic matrix preceding the mineralized tissue [47–52]. During mineralization, these bioactive factors become embedded and immobilized in the dentin matrix [47, 53]. Whereas proteins and growth factors in an active form have a short half-life, binding to extracellular matrix components may be required to maintain their bioactivity by protecting them from proteolytic degradation and thus prolonging their life span. Among growth-factor-binding compounds are proteoglycans, mainly heparan sulfate [53], furthermore specific binding proteins [54], glycoproteins such as fibronectin [55], or different types of collagen [56, 57].

As there is no turnover in dentin extracellular matrix, bioactive regulatory molecules can be reactivated much later in life upon release from their bond. Organic acids or chelating agents such as EDTA (ethylenediaminetetraacetic acid) are suitable for dentin demineralization. EDTA acts as a potent chelator which binds and withdraws calcium ions from solution and thus alters the balance between binding and release of ions of the hydroxyapatite crystal lattice. A variety of growth factors have been shown to be present in the EDTA-soluble fraction of demineralized human dentin extracellular matrix, including transforming growth factor β1 (TGF-β1), fibroblast growth factor 2 (FGF-2), bone morphogenetic protein 2 (BMP-2), platelet-derived growth factor (PDGF), placenta growth factor (PIGF), epidermal growth factor (EGF), but also angiogenic factors such as vascular endothelial growth factor (VEGF) [47–52]. These molecules are effective at very low concentrations and elicit cellular responses still at picogram concentrations, modifying immunodefense, angiogenesis, cell recruitment, proliferation, and differentiation as well as mineralization [58–61].

As described previously, growth factors in natural extracellular matrix are bound to ECM elements. This feature can be mimicked by synthetic matrices, for example, via binding to heparin. This negatively charged glycosaminoglycan can be incorporated into scaffold materials and in turn bind to growth factors such as TGF-β1, FGF-2, or VEGF. Binding and slow release of these growth factors from synthetic scaffolding materials has been demonstrated and can be taken advantage of [62, 63].

18.6 Cell-Based vs. Cell-Free Approaches to Dental Pulp Engineering

It is accepted to date that after transplantation of stem cells into the root canal, dental pulp tissue engineering is possible [4–7, 34, 35]. If the goal is to develop a clinical procedure of stem cell transplantation for pulp regeneration, the choice of scaffold material seems to be of secondary importance, as we can utilize the stem cells' inherent competence to form pulp tissue, especially in contact with the appropriate matrix, the dentin. However, cell-based tissue engineering of dental pulp is afflicted with several problems. At the time point and age when dental stem cells are available from a patient (loss of primary teeth, removal of wisdom teeth), therapies to regenerate dental pulp are usually not necessary yet. Thus, stem cells need to be stored, but banking of stem cells as well as expansion in culture before

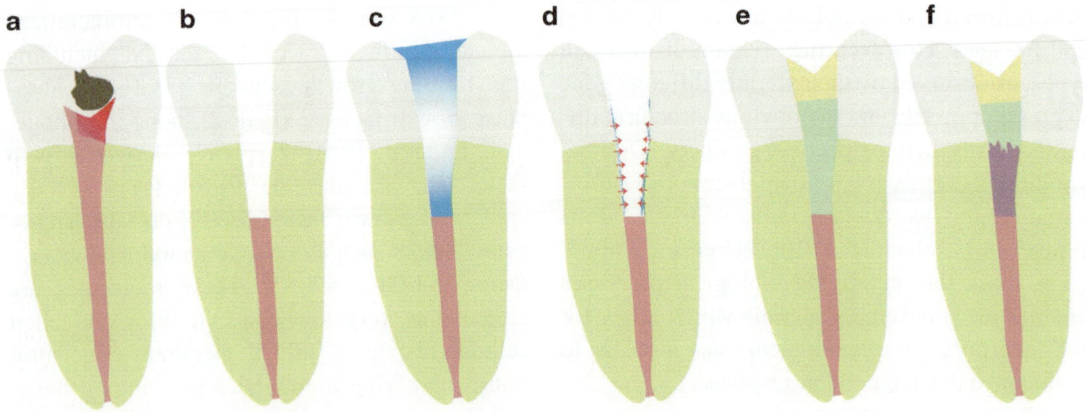

Fig. 18.1 Clinical procedure for dental pulp regeneration using stem cell homing. (**a**) Irreversible pulpitis. (**b**) Pulpotomy. (**c**) EDTA conditioning. (**d**) Release of growth factors. (**e**) Insertion of biomaterial. (**f**) Regeneration (Used with permission of Elsevier. Galler et al. [99])

transplantation is costly. Even if money issues do not matter, it can barely be envisioned to perform a transplantation procedure in a dental office but has to be accomplished in a hospital where GMP guidelines are followed. Generally, it appears that the benefit does not outweigh the cost and effort.

A different approach to pulp regeneration is cell-free regeneration or cell homing. In that case, a scaffold material in combination with growth factors is supposed to recruit resident stem cells and attract them to populate the scaffold material, proliferate, differentiate, and form a three-dimensional tissue [27]. In that case, cells could be recruited either from remnant vital pulp tissue after pulpotomy or, if the pulp tissue is lost, from the periapical region. In the case of cell homing, the scaffold material plays a crucial role and should be tailor-made for this specific application. The envisioned clinical procedure for dental pulp regeneration using stem cell homing is depicted in Fig. 18.1.

Ideally, the scaffold material should most closely resemble the cells' physiological environment – natural extracellular matrix. The ECM acts as a structural support, but its role goes far beyond this. It is a nanostructured environment that provides the biochemical cues to modulate cellular behavior and reinforce a particular phenotype. Furthermore, the ECM is dynamic; it can be selectively degraded and remodeled by the cells living within it. Polymers like PLA have the advantage of being biodegradable, biocompatible, inexpensive, and easy to prepare. However, they lack the chemical information that can be found in the ECM physiologically. On the other hand, collagen offers the chemical and structural information of the ECM but is difficult to customize for specific applications. Because of its biological origin, purity and immune reaction can be an issue. An ideal scaffold should combine the best properties of each of these groups of biomaterials. These would be structurally similar to ECM at the nanoscale, be able to present complex molecular information to the cells, and be easy to modify for specific applications. To address these deficiencies novel synthetic matrices are developed for tissue engineering. Among these, peptide-based nanofibers are an example of a tunable, ECM-like matrix and are particularly promising due to their ease of synthesis, chemical diversity, and high control over various aspects of material behavior [25, 62, 64–66]. Regarding dentin–pulp-complex engineering, the scaffold should allow us to address the particular challenges of this approach, including injectability into the root canal, contamination control, vascularization and innervation of a long and narrow space, the incorporation of growth and differentiation factors relevant to odontoblast differentiation, the support of mineral formation, and the possibility to create and insert primarily cell-free matrices, which are capable of recruiting resident stem cells in the respective tissues.

18.7 Desirable Properties in a Scaffold

18.7.1 Biodegradability

Regardless of whether a cell-based or cell-free approach to dental pulp tissue engineering is chosen, this application requires a biodegradable scaffold. Biodegradability of a scaffolding system has to closely match the cells' ECM synthesis rate. Too rapid a rate of degradation will result in dissolution of the material before the living tissue has had an opportunity to fill in the defect. In contrast, if degradation occurs too slowly, unintended reactions to the matrix may occur. For example, ECM buildup may take place only around the cells and interfere with cellular functions. A number of factors may affect the degradation rate, including cross-linking density, mechanism of degradation, susceptibility of the substrate to degradation, monomer concentration, cell type, and cell seeding density. Some systems like hydrogels which are formed via physical or ionic cross-links might dissolve by reversing the gelation process. Alginate gels degrade as divalent Ca^{2+} ions, which cross-link polymer chains, are exchanged with monovalent ions as new tissue is formed. Higher control over biodegradability can be achieved by incorporation of degradation sites, which allow for enzyme-mediated degradation, hydrolysis, or a combination of both. Degradation kinetics can be altered by changing the material's composition, the number of linkages, or their chemistry. The degradation rate of inorganic calcium phosphate, which has been widely used for bone tissue engineering, can be adjusted using biphasic materials and varying the ratio of nondegradable hydroxyapatite (HA) and degradable tricalcium phosphate (TCP) [14]. Synthetic polymers such as PLA or PGA degrade by hydrolysis, which makes it difficult to control the time frame of degradation. Ester linkages can be incorporated in the backbone or cross-links of the material. Both the molecular weight of the monomer and the type of polyester influence the degradation rate [67]. Alternatively, materials can be designed to allow for cell-mediated degradation, which enables cells to degrade the matrix locally, thus facilitating cell spreading and remodeling. Hydrogels fabricated from natural biopolymers such as collagen, fibrin, or hyaluronan are naturally subject to cell-mediated degradation. Incorporation of enzyme-cleavable amino acid sequences in synthetic polymers can mimic this effect. Since the amino acids flanking the scissile bond determine the specificity of the site [68], this strategy provides means to control the susceptibility to degradation. Furthermore, cleavage sites can be designed according to the cell type. In dental pulp, for example, the most abundant protease is matrix metalloproteinase-2 (MMP-2), a gelatinase capable of degrading various ECM components, which is essential for tissue remodeling, growth factor release, and regeneration. In previous work, a hexapeptide containing the MMP-2 consensus cleavage motif LRG was successfully incorporated into the peptide sequence. The degradation of hydrogels resulting from this enzyme-cleavable peptide by MMP-2 was confirmed in vitro, and cell culture studies demonstrated that the cells started migrating into the hydrogel matrix only if the cleavage site was present [69].

18.7.2 Cell–Matrix Interactions

To date, biomaterials can be designed, synthesized, and modified in numerous ways to promote selective interactions with cells. Through scaffold-mediated cues, extracellular events can be triggered, which in turn can induce a desired effect intracellularly. Through these interactions, we can control cell adhesion, proliferation, migration, differentiation, matrix synthesis, and, ultimately, tissue formation. The first step of cell–matrix interaction, cell adhesion, can be mediated by various short peptide motifs, which mimic ligands on molecules abundant in natural ECM. It is particularly attractive to incorporate these motifs into tissue engineering scaffolds. Sequences derived from fibronectin include the integrin-binding tripeptide RGD (arginine–glycine–aspartic acid) [70], which has been used for tissue engineering scaffolds countless times, furthermore REDV (arginine–glutamic acid–aspartic acid–valine) [71], PHSRN (proline–histidine–serine–arginine–asparagine) [72], and KNEED (lysine–arginine–glutamic acid–aspartic acid) [73]. Similarly, short peptide motifs can be found in laminin [74] and collagen [75]. Other sequences to be considered include short heparin-binding sequences, which can be utilized to link growth factors to the matrix. Heparin or heparan sulfate can bind growth factors in the ECM, protect them from rapid proteolysis, and release them slowly in response to cell-mediated matrix degradation [53]. Consensus heparin-binding sequences have been described as XBBXBX or XBBBXXBX (where X is a hydrophobic amino acid and B a basic amino acid residue) [76]. Using this indirect mechanism of binding, various growth factors can be incorporated into scaffolding systems including TGF-β1 and FGF-2, which might be particularly interesting for the work with dental stem cells (see section on growth factors). Heparan-sulfate-mediated incorporation or growth factors can furthermore be utilized to address another important aspect in tissue engineering, namely, the supply of implanted cells with nutrients through new blood vessel formation and connection to the existing vascular network. Stimulation of vasculogenesis via heparin-bound vascular endothelial growth factor (VEGF) has been demonstrated in in vivo applications [64]. Binding of TGF-β1, FGF-2, and VEGF into self-assembling peptide hydrogels via heparin, slow release, and bioactivity after release have been demonstrated [62]. Dental stem cells seeded into growth-factor-laden hydrogels rendered a pulp-like tissue in dentin cylinders after subcutaneous transplantation into immunocompromised mice for 4 weeks [62]. Another strategy well worth exploring for growth factor delivery in dental tissue engineering might be the use of recombinant biopolymers. Sustained release of growth factors such as BMPs from elastin-like polymer nanoparticles [77], poly(lactic) acid [78], collagen [79], or dextran-derived microspheres [80] has been demonstrated, which stimulated the mineralization activity of bone-derived cells. Recent work on dental pulp stem cells demonstrated increased proliferation and induction of chemotaxis after controlled release of TGF-β1 and FGF-2 from biodegradable polymer microspheres [81].

18.7.3 Viscoelastic Properties

In living tissues, the elastic moduli span several orders of magnitude, ranging from 100 Pa in the brain to 950 kPa in cartilage or tendon [82]. Matrix stiffness is increasingly appreciated as cells noticeably alter their adhesion behavior, morphology, and gene expression profile when cultured on chemically equivalent surfaces with different rigidities [83]. Specific anchorage points termed focal adhesions allow the cells to sense and respond to the substrate's resistance. The impact of proteolytic resistance and matrix stiffness on cell spreading are interconnected, as demonstrated in experiments on a PEGylated fibrin material displaying a range of biochemical and physical properties. Initial cell spreading is dependent on both stiffness and proteolytic susceptibility, whereas the extent of matrix remodeling and compaction as cells adhere and exert tractional forces is mainly dictated by the stiffness [84]. In general, cells tend to migrate from softer to stiffer environments, a behavior termed durotaxis [85]. Whereas they appear most motile at intermediate stiffness, increased matrix moduli generally stimulate cellular differentiation, where the optimum has to be established individually for each cell type.

A number of biomaterials offer the possibility of changing the matrix modulus without changing the chemical properties, including self-assembling peptides [86], fibrin [84], and polyethylene glycol [87]. A fairly simple way of increasing matrix stiffness is by increasing the material concentration. Furthermore, gradual stiffening of collagen matrices via glycation [88] or lysyl oxidase [89] has been described. For self-assembling peptides, relatively minor changes in fiber surface chemistry by changing amino acid residues can increase the rigidity, and highest moduli can be achieved by induction of covalent bond formation via cysteine disulfide bonds, which allows for an increase of stiffness up to 60-fold [86].

Another aspect that might be considered specifically for applications in the pulp chamber and root canal is shear-thinning behavior and shear recovery. In order to load a biomaterial into a syringe and inject it through a needle, these properties may be desirable. Several self-assembling peptide systems offer this feature, which can be confirmed via oscillatory rheometry. If a material undergoes sheer recovery, the storage modulus, a measure for gel stiffness, recovers to nearly 100 % of the initial value within a relatively short time period of <1 min [90].

18.7.4 Biomineralization

During the synthesis of both dentin and enamel, an organic matrix precedes the mineral phase produced by ameloblasts and odontoblasts. Non-collagenous proteins play a key role in the mineralization process, where negatively charged surfaces and phosphorylated serine residues attract Ca^{2+}, initiate crystal growth, and control the orientation and elongation of the hydroxyapatite crystals. The main players are called SIBLINGs (*s*mall *i*ntegrin *b*inding *l*igand, *N*-linked *g*lycoprotein), which include bone sialoprotein (BSP), osteopontin (OPN), dentin sialophosphoprotein (DSPP), dentin matrix protein-1 (DMP-1), and matrix extracellular phosphoprotein (MEPE). These organic components commonly found in the matrix of mineralizing tissues share the presence of the integrin-binding peptide RGD to mediate cell adhesion. Furthermore, they contain multiple serine-rich domains in β-sheet conformation for phosphorylation and display a high density of negative charge to initiate crystal nucleation [91, 92]. Short peptide sequences derived from DMP-1 have been identified, which display Ca-induced self-assembly into β-sheet structures and induce hydroxyapatite crystal growth [93]. Peptide fragments of MEPE injected into rat calvaria stimulated new bone formation and increased osteoblast proliferation and alkaline phosphatase activity in vitro [94]. Features of these mineralization-inducing peptide motifs can be programmed into scaffolds in order to influence kinetic and structural aspects of crystal formation. Self-assembling peptides offer the advantage that mineralization domains such as phosphorylated serines and acidic domains can be incorporated into the peptide monomers during synthesis. Groundbreaking work demonstrated that peptide-amphiphile molecules can be functionalized by incorporation of phosphorylated serine residues, which enables Ca^{2+} binding, HA crystal nucleation, and growth along the fiber long axis [25]. Previous work demonstrated mineral nucleation along the nanofibers of self-assembling peptides in the presence of dental stem cells [90]. Utilizing these motifs for dental tissue engineering might facilitate the generation of the mineralized component of the dentin–pulp complex.

18.7.5 Contamination Control

Pre- or perioperative contamination and biomaterial-associated infections pose a serious problem in tissue engineering in general. This is particularly true in the oral cavity, as microorganisms are the major cause for dental tissues in the oral cavity. For future clinical applications, local disinfection, e.g., with chlorhexidine, in combination with an antibacterial effect or prolonged release of an antibacterial drug from the scaffold might enable contamination control and allow for optimized healing and regeneration.

It is known that certain low-molecular-weight peptides show antibacterial activity [95]. High-throughput approaches allow the screening of large numbers of appropriate peptide sequences to identify additional motifs. Such sequences can be synthesized and incorporated into scaffold materials.

Peptide or protein hydrogels can incorporate antimicrobial activity, for example, via lysine-rich surfaces, which enable electrostatic interaction of the peptide with the negatively charged bacterial surface, leading to disruption of the bacterial membrane [96]. Whereas the hydrogel exhibits antimicrobial activity against both gram-positive and gram-negative strains, cell culture experiments showed selective toxicity to bacterial rather than mammalian cells. High-throughput approaches can be utilized to screen large numbers of peptide sequences and identify additional motifs with antibacterial activity [97]. A different approach might be the incorporation of antibiotics into scaffolds for dental tissue engineering. A novel drug loading technique was described recently, where the antibiotic vancomycin was incorporated into a polymeric hydrogel, with a sustained release profile, activity of vancomycin after the loading procedure, and promising results after in vitro testing with osteoblast cell cultures [98]. Attachment or encapsulation of antibacterial drugs into bioengineering scaffolds could provide a means to circumvent systemic effects of antibiotics that are frequently administered to support dental treatment.

18.8 Conclusion and Future Perspective

Scaffold biomaterials might play an important role for future approaches to dental pulp tissue engineering. We are in need for customized, multifunctional delivery systems, which (1) accommodate cells or assist in the homing of resident stem cells, (2) bind and prolong the bioavailability of several different growth factors and bioactive molecules, (3) establish temporally controlled release profiles and concerted release kinetics of different molecules, (4) provide possibilities of contamination control, (5) allow for connection to the vasculature, and (6) can be inserted into small defects. Interdisciplinary approaches and close communication between material scientists, biologists, and clinicians are necessary to develop systems capable of all of these. Optimized scaffolding systems might contribute toward developing therapeutic strategies for regenerative endodontic approaches in the near future.

References

1. Gronthos S, Mankani M, Brahim J, Robey PG, Shi S. Postnatal human dental pulp stem cells (DPSCs) in vitro and in vivo. Proc Natl Acad Sci U S A. 2000;97:13625–30.
2. Miura M, Gronthos S, Zhao M, Lu B, Fisher LW, Robey PG, et al. SHED: stem cells from human exfoliated deciduous teeth. Proc Natl Acad Sci U S A. 2003;100:5807–12.
3. Prescott RS, Alsanea R, Fayad MI, Johnson BR, Wenckus CS, Hao J, et al. In vivo generation of dental pulp-like tissue by using dental pulp stem cells, a collagen scaffold, and dentin matrix protein 1 after subcutaneous transplantation in mice. J Endod. 2008;34:421–6.
4. Iohara K, Zheng L, Ito M, Ishizaka R, Nakamura H, Into T, et al. Regeneration of dental pulp after pulpotomy by transplantation of CD31(-)/CD146(-) side population cells from a canine tooth. Regen Med. 2009;4:377–85.
5. Nakashima M, Iohara K. Regeneration of dental pulp by stem cells. Adv Dent Res. 2011;23:313–9.
6. Cordeiro MM, Dong Z, Kaneko T, Zhang Z, Miyazawa M, Shi S, et al. Dental pulp tissue engineering with stem cells from exfoliated deciduous teeth. J Endod. 2008;34:962–9.
7. Sakai VT, Zhang Z, Dong Z, Neiva KG, Machado MA, Shi S, et al. SHED differentiate into functional odontoblasts and endothelium. J Dent Res. 2010;89:791–6.
8. Ferreira AM, Gentile P, Chiono V, Ciardelli G. Collagen for bone tissue regeneration. Acta Biomater. 2012;8:3191–200.
9. Badylak SF, Freytes DO, Gilbert TW. Extracellular matrix as a biological scaffold material: structure and function. Acta Biomater. 2009;5:1–13.
10. Glowacki J, Mizuno S. Collagen scaffolds for tissue engineering. Biopolymers. 2008;89:338–44.
11. Boden SD, Zdeblick TA, Sandhu HS, Heim SE. The use of rhBMP-2 in interbody fusion cages. Definitive evidence of osteoinduction in humans: a preliminary report. Spine. 2000;25:376–81.
12. Fleming Jr JE, Cornell CN, Muschler GF. Bone cells and matrices in orthopedic tissue engineering. Orthop Clin North Am. 2000;31:357–74.
13. Muschler GF, Negami S, Hyodo A, Gaisser D, Easley K, Kambic H. Evaluation of collagen ceramic composite graft materials in a spinal fusion model. Clin Orthop Relat Res. 1996;328:250–60.
14. Van Bitterswijk C, editor. Tissue engineering. San Diego: Elsevier; 2008.

15. Hutmacher DW, Goh JC, Teoh SH. An introduction to biodegradable materials for tissue engineering applications. Ann Acad Med Singapore. 2001;30:183–91.
16. Zhu J. Bioactive modification of poly(ethylene glycol) hydrogels for tissue engineering. Biomaterials. 2010;31:4639–56.
17. Chan G, Mooney DJ. New materials for tissue engineering: towards greater control over the biological response. Trends Biotechnol. 2008;26:382–92.
18. Langer R, Tirrell DA. Designing materials for biology and medicine. Nature. 2004;428:487–92.
19. Ahmed TA, Dare EV, Hincke M. Fibrin: a versatile scaffold for tissue engineering applications. Tissue Eng Part B Rev. 2008;14:199–215.
20. Ferdous Z, Grande-Allen KJ. Utility and control of proteoglycans in tissue engineering. Tissue Eng. 2007;13:1893–904.
21. Masters KS, Shah DN, Leinwand LA, Anseth KS. Crosslinked hyaluronan scaffolds as a biologically active carrier for valvular interstitial cells. Biomaterials. 2005;26:2517–25.
22. Nguyen MK, Lee DS. Injectable biodegradable hydrogels. Macromol Biosci. 2010;10:563–79.
23. Coviello T, Matricardi P, Alhaique F. Drug delivery strategies using polysaccharidic gels. Expert Opin Drug Deliv. 2006;3:395–404.
24. Nguyen KT, West JL. Photopolymerizable hydrogels for tissue engineering applications. Biomaterials. 2002;23:4307–14.
25. Hartgerink JD, Beniash E, Stupp SI. Self-assembly and mineralization of peptide-amphiphile nanofibers. Science. 2001;294:1684–8.
26. Nicodemus GD, Bryant SJ. Cell encapsulation in biodegradable hydrogels for tissue engineering applications. Tissue Eng Part B Rev. 2008;14:149–65.
27. Chen FM, Wu LA, Zhang M, Zhang R, Sun HH. Homing of endogenous stem/progenitor cells for in situ tissue regeneration: promises, strategies, and translational perspectives. Biomaterials. 2011;32:3189–209.
28. Villar CC, Cochran DL. Regeneration of periodontal tissues: guided tissue regeneration. Dent Clin N Am. 2010;54:73–92.
29. Olsson H, Petersson K, Rohlin M. Formation of a hard tissue barrier after pulp cappings in humans. A systematic review. Int Endod J. 2006;39:429–42.
30. Parirokh M, Torabinejad M. Mineral trioxide aggregate: a comprehensive literature review–part III: clinical applications, drawbacks, and mechanism of action. J Endod. 2010;36:400–13.
31. Gloria A, De Santis R, Ambrosio L. Polymer-based composite scaffolds for tissue engineering. J Appl Biomater Biomech. 2010;8:57–67.
32. Mooney DJ, Powell C, Piana J, Rutherford B. Engineering dental pulp-like tissue in vitro. Biotechnol Prog. 1996;12:865–8.
33. Bohl KS, Shon J, Rutherford B, Mooney DJ. Role of synthetic extracellular matrix in development of engineered dental pulp. J Biomater Sci Polym Ed. 1998;9:749–64.
34. Huang GT, Yamaza T, Shea LD, Djouad F, Kuhn NZ, Tuan RS, et al. Stem/progenitor cell-mediated de novo regeneration of dental pulp with newly deposited continuous layer of dentin in an in vivo model. Tissue Eng Part A. 2010;16:605–15.
35. Galler KM, D'Souza RN, Federlin M, Cavender AC, Hartgerink JD, Hecker S, et al. Dentin conditioning codetermines cell fate in regenerative endodontics. J Endod. 2011;37:1536–41.
36. Boontheekul T, Kong HJ, Mooney DJ. Controlling alginate gel degradation utilizing partial oxidation and bimodal molecular weight distribution. Biomaterials. 2005;26:2455–65.
37. Jiang T, Kumbar SG, Nair LS, Laurencin CT. Biologically active chitosan systems for tissue engineering and regenerative medicine. Curr Top Med Chem. 2008;8:354–64.
38. D'Souza RN, Happonen RP, Ritter NM, Butler WT. Temporal and spatial patterns of transforming growth factor-beta 1 expression in developing rat molars. Arch Oral Biol. 1990;35:957–65.
39. D'Souza RN, Flanders K, Butler WT. Colocalization of TGF-beta 1 and extracellular matrix proteins during rat tooth development. Proc Finn Dent Soc. 1992;88 Suppl 1:419–26.
40. Nakashima M, Nagasawa H, Yamada Y, Reddi AH. Regulatory role of transforming growth factor-beta, bone morphogenetic protein-2, and protein-4 on gene expression of extracellular matrix proteins and differentiation of dental pulp cells. Dev Biol. 1994;162:18–28.
41. Iohara K, Nakashima M, Ito M, Ishikawa M, Nakasima A, Akamine A. Dentin regeneration by dental pulp stem cell therapy with recombinant human bone morphogenetic protein 2. J Dent Res. 2004;83:590–5.
42. Nakashima M. Induction of dentin formation on canine amputated pulp by recombinant human bone morphogenetic proteins (BMP)-2 and -4. J Dent Res. 1994;73:1515–22.
43. Six N, Decup F, Lasfargues JJ, Salih E, Goldberg M. Osteogenic proteins (bone sialoprotein and bone morphogenetic protein-7) and dental pulp mineralization. J Mater Sci Mater Med. 2002;13:225–32.
44. Almushayt A, Narayanan K, Zaki AE, George A. Dentin matrix protein 1 induces cytodifferentiation of dental pulp stem cells into odontoblasts. Gene Ther. 2006;13:611–20.
45. Alliot-Licht B, Bluteau G, Magne D, Lopez-Cazaux S, Lieubeau B, Daculsi G, et al. Dexamethasone stimulates differentiation of odontoblast-like cells in human dental pulp cultures. Cell Tissue Res. 2005; 321:391–400.
46. Couble ML, Farges JC, Bleicher F, Perrat-Mabillon B, Boudeulle M, Magloire H. Odontoblast differentiation of human dental pulp cells in explant cultures. Calcif Tissue Int. 2000;66:129–38.
47. Smith AJ, Scheven BA, Takahashi Y, Ferracane JL, Shelton RM, Cooper PR. Dentine as a bioactive extracellular matrix. Arch Oral Biol. 2012;57:109–21.

48. Zhang R, Cooper PR, Smith G, Nör JE, Smith AJ. Angiogenic activity of dentin matrix components. J Endod. 2011;37:26–30.
49. Roberts-Clark DJ, Smith AJ. Angiogenic growth factors in human dentine matrix. Arch Oral Biol. 2000;45:1013–6.
50. Finkelman RD. Growth factors in bones and teeth. J Calif Dent Assoc. 1992;20:23–9.
51. Finkelman RD, Mohan S, Jennings JC, Taylor AK, Jepsen S, Baylink DJ. Quantitation of growth factors IGF-I, SGF/IGF-II, and TGF-beta in human dentin. J Bone Miner Res. 1990;5:717–23.
52. Smith AJ, Matthews JB, Hall RC. Transforming growth factor-beta1 (TGF-beta1) in dentine matrix. Ligand activation and receptor expression. Eur J Oral Sci. 1998;106 Suppl 1:179–84.
53. Baker SM, Sugars RV, Wendel M, Smith AJ, Waddington RJ, Cooper PR, et al. TGF-beta/extracellular matrix interactions in dentin matrix: a role in regulating sequestration and protection of bioactivity. Calcif Tissue Int. 2009;85:66–74.
54. Dreyfuss JL, Regatieri CV, Jarrouge TR, Cavalheiro RP, Sampaio LO, Nader HB. Heparan sulfate proteoglycans: structure, protein interactions and cell signaling. An Acad Bras Cienc. 2009;81:409–29.
55. Rahman S, Patel Y, Murray J, Patel KV, Sumathipala R, Sobel M, et al. Novel hepatocyte growth factor (HGF) binding domains on fibronectin and vitronectin coordinate a distinct and amplified Met-integrin induced signalling pathway in endothelial cells. BMC Cell Biol. 2005;6:8.
56. Paralkar VM, Vukicevic S, Reddi AH. Transforming growth factor beta type 1 binds to collagen IV of basement membrane matrix: implications for development. Dev Biol. 1991;143:303–8.
57. Somasundaram R, Ruehl M, Tiling N, Ackermann R, Schmid M, Riecken EO, et al. Collagens serve as an extracellular store of bioactive interleukin 2. J Biol Chem. 2000;275:38170–5.
58. Casagrande L, Demarco FF, Zhang Z, Araujo FB, Shi S, Nör JE. Dentin-derived BMP-2 and odontoblast differentiation. J Dent Res. 2010;89:603–8.
59. He H, Yu J, Liu Y, Lu S, Liu H, Shi J, et al. Effects of FGF2 and TGFbeta1 on the differentiation of human dental pulp stem cells in vitro. Cell Biol Int. 2008;32:827–34.
60. Kalyva M, Papadimitriou S, Tziafas D. Transdentinal stimulation of tertiary dentine formation and intratubular mineralization by growth factors. Int Endod J. 2010;943:382–92.
61. Melin M, Joffre-Romeas A, Farges JC, Couble ML, Magloire H, Bleicher F. Effects of TGFbeta1 on dental pulp cells in cultured human tooth slices. J Dent Res. 2000;79:1689–96.
62. Galler KM, Hartgerink JD, Cavender AC, Schmalz G, D'Souza RN. A customized self-assembling peptide hydrogel for dental pulp tissue engineering. Tissue Eng Part A. 2012;18:176–84.
63. Rajangam K, Behanna HA, Hui MJ, Han X, Hulvat JF, Lomasney JW, et al. Heparin binding nanostructures to promote growth of blood vessels. Nano Lett. 2006;6:2086–90.
64. Silva GA, Czeisler C, Niece KL, Beniash E, Harrington DA, Kessler JA, et al. Selective differentiation of neural progenitor cells by high-epitope density nanofibers. Science. 2004;303:1352–5.
65. Zhang S. Fabrication of novel biomaterials through molecular self-assembly. Nat Biotechnol. 2003;21: 1171–8.
66. Hartgerink JD, Beniash E, Stupp SI. Peptide-amphiphile nanofibers: a versatile scaffold for the preparation of self-assembling materials. Proc Natl Acad Sci U S A. 2002;99:5133–8.
67. Sawhney AS, Pathak CP, Hubbell JA. Interfacial photopolymerization of poly(ethylene glycol)-based hydrogels upon alginate-poly(l-lysine) microcapsules for enhanced biocompatibility. Biomaterials. 1993;14:1008–16.
68. Turk BE, Huang LL, Piro ET, Cantley LC. Determination of protease cleavage site motifs using mixture-based oriented peptide libraries. Nat Biotechnol. 2001;19:661–7.
69. Galler KM, Aulisa L, Regan KR, D'Souza RN, Hartgerink JD. Self-assembling multidomain peptide hydrogels: designed susceptibility to enzymatic cleavage allows enhanced cell migration and spreading. J Am Chem Soc. 2010;132:3217–23.
70. Ruoslahti E. RGD and other recognition sequences for integrins. Annu Rev Cell Dev Biol. 1996;12: 697–715.
71. Massia SP, Hubbell JA. Vascular endothelial cell adhesion and spreading promoted by the peptide REDV of the IIICS region of plasma fibronectin is mediated by integrin alpha 4 beta 1. J Biol Chem. 2002;267:14019–26.
72. Aota S, Nomizu M, Yamada KM. The short amino acid sequence Pro-His-Ser-Arg-Asn in human fibronectin enhances cell-adhesive function. J Biol Chem. 1994;269:24756–61.
73. Wong JY, Weng Z, Moll S, Kim S, Brown CT. Identification and validation of a novel cell-recognition site (KNEED) on the 8th type III domain of fibronectin. Biomaterials. 2002;23:3865–70.
74. Yamada Y, Kleinman HK. Functional domains of cell adhesion molecules. Curr Opin Cell Biol. 1992;4:819–23.
75. Bhatnagar RS, Qian JJ, Gough CA. The role in cell binding of a beta-bend within the triple helical region in collagen alpha 1 (I) chain: structural and biological evidence for conformational tautomerism on fiber surface. J Biomol Struct Dyn. 1997;14:547–60.
76. Cardin AD, Weintraub HJ. Molecular modeling of protein-glycosaminoglycan interactions. Arteriosclerosis. 1989;9:21–32.
77. Bessa PC, Machado R, Nürnberger S, Dopler D, Banerjee A, Cunha AM, et al. Thermoresponsive self-assembled elastin-based nanoparticles for delivery of BMPs. J Control Release. 2010;142:312–8.
78. Ji Y, Xu GP, Yan JL, Pan SH. Transplanted bone morphogenetic protein/poly(lactic-co-glycolic acid)

delayed-release microcysts combined with rat micromorselized bone and collagen for bone tissue engineering. J Int Med Res. 2009;37:1075–87.
79. Zhao Y, Zhang J, Wang X, Chen B, Xiao Z, Shi C, et al. The osteogenic effect of bone morphogenetic protein-2 on the collagen scaffold conjugated with antibodies. J Control Release. 2010;141:30–7.
80. Chen FM, Wu ZF, Sun HH, Wu H, Xin SN, Wang QT, et al. Release of bioactive BMP from dextran-derived microspheres: a novel delivery concept. Int J Pharm. 2006;307:23–32.
81. Mathieu S, Jeanneau C, Sheibat-Othman N, Kalaji N, Fessi H, About I. Usefulness of controlled release of growth factors in investigating the early events of dentin-pulp regeneration. J Endod. 2013;39:228–35.
82. Levental I, Georges PC, Janmey PA. Soft biological materials and their impact on cell function. Soft Matter. 2007;3:299–306.
83. Nemir S, West JL. Synthetic materials in the study of cell response to substrate rigidity. Ann Biomed Eng. 2010;38:2–20.
84. Dikovsky D, Bianco-Peled H, Seliktar D. The effect of structural alterations of PEG-fibrinogen hydrogel scaffolds on 3-D cellular morphology and cellular migration. Biomaterials. 2006;27:1496–506.
85. Wells RG. The role of matrix stiffness in regulating cell behavior. Hepatology. 2008;47:1394–400.
86. Aulisa L, Dong H, Hartgerink JD. Self-assembly of multidomain peptides: sequence variation allows control over cross-linking and viscoelasticity. Biomacromolecules. 2009;10:2694–8.
87. Khatiwala CB, Peyton SR, Metzke M, Putnam AJ. The regulation of osteogenesis by ECM rigidity in MC3T3-E1 cells requires MAPK activation. J Cell Physiol. 2007;211:661–72.
88. Girton TS, Oegema TR, Grassl ED, Isenberg BC, Tranquillo RT. Mechanisms of stiffening and strengthening in media-equivalents fabricated using glycation. J Biomech Eng. 2000;122:216–23.
89. Elbjeirami WM, Yonter EO, Starcher BC, West JL. Enhancing mechanical properties of tissue-engineered constructs via lysyl oxidase crosslinking activity. J Biomed Mater Res A. 2003;66:513–21.
90. Galler KM, Cavender A, Yuwono V, Dong H, Shi S, Schmalz G, et al. Self-assembling peptide amphiphile nanofibers as a scaffold for dental stem cells. Tissue Eng Part A. 2008;14:2051–8.
91. George A, Sabsay B, Simonian PA, Veis A. Characterization of a novel dentin matrix acidic phosphoprotein. Implications for induction of biomineralization. J Biol Chem. 2003;268:12624–30.
92. Tye CE, Rattray KR, Warner KJ, Gordon JA, Sodek J, Hunter GK, et al. Delineation of the hydroxyapatite-nucleating domains of bone sialoprotein. J Biol Chem. 2003;278:7949–55.
93. He G, Dahl T, Veis A, George A. Nucleation of apatite crystals in vitro by self-assembled dentin matrix protein 1. Nat Mater. 2003;8:552–8.
94. Hayashibara T, Hiraga T, Yi B, Nomizu M, Kumagai Y, Nishimura R, et al. A synthetic peptide fragment of human MEPE stimulates new bone formation in vitro and in vivo. J Bone Miner Res. 2004;19:455–62.
95. Brennan EP, Reing J, Chew D, Myers-Irvin JM, Young EJ, Badylak SF. Antibacterial activity within degradation products of biological scaffolds composed of extracellular matrix. Tissue Eng. 2006;12:2949–55.
96. Salick DA, Kretsinger JK, Pochan DJ, Schneider JP. Inherent antibacterial activity of a peptide-based beta-hairpin hydrogel. J Am Chem Soc. 2007;129:14793–9.
97. Xie Q, Matsunaga S, Wen Z, Niimi S, Kumano M, Sakakibara Y, et al. In vitro system for high-throughput screening of random peptide libraries for antimicrobial peptides that recognize bacterial membranes. J Pept Sci. 2006;12:643–52.
98. Zhang LF, de Yang J, Chen HC, Sun R, Xu L, Xiong ZC, et al. An ionically crosslinked hydrogel containing vancomycin coating on a porous scaffold for drug delivery and cell culture. Int J Pharm. 2008;353:74–87.
99. Galler KM, Eidt A, Schmalz G. Cell-free approaches for dental pulp tissue engineering. J Endod. 2014;40:41.
100. Kim NR, Lee DH, Dhung P-H, Yang H-C. Distinct differentiation properties of human dental pulp cells on collagen, gelatin, and chitosan scaffolds. Oral Surg Oral Med Oral Pathol Endod. 2009;108(5):e94–100.

Regenerative Endodontics: Regeneration or Repair? 19

Stephane Simon and Michel Goldberg

19.1 Introduction

Pulp capping leading to induction of dentin bridge formation has been recognized for more than 70 years [1]. The beneficial effects of pulp capping after $Ca(OH)_2$ capping have been reported in the pioneer works of Hermann [2] and analyzed by Schröder [3]. Clinical results obtained by this method are encouraging and tend to motivate the clinician maintaining pulp vitality as long as possible. Doing so the aim was to enhance tooth longevity and avoid endodontic treatments. Unfortunately, robust clinical trials are lacking [4]. The data which have been published are contradictory and therefore do not offer clear guidance for this technique. A recent systematic review reported success rates for maintaining pulp vitality after 3 years of 72.9 % for pulp capping and 99.4 % for partial pulpotomy [5]. However, a recent randomized clinical trial by Bjorndal et al. yielded much lower success rates: 31.8 % for pulp capping and 34.5 % for partial pulpotomy [6].

After careful elimination of the soft carious dentin, presumably restorative dental cement avoids further bacterial contamination and allows mechanical protection of the cavity liner containing $Ca(OH)_2$. The alkaline calcium hydroxide burns the superficial exposed layer of the pulp. Under the scar, pulp cells differentiate into pre-odontoblasts, elongate, and become functional. They secrete a fibrous collagen-rich extracellular matrix (ECM), which underwent mineralization (Fig. 19.1a–d). The early formation of a dentin bridge isolates the superficial pulp from the cavity. This construction is gradually becoming thicker to the detriment of the dental pulp. However, the dentin bridge is not homogeneous, leaving some non-mineralized tunnels containing pulp remnants and cell lacunae. The partial closure of the pulp exposure favors pulp superficial healing and regeneration. In contrast, the dentin bridge does not seal totally the pulp, and bacteria may diffuse between the cavity and the pulp. This is consequently why pulp reinfection and degradation occur (Figs. 19.2a–c, 19.3, 19.4, 19.5, and 19.6a, b).

Significant progress in the prevention and treatment of pulp and periradicular disease has led to an intensified focus on the ability of the dentin-pulp complex to repair itself and regenerate mineralized tissue. Recent advances in biotechnology and translational research offer hope of providing new treatment modalities that can protect the vital pulp, enable manipulation of

S. Simon, DDS, PhD
Department of Oral Biology and Endodontics,
Dental School of the University of Paris Diderot,
Hoptial de la Pitié Salpétrière,
5 rue de Garancière, Paris 75006, France
e-mail: stephane.simon@univ-paris-diderot.fr

M. Goldberg, MD, DDS (✉)
Department of Oral Biology,
Institut National de la Santé et de la
Recherche Médicale, Université Paris Descartes,
45 Rue des Saints Pères, Paris 75006, France
e-mail: mgoldod@gmail.com,
michel.goldberg@parisdescartes.fr

M. Goldberg (ed.), *The Dental Pulp*,
DOI 10.1007/978-3-642-55160-4_19,

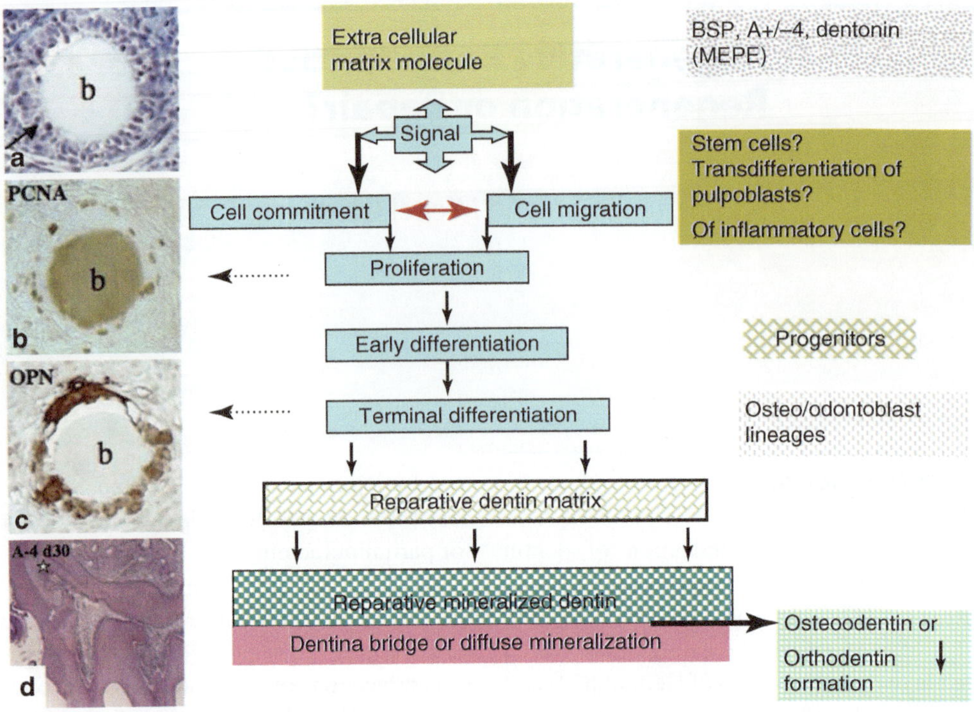

Fig. 19.1 (**a–d**) Schematic cascade leading from cell commitment to differentiation in osteoblasts or odontoblasts and the formation of a reparative dentin bridge (*asterisk*), *b* bead

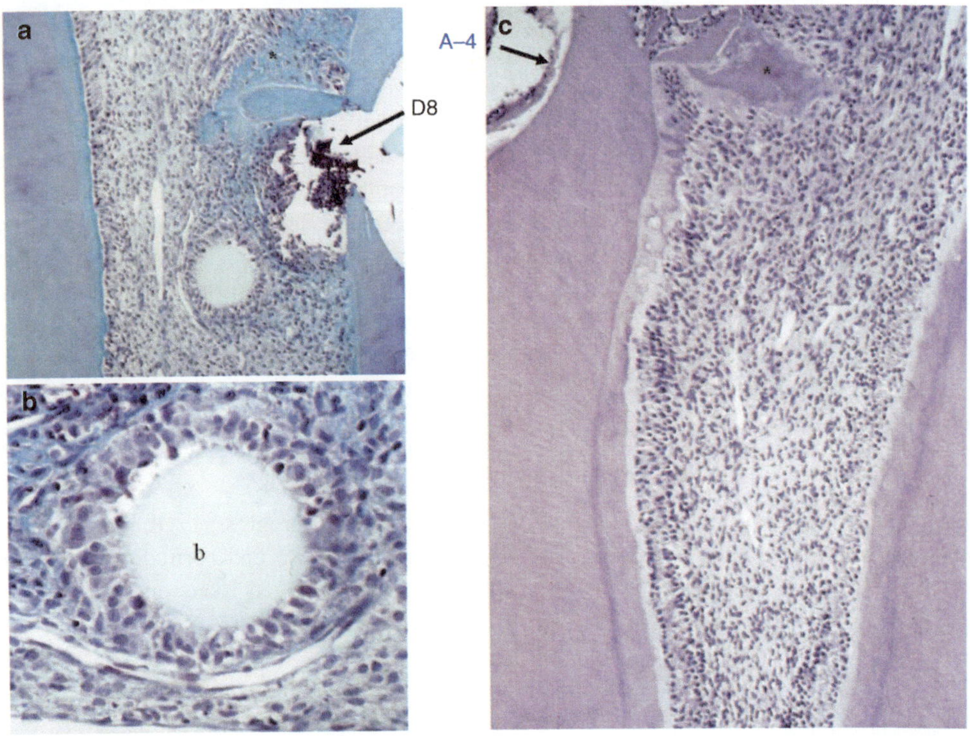

Fig. 19.2 Eight days after the implantation of A-4, pulp cells are committed around agarose beads (**a**) and start to differentiate into osteo-/odontoblasts (**b**) (reparative dentin). After 15 days (**c**), mineralization loci begin to be formed in the crown part (*asterisk*), whereas in the root reactionary dentin is formed along the root canal lumen behind a calciotraumatic line, *b* bead

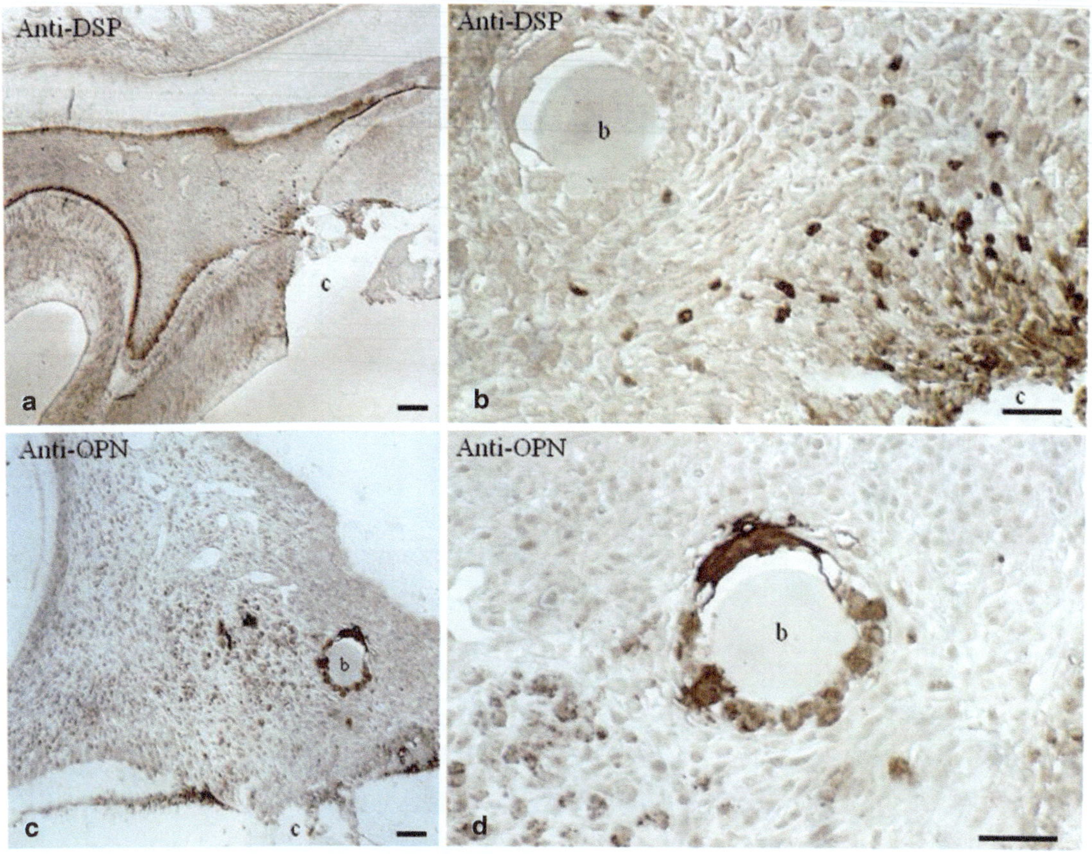

Fig. 19.3 (**a**, **b**) Anti-dentin sialoprotein labeling. No labeling is detected around the bead (carrier), but near the pulp exposure DSP-labeled cells are stained positively. They will be implicated in reparative dentin formation. In (**c**, **d**), anti-osteopontin labeling is especially dense around the beads (osteoblast labeling), *b* bead; *c* cavity

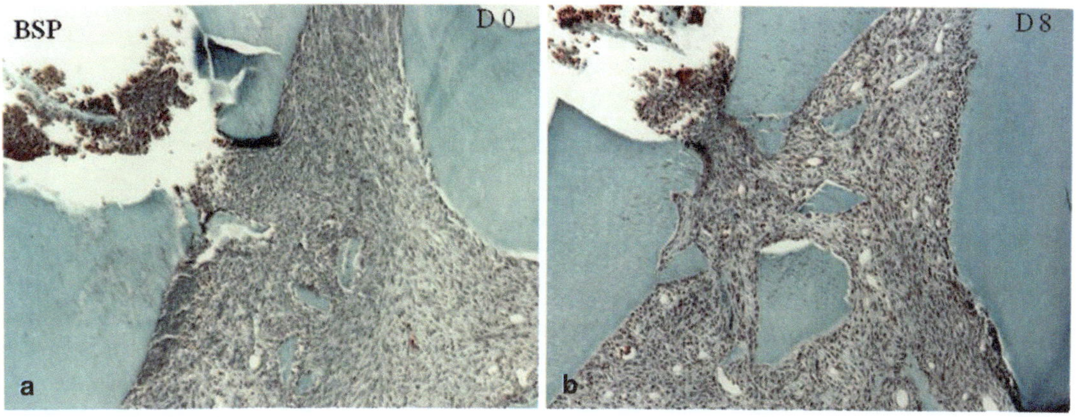

Fig. 19.4 After implantation of bone sialoprotein (*BSP*), no reaction is detected at day 0–day 8 (**a**, **b**). A slight inflammatory process is seen in the pulp horn, due mostly to the dentin debris pushed in the pulp. At day 15 (**c**), mineralization starts to be formed around dentin debris; and a solid dentin bridge is formed at day 30 (**d**), occluding homogeneously the pulp exposure

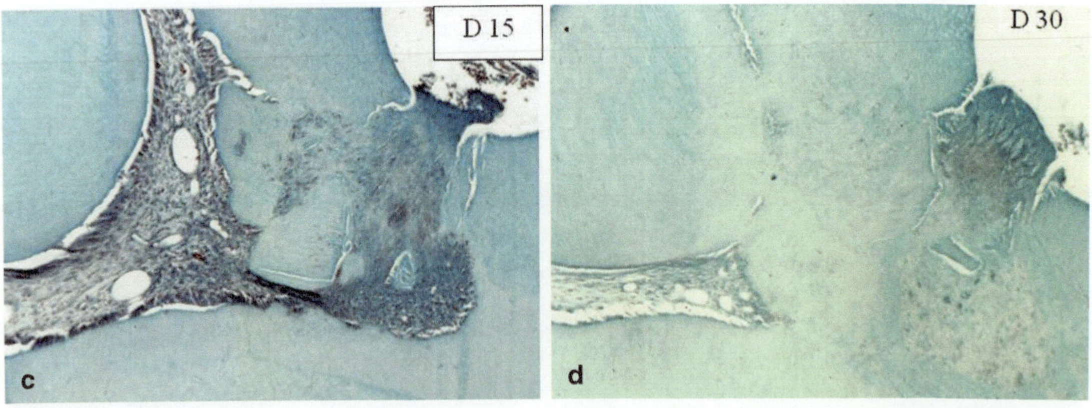

Fig. 19.4 (continued)

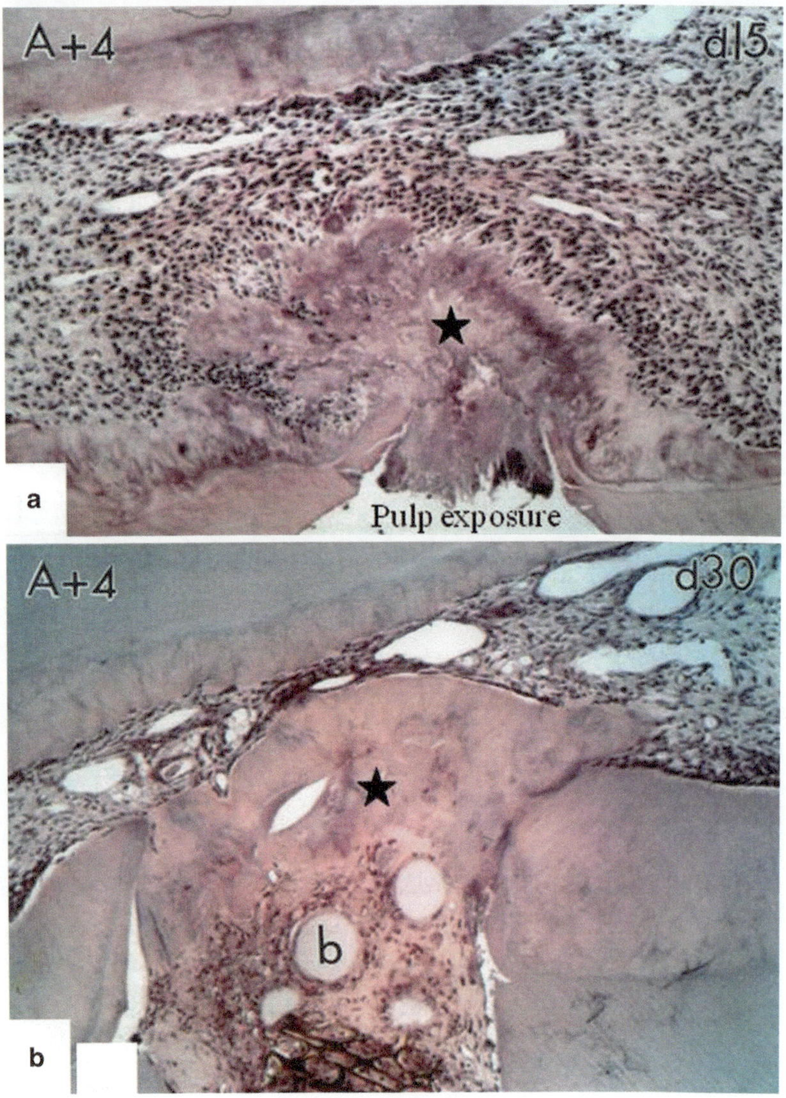

Fig. 19.5 Implantation of A+4 in the pulp. After 15 days (**a**), the pulp exposure is partially closed by a thin dentin bridge. After 30 days (**b**), a thick and homogeneous dentin bridge is formed (*asterisk*). The dental pulp is totally protected from bacterial contamination

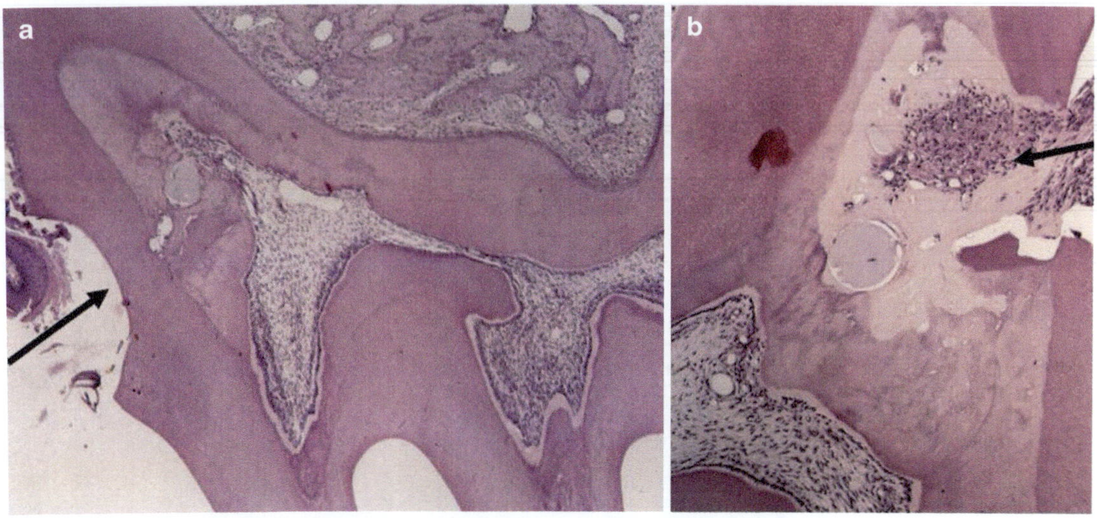

Fig. 19.6 (**a**) Reactionary dentin (*arrow*) filling the upper part of the root. (**b**) Reparative dentin (*arrow*), closing the pulp exposure. Agarose bead and cell debris are located in the mineralized tissue

reactionary dentinogenesis, and stimulate revascularization of an infected root canal space. Indeed, the volume of the mature pulp is very small (less than 100 µl), so it might seem that regeneration of such a tiny island of tissue could be achieved relatively easily. Unfortunately, this is not the case. The pulp is a highly specialized connective tissue enclosed in a mineralized shell with a limited blood supply and thus poses a unique challenge for the design and development of new therapeutic strategies.

Attempts were made to use dentin extracellular matrix molecules. Rutherford et al. [7, 8] have implanted BMP-7 (OP-1) in the pulp. Osteodentin was formed filling totally the root canal. Better results were obtained with bone sialoprotein (BSP), and a solid homogeneous dentin bridge occluded totally the pulp exposure (Fig. 19.4a–d). The biological effects of a few other ECM molecules were evaluated. This was the case for Dentonin, a 100 amino acid long peptide from MEPE, and for A+4 and A−4, two amelogenin gene splice products. Each molecule contributed to the regeneration of a superficial pulp [9] (Figs. 19.5a, b and 19.6a, b).

It was shown that around the agarose beads used as carrier for bioactive molecules, pulp cells were committed and recruited and they differentiate toward an osteo-/odontogenic lineage. The proliferating cell nuclear antigen (PCNA), a characteristic labeling indicative for cell division, labeled the cell nuclei. Pulp cells migrate, underwent an early differentiation forming a ring around and close to the surface of the carrier agarose beads. They were labeled firstly as mesenchymal cells, then becoming osteopontin positive. Later dentin sialoprotein labeled positively the cells implicated in the formation of a reparative dentin matrix, which further become mineralized. The cascade of differentiation leads to the formation of ortho- or osteodentin (Figs. 19.6a, b and 19.7a, b). These biological approaches allow a better understanding of what is occurring during pulp capping or regeneration.

Regenerative endodontics involves two alternatives:

1. Dentin-pulp complex regeneration (which could also be called dentin-odontoblast complex regeneration), which involves the preservation of pulp vitality and pulp capping
2. Dental pulp regeneration, which is the regeneration of vital tissue in an infected root canal space

19.2 Dentin-Pulp Complex Regeneration

Clinically, the aims of such treatments are to keep the pulp vital and maintain its homeostatic functions so as to avoid pulpectomy or tooth

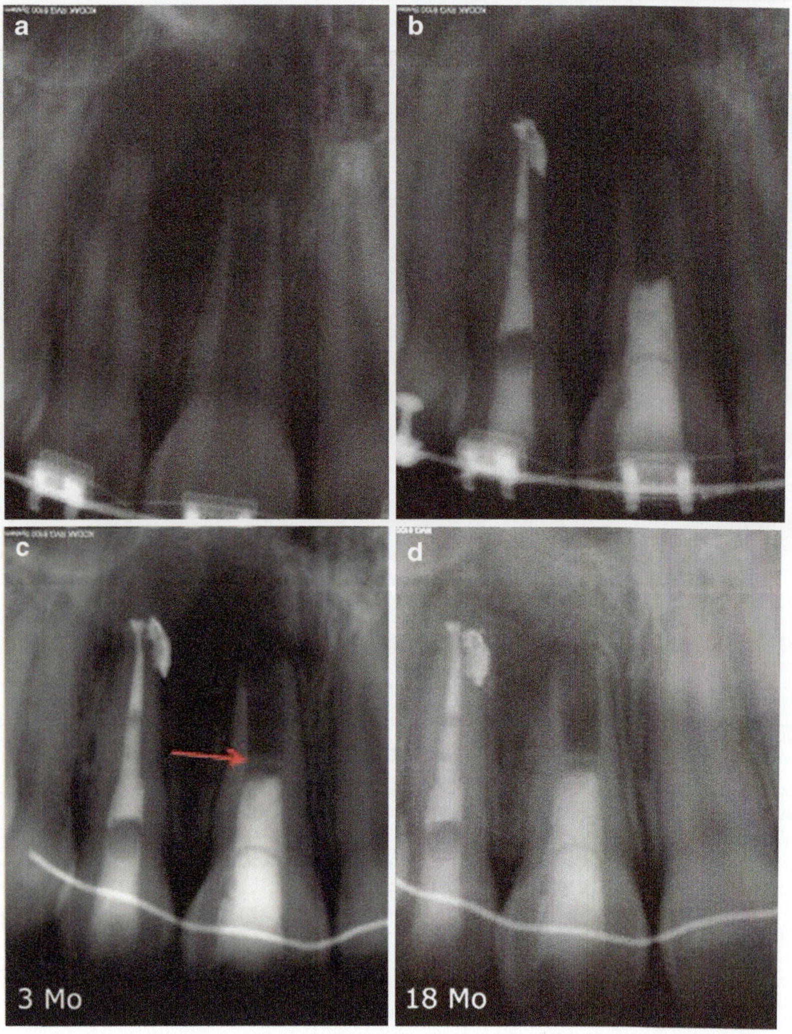

Fig. 19.7 Endodontic treatment by revascularization on tooth #8 on a 16-year-old girl and with conventional approach on tooth #7 (**a**, **b**). Note the formation of a mineralized barrier at low distance from coronal filling material (mineral trioxide aggregate) (*arrow*) at an early stage of healing (3 months postoperative) (**c**). At an 18-month recall (**d**), the bone healing is complete and the mineral barrier still present. Nevertheless, no root lengthening, neither apexogenesis, was noticeable

extraction. Treatment success is based on symptoms reported by the patient as well as the results of relatively rudimentary tests such as thermal tests, electrical tests, tenderness to percussion or palpation, and radiographic assessment. It is well established that clinical signs and symptoms do not correlate with the histological status of the tooth [10]. More advanced assessment using biological research methods allows the investigator to analyze histological structure, cellular behavior, and immunological/inflammation status of the tissue, thus providing more specific indications of pulp responses which may not be evident in clinical studies. To study physiological and reparative processes in the pulp, in vitro experiments can be conducted with immortalized cells or primary cell cultures. However, such studies are limited because they can only mimic biological processes such as mineral production but cannot determine if the tissue produced is actually dentin or another form of mineralized tissue. Nonetheless, these experiments can be supported with RT-PCR or microarray data to identify the likely gene expression profile which induced the mineral production. Confidence in the tissue regeneration results is then enhanced when phenotypic markers associated with dentin production are observed.

19.3 Regeneration/Repair and Remodeling

The bone undergoes constant remodeling, with turnover rates of less than a few months. Thus, gradually, newly secreted bone tissue will merge with older tissue such that the new and old become homogeneous. Some exceptions occur, such as in pseudarthrosis, in which the new non-resident tissue should be considered reparative tissue and is histologically and ultrastructurally distinctly different from that formed through regeneration. Remodeling can also lead to destruction of regenerated tissue (partial or complete) in cases where the tissue is not biologically compatible with its microenvironment [11].

In contrast to the bone, in the tooth, remodeling of dentin does not occur and newly formed tissue will never be replaced. Histologically, tertiary dentin can appear to be similar to secondary dentin; however, it is never truly the same nor continuous with preexisting dentin.

The dentin-pulp interface may be considered a dentin-odontoblast complex, because dentin is uniquely penetrated by odontoblast processes, which form an intimate union with the cohesive and intermingling odontoblast palisade beneath. This layer acts like a membrane, separating itself from the pulp underneath by the acellular layer of Weil. Breakage of this odontoblast membrane by caries, trauma, or iatrogenic damage exposes the pulp tissue, leaving it unprotected and vulnerable.

At present, determination of the origin of tissue secreted by the processes of repair/regeneration generally relies on identification of cellular markers (usually proteins) left by the cells that produced the tissue. Based on the presence of such proteins as well as the cellular behavior (especially biomineralization) and structure of the newly generated tissue, conclusions can be drawn about the origin of these dental tissue secretions. Nevertheless, owing to a lack of known specific molecular markers of newly secreted tissues, the implicated cells are given the suffix "-like" (e.g., osteoblast-like, odontoblast-like). This approach makes it possible to distinguish between normal tissue and altered tissue, suggesting that the latter is formed from reparative processes rather than regenerative ones. In the field of tooth biology, it is conventional to consider the mineralized tissue secreted by dental cells to be dentin; however, very few experiments have sought to characterize the type of mineral produced during synthesis of tertiary dentin. In a previous experiment, we showed by X-ray analysis that the crystal structure of reparative dentin formed after pulp capping was close to that of orthodentin, but differed in terms of protein levels [12]. There is a lack of understanding of the precise nature of mineralized tissue to the extent that we are not even certain of any formal differences, which may exist between dentin and bone, for example. Such knowledge would make it possible to describe the precise nature and composition of secreted tissue and identify the healing process as regenerative or reparative.

Under specific pathophysiological conditions, cells other than odontoblasts may secrete mineralized tissue. For example, pulp stones and other intra-pulp mineral tissue accumulations may form in the presence of chronic inflammation [13, 14]. Clinically, it is possible to distinguish morphologically between orthodentin and pulp mineralization, but in vitro, it is much more difficult to distinguish pathological tissue from true dentin. In vitro, under appropriate experimental conditions, it is possible to demonstrate the production of mineralized tissue, but impossible to identify the precise nature of this tissue. A better understanding of the ultrastructure of the mineral produced in these different pathophysiological situations would allow us to identify, in every in vitro situation, the true origin of the mineralized tissue, for example, whether it was from a phenotypically odontoblast cell or induced by an inflammatory process. In other tissues/organs, for example, the urinary system [15], it has been demonstrated that there is a correlation between the ultrastructure or chemical composition of the mineral and the etiology of its secretion.

Whereas the synthesis of dentin-like tissue by odontoblast-like cells can be considered a regenerative process, ectopic biomineralization would be considered more as a reparative one. The experimental conditions used in in vitro and in vivo experiments are usually designed to investigate reparative processes rather than true processes of regeneration (e.g., progenitor cell migration, recruitment, differentiation).

The driving factor for material-induced dentin bridge formation is unknown. No published data decisively demonstrate that the dentin bridge secreted upon contact with a bioactive material is related to a true process of odontoblast differentiation. It has long been suggested that pulp-capping agents initially induce tissue irritation. It may be that the initial inflammatory reaction provoked by contact of the biomaterial with the dental pulp induces this reparative process and mineral formation [16].

Owing to our lack of knowledge, especially with regard to the structure and ultrastructure of the mineralized tissue produced from pulp capping, it would be wise to regard such therapeutically induced wound healing as a process of repair of the dentin-pulp complex, rather than as a process of regeneration of the dentin-odontoblast complex.

19.4 Dental Pulp Regeneration or Root Canal Revascularization

The therapeutic strategies discussed so far for inducing wound healing of the dental pulp tissue are aimed at limiting tissue degeneration and enabling the rest of the pulp tissue to remain vital. If this is not possible due to severe pulp damage or the pulp being badly inflamed or necrotic, it is likely that its preservation will not be achieved. Under such conditions, the clinician has to perform a pulpectomy and disinfect the entire canal system and then fill in the canal to prevent recontamination by bacteria.

Although current root canal therapeutic approaches provide reliable outcomes, it appears that de novo synthesis of pulp or connective tissue inside the root canal system itself might be a better approach for endodontic treatment in the future.

Treatment of an empty canal with a regenerative strategy is a true technical challenge. It requires numerous simultaneous objectives to be met in a difficult microenvironment. In terms of the basic cellular processes that would need to be harnessed in order to engineer such tissue, research and validation are required for the type of scaffold used, the source and subsequent recruitment of stem cells, and the correct signaling molecules to induce the molecular responses required for tissue development and maturation and neovascularization.

The first attempts to carry out root canal revascularization were made in the 1960s [17]. The principal objective of this treatment is to regenerate dental pulp tissue de novo. One of the biggest limitations of this approach early on was that the only possible source of viable cells was from inducing bleeding into the root canal space. This meant that these cells were derived from circulating cells or cementum or periodontal ligament or alveolar bone and therefore were not of pulp origin.

Although first described in the 1960s, it was not until 2001 that the goal of root canal revascularization gained renewed interest. This has produced an interesting debate to determine if this tissue is generated by repair or regenerative processes [18]. It has been proposed that stem cells from the apical papilla (SCAP) could be introduced into the canal by disorganizing the apical papilla tissue with an endodontic file and letting the SCAP be carried into the canal space by the forming blood clot.

Despite the publication of a significant number of case studies, little is known about the cellular processes involved in this therapeutic approach. Most cases show examples of revascularization of the pulp space where there is a preexisting lesion of endodontic origin. A number of these publications demonstrate completion of apexogenesis, which ceased because of pulp necrosis, increased root-end dentin thickness, and reduction of the volume of root canal space. These observations tend to support regeneration of a dental pulp-like tissue inside the root canal, with peripheral cells showing dentinogenic capability. However, this treatment is not always successful, as some case reports have described the teeth having had to be extracted and subsequent histological analysis carried out.

The first histological observations of regenerated tissue using the revascularization technique inside a tooth were made on a dog model [19]. The results clearly show, inside the root canal space, dentin walls covered with a layer of cementum, a *neo*-ligament, and an osteoid structure. More recently, histological analysis of the teeth treated by simple revascularization [20] or filling with platelet-rich plasma [21] showed a mineralized layer was deposited on the radicular walls. This newly formed tissue appeared to be of periodontal rather than pulp origin and thus was not formed by dentinogenesis. In this case, the recruited progenitors migrating from the apical papilla or from the surrounding periradicular tissues would have differentiated into cells of periodontal origin. Radiographic analysis of this treatment thus far may have misled us into thinking that the mineralized tissue was dentin rather than cementum, as shown by recent histological analysis. If this evidence becomes more established, then the process of apexogenesis becomes difficult to describe when formed by revascularization. Nevertheless, similar treatments with apexification are described in the literature [22, 23]. Apical closure with a periodontal structure might be possible, and it is besides what was for a long time required in the technique of apexification.

To determine if this treatment modality can be considered efficacious, it is important to clarify the therapeutic objective. If it is to induce healing of the periapical tissue, stimulate bone regeneration, and render the patient free from any signs or symptoms, then it could be termed a success. However, if the objective is to regenerate a pulp tissue ad integrum, then this treatment must be regarded as a failure. In other words, it would be a clinical success, but a biological failure.

The objectives of endodontic treatment in an infected tooth are firstly to disinfect the root canal system and secondly to prevent reinfection over time. Both of these objectives can lead to favorable bone healing and regeneration of the periradicular tissues. Filling the root canal space with a biological tissue does not have the disadvantages of filling the root canal with a synthetic material, such as potential loss of seal and toxicity. However, it has the huge advantage of rendering the root canal immune competent so it can defend itself from bacterial contamination, like normal pulp tissue does (Fig. 19.7a, b).

Revascularization is an interesting technique, which allows the pulp space to be filled with a vital tissue. This tissue is different from that which was initially present in the canal and will never be modified to reform dentin. Among the 132 cases published at the date of writing this manuscript, only one shows the presence of a dental pulp inside a treated canal. It shows a true palisade of odontoblasts and a well-organized dental pulp tissue. An important difference concerning this latter case compared to most other published revascularization cases is that the tooth experienced pulpitis and not necrosis. Thus, the odontoblast layer was mostly preserved and the treatment consisted of disorganizing the remaining pulp tissue without completely destroying it. Although the dentin-odontoblast complex is highly specialized and difficult to regenerate, with the resident cells still present, it was possible for the pulp tissue to reorganize, regenerate, and finally preserve itself.

The precise meaning of "regeneration" requires that pulp tissue reform in the vacant root canal space and exhibit normal homeostatic function. If this definition is accepted, then none of our present therapeutic strategies should be considered regenerative as they do not fulfill these requirements. Instead, the treatments should be regarded as reparative strategies only. If the demand is for generation of a vital biological tissue within the vacant space, then the technique of revascularization would fit that criterion.

These considerations are obviously important but remain as scientific problems in need of solutions. Clinically, these therapeutic strategies have their place even if clear indications and contraindications have not yet been determined.

Beyond the semantics of definitions, many questions remain about how a vital tissue can form in a vacant biological space. The presence of stem cells in a revascularized canal was clearly shown recently [24, 25]. The most plausible hypothesis to explain this would be that recruitment of SCAP and their subsequent migration are critical to the formation of this new tissue; however, the origin of these cells remains debatable.

At this point these assumptions suggest that the clinical indications of revascularization treatment should be confined to immature teeth.

If progenitor cells could be recruited from a niche other than the apical papilla, the indications of treatment could be extended to mature teeth. If progenitor cell niches lie inside periradicular tissues, then their recruitment to inside the root canal could be plausible. This may explain why the newly formed tissue is closer to a periodontal tissue rather than a pulp one. If proven to be successful, then treatment of mature teeth using this method may be possible.

Acknowledgments to the contributors of the figures chapters 1, 2, 8, 10, 13, 16, and 19 Hervé Lesot, Sylvie Lecolle, Dominique Septier, Azumi Hirata, Arnaud Marchadier, Françoise Escaig, Sasha Dimitrova-Nakov, David Montero, Ngampis Six, Frank Decup, Yassine Harichane.

References

1. Zander HA. Reaction of the pulp to calcium hydroxide. J Dent Res. 1939;18(4):373–9.
2. Hermann BW. Dentin obliteration der wurzelkanale nach behandlung mit calcium. Zahnärzl Bdsch. 1930;39:887–99.
3. Schröder U. Effects of calcium hydroxide-containing pulp-capping agents on pulp cell migration, proliferation, and differentiation. J Dent Res. 1985;64(Spec Iss):541–8.
4. Miyashita H, Worthington HV, Qualtrough A, Plasschaert A. Pulp management for caries in adults: maintaining pulp vitality. Cochrane Database Syst Rev. 2007;2:CD004484.
5. Aguilar P, Linsuwanont P. Vital pulp therapy in vital permanent teeth with cariously exposed pulp: a systematic review. J Endod. 2011;37(5):581–7.
6. Bjorndal L, Reit C, Bruun G, Markvart M, Kjaeldgaard M, Nasman P, et al. Treatment of deep caries lesions in adults: randomized clinical trials comparing stepwise vs. direct complete excavation, and direct pulp capping vs. partial pulpotomy. Eur J Oral Sci. 2010;118(3):290–7.
7. Rutherford RB, Waahle J, Tucker M, Rueger D, Charrette M. Induction of reparative dentine formation in monkeys by recombinant human osteogenic protein-1. Arch Oral Biol. 1993;38:571–6.
8. Rutherford RB, Spangberg L, Tucker M, Rueger D, Charrette M. The time-course of the induction of reparative dentine formation by recombinant human osteogenic protein-1. Arch Oral Biol. 1994;39:833–8.
9. Goldberg M, Six N, Chausssain C, DenBesten P, Veis A, Poliard A. Dentin extracellular matrix molecules implanted into exposed pulps generate reparative dentin: a novel strategy in regenerative dentistry. J Dent Res. 2009;88:396–9.
10. Dummer PM, Hicks R, Huws D. Clinical signs and symptoms in pulp disease. Int Endod J. 1980;13(1):27–35.
11. Leucht P, Kim J-B, Amasha R, James AW, Girod S, Helms JA. Embryonic origin and Hox status determine progenitor cell fate during adult bone regeneration. Development. 2008;135(17):2845–54.
12. Simon S, Cooper P, Smith A, Picard B, Ifi CN, Berdal A. Evaluation of a new laboratory model for pulp healing: preliminary study. Int Endod J. 2008;41(9):781–90.
13. Sundell JR, Stanley HR, White CL. The relationship of coronal pulp stone formation to experimental operative procedures. Oral Surg Oral Med Oral Pathol. 1968;25(4):579–89.
14. Goga R, Chandler NP, Oginni AO. Pulp stones: a review. Int Endod J. 2008;41(6):457–68.
15. Dessombz A, Méria P, Bazin D, Daudon M. Prostatic stones: evidence of a specific chemistry related to infection and presence of bacterial imprints. PLoS One. 2012;7(12):e51691.
16. Cooper PR, McLachlan JL, Simon S, Graham LW, Smith AJ. Mediators of inflammation and regeneration. Adv Dent Res. 2011;23(3):290–5.
17. Ostby BN. The role of the blood clot in endodontic therapy. An experimental histologic study. Acta Odontol Scand. 1961;19:324–53.
18. Trope M. Letters to editor: reply. J Endod. 2008;34(5):511.
19. Thibodeau B, Teixeira F, Yamauchi M, Caplan DJ, Trope M. Pulp revascularization of immature dog teeth with apical periodontitis. J Endod. 2007;33(6):680–9.
20. Shimizu E, Ricucci D, Albert J, Alobaid AS, Gibbs JL, Huang GT-J, et al. Clinical, radiographic, and histological observation of a human immature permanent tooth with chronic apical abscess after revitalization treatment. J Endod. 2013;39:1078–83.
21. Martin G, Ricucci D, Gibbs JL, Lin LM. Histological findings of revascularized/revitalized immature permanent molar with apical periodontitis using platelet-rich plasma. J Endod. 2012;39:138–44.
22. Simon S, Rilliard F, Berdal A, Machtou P. The use of mineral trioxide aggregate in one-visit apexification treatment: a prospective study. Int Endod J. 2007;40(3):186–97.
23. Nosrat A, Li KL, Vir K, Hicks ML, Fouad AF. Is pulp regeneration necessary for root maturation? J Endod. 2013;39(10):1291–5.
24. Regenerative endodontics – AAE [Internet]. [cited 2011 Nov 14]. Available from: http://www.aae.org/publications-and-research/research/regenerative-database.aspx.
25. Lovelace TW, Henry MA, Hargreaves KM, Diogenes A. Evaluation of the delivery of mesenchymal stem cells into the root canal space of necrotic immature teeth after clinical regenerative endodontic procedure. J Endod. 2011;37(2):133–8.

Index

M. Goldberg (ed.), *The Dental Pulp*,
DOI 10.1007/978-3-642-55160-4, © Springer-Verlag Berlin Heidelberg 2014

GPSR Compliance

The European Union's (EU) General Product Safety Regulation (GPSR) is a set of rules that requires consumer products to be safe and our obligations to ensure this.

If you have any concerns about our products, you can contact us on ProductSafety@springernature.com

In case Publisher is established outside the EU, the EU authorized representative is:

Springer Nature Customer Service Center GmbH
Europaplatz 3
69115 Heidelberg, Germany

Batch number: 09430315

Printed by Printforce, the Netherlands